3D Engineered Polysaccharides

Mahmoud H. Abu Elella · Riham R. Mohamed ·
Nedal Y. Abu-Thabit
Editors

3D Engineered Polysaccharides

Fabrication Methods and Environmental Applications

Springer

Editors
Mahmoud H. Abu Elella
Department of Chemistry
Faculty of Science, Cairo University
Giza, Egypt

Riham R. Mohamed
Department of Chemistry
Faculty of Science, Cairo University
Giza, Egypt

Nedal Y. Abu-Thabit
Department of Chemical Engineering
Jubail Industrial College
Jubail Industrial City, Saudi Arabia

ISBN 978-3-032-26153-3 ISBN 978-3-032-26154-0 (eBook)
https://doi.org/10.1007/978-3-032-26154-0

This Springer imprint is published by the registered company Springer Nature Switzerland AG
The registered company address is: Gewerbestrasse 11, 6330 Cham, Switzerland

Contents

Introduction to 3D Engineered Polysaccharide Formulations

An Overview of Polysaccharides: Classifications and Applications

Indu Kumari, Ratnesh Das, and Nutan Shukla

Abstract Polysaccharides have attracted much attention owing to their immense biological and pharmacological usage such as in medical, chemical, nanotechnology, pharmaceutical, material science and food industry. Polysaccharides being one of the essential macromolecules, can be obtained from various sources viz., plants, microorganisms, algae and animals. The diverse applications of polysaccharides in various pharmaceutical or biopharmaceutical fields is mainly due to their physicochemical characteristics that makes them prone to physical and chemical modifications to enhance their properties. By examining the multifaceted aspects of polysaccharides, the present chapter endeavours to give insight into their significance in diverse fields, covering the classification and most recent updated applications. Additionally, the chemistry and structure of various polysaccharides in nature have been addressed. Finally, future trends of the polysaccharides for usage in various fields, including tissue engineering is summarized.

Keywords Polysaccharides · Classification of polysaccharides · Sources · Pharmaceutical and biological usage

Abbreviations

GAGs	Glycosaminoglycans
GG	Guar gum
CPSs	Capsular polysaccharides
LPSs	Lipopolysaccharides

I. Kumari
Department of Applied Sciences, CT Group of Institutions, Shahpur, Jalandhar, Punjab, India

R. Das (✉)
Department of Chemistry, Dr Harisingh Gour Central University, Sagar, Madhya Pradesh, India
e-mail: ratneshdas1@gmail.com

N. Shukla
The Smart Materials Research Institute, Southern Federal University, Rostov-On-Don, Russia

M. H. Abu Elella et al. (eds.), *3D Engineered Polysaccharides*,
https://doi.org/10.1007/978-3-032-26154-0_1

EPSs	Exopolysaccharides
NDP	Nucleoside diphosphate
UDPG	Uridine diphosphate glucose
Wzz	An integral inner membrane protein
Wzy	An O-antigen polymerase
PCR	Polymerase chain reaction
DNA	Deoxyribonucleic acid
mRNA	Messenger ribonucleic acid
NO	Nitric oxide
HIV	Human immunodeficiency virus
HSV	Herpes simplex virus
IL-2	Interleukin-2
IFN	Interferon gamma
TLR3/4	Toll-like receptor 3
COVID-19	Coronavirus disease 2019
SARS-CoV-2	Severe acute respiratory syndrome coronavirus 2
PSP	Polysaccharide-peptide complex
PSK	Polysaccharide-protein complex

1 Introduction

Polysaccharides are the most abundant naturally available macromolecular polymers that are derived from the various renewable natural sources viz., plants, algae, animals, bacteria, fungi and microorganisms [1]. In addition to other biomolecules like proteins and nucleotides, polysaccharides are a vital component for the functionality and specificity of the species in the biological system. Polysaccharides own exceptional physical, chemical and biological characteristics. Regardless the wide distribution of polysaccharides in nature, they have received very little popularity. Owing to the diverse functions of polysaccharides such as, structural and energy reserves, they are also referred as carbohydrates [2]. Polysaccharides are utilized by plants, animals, microorganisms and algae to store energy and structural support [3]. In various photosynthetic organisms, they are synthesized by using natural carbon-capture method. Unlike proteins and nucleic acids, the cellular structure does not follow any particular pattern to synthesize polysaccharides. Thus, their synthesis result into the formation of various microstructures whose composition depends on the type of cells or microorganisms [3]. Polysaccharides are also known as glycans. In the cells of living organisms, a dense and complex layer of polysaccharides cover the plasma membrane. They can be connected to proteins or lipids for protein folding (i.e. glycosylation) and multifunctional biomacromolecules synthesis. Additionally, polysaccharides play significant role in various biological processes including differentiation, cell adhesion, cell recognition, signalling and microbial attachment [3]. Considering the bioactivity of polysaccharides, various polysaccharides constituting plants, algae and fungi have gained medicinal applications [4]. They have also been used for clinical purposes. For instance, in 1988, polysaccharides were used as immunomodulating agents with antiviral characteristics [5]. Polysaccharides and their derivations are also used in food industry [6].

Polysaccharides are constituents of monosaccharide units that are joined in a biopolymeric chain via glycosidic bond [7]. The monosaccharide unit number in a polysaccharide is referred as degree of polymerization that changes with the type of polysaccharide. Hence, polysaccharides with different degree of polymerization lead to different molecular weights. The molecular weight of polysaccharides is mostly high that varies with the monosaccharide residues or units involved in their chemical structure [7]. Therefore, polysaccharides are large carbohydrate molecules and their conversion into small polysaccharides, disaccharides or monosaccharides can be performed by the hydrolysis of the glycosidic bonds. The medium of hydrolysis can be acidic, basic, thermal or enzymatic. This chemical modification is due to different reactive functional groups in the chemical structure of polysaccharides such as amino, hydroxyl and carboxylic acid groups.

Considering the source, polysaccharides can be distinguished into four major types, i.e. Polysaccharides of plant, algal, animal and microorganisms. The diversity and complexity of the polysaccharides can be determined using the nature of the monosaccharide unit i.e. linear or branched chain. Branches can either be short on a linear backbone or branch-on-branch structure. Branches can either be short, isolated, or positioned to produce bush-like structures, a branch-on branch structure or clustered [8]. Natural polysaccharides can be obtained from various sources such as algae origin (viz., alginate), plant origin (viz., pectin and gums), microbial origin (viz., dextran) and animal origin (viz., chitosan) [9]. The function of polysaccharides is based on its structure. Linear molecules, such as cellulose and chitin, are strong and rigid wherein branched polymers are utilized for energy storage due to their insolubility in water.

Keeping monosaccharide units (i.e. chemical structure) in mind, polysaccharide can be distinguished into homopolysaccharide (includes same monosaccharides) and heteropolysaccharide (includes different monosaccharides). Considering the functionality, homopolysaccharides can be further categorized into three types, i.e. storage, structural and gel-forming polysaccharides (Fig. 1). For instance, Benalaya et al. [9] represented the classification of polysaccharides on the basis of physiological characteristics. The food reserve material of plants i.e. starch is widely available as main storage polysaccharide in foods; such as in wheat, maize, potato, rice, legumes, and other vegetables [10]. In contrast to storage polysaccharides, structural homopolysaccharides preserve mechanical shape and support to the cells. For instance, wood, plants and plant cell walls contain cellulose and lignin. Chitin is a structural homopolysaccharide present on the exoskeleton of crustaceans [9].

Heteropolysaccharides can also be categorized into nitrogen and non-nitrogen molecule. For instance, nitrogen heteropolysaccharides are glycosaminoglycans (available in different tissues and intracellular cells) and non-nitrogen heteropolysaccharides are agar (found in microalgae, seaweeds and coral reefs), gums (seeds, plant exudates and microorganisms), hemicelluloses (plant cell walls), and pectin (plant

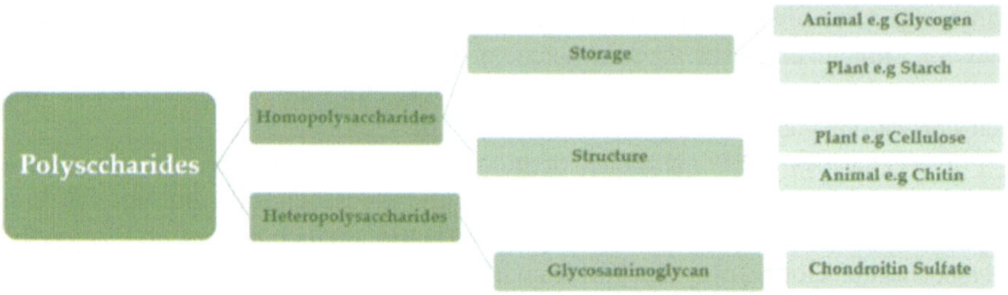

Fig. 1 Classification of polysaccharides based on physiological characteristics [9]. From Ref. [9], used under a Creative Commons CC BY license

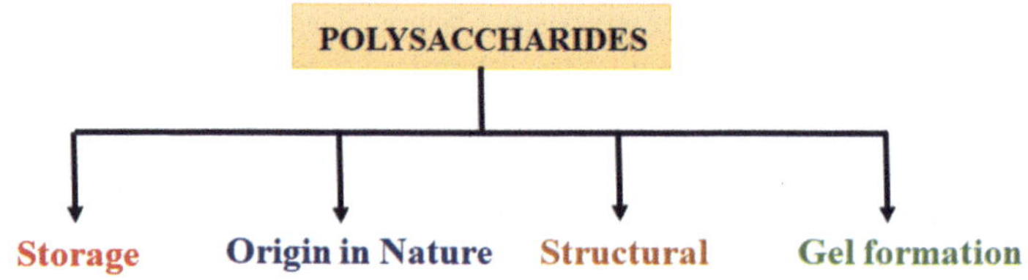

Fig. 2 General classification of polysaccharides

cell walls, citrus fruits, carrot, apple, and beet) [10]. Overall polysaccharides can be categorized into four categories i.e. storage, structural, origin in nature and gel formation that have been discussed in detail in the following sections (Fig. 2). The classification and chemical properties of polysaccharides have also been discussed by Mohammed et al. in their review [11]. Sharma et al. [12] reported the significance, classification, properties, and methods of preparation of polysaccharides as well.

At present, extensive research has been conducted on polysaccharides. They have been utilized as safer, biocompatible and biodegradable encapsulating materials. These materials can support drug solubility, bioavailability, delay or control drug release and augment active/passive targeting of bioactive agents to the required sites [13]. The tendency of polysaccharides to interact with tissues and cells shows some characteristics the body recognises and thus does not consider them as foreign material. Also, Zahra et al. [14] reported the various characteristics of renewable polysaccharides, their classification, and their contribution towards environmental sustainability. They have discussed environment friendly extraction methods to control chemical pollution. In another review, the categorization of biopolymers-polynucleotides, poly-peptides and polysaccharides-based monomeric units is documented by Biswas et al. [15]. They mainly highlighted the applications of biopolymers in drug delivery, tissue engineering, wound healing, and biosensors. Koyyada et al. [16] have documented the role of natural gum polysaccharides in drug delivery and tissue engineering. Salave et al. [17] presented insightful approaches for developing implants by using polysaccharides. Another review by Jabeen et al. [18] covers the role of polysaccharide based biopolymers in the field of biomedical science.

Considering the vital applications of polysaccharides, the current chapter provides classification, properties, and most recent updated applications of polysaccharides. Additionally, the chemistry of various polysaccharides in nature have been addressed.

2 Classification of Polysaccharides

Polysaccharides are the third major type of biopolymers (carbohydrates) that play an important role in the blood coagulation, immune system, fertilization, pathogenesis prevention, and therapeutic effectiveness [19]. Polysaccharides have noticeable impact in the biological functions viz., structural support, energy storage and cell signal transduction.

2.1 On the Basis of Structural Difference and Functions

2.1.1 Storage Polysaccharides

Storage polysaccharides also known as reserve polysaccharides, stores sugar molecules without creating an osmotic problem. Storage polysaccharides play different role in fungi, plants and animals. Few storage polysaccharides are starch and glycogen. Glucose is the major molecule that is responsible for transferring energy to the living cells. Recently, Hu et al. [20] discussed the effects of the molecular structures of starch and glycogen on human health and examined their role in pathological conditions. Owing to essential characteristics, the starch is known to be favourable natural polysaccharides. It is made up of linear a-amylose and branched amylopectin materials [20]. Starch is found in the chloroplasts of plant cells. The a-Amylose polymer contains many glucose monomers joined via a-(1, 4) bonds wherein the amylopectin molecule contains a-(1, 4) linked glucose residues having a-(1, 6) branch points [20]. The main application of starch is in food processing.

Another major carbohydrate which stores energy is glycogen. Both starch and glycogen consists of highly branched structure, except glycogen contains irregular spacing between branch points. This irregularity in structure is responsible for its high solubility in water and low viscosity in aqueous solution [20]. Despite the similarities with other similar polysaccharides such as amylopectin, glycogen is more complicated and branched in structure. In case of animals, only small amount of glycogen is generally being stored that is responsible for permanent supply of dissolved glucose. In order to preserve large amount of storage of reserves, fats are converted to lipids in animals [21]. Also, they are liquid in nature in the cells that makes them more companionable with the body actions. The storage polysaccharides are sub classified into further types as shown in Fig. 3.

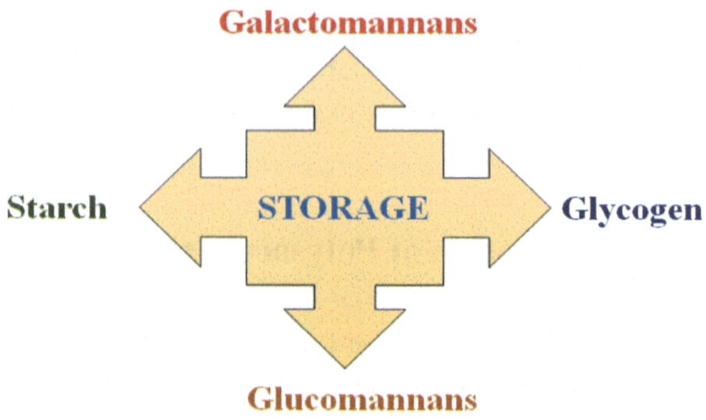

Fig. 3 Classification of storage polysaccharides

2.1.2 Structural Polysaccharides

Structural polysaccharides, as name implies, supports the formation of structures of different molecules. The essential structural polysaccharides are present in the main part of cell wall of plants, fungi and eukaryotic osmotrophic organisms [10]. The cell wall of these organisms is flexible and resistant to support their body via opposing the osmotic pressure of the cell and therefore prevents hydrostatic skeleton. The structural polysaccharides are represented in Fig. 4.

Main examples of structural polysaccharides are chitin and cellulose. Generally, cellulose found in plant cell walls while arthropod exoskeletons and fungal cell walls both comprise of chitin [11, 22]. Chitin is a type of linear polysaccharide found in animals. It is generally available in fungi's mycelia and spores, as well as the exoskeletons of insects and invertebrates [9]. The acetyl and amino groups in chitin are responsible for its hydrophobic nature. It is soluble only in specialized solvents such as dimethylacetamide with 5% lithium chloride [9].

Another structural polysaccharide i.e. cellulose is present in the plant cell walls and almost half of the biomass in photosynthetic species. It is an essential structural polysaccharide that includes a linear chain of D-glucopyranosyl units connected via ß-1,4 -glycosidic bonds The chemical structure of cellulose has been discussed by Benalaya et al. [9] (Fig. 5). It is the most abundant biopolymer [18]. Main sources of cellulose include cotton, jute, wood and flax. Other sources are bacteria, algae and marine animals of Ascite family [23, 24]. The chemical structure of cellulose includes three hydroxyl groups excepting terminal ends. The C-1 and C-4 carbon atoms of adjacent glucose units forms ß-1,4-glycosidic bond resulting in a linear

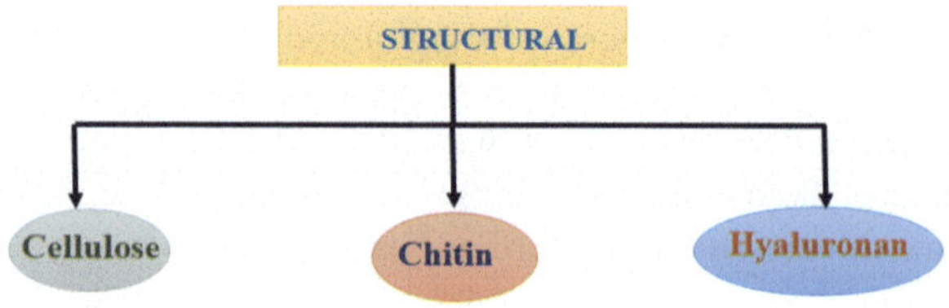

Fig. 4 Classification of structural polysaccharides

chain of glucose molecules having hydroxyl groups on each glucose unit [18]. The a-1,4-glycosidic bonds in starch makes it more easily to be digested by animals, including humans. However, the ß-1,4-glycosidic bonds in cellulose show intra- and inter-molecular hydrogen bonding that increase the linear integrity of polymers. This makes it more resistant to digestion and can be used as an important source of dietary fiber [18, 25]. The structure of cellulose have great role on its reactivity. However, it has also few disadvantages such as poor solvent solubility, lack of thermoplasticity and low crease resistance [18]. Controlled chemical and physical surface modification of cellulose can be used to overcome these disadvantages. Cellulose can be modified to change its crystalline nature and make it water soluble, non-toxic, biodegradable and thus have applications in medicinal field [18]. For example, cellulose esters acquire better solubility in organic solvents as compared to cellulose and hence possess applications in drug delivery systems.

Along with cellulose, hemicellulose is also found in plant cell walls and second most abundant renewable resource after cellulose [26]. The main difference lies between hemicellulose and cellulose is their chemical heterogenous nature. Hemicellulose is an important polysaccharide component of all land plants, freshwater plants, and some seaweeds. The distinct features of hemicellulose such as environment-friendliness, renewability, and biodegradability lead to synthesize various hemicellulose-based functional polymeric materials [26]. It has gained attention in many industrial fields such as biological implants, food packaging, bio devices, and medicinal products [18]. It is heteropolysaccharide comprises different monosaccharide units like pentoses, hexoses, acetylated sugars, and uronic acids. The variety of these sugar monomer units produces different types of hemicellulose viz., Glucomannan and Galactoglucomannan [27]. In glucomannan, glucose and mannose units have been linked by 1,4-glycosidic bonds in a linear manner. A small branching via

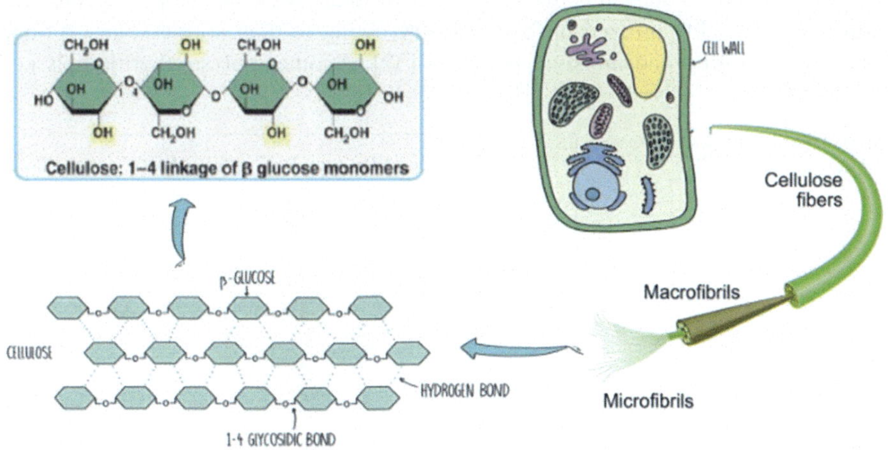

Fig. 5 Schematic structure of cellulose [9]. From Ref. [9], used under a Creative Commons CC BY license

1,6-glycosidic linkages is also present in it [28]. Galactoglucomannan is a modified version of glucomannan chain wherein the mannose units have been partially substituted by both galactose units through 1,6-glycosidic linkages and acetyl groups at C2 and C3 hydroxyl groups of the mannose it [28].

2.1.3 Polysaccharide Gels

Polysaccharide gels are widespread in daily life activities and thus possess diverse applications across different fields. A gel polysaccharide is a combination of two phases wherein polysaccharide acts as a continuous phase and liquid as a disperserd phase [29]. It is kind of colloidal system. Dispersed phase can be either water or any organic compound such as methanol, ethanol, etc. Gels are basically intermediate between solids and liquids; it has same density as liquids however structure is more close to the solids [29]. The amount of liquid is more in gel polysaccharides. The best known polysaccharide gel is agar. Apart from agar, another gel polysaccharide are mucopolysaccharides which are carbohydrates containing amino sugars and uronic acids. These are also referred as glycosaminoglycans (GAGs) [30]. These are generally found in the mucus and fluid present around the joints. The most common examples of mucopolysaccharide are chondroitin-6-sulfate, heparin sulfate and keratin sulfate [31]. Some types of gels possess the special property called thixotropy in which they can change their state. For instance, they can become fluid on agitation while becomes solid in the static state. The formation of gel is called gelatin. Some materials (such as aerosols) that possess low density, high porosity and excellent thermal insulation can be formed via changing the liquid with gas in gel polysaccharides. Figure 6 represents the classification of polysaccharides gels.

Comprehensive studies have been documented on the preparation of polysaccharide gels i.e. gelatin mechanism [29, 32]. For instance, well known polysaccharide hydrogels are highly utilized in food industry owing to their softness, absorption, elasticity, flexibility and moisture retention [32]. Another polysaccharide gels i.e.

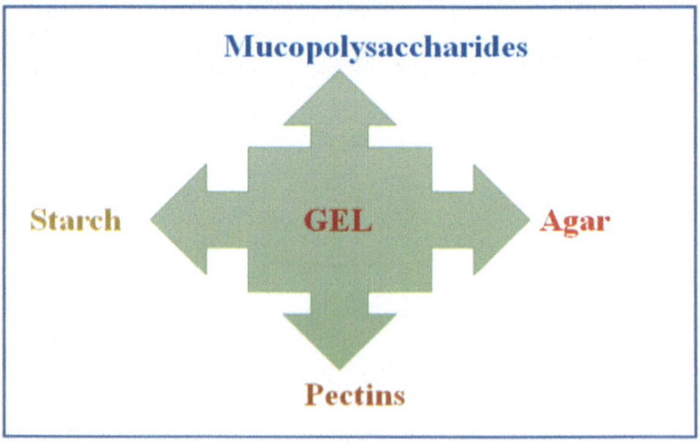

Fig. 6 Classification of gel polysaccharides

polysaccharide aerogels with porous structure and wide surface area found applications in biomedical science [29]. In the last decades, stimuli-responsive gels, cell-compatible gels, and self-healing gels have attracted attention broad wide [30]. For example, protein–polysaccharide gels are gaining popularity owing to their tuning gel properties [33]. These gels can be formed by mixing gelling polysaccharides and non-gelling proteins, gelling proteins and non-gelling polysaccharides, or gel as both agents [34]. In literature, different factors are reported that can trigger the gelatin of polysaccharides such as acid induction [35], salt induction [36] and alcohol induction [37]. These induction methods produce different polysaccharide gel processes and thus yield final gels with varied characteristics.

2.2 On the Basis of Their Origin in Nature

Due to the wide range of properties such as non-toxicity, abundance and biocompatibility, the natural polysaccharides have gained much attention for their consumption in green materials. The natural polysaccharides possess brilliant biological functions that are responsible for their wide applications in the medicinal field. According to origin in nature, polysaccharides can be classified into four main categories, which are Polysaccharides of plant, algal, animal and microorganisms [7, 38]. Plant polysaccharides are generally available in food plants, medicanical plants and marine algae [38]. Animal polysaccharides are widely distributed among mammals, birds, insects and marine organisms [38]. Microbial polysaccharides are commonly found in fungi, bacteria and algae [38].

2.2.1 Polysaccharides from Plants

Plant polysaccharides exhibits a wide range of applications such as, antitumor, anti-inflammatory, tissue support, anticoagulant, nutrient storage and immune enhancement [7]. Xu et al. [7] reported various plant polysaccharides exhibiting these characteristics. Depending on location within the cell, plant polysaccharides are distinguished into three major categories viz., intracellular, cell wall and extracellular [7]. Amongst these, cell wall polysaccharides exhibit extensive medicinal importance. As already discussed, starch is one of the most favourable natural polymers due to its essential properties. This polysaccharide is the main constituent of plant food items and stored in the chloroplasts of plant cells [39]. The main application of starch is in food processing as a thickener and stabilizer. Cellulose, hemicellulose and pectin are famous examples of plant cell wall [40]. Cellulose is the most abundant, renewable and ecological. It is the chief structural element of the plant system. It is widely available thus, considered as a maintainable feedstock of energy source. It possesses various applications viz., paper, cloth, film, coating, tissues, medicines, polymeric materials [11].

On the basis of sugar groups, hemicellulose can be further categorized into mannan, xylan and xyloglucan [7, 18]. Polysaccharides can be easily extracted from numerous plant sources including tea, brown algae, mistletoe and echinacea (the medicinal tree of Mali). These sources act as raw materials. Many medicinal plants are utilized to extract polysaccharides with potential therapeutic effects against viruses, cancer and immune modulation. Examples are Glycyrrhiza (Licorice), Morinda officinalis, Astragali radix, Dendrobium officinale kimura et migo, Codonopsis pilosula, [7, 41].

Apart from these, Guar gum (GG) is also a natural polysaccharide. GG belong to the Leguminosae family and is extracted from the seed endosperm of the plant Cyamopsis tetragonoloba (guar kernels) [41]. Owing to easy availability, nontoxicity, biodegradability and low cost, GG is widely used in high performance applications such as biomedical, explosive, cosmetic, food, textile, paper, etc. In the chemical structure of GG, ß-D-mannopyranosyl units are connected with a branched a-D-galactopyranose unit at 6 position that have mannose: galactose ratio of approximately 2:1 [41]. Hence, great amount of research on plant polysaccharides is gaining widespread attention in recent years.

2.2.2 Polysaccharides from Algal Origin

Algal polysaccharides have garnered attention in the research community owing to their widespread bioactivities such as, antiviral, antidiabetic, antioxidant, anticancer and immunoregulation [42]. The significance of algal polysaccharides in terms of their extraction, purification and potential use has been well documented in literature [42, 43]. The main advantage of using algal polysaccharides is to make the whole process renewable and recyclable. This leads to biodegradable, biocompatible and sustainable environment [42].

Most common algal polysaccharides are alginates, laminarin, fucoidan, carrageenan and ulvan that act as anti-inflammatory, anticancer, antioxidant, anticoagulant, immunostimulatory and antiproliferative agents [44–46]. These algae-derived polysaccharides possess exclusive structural features that make them valuable functional components in the food industry. Alginate is the most common linear polysaccharide and abundantly present in the nature. It is eco-friendly and made from the cell wall of brown algae. It is a copolymer of ß-D-mannuronic acid and a-l-guluronic acid monomer units [47]. Owing to exceptional properties such as, water solubility, low toxicity, biocompatibility, low cost and gelling ability, alginates possess wide applications in different fields viz., medical, food additives, cosmetics, antibacterial films, etc. [47]. Three main species of brown algae are used for commercial uses to extract alginates i.e. Laminaria hyperborean, Ascophyllum nodosum, and Macrocystis pyrifera.

Apart from alginate, carrageenan is also naturally occurring linear sulfonated polysaccharide. It is synthesized by alkaline extraction of red seaweed. It is a copolymer of (1,3)-D-galactopyranose and (1,4)-3,6-anhidro-a-D-galactopyranose with sulfate groups [42]. The hydroxyl and sulphate groups are responsible for the hydrophilic nature of carrageenan. Due to the properties such as biodegradability, biocompatibility and gelling behaviour, carrageenan is widely utilized as a stabilizer and a structure provider in food processing [42]. Wu et al. [42] analysed the structural properties of brown, red, and green algal polysaccharides and studied their roles in diabetes, cancer and cardiovascular diseases.

2.2.3 Polysaccharides from Animals

Animal polysaccharides are polysaccharides found in animal tissues and organs. In nature, animal polysaccharides are available in shrimps, sea cucumbers, loaches and velvet antlers. The animal polysaccharides are mainly constitutes of hyaluronic acid, heparin, chitin and chondroitin [48]. The most popular natural polysaccharides that obtained from animal origin are chitin and chitosan. Chitin is the abundantly found natural polysaccharide that constitutes of ß-(1,4) linked D-glucosamine and randomly positioned N-acetylglucosamine groups [49]. Chitosan is the derivative of chitin and thus unique bio-based polymer. While chitin can be extracted through fermentation technique, chitosan is obtained via partial deacetylation of chitin using a chemical method or by enzymatic hydrolysis [50]. Chitin is available in the shells of crabs, lobsters, shrimps and insects whereas, chitosan forms the exoskeleton of arthropod [49]. Due to biocompatibility, biodegradability and non-toxicity, both chitin and chitosan are widely used for biomedical applications such as tissue engineering, drug delivery, wound healing etc.

Hyaluronic acid is a biodegradable natural polysaccharide. It is a linear GAG composed of repeating disaccharide unit of D-glucuronic acid and N-acetyl-D-glucosamine [51]. It is mainly exist in the heads, skin, eyeballs of animals and umbilical cords. It is highly bio-compatible, nonimmunogenic and does not encourage chronic inflammation. It helps in the lubrication of bones and joints and protecting cells [51].

Another animal polysaccharide, Heparin is widely available in the animal blood, muscles, thymus and internal organs (liver, heart, spleen, kidneys and lungs) [52]. It is a sulfur containing mucopolysaccharide and exhibits various pharmacological effects including unblocking blood vessels and regulating blood lipids [52]. Thus, animal polysaccharides possess numerous potential applications in pharmaceutical field.

2.2.4 Polysaccharides from Microorganisms

Microbial polysachharides are the natural polysaccharides that are obtained from microorganisms [38]. These polymers are obtained from microbial cell and consist of high molecular weight. These polysaccharides are non-toxic modules exhibits different physical properties such as, elasticity, mechanical strength, viscosity and biocompatibility [53]. The repeating unit of microbial polysaccharides are made from groups of 10 or less that contains 3 to 7 homo or hetero-monosaccharides joined in a linear manner. Owing to their intriguing and alluring properties, they are utilized as bio-absorbents, bio-flocculants and drug delivery agents in the food, cosmetic, and pharmaceutical fields. The common monosaccharides included in this connection are uronic acids, amino sugars, hexose etc. [38]. The microbes used in the preparation of different types of microbial polysaccharides include extensive variety of bacteria, yeast and fungi. The most common microbial polysaccharides are dextran, Pullulan and Xanthan gum.

Microbial polysaccharides are divided into three main groups i.e. capsular polysaccharides (CPSs), lipopolysaccharides (LPSs) and exopolysaccharides (EPSs). Like plant polysaccharides, microbial polysaccharides can also be categorized on the basis of their morphological localization i.e. intracellular, extracellular, and cell wall polysaccharides [38]. The xanthan gum (extracellular microbial polysaccharide) was first extracellular polysaccharide that was first successfully extracted and studied in 1952 from *Xanthomonas* black rot in cabbage [54]. This polysaccharide is still getting attention from the scientific community worldwide. Xanthan gum is heteropolysaccharide wherein D-mannose, D-glucose, and D-glucuronic acid are connected in a molecular ratio of 3:3:2 [54]. It is synthesised by aerobic fermentation of glucose and sucrose using bacterial species named *Xanthomonas campestris*. It is widely used as food additive and thickener [54]. Another example, Dextran is homopolysaccharide wherein glucose acts as a monomer unit. It is odourless, non-toxic and tasteless [55]. It has been found in wine as a microbial product by scientist Louis Pasteur, however, it can be manufactured using sucrose at large scale through fermentation process with the aid of enzyme i.e. *Leuconostocmes enteroides* or *Streptococcus mutans* [55]. It is widely used in food, pharmaceuticals and agricultural fields. Similar to dextran, Pollulan is also tasteless, non-toxic and odourless biodegradable homopolysaccharide wherein maltotriose acts as a repeating unit. It can be prepared by fermentation process by using the fungus *Aureobasidium* pullulans i.e. also known as black yeast [56]. It was discovered by Bernier in 1958. It is greatly used in food and pharmaceutical area. It is also used as a binder and thickener.

In addition to this, various marine microbes produce polysaccharides as marine microbial resources are abundant [7, 57]. Thus, marine microbial polysaccharides are gaining attention due to their unique functions. They are mainly extracellular in nature that exhibits scientific importance and commercial value [57].

3 Synthesis of Polysaccharides

The synthesis of polysaccharides can be done by two methods i.e. natural and chemical. While natural method is performed within living organisms and depends on enzymatic catalysis, chemical synthesis is conducted in the research laboratory setting. Natural synthesis is efficient, specific method that involves only mild conditions [58]. However, chemical method requires designed planned routes and controlled reaction conditions to synthesise polysaccharides [59].

3.1 Natural Synthesis

The natural method of preparation of polysaccharides follows various pathways such as glycogen, nucleotide sugar, synthesis, cleavage and modification [60]. The crucial route for polysaccharide synthesis is the nucleotide sugar pathway. It includes sugar nucleic acids combine with other molecules and form polysaccharide via phosphorylation and glycosyltransferase. Plant polysaccharides are formed by nucleotide sugars whereas nucleoside diphosphate (NDP) produces sugar complexes by chemically activated phosphoester bonding [61]. The incorporation of specific monomers from NDP is catalysed by glycosyltransferases. This produces polysaccharide chain and finally forms complete polysaccharides. Another significant pathway i.e. the glycogen route encompasses the conversion of glucose into uridine diphosphate glucose (UDPG). Then, UDPG is transferred to the non-reducing end of the glycogen primer and connected with a-1,4 glycosidic bonds. It leads to the formation of the main chain using glycogen synthase. This leads to the further chain elongation and glycogen [62]. Extracellular polysaccharide synthesis resembles the glyconucleic acid route that involves monosaccharide binding to nucleotides and produces precursor nucleotide-activated sugars [7]. Here also, these precursors converted into starting units using enzyme catalysis, gets elongated and undergo polymerization reactions to form repeating units. These repeating units are then finally linked consequently connected and transported out of the cell through lipid carriers [7].

In addition to these natural synthesis pathways, cleavage and modification pathways also play a vital role for polysaccharide synthesis [62]. In these methods, the monomeric sugars are joined to form polysaccharide chains by using polymerase. For example, Wiseman et al. [63] reported that the length of polysaccharide chains can be regulated by polysaccharide copolymerase Wzz whereas polymerase Wzy polymerizes oligosaccharides into longer polysaccharide chains. Another study by Islam et al. [64] further supported this report and also highlighted the importance of close position of polymerase Wzy and copolymerase Wzz. These natural synthesis routes ensure the normal functioning of polysaccharides in organisms.

3.2 Chemical Synthesis

Owing to their complex structure and microscopic heterogeneity, the separation and purification of natural polysaccharides becomes difficult. The mass production of natural polysaccharides using PCR technology is not possible since no set template is available. In the same way, formation of polysaccharides with consistent structure is also partially attainable using chemical method [65]. However, the limitation of chemical methods is that they only yield short polysaccharide fragments due to the complexity of polysaccharide structures. In literature, it is reported that the imitation of complicated effects of natural polysaccharides for short sugar chains is very challenging [66]. Petitou et al. [67] reported the chemical synthesis of small molecule heparin that inhibits thrombin activity. However, the inhibitory consequence of chemically synthesized pentadecaccharide remains weak. Considering literature, it becomes necessary to synthesize glycan compounds having same size and chemical complexity as natural polysaccharides.

The glycosylation reaction is one of the most important reactions in the chemical synthesis of polysaccharides. It entails the coupling reactions between sugar units. On the basis of nucleophilic reactive groups present in the glycosyl acceptor, glycosylation reactions are divided into oxygen, nitrogen and carbon glycosylation reactions. The resulting reaction products are known as oxygen, nitrogen and carbon glycosides [68]. In polysaccharide synthesis, the arrangement of polysaccharides is another main problem. The sugar units are effectively arranged to give complex polysaccharide structures by using existing glycosylation methods. These days, two main approaches i.e. linear and convergent methods have been followed for assembling polysaccharides [69]. These approaches either involve the adjustment of end group substituents or the elimination of temporary protecting groups on the sugar units. Thus, results in the complicated and time-consuming nature of polysaccharide synthesis. This makes difficult to isolate and purify intermediates and therefore, decreasing the efficiency of polysaccharide synthesis. To overcome these challenges, few new strategies have been adopted by scientific community including automated solid-phase synthesis [70]. However, the preparation of complex long chain polysaccharides due to the structural diversity, chemical complexity and difficult isolation is still a major issue. Thus, it becomes a significant problem to synthesize polysaccharide using efficient chemical process. Subsequently, designing and manufacturing tailor-made polysaccharides has proven to be quite challenging.

3.2.1 Tailor-Made Polysaccharides by Using Chemical Modification Process

The chemical and conformational characteristics of polysaccharides can be responsible for altering or reversing their properties. Therefore, tailoring polysaccharides becomes significant to changing their characteristics for the purpose of obtaining polymeric materials with specific applications in biomedical field [3]. Over the past few decades, tailor-made polymers are getting lot of attention worldwide. The purpose behind this concept of tailor-made polysaccharides has been encouraged from polymerization reactions [3]. This allows scientists to design and prepare polysaccharides with various features.

Modification techniques have been utilized in order to selectively alter the characteristics of polysaccharides. Generally, physical, chemical and biological modifications are followed to increase the functional property of polysaccharides [3]. Amongst these, chemical modifications effectively support the structural and functional characteristics of polysaccharides [3]. This includes methods viz., carboxymethylation, chelation, selenization, sulfation, phosphorylation that significantly improved the properties of polysaccharide [3]. Using these chemical modification methods, the functional groups can be replaced or introduced into polysaccharides under particular conditions, thus their biological applications can be improved. For instance, sulfur groups can be introduced in polysaccharides polymer chains by sulfation modification method [71]. In this method, a known amount of polysaccharides in a solvent is dissolved with specific reagent under specific conditions. This can replace the hydroxyl group with a sulfate group. Owing to its ease, non-toxicity and low cost, this method is very popular for chemical modification of polysaccharides [71]. Similarly, phosphorylation modification method can enhance the biological activity of polysaccharides by replacing the hydroxyl group with phosphate group in the polysaccharide polymerization chain [72].

Selenization modification method is another known method for polysaccharide modification. In this, selenium group can be introduced into the polymer chain of polysaccharides and form selenopolysaccharides [73]. Many studies reported the low toxicity, higher bioavailability, antioxidant, antitumor, hypoglycemic, immunomodulatory characteristics of selenopolysaccharides [73, 74]. These specific properties of selenopolysaccharides make them better than traditional selenium compounds. In the same way, acetylation modification converts hydroxyl groups into acetyl groups through esterification reactions [75]. Since, acetyl groups can stretch and change the spatial arrangement of polysaccharide chains, thus more polysaccharide hydroxyl groups are exposed for chemical reactions. This increases the water solubility of polysaccharides. Yazdi et al. [3] discussed the properties and applications of modified polysaccharides in detail. They have highlighted various chemical modification methods for tailor-made polysaccharides such as, esterification, etherification, amidation, oxidation, nucleophilic displacement reaction and grafting reactions.

In addition to the aforementioned chemical techniques, other modifications such as, methylation, sulfonylation, iodination, desulfation can also be followed for the chemical modification of polysaccharides [76]. These methods can improve the chemical, physical and pharmacological properties of polysaccharides. However, more research is required to gain insights into the structural and functional properties of modified polysaccharides.

4 Polysaccharide Degradation Methods

The gastrointestinal systems of animals cannot readily absorb polysaccharides due to their high molecular weight and low penetration power [77]. Thus, polysaccharides should follow degradation methods to perform various functions. Polysaccharide degradation process entails the breaking of enzymes or other degrading substances into smaller molecules or monosaccharides. Two primary methods i.e. enzymatic degradation and microbial transformation are known to degrade the polysaccharides. The first method, enzymatic degradation uses the particular biological enzymes to break the polysaccharides [78]. In this mechanism, the enzymes firstly join the polysaccharide substrate noncovalently and forms enzyme–substrate complex. Then, the enzyme hydrolyses the substrate using a particular active site. Once the product is formed via enzyme degradation of the substrate, the enzyme then goes to new substrate to carry new reaction. Since, the mechanism depends on the chemical properties of the substrate and structure of the enzyme, thus each degradation mechanism of polysaccharides can vary. In literature, various studies are reported on the enzymatic degradation of polysaccharides. For instance, Kobayashi et al. [79] acknowledged a variety of polysaccharide hydrolysing enzymes in marine species. This makes them to absorb nutrients from plant debris and perform enzymatic mechanisms at low temperatures. Another study by Zeng et al. [80] found that the degradation of peach gum polysaccharides can be performed through glycohydrolase and fungal galactosidase.

Another significant route for polysaccharide degradation is microbial transformation. In this method, some microorganisms can break down complex polysaccharides into small molecules or monosaccharides with the release of specific enzymes. These enzymes help the host to absorb and utilize the polysaccharides [3]. For instance, in case of ruminants, *Ruminococcus* spp. can convert polysaccharides into various nutrients [81]. In another study, bacteria belonging to the order *Actinomycetes* sufficiently degrade polysaccharides in soil [82].

5 Applications of Polysaccharides

5.1 Role of Polysaccharides in Living Organisms

In living organisms, energy storage is an essential physiological function that preserves the excess nutrients and release when physiologically required [83]. This bioenergy is stored in organisms in the form of proteins, polysaccharides and fats. In polysaccharides, glycogen acts as a main energy storage molecule for some animals and microorganisms [84]. It helps in maintaining energy metabolism homeostasis. As already discussed, starch and cellulose are essential energy-storing polysaccharides found in plants [3]. The digestion of cellulose is difficult in humans however; it acts as a useful dietary fiber. Polysaccharides also play a vital role in maintaining the tissue cell structure. They are known to be integral part of the cell wall of plants. In animals, the strength and elasticity of bones rely on polysaccharides [3].

Polysaccharides also help in cell recognition by identify and differentiating different cell types. For instance, Dots et al. [85] reported the blood group antigens on the surface of red blood cells that play a key role in blood group matching and transfusion. Polysaccharides show significant impact in cellular signalling and cytopathic information [3]. Reports have shown their property as signal amplifiers in disease detection systems. For instance, polysaccharide based electrochemical biosensor is used to identify nonsmall cell lung cancer biomarkers by amplifying the electrochemical signal [86]. These properties of polysaccharides can help to in the development of specific drugs that can offer more effective therapeutic effects.

5.2 *Biomedical Applications*

Owing to the extensive properties of polysaccharides viz., biocompatibility, biodegradability, non-immunogenicity, antimicrobial, antioxidant, improved solubility and stability, polysaccharides have been considered as promising candidates for applications in many biomedical applications [3]. Additionally the low cost of polysaccharides made them better choice for many biomedical applications. Recently, various significant biological properties of polysaccharides have been investigated such as, antitumor, antioxidant, anti-inflammatory, antiaging, antibacterial and immunoregulatory effects.

5.2.1 Antioxidative Applications

Reactive oxygen species degrades human body via attacking the macromolecules (including proteins, DNA and membrane lipids) and leads to various health problems such as cancer and inflammatory diseases. Oxygen species decreases antioxidant tendency that produces physiological and pathological responses [87]. In order to reduce the effect of these hazardous agents, antioxidative agents have been used to enhance the human health. However, these antioxidants may have drawbacks such as, carcinogenesis and liver damages. Considering this, polysaccharide derived from plants exhibit strong antioxidative properties and can protect human body against free radicals and lowers the problems of many diseases [87]. These polysaccharides have wide range of applications such as antitumor and antioxidant activities. Many polysaccharides have been known to prevent cardio and cerebrovascular disease caused by free radicals.

Polysaccharides follow scavenging of free radicals and thus exert antioxidant effects. For instance, lignified okra polysaccharide is reported to protect cells by inhibiting mRNA and reducing NO secretion [88]. Polysaccharides improve the antioxidant tendency of living organisms using antioxidant enzymes such as glutathione peroxidase and superoxide dismutase [89]. They also enhance the stability of antioxidant substances by increasing the time of their antioxidant activity. For instance, soluble *soyabean* polysaccharides with wheat alcohol soluble protein nanoparticles, shared with curcumin are known to increase their constancy and antioxidant characteristics [90].

5.2.2 Antitumor and Immunomodulatory Applications

Since tumor is a one of the major health issue worldwide that is causing deaths [3]; thus researches have been made to inhibit the growth of tumor. Many antitumor activities are found to exist in various plants and marine polysaccharides. Large variety of natural polysaccharides has been considered as effective antitumor agents [91]. Polysaccharide-protein conjugate exhibits antitumor activity and support the function of chemo related drugs. Many studies have been documented on the immunosuppressive activity of polysaccharides against tumor growth [3]. These studies include induction of tumor cell apoptosis, autophagy, immune regulation and antitumor cell metastasis.

Recently, the autophagy-mediated antitumor activity of polysaccharides is widely utilized to regulate tumorigenesis and therapy [91]. Polysaccharides such as those available in *Ganoderma lucidum*, and *L. fructus*, inhibit the existence and development of tumor cells through the immonumodulatory mechanisms [92, 93]. These mechanisms enhance the killing tendency of natural killer cells and macrophages [92]. Also, polysaccharides are well known for the prevention and of tumor metastasis and recurrence. Despite the variation in the chemical structures and conjugated parts, mushrooms are well known example for many therapeutic applications and also acts as food and medicinal agents. Considering this, natural polysaccharides resulting from mushrooms have been significantly acknowledged for their antitumor applications. For instance, polysaccharides derived from mushrooms include *Ganoderma lucidum* is reported for in vitro immune activation, and antitumor activity on the breast cancer cells [92]. Polysaccharides obtained from mushrooms show antitumor effects through its direct tumor-inhibitory effect. Polysaccharides derived from nanoparticles are also known to expressively accumulate in tumor tissues. Various studies demonstrated the antioxidant property of polysaccharides that are directly connected to the antitumor mechanism of polysaccharides [3]. These reports show the polysaccharides having antioxidant characteristics can exhibit antitumor effects [94]. Therefore, polysaccharides exhibit favourable therapeutic capacity by manifesting antitumor effects via multiple mechanisms.

5.2.3 Anti-Inflammatory and Antiviral Activity

Inflammatory reaction is a basic physiological response of tissue damage or infection. Currently, polysaccharides having anti-inflammatory characteristics are studied to have modulating pro- and anti-inflammatory mediators and cell signalling [95]. Polysaccharides can either increase the release of anti-inflammatory mediators or control pro-inflammatory mediators. Then, they can suppress the inflammatory response and prevent issue damage [96]. Several reports have investigated the anti-inflammatory properties of polysaccharides through the regulation and inhibiting of inflammation related signalling pathways [97].

Viral infections have shown significant risks to human and animal health. Polysaccharides especially sulphated polysaccharides have been well known for their antiviral properties. For instance, sulphated polysaccharides from seaweeds possess inhibitory effect against the replication of enveloped viruses including human immunodeficiency virus (HIV), herpes simplex virus (HSV), dengue virus, and respiratory syncytial virus [98]. Chinese traditional medicine have been widely used as antiviral agents due to their tendency to enhance and support the immune system via activating macrophagocytes to indorse their phagocytic capacity and persuade the secretion of IL-2, IFN, and antibodies [99]. Also, Inulin polysaccharides showed its antiviral characteristic by inhibiting the expression of TLR3/4 mRNA. It is noteworthy to mention that some factors such as polysaccharide concentration, structure and host cell type can impact the antiviral property of polysaccharides. Hence, this area needs further exploration to understand the antiviral mechanisms of different polysaccharides.

5.3 Drug Delivery and Tissues Engineering

Polysaccharides are greatly used in pharmaceutical industries due to its extensive properties. The major area of pharmaceutical industry using polysaccharide applications is focussed on drug delivery systems to support drug delivery and bioactive materials [3]. Polysaccharides have been greatly used to support the immune system, to stabilize therapeutics and their widespread and diverse tissue-specific targeting ability (or tissue engineering) [100]. The reason behind polysaccharides being utilized in drug and vaccine delivery is their tendency to get shaped and modified for specific functions.

Many studies have been reported in the literature where polysaccharides are broadly used in drug delivery. For instance, microorganisms-derived polysaccharides such as, xanthan gum, gellan gum, and scleroglucan have been studied in drug delivery wherein their absorption can be enhanced by loading them in bioadhesive polysaccharides nanoparticle carriers [101]. Additionally, chitosan and chitin are known to be low-immunogenic and tissue compatible polysaccharides and acquire applications in tissue engineering, bone regeneration, in drugs and vaccine delivery. Other polysaccharides i.e. hydrogels based on cross linked polysaccharides are also potential candidates for drug delivery systems and tissue engineering [101]. The hyper-branched polysaccharides are great choice for gene delivery nanoparticles. Recently, polysaccharides based nanoparticles are gaining much interest as nanometric carriers for drug delivery systems. For instance, Kotenkova et al. [102] discussed polysaccharide-based nanoformulations with natural antimicrobials. Polysaccharide-based nanoformulations exhibit diverse benefits in different area such as drug and gene delivery, tissue engineering, cancer therapy, food preservation, food packaging, etc. Polysaccharide-based nanoformulations can be formed via trapping, encapsulating or conjugating the bioactive compounds (Fig. 7) [102]. Polysaccharide nanocarriers can be prepared using different methods such as self assembly, cross-linking, ionic gelation, emulsification, desolvation, etc. [102]. This choice is entirely depends on the desired nanoparticle and polysaccharide properties. These methods are well discussed by Kotenkova et al. [102] (Fig. 8).

Polyelectrolyte complexes based on cross-linked ionic polysaccharides offers application in site-specific drug delivery, targeting, and biomarker-activated drug release. Emmanuel et al. [103] discussed drug delivery properties of some polyelectrolyte complexes based on ionically cross-linked polysaccharides. Their properties are facilitated by the diverse responsive mechanisms of polysaccharides to various external and internal stimuli. Efthimiadou et al. [100] reviewed the utilization of

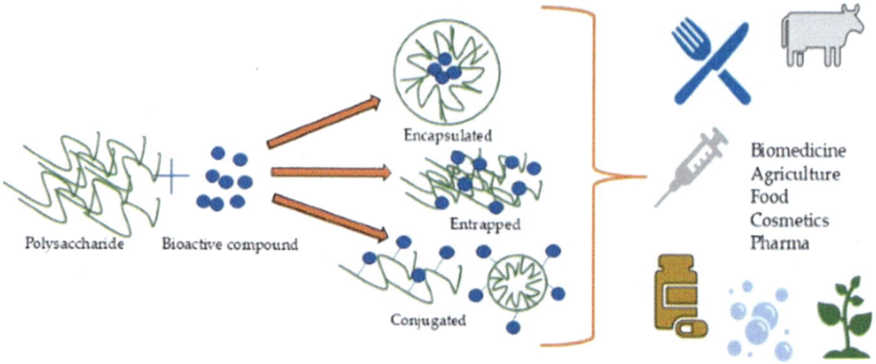

Fig. 7 Polysaccharide-based nanoformulations and their various applications [102]. From Ref. [102], used under a Creative Commons CC BY license

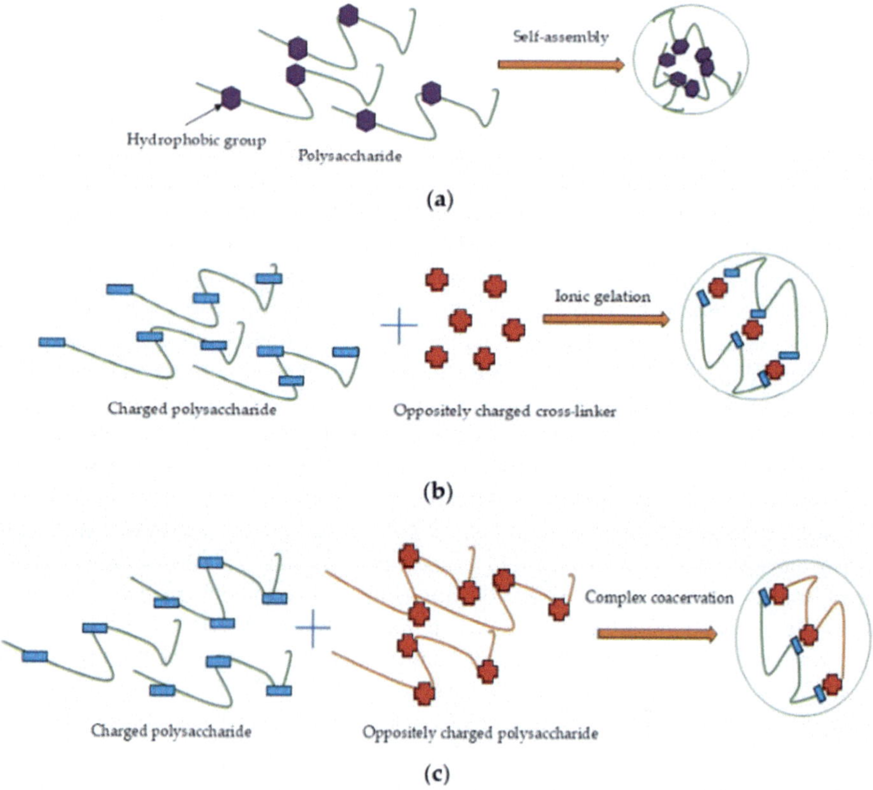

Fig. 8 Methods of synthesis of polysaccharide-based nanoformulations [102]. From Ref. [102], used under a Creative Commons CC BY license

biopolymer-based nanocarriers along with different factors such as pH, thermo- or redox sensitivity for targeted drug delivery applications. Encapsulation is applicable for drug and diagnostic delivery systems in the medical field. The wide properties of polysaccharides make them excellent choice for encapsulation of drugs and diagnostics. For example, Ndidi et al. [104] demonstrated the recent and advanced applications of polysaccharides and proteins as nanocarrier materials for encapsulation of diagnostics and therapeutics.

5.4 In Food Industry

Polysaccharides have been greatly utilized in the food industry as they are abundant in nature. Incorporation of polysaccharides in food processing can impact the physiological characteristics of food items, including their stability, texture, nutritional value and rheological properties [105]. Also, their incorporation increases the flavour and taste of food and promotes digestive health. For instance, many polysaccharides including starch, cellulose, chitosan, galactomannans, inulins, pectins, agars, alginates, xanthans and gums are significant source of nutrition and found in abundance [3]. These polysaccharides are considered as vital dietary components that are rich in macronutrients. Their remarkable properties such as, stabilizing, thickening, gelling, and emulsifying agents can be utilized as important food additives [3]. In biological processes, nutritional functions act as energy storage for metabolism and building material. For instance, starch and glycogen are responsible for metabolism in plants and animals, respectively.

Seed-derived polysaccharides can also be utilized as preservatives that enhance the food safety and durability [106]. Their incorporation can significantly inhibit the precipitation and sedimentation of proteins in food products thus, maintains the stability and quality. This application of polysaccharides can help to control body weight and chronic diseases. Basically, they form a colloidal structure that shields external environmental factors and thus increases the shelf life of food items [3].

6 Role of Polysaccharides in Coronavirus Disease (COVID-19)

Certainly, the coronavirus disease 2019 (COVID-19) outbreak has had a catastrophic effect on the world and became a major health challenge. It has become global health crisis with continuous searching for safe and effective therapeutic agents and vaccines [107]. Owing to the broad antiviral characteristics, polysaccharides have been considered as potential candidates for the prevention of the virus and as a therapeutic agent and carriers. Traditional Chinese medicine, GAGs and marine polysaccharides are well known for their potent anti-coronavirus properties. The sulfated polysaccharides inhibit the entry of the virus to the host cell via hindering the positive charge of the pathogen surface receptors. For instance, in vitro studies have shown that sulfated polysaccharides holds tightly the S-protein of SARS-CoV-2 and this interference has a potent inhibitory effect on viral infection [108]. Also, the cationically modified chitosan has been shown the major inhibition against the human coronavirus that indicate its inhibitory effect against low-pathogenic human coronaviruses.

Traditional Chinese medicine possesses potential anti- SARS-CoV-2 active compounds [109]. The major complication during COVID-19 that has been noticed was the vulnerability of the infected patients to the bacterial secondary infection wherein polysaccharides play an essential role. For instance, *Glycyrrhiza* polysaccharide exhibited antimicrobial activity by hindering the growth of *B. cereus, Staphylococcus aureus*, *E. Aerogens*, and *E. coli.*

7 Limitations and Challenges

Polysaccharides are the most abundant biopolymers that have gained much attention recently. Owing to the extensive properties such as, economical cheaper, less toxic and chemically modified, polysaccharides have a wide range of applications in different fields. Recently, polysaccharides acquire interest in the area of tissue engineering. Polysaccharide derived from mushrooms has gained attention in cancer therapy such as polysaccharide-peptide complex (PSP) and polysaccharide-protein complex (PSK). These polysaccharides can induce the immune response and shows the antitumor effect. Owing to the biodegradability and biocompatibility, polysaccharides have been greatly used to control the delivery of some drugs. The different functional groups present in the chemical structure of polysaccharides can be further modified and assist as a specific biological tools in different fields such as, tissue engineering, drug delivery systems and pharmacological activities.

As the polysaccharides are generally derived from natural sources, thus, there are few limitations and challenges linked with the polysaccharides in pharmaceutical and biological areas such as microbial contamination, thickening and uncontrolled rate of hydration. However, these limitations can be overcome through some modifications such as, grafting and cross-linking. Since, polysaccharides are naturally available in complex form with other compounds such as proteins and lipids; their isolation needs proficient and best approaches to avoid co-extraction and contamination with other compounds. Lastly, the determination and full knowledge of the structural-function relationship of polysaccharides may open way to new opportunities in pharmaceutical area.

8 Conclusion

Polysaccharides being a complex collection of biopolymers are generally derived from plant, animal, microalgal sources. These are made from monosaccharide units connected via O-glycosidic linkages. The present chapter comprises the main characteristics, synthesis, degradation and potential applications of polysaccharides. Polysaccharides possess several advantages over other synthetic polymers such as high abundance, low cost, stability, biocompatibility and prone to chemical modifications. Owing to these properties, polysaccharides have wide applications in different fields such as, pharmaceutical fields, drug delivery systems, tissue engineering and vaccine area. Additionally, polysaccharide can be applied biologically in other fields such as, like anti-tumor, anti-virus, anti-inflammatory, immunoregulatory and antioxidation activity. Despite the wide variety of applications, extensive investigation and elucidation of the structural-function relationship of polysaccharides are crucial to gain more insight into their precise mechanisms and entirely explore their future applications.

Conflict of Interest Authors declare no conflict of interest.

References

1. Darge HF, Andrgie AT, Tsai HC, Lai JY. Polysaccharide and polypeptide based injectable thermo-sensitive hydrogels for local biomedical applications. Int J Biol Macromol. 2019;15(133):545–63.
2. Zhang Y, Wang F. Carbohydrate drugs: current status and development prospect. Drug Discoveries Therapeutics. 2015;9(2):79–87.
3. Khodadadi Yazdi M, Seidi F, Hejna A, Zarrintaj P, Rabiee N, Kucinska-Lipka J, Saeb MR, Bencherif SA. Tailor-made polysaccharides for biomedical applications. ACS Appl Bio Mater. 2024;7(7):4193–230.
4. Liu Y, Huang G. Extraction and derivatisation of active polysaccharides. J Enzyme Inhib Med Chem. 2019;34(1):1690–6.
5. Mizumoto K, Sugawara I, Ito W, Kodama T, Hayami M, Mori S. Sulfated homopolysaccharides with immunomodulating activities are more potent anti-HTLV-III agents than sulfated heteropolysaccharides. Jpn J Exp Med. 1988;58(3):145–51.
6. Yazdi MK, Sajadi SM, Seidi F, Rabiee N, Fatahi Y, Rabiee M, Dominic CM, Zarrintaj P, Formela K, Saeb MR, Bencherif SA. Clickable polysaccharides for biomedical applications: a comprehensive review. Prog Polym Sci. 2022;1(133): 101590.
7. Xu BW, Li SS, Ding WL, Zhang C, Rehman MU, Tareen MF, Wang L, Huang SC. From structure to function: a comprehensive overview of polysaccharide roles and applications. Food Frontiers. 2025;6(1):15–39.
8. BeMiller JN. Gums and related polysaccharides. In: Glycoscience. Heidelberg: Springer; 2008. p. 1513–1533.
9. Benalaya I, Alves G, Lopes J, Silva LR. A review of natural polysaccharides: sources, characteristics, properties, food, and pharmaceutical applications. Int J Mol Sci. 2024;25(2):1322.
10. Delgado LL, Masuelli M. Polysaccharides: concepts and classification. Evol Polym Technol J. 2019;2(2): 180013.

11. Mohammed AS, Naveed M, Jost N. Polysaccharides; classification, chemical properties, and future perspective applications in fields of pharmacology and biological medicine (a review of current applications and upcoming potentialities). J Polym Environ. 2021;29(8):2359–71.
12. Sharma S, Bhende M. An overview: non-toxic and eco-friendly polysaccharides—its classification, properties, and diverse applications. Polym Bull. 2024;81(14):12383–429.
13. Ngwuluka NC, Abu-Thabit NY, Uwaezuoke OJ, Erebor JO, Ilomuanya MO, Mohamed RR, Soliman SM, Elella MH, Ebrahim NA. Natural polymers in micro-and nanoencapsulation for therapeutic and diagnostic applications: part I: lipids and fabrication techniques. In: Nano-and microencapsulation-techniques and applications 2020 Dec 29. IntechOpen.
14. Behrooznia Z, Nourmohammadi J. Polysaccharide-based materials as an eco-friendly alternative in biomedical, environmental, and food packaging. Giant. 2024;1(19): 100301.
15. Biswas MC, Jony B, Nandy PK, Chowdhury RA, Halder S, Kumar D, Ramakrishna S, Hassan M, Ahsan MA, Hoque ME, Imam MA. Recent advancement of biopolymers and their potential biomedical applications. J Polym Environ. 2022;30(1):51–74.
16. Koyyada A, Orsu P. Natural gum polysaccharides as efficient tissue engineering and drug delivery biopolymers. J Drug Deliv Sci Technol. 2021;1(63): 102431.
17. Salave S, Rana D, Sharma A, Bharathi K, Gupta R, Khode S, Benival D, Kommineni N. Polysaccharide based implantable drug delivery: development strategies, regulatory requirements, and future perspectives. Polysaccharides. 2022;3(3):625–54.
18. Jabeen N, Atif M. Polysaccharides based biopolymers for biomedical applications: a review. Polym Adv Technol. 2024;35(1): e6203.
19. Luo M, Zhang X, Wu J, Zhao J. Modifications of polysaccharide-based biomaterials under structure-property relationship for biomedical applications. Carbohyd Polym. 2021;15(266): 118097.
20. Hu X, Gilbert RG. Molecular structural insights into starch and glycogen: impact on health. Carbohyd Polym. 2025;25: 123958.
21. Neoh GK, Tan X, Chen S, Roura E, Dong X, Gilbert RG. Glycogen metabolism and structure: a review. Carbohyd Polym. 2024;15(346): 122631.
22. Imam, Green, Zaidi. Biopolymers utilizing nature's advances materials. ACS Series 723; 1997.
23. Seddiqi H, Oliaei E, Honarkar H, Jin J, Geonzon LC, Bacabac RG, Klein-Nulend J. Cellulose and its derivatives: towards biomedical applications. Cellulose. 2021;28(4):1893–931.
24. Marinho E. Cellulose: a comprehensive review of its properties and applications. Sustain Chem Environ. 2025;16: 100283.
25. Lin Q, Huang Y, Yu W. Effects of extraction methods on morphology, structure and properties of bamboo cellulose. Ind Crops Prod. 2021;1(169): 113640.
26. Rao J, Lv Z, Chen G, Peng F. Hemicellulose: structure, chemical modification, and application. Prog Polym Sci. 2023;1(140): 101675.
27. Zhu Y, Cai Y, Cao X, Li P, Liu D, Ye S, Xu Z, Shen B, Liao Q, Li H, Xie Z. Structural analysis and immunomodulatory activity of a homopolysaccharide isolated from Parabacteroides distasonis. Arab J Chem. 2022;15(5): 103755.
28. Katsuraya K, Okuyama K, Hatanaka K, Oshima R, Sato T, Matsuzaki K. Constitution of konjac glucomannan: chemical analysis and 13C NMR spectroscopy. Carbohyd Polym. 2003;53(2):183–9.
29. Li K, Liu X, Jiang F, Zhang B, Qiao D, Xie F. In the process of polysaccharide gel formation: a review of the role of competitive relationship between water and alcohol molecules. Int J Biol Macromol. 2024;1(281): 136398.
30. Melrose J. Glycosaminoglycans, instructive biomolecules that regulate cellular activity and synaptic neuronal control of specific tissue functional properties. Int J Mol Sci. 2025;26(6):2554.
31. Bedini E, Corsaro MM, Fernández-Mayoralas A, Iadonisi A. Chondroitin, dermatan, heparan, and keratan sulfate: structure and functions. In: Extracellular sugar-based biopolymers matrices. Cham: Springer International Publishing; 2019. p. 187–233.

32. Duceac IA, Stanciu MC, Nechifor M, Tanasa F, Teaca CA. Insights on some polysaccharide gel type materials and their structural peculiarities. Gels. 2022;8(12):771.
33. Cao Y, Mezzenga R. Design principles of food gels. Nature Food. 2020;1(2):106–18.
34. Makshakova ON, Zuev YF. Interaction-induced structural transformations in polysaccharide and protein-polysaccharide gels as functional basis for novel soft-matter: a case of carrageenans. Gels. 2022;8(5):287.
35. Li W, Fang K, Yuan H, Li D, Li H, Chen Y, Luo X, Zhang L, Ye X. Acid-induced Poria cocos alkali-soluble polysaccharide hydrogel: gelation behaviour, characteristics, and potential application in drug delivery. Int J Biol Macromol. 2023;1(242): 124383.
36. Patel P, Mujmer K, Aswal VK, Gupta S, Thareja P. Structure, rheology, and 3D printing of salt-induced ?-carrageenan gels. Mater Today Commun. 2023;1(35): 105807.
37. Chen M, Yu P, Ao C, Zhang M, Xing J, Ding C, Xie J, Li J. Ethanol-induced responsive behavior of natural polysaccharide hydrogels. Ind Eng Chem Res. 2022;61(35):13145–53.
38. Song Y, Li S, Gong H, Yip RC, Chen H. Biopharmaceutical applications of microbial polysaccharides as materials: a review. Int J Biol Macromol. 2023;1(239): 124259.
39. Bojarczuk A, Skapska S, Khaneghah AM, Marszalek K. Health benefits of resistant starch: a review of the literature. J Functional Foods. 2022;1(93): 105094.
40. Ruprecht C, Blaukopf M, Pfrengle F. Synthetic fragments of plant polysaccharides as tools for cell wall biology. Curr Opin Chem Biol. 2022;1(71): 102208.
41. Sharahi M, Bahrami SH, Karimi A. A comprehensive review on guar gum and its modified biopolymers: their potential applications in tissue engineering. Carbohyd Polym. 2025;1(347): 122739.
42. Wu X, Guo Y, Dai C, Zhao C. The pharmacological potential of algal polysaccharides in food applications and chronic disease management. Future Pharmacol. 2025;5(2):29.
43. Nigam S, Singh R, Bhardwaj SK, Sami R, Nikolova MP, Chavali M, Sinha S. Perspective on the therapeutic applications of algal polysaccharides. J Polym Environ. 2022;30(3):785–809.
44. Tziveleka LA, Ioannou E, Roussis V. Ulvan, a bioactive marine sulphated polysaccharide as a key constituent of hybrid biomaterials: a review. Carbohyd Polym. 2019;15(218):355–70.
45. Oliveira C, Ferreira AS, Novoa-Carballal R, Nunes C, Pashkuleva I, Neves NM, Coimbra MA, Reis RL, Martins A, Silva TH. The key role of sulfation and branching on fucoidan antitumor activity. Macromol Biosci. 2017;17(5):1600340.
46. Patel S. Therapeutic importance of sulfated polysaccharides from seaweeds: updating the recent findings. 3 Biotech. 2012;2(3):171–85.
47. Rashedy SH, Abd El Hafez MS, Dar MA, Cotas J, Pereira L. Evaluation and characterization of alginate extracted from brown seaweed collected in the Red Sea. Appl Sci. 2021;11(14):6290.
48. Liu YY, Ding YF, Sui HJ, Liu W, Zhang ZQ, Li F. Pilose antler (Cervus elaphus Linnaeus) polysaccharide and polypeptide extract inhibits bone resorption in high turnover type osteoporosis by stimulating the MAKP and MMP-9 signaling pathways. J Ethnopharmacol. 2023;25(304): 116052.
49. Satitsri S, Muanprasat C. Chitin and chitosan derivatives as biomaterial resources for biological and biomedical applications. Molecules. 2020;25(24):5961.
50. Gal MR, Rahmaninia M, Hubbe MA. A comprehensive review of chitosan applications in paper science and technologies. Carbohyd Polym. 2023;1(309): 120665.
51. Vasvani S, Kulkarni P, Rawtani D. Hyaluronic acid: a review on its biology, aspects of drug delivery, route of administrations and a special emphasis on its approved marketed products and recent clinical studies. Int J Biol Macromol. 2020;15(151):1012–29.
52. Hao C, Xu H, Yu L, Zhang L. Heparin: an essential drug for modern medicine. Prog Mol Biol Transl Sci. 2019;1(163):1–9.
53. Bhattacharya M, Samanta D. Microbial polysaccharides and their classification: a brief study on fundamental biochemical properties.
54. Kumar A, Rao KM, Han SS. Application of xanthan gum as polysaccharide in tissue engineering: a review. Carbohyd Polym. 2018;15(180):128–44.
55. Lanvers P, Broeker J, Schmid J. Clinical-type dextran: a review on process and enzyme engineering strategies to control molecular weight distribution. Carbohyd Polym. 2025;22: 123783.

56. de Souza CK, Ghosh T, Lukhmana N, Tahiliani S, Priyadarshi R, Hoffmann TG, Purohit SD, Han SS. Pullulan as a sustainable biopolymer for versatile applications: a review. Mater Today Commun. 2023;1(36): 106477.
57. Xu SY, Huang X, Cheong KL. Recent advances in marine algae polysaccharides: isolation, structure, and activities. Mar Drugs. 2017;15(12):388.
58. Zhao L, Ma Z, Yin J, Shi G, Ding Z. Biological strategies for oligo/polysaccharide synthesis: biocatalyst and microbial cell factory. Carbohyd Polym. 2021;15(258): 117695.
59. Li R, Yu H, Chen X. Recent progress in chemical synthesis of bacterial surface glycans. Curr Opin Chem Biol. 2020;1(58):121–36.
60. Yang S, Yang X, Zhang H. Extracellular polysaccharide biosynthesis in Cordyceps. Crit Rev Microbiol. 2020;46(4):359–80.
61. Nag A, Karpinets TV, Chang CH, Bar-Peled M. Enhancing a Pathway-Genome Database (PGDB) to capture subcellular localization of metabolites and enzymes: the nucleotidesugar biosynthetic pathways of Populus trichocarpa. Database. 2012 Jan 1;2012:bas013.
62. Thompson AJ, Cao L, Ma Y, Wang X, Diedrich JK, Kikuchi C, Willis S, Worth C, McBride R, Yates JR, Paulson JC. Human influenza virus hemagglutinins contain conserved oligomannose N-linked glycans allowing potent neutralization by lectins. Cell Host Microbe. 2020;27(5):725–35.
63. Wiseman B, Nitharwal RG, Widmalm G, Högbom M. Structure of a full-length bacterial polysaccharide co-polymerase. Nat Commun. 2021;12(1):369.
64. Islam ST, Huszczynski SM, Nugent T, Gold AC, Lam JS. Conserved-residue mutations in Wzy affect O-antigen polymerization and Wzz-mediated chain-length regulation in Pseudomonas aeruginosa PAO1. Sci Rep. 2013;3(1):3441.
65. Li BH, Ye XS. Recent advances in glycan synthesis. Curr Opin Chem Biol. 2020;1(58):20–7.
66. Tanaka H, Kawai T, Adachi Y, Hanashima S, Yamaguchi Y, Ohno N, Takahashi T. Synthesis of ß (1, 3) oligoglucans exhibiting a Dectin-1 binding affinity and their biological evaluation. Bioorg Med Chem. 2012;20(12):3898–914.
67. Petitou M, Hérault JP, Bernat A, Driguez PA, Duchaussoy P, Lormeau JC, Herbert JM. Synthesis of thrombin-inhibiting heparin mimetics without side effects. Nature. 1999;398(6726):417–22.
68. Inuki S, Tabuchi H, Matsuzaki C, Yonejima Y, Hisa K, Kimura I, Yamamoto K, Ohno H. Chemical synthesis and evaluation of exopolysaccharide fragments produced by Leuconostoc mesenteroides strain NTM048. Chem Pharm Bull. 2022;70(2):155–61.
69. Liu M, Qin X, Ye XS. Glycan assembly strategy: from concept to application. Chem Rec. 2021;21(11):3256–77.
70. Zhou S, Huang G. Preparation, structure and activity of polysaccharide phosphate esters. Biomed Pharmacother. 2021;1(144): 112332.
71. Chen F, Huang G, Yang Z, Hou Y. Antioxidant activity of Momordica charantia polysaccharide and its derivatives. Int J Biol Macromol. 2019;1(138):673–80.
72. Zhang M, Su N, Huang Q, Zhang Q, Wang Y, Li J, Ye M. Phosphorylation and antiaging activity of polysaccharide from Trichosanthes peel. J Food Drug Anal. 2017;25(4):976–83.
73. Gao P, Bian J, Xu S, Liu C, Sun Y, Zhang G, Li D, Liu X. Structural features, selenization modification, antioxidant and anti-tumor effects of polysaccharides from alfalfa roots. Int J Biol Macromol. 2020;15(149):207–14.
74. Huang S, Yang W, Huang G. Preparation and activities of selenium polysaccharide from plant such as Grifola frondosa. Carbohyd Polym. 2020;15(242): 116409.
75. Ren YY, Sun PP, Ji YP, Wang XT, Dai SH, Zhu ZY. Carboxymethylation and acetylation of the polysaccharide from Cordyceps militaris and their a-glucosidase inhibitory activities. Nat Prod Res. 2020;34(3):369–77.
76. Staudacher E. Methylation–an uncommon modification of glycans. Biol Chem. 2012;393(8):675.
77. Xia T, Liu CS, Hu YN, Luo ZY, Chen FL, Yuan LX, Tan XM. Coix seed polysaccharides alleviate type 2 diabetes mellitus via gut microbiota-derived short-chain fatty acids activation of IGF1/PI3K/AKT signaling. Food Res Int. 2021;1(150): 110717.

78. Wu Q, Qin D, Cao H, Bai Y. Enzymatic hydrolysis of polysaccharide from Auricularia auricula and characterization of the degradation product. Int J Biol Macromol. 2020;1(162):127–35.
79. Kobayashi H, Nagahama T, Arai W, Sasagawa Y, Umeda M, Hayashi T, Nikaido I, Watanabe H, Oguri K, Kitazato H, Fujioka K. Polysaccharide hydrolase of the hadal zone amphipods Hirondellea gigas. Biosci Biotechnol Biochem. 2018;82(7):1123–33.
80. Zeng S, Long J, Sun J, Wang G, Zhou L. A review on peach gum polysaccharide: hydrolysis, structure, properties and applications. Carbohyd Polym. 2022;1(279): 119015.
81. La Reau AJ, Suen G. The Ruminococci: key symbionts of the gut ecosystem. J Microbiol. 2018;56(3):199–208.
82. Yeager CM, Gallegos-Graves LV, Dunbar J, Hesse CN, Daligault H, Kuske CR. Polysaccharide degradation capability of Actinomycetales soil isolates from a semiarid grassland of the Colorado Plateau. Appl Environ Microbiol. 2017;83(6):e03020-e3116.
83. Cifuente JO, Comino N, Trastoy B, D'Angelo C, Guerin ME. Structural basis of glycogen metabolism in bacteria. Biochem J. 2019;476(14):2059–92.
84. Ellingwood SS, Cheng A. Biochemical and clinical aspects of glycogen storage diseases. J Endocrinol. 2018;238(3):R131–41.
85. Dotz V, Wuhrer M. Histo-blood group glycans in the context of personalized medicine. Biochimica et Biophysica Acta (BBA)-General Subjects. 2016;1860(8):1596–607.
86. Zhao L, Yang H, Zheng X, Li J, Jian L, Feng W, Kong J. Dual signal amplification by polysaccharide and eATRP for ultrasensitive detection of CYFRA 21–1 DNA. Biosens Bioelectron. 2020;15(150): 111895.
87. Zhu H, Tian LI, Zhang L, Bi J, Song Q, Yang H, Qiao J. Preparation, characterization and antioxidant activity of polysaccharide from spent Lentinus edodes substrate. Int J Biol Macromol. 2018;1(112):976–84.
88. Liu Y, Ye Y, Hu X, Wang J. Structural characterization and anti-inflammatory activity of a polysaccharide from the lignified okra. Carbohyd Polym. 2021;1(265): 118081.
89. Peng L, Tang S, Li H, Wang Q, Long T, Zhang H, Wu Q, Wang Y, Liu L, Tang X, Li Z. Angelica sinensis polysaccharide suppresses the aging of hematopoietic stem cells through Sirt1/FoxO1 signaling. Clin Lab. 2022;68(5).
90. Guo S, Zhao Y, Luo S, Mu D, Li X, Zhong X, Jiang S, Zheng Z. Encapsulation of curcumin in soluble soybean polysaccharide-coated gliadin nanoparticles: interaction, stability, antioxidant capacity, and bioaccessibility. J Sci Food Agric. 2022;102(12):5121–31.
91. Zhang Y, Liu Y, Zhou Y, Zheng Z, Tang W, Song M, Wang J, Wang K. Lentinan inhibited colon cancer growth by inducing endoplasmic reticulum stress-mediated autophagic cell death and apoptosis. Carbohyd Polym. 2021;1(267): 118154.
92. Yu S, Dong X, Ji H, Yu J, Liu A. Antitumor activity and immunomodulation mechanism of a novel polysaccharide extracted from Polygala tenuifolia Willd. evaluated by S180 cells and S180 tumor-bearing mice. International Journal of Biological Macromolecules. 2021;192:546–56.
93. Wang Y, Sun M, Jin H, Yang J, Kang S, Liu Y, Yang S, Ma S, Ni J. Effects of Lycium barbarum polysaccharides on immunity and the gut microbiota in cyclophosphamideinduced immunosuppressed mice. Front Microbiol. 2021;6(12): 701566.
94. Guo X, Wang Y, Qin Y, Shen P, Peng Q. Structures, properties and application of alginic acid: a review. Int J Biol Macromol. 2020;1(162):618–28.
95. Bai Y, Jia X, Huang F, Zhang R, Dong L, Liu L, Zhang M. Structural elucidation, antiinflammatory activity and intestinal barrier protection of longan pulp polysaccharide LPIIa. Carbohyd Polym. 2020;15(246): 116532.
96. Jen CI, Su CH, Lu MK, Lai MN, Ng LT. Synergistic anti-inflammatory effects of different polysaccharide components from Xylaria nigripes. J Food Biochem. 2021;45(4): e13694.
97. Wang P, Chi L, Zhang Z, Zhao H, Zhang F, Linhardt RJ. Heparin: An old drug for new clinical applications. Carbohyd Polym. 2022;1(295): 119818.
98. Jiao G, Yu G, Zhang J, Ewart HS. Chemical structures and bioactivities of sulfated polysaccharides from marine algae. Mar Drugs. 2011;9(2):196–223.

99. Chen Y, Yao F, Ming K, Wang D, Hu Y, Liu J. Polysaccharides from traditional Chinese medicines: extraction, purification, modification, and biological activity. Molecules. 2016;21(12):1705.
100. Efthimiadou EK, Theodosiou M, Toniolo G, Abu-Thabit NY. Stimuli-responsive biopolymer nanocarriers for drug delivery applications. Stimuli Responsive Polym Nanocarriers Drug Deliv Appl. 2018;1(1):405–32.
101. Ganesan O, Kiwanuka H, Hamaguchi R, Orgill DP. A review of regenerative medicine and tissue engineering with a focus on wound healing and anti-aging. Frontiers Surgery. 2025;5(12):1504563.
102. Kotenkova E, Kotov A, Nikitin M. Polysaccharide-based nanocarriers for natural antimicrobials: a review. Polymers. 2025;17(13):1750.
103. Emmanuel BD, Abu-Thabit NY, Ngwuluka NC. Responsive polyelectrolyte complexes based on natural polysaccharides for drug delivery applications. Stimuli Responsive Polym Nanocarriers Drug Deliv Appl. 2018;1(1):267–87.
104. Ngwuluka NC, Abu-Thabit NY, Uwaezuoke OJ, Erebor JO, Ilomuanya MO, Mohamed RR, Soliman SM, Elella MH, Ebrahim NA. Natural polymers in micro-and nanoencapsulation for therapeutic and diagnostic applications: part II-polysaccharides and proteins. Nano-and microencapsulation-techniques and applications. 2021 Jan 4.
105. Wang Z, Wang L, Yu X, Wang X, Zheng Y, Hu X, Zhang P, Sun Q, Wang Q, Li N. Effect of polysaccharide addition on food physical properties: a review. Food Chem. 2024;15(431): 137099.
106. Abbasi A, Sabahi S, Bazzaz S, Tajani AG, Lahouty M, Aslani R, Hosseini H. An edible coating utilizing Malva sylvestris seed polysaccharide mucilage and postbiotic from Saccharomyces cerevisiae var. boulardii for the preservation of lamb meat. Int J Biol Macromolecules. 2023;246:125660.
107. Naveed M, Uddin S, Ishaq SE, Ahmad T. Various evidence-based hypothetical and experimental treatment approaches and their effectiveness against COVID-19 worldwide: a comprehensive literature review. Eurasian J Med Oncol. 2020;4(4):265–85.
108. Kwon PS, Oh H, Kwon SJ, Jin W, Zhang F, Fraser K, Hong JJ, Linhardt RJ, Dordick JS. Sulfated polysaccharides effectively inhibit SARS-CoV-2 in vitro. Cell Discovery. 2020;6(1):50.
109. Chen X, Han W, Wang G, Zhao X. Application prospect of polysaccharides in the development of anti-novel coronavirus drugs and vaccines. Int J Biol Macromol. 2020;1(164):331–43.
110. Ji X, Liu N, Huang S, Zhang C. A comprehensive review of licorice: the preparation, chemical composition, bioactivities and its applications. Am J Chin Med. 2024;52(03):667–716.

Polysaccharide Hydrogel-Based 3D Printing Technology

Muhammad Humais Yamin, Qudsia Kanwal, Muhammad Nawaz Rasool, Malaika, Muhammad Hanif, and Muhammad Hamza

Abstract This chapter explores the evolution and application of polysaccharide hydrogel materials in additive manufacturing technology. It also highlights 3D-printed polysaccharides adaptability, biocompatibility, and sustainability across various biomedical domains like drug delivery, tissue engineering, and regenerative medicines. Polysaccharide-based hydrogels, sourced from natural materials e.g., starch, chitosan, alginate, and cellulose, have discrete benefits, counting facile functionalization, tunable properties, and biodegradability. The chapter examines various fabrication techniques, namely extrusion-based and inkjet-based 3D printing technologies. It highlights the limitations associated with their processing or fabrication, including maintaining structural integrity and optimizing print resolution. The capabilities of polysaccharide hydrogels in fabricating intricate, hierarchical structures with exceptional accuracy and application in developing scaffolds for cell culture and tissue regeneration are amazing. This chapter also covers the recent breakthroughs in enhancing these hydrogels' mechanical strength, stability, and bioactivity via crosslinking techniques and composite formations. The future possibilities and current research initiatives focused on improving the performance and broadening the applications of polysaccharide hydrogel-based 3D printing are examined, establishing them as a potential material for next-generation biomedical and industrial applications.

Keywords 3D Printing · Polysaccharide hydrogels · Crosslinking · Chemical modification · Biocompatibility · Mechanical properties · Tissue engineering · Drug delivery · Regenerative medicine

---

Muhammad Humais Yamin, Qudsia Kanwal are contributed equally

M. H. Yamin
Department of Physics, University of Gujrat, Gujrat, Pakistan

Q. Kanwal · M. N. Rasool · Malaika
Department of Chemistry, The University of Lahore, Lahore, Pakistan

M. Hanif · M. Hamza (✉)
Additive Manufacturing Institute, Shenzhen University, Shenzhen, China
e-mail: muhammadhamza7596173@gmail.com

M. H. Abu Elella et al. (eds.), *3D Engineered Polysaccharides*,
https://doi.org/10.1007/978-3-032-26154-0_2

Abbreviations

3D	Three Dimensional
AgNPs	Silver Nanoparticles
BSA	Bovine Serum Albumin
BTB	Bromothymol Blue
Ca^{2+}	Calcium Ion
$CaCO_3$	NPs-GDL Calcium Carbonate Nanoparticles–Gluconic Delta-Lactone
CAD	Computer-Aided Design
Cap	Calcium Phosphate
ChNF	Chitin Nanofibers
CMC	Carboxymethyl Cellulose
CO_2	Carbon Dioxide
–COOH	Carboxyl Group
CT	Computed Tomography
CuAAC	Copper-Catalyzed Azide–Alkyne Cycloaddition
CuO	Copper (II) Oxide
DIW	Direct Ink Writing
DLP	Digital Light Processing
ECM	Extracellular Matrix
E-Jet	Electrohydrodynamic Jet
FDM	Fused Deposition Modeling
G	Guluronate
G'	Storage Modulus
G"	Loss Modulus
GelMA	Gelatin Methacryloyl
GO	Graphene Oxide
GO/CS	Graphene Oxide/Chitosan
H_2O_2	Hydrogen Peroxide
H_3PO_4	Phosphoric Acid
HA	Humic Acid
HAMA	Hyaluronic Acid Methacryloyl
K^+	Potassium Ion
Kda	Kilodalton
M	Mannuronate
MBG	Mesoporous Bioactive Glass
MR	Methyl Red
MRI	Magnetic Resonance Imaging
Na^+	Sodium Ion
$-NH_2$	Amino Group
NPs	Nanoparticles
–OH	Hydroxide Group
PEGDA	Poly(Ethylene Glycol) Diacrylate
PLA	Polylactic Acid

PVA	Polyvinyl Alcohol
ROS	Reactive Oxygen Species
SA	Sodium Alginate
SIS	Small Intestinal Submucosa
SLA	Stereolithography
TCNF	TEMPO-Oxidized Cellulose Nanofibers
TEMPO	(2,2,6,6-Tetramethylpiperidine-1-Oxyl Radical)
TPMS	Triply Periodic Minimal Surface
UF	Ultrafiltration
UV	Ultra Violet

1 Introduction

Additive manufacturing has become a groundbreaking technology in various fields, especially in biomedical engineering, environmental and materials science. In this technological advancement, polysaccharide hydrogels have received considerable attention as a better material with excellent biocompatibility, adjustable mechanical properties, and environmental compatibility. The chapter discusses polysaccharide hydrogel materials have been developed and used in additive manufacturing and highlights their versatility, bio-compatibility [1], and sustainability to find applications in tissue engineering as well as pharmaceutical delivery systems and cell regeneration [2].

The polysaccharides are based on natural materials such as starch, chitosan, alginate, and cellulose, and they convey distinct advantages in terms of 3D fabrication. These renewable materials are intrinsically biodegradable and superbly biocompatible coupled with broad spectrum of chemical modification and property tailoring capabilities. This chapter addresses the manufacturing technologies used with these hydrogels with a specific emphasis on extrusion-based and inkjet printing technologies [3–5]. They enable the creation of sophisticated, multi-scale structures with precision at the micron scale, and have proven effective in creating complex cellular structures and in regenerative tissue scaffolding [6].

Chemical modification and crosslinking approaches complement the functional modification of polysaccharide hydrogels to critically improve their mechanical strength, structural integrity, and biologic activity [7, 8]. Recent developments in crosslinking procedures and composite materials technology have produced hydrogels with superior performance attributes, which have greatly improved their potential uses [9–12]. The technical challenges associated with the processing of these materials are also discussed which includes the maintenance of structural fidelity and optimal resolution of printed structures [13].

Polysaccharide hydrogels demonstrate outstanding promise as biomimetic surfaces in tissue engineering applications.

These materials are precisely engineered to replicate the structural and biochemical properties of native ECM [14], thereby creating optimal microenvironments for cellular growth and specialization. The current investigations specifically evaluate polysaccharide implementation in regenerative scaffolds for osseous, chondral, and neural tissue applications [15]. The capacity to precisely tailor scaffold morphology and mechanical characteristics through additive manufacturing techniques provides unique opportunities for developing personalized medical implants and tissue constructs [6, 16].

Polysaccharide hydrogels have exhibited significant potential beyond tissue engineering, particularly in drug delivery systems where their ability to encapsulate and controllably release therapeutic agents enables sustained release kinetics and enhanced drug bioavailability [17]. The chapter explores the application of polysaccharides-based hydrogels for diabetes treatment through insulin delivery, targeted cancer therapy, and ocular drug administration, with their tunable properties allowing adaptation to various administration routes and release profiles [18]. Additionally, this chapter highlights environmental and industrial applications of polysaccharides, including eco-friendly filtration systems [19], biodegradable packaging solutions [20], and slow-release fertilizers [21], where 3D printing technology helps the fabrication of complex structures with precisely controlled material properties for water purification, food packaging, and agricultural applications.

Polysaccharide hydrogels offer several advantages for 3D printing applications, although many challenges persist, including the requirement to improve mechanical properties, optimize printing processes for higher resolution and structural integrity [22], and address regulatory and ethical considerations in biomedical implementations. Looking ahead, the field shows promising potential through developing smart responsive hydrogels, creating hybrid materials, and adopting sustainable manufacturing practices [23]. The integration of polysaccharide hydrogels with 3D printing technology has created transformative opportunities across biomedical, environmental, and industrial sectors by enabling precise functional tailoring through chemical modifications and advanced fabrication techniques [24]. These capabilities of these materials as key components in next-generation solutions for tissue engineering, drug delivery systems, and sustainable material development, with ongoing research advancing their potential to address critical challenges in modern science and technology.

2 Polysaccharides as Hydrogel Materials

2.1 *Types of Polysaccharides Used in Hydrogels*

2.1.1 Starch

Starch is a neutral polysaccharide produced by plants like maize, rice, and wheat, serving as their primary energy reserve. It consists of two main components: amylose, a linear polymer of (1,4)-linked α-d-glucan (10^5 g·mol^{-1}), and amylopectin, a branched form of α-d-glucan (10^6–10^7 g·mol^{-1}). At room temperature, starch is insoluble in water and forms a suspension [25]. However, upon heating, the granules swell and gelatinize, which facilitates the separation of amylose from amylopectin and creates a continuous phase around the swollen granules. When cooled, the amylose phase separates, resulting in a stable gel that is biocompatible, making it suitable for use in biomedical applications, such as drug delivery systems [24]. To date, starch has not been explored for developing cell-laden bioinks for 3D bioprinting, possibly due to the requirement for heat treatments to dry and preserve the structural integrity of starch-based prints [26].

2.1.2 Chitosan

The shells of crustaceans, such as shrimp and crabs are significant sources of chitin, the second most abundant polymer after cellulose. Chitin is a polysaccharide made of β-(1,4)-linked N-acetyl-D-glucosamine units, primarily found in the exoskeletons of crustaceans, as well as in insects and fungal cell walls [27]. It is biodegradable, non-toxic, and known for its antibacterial properties, which make it valuable in various biomedical applications, including drug delivery and tissue engineering. A notable derivative of chitin is chitosan, produced through the enzymatic or chemical deacetylation of chitin, typically exceeding 50% deacetylation. Chitosan has a molecular weight ranging from 50 to 2000 kDa. Its antimicrobial effects are partly due to its positive charge and its solubility in slightly acidic solutions because of the deacetylation [28]. Multivalent anions (glycerophosphate and tripolyphosphate) may also be used to form gels along with chitosan. It is biodegradable, biocompatible, and low immunogenicity, hence, a potential source of numerous studies in the application of the substance within the field of drug delivery and wound healing [2].

2.1.3 Alginate

Alginate is a brown algae (Phaeophyceae) polysaccharide, with a polyanionic, water-soluble linear structure. Its polymeric structures are composed of mannuronate (M) and guluronate (G) units with different proportions and molecular weight of 10–600 kDa, depending on the source of the algae [29]. Ionotropic gelation of divalent cations such as Ca^{2+} can be used to rapidly and at mild conditions to form hydrogels of alginate. This gelation mechanism is based on the "egg-box" model with Ca^{2+} ions entrap into the cavities formed by the cooperative structure of the surrounding G units [30]. Alginate hydrogels are affected by the concentration and molecular weight of polymer and the amount of G units, affecting the properties of the polymer, such as viscosity, elasticity, and pore size. One of the best-researched 3D bioprinting systems are Ca^{2+}-alginate hydrogels because they are highly tunable and printable. Nevertheless, recent research emphasized overcoming some biological and mechanical drawbacks of alginate-based hydrogels. It is important to note that, unlike cell-binding receptors, alginate is quite inert and therefore, does not facilitate cell adhesion and growth in bioinks [3]. Moreover, the non-uniformity of degradation rates may cause eventual mechanical instability in 3D bioprinted structures. To address these issues, continuous research efforts are being conducted to make changes to the alginate hydrogel to make it more efficient.

2.1.4 Cellulose and Derivatives

Plant cell walls are mainly composed of cellulose, which is the most common polysaccharide on earth. It consists of β-D-glucopyranose units connected by β-(1,4) glycosidic bonds, and its degree of polymerization is variable, ranging about 10,000 in the case of the native cellulose and about 15,000 in the case of cellulose in cotton fibres. In other plant fibers, the level of polymerization lies between 800 and 10,000 basing on the extraction methods to be applied. Cellulose is rich in free hydroxyl groups that form a high density of intra and intermolecular hydrogen bonds, making it easy to form strong cellulose fibers. These fibers are comprised of crystalline and amorphous areas located alternately [31]. There are many uses of cellulose fibers such as paper, textiles, and composite production. Nonetheless, cellulose is insoluble in many typical solvents due to the strong hydrogen bonding network [7]. To overcome this problem, a range of properties and applications of cellulose derivatives can be developed by chemical restructuring. Examples include carboxymethyl cellulose (CMC), methyl cellulose, ethyl cellulose, hydroxyethyl cellulose, and hydroxypropyl methyl cellulose, which are produced by treating cellulose fibers with alkyl halides in an alkaline medium. The derivatives can be custom-made with different degrees of substitutions, which have different features. The recent research revealed the possibility of using cellulose derivatives to create bioinks, with emphasis on CMC. Other derivatives such as methyl cellulose, hydroxyethyl cellulose, and hydroxypropyl methyl cellulose are also finding applications in such a manner.

2.2 *Properties of Polysaccharide Hydrogels*

2.2.1 Biocompatibility

In 3D bioprinting tissue engineering, preservation of cell viability, proliferation, and differentiation ability in bioinks is very important. Hydrogels made of polysaccharides are extensively investigated because of their biocompatibility and biodegradability that helps to reduce cytotoxicity and guarantees a smooth integration of the hydrogel with the host tissue [1]. Recent developments have redefined biocompatibility, not only in terms of lack of harmful effects, but also how it serves as a positive stimulus to cellular functions, and even tissue regeneration. Natural polysaccharides (alginate, chitosan, and cellulose, in particular) are also beneficial as bioinks since they are more reminiscent of the extracellular matrix (ECM), which creates more friendly microenvironment conditions in cells. Nevertheless, polysaccharide hydrogels have many limitations like poor mechanical strength and possible immune responses, which need to be utilized to better their use in bio-printing practices.

2.2.2 Biodegradability

The biodegradability of polysaccharide-based hydrogels, enabling them to degrade spontaneously during enzyme or microbial activity, makes them highly prized in biomedical applications and in environmental fields. Biodegradation is the process of breaking down polymeric structures into simpler chemical structures, and it depends on the cross-linking density, chemical modification, and environmental factors [32]. To best utilize them it is critical to know how these factors influence the degradation kinetics. The breakdown of the polysaccharide hydrogels is highly variable in that, days to years, it depends on the chemical composition of the polysaccharide gel. Examples include chitosan which decomposes in weeks to months, alginate which decomposes in weeks to months, and cellulose derivatives which last months to years. Biodegradability can be manipulated to suit a given need and application in polysaccharide hydrogel. Control methods of biodegradability include chemical modification [9]. The more the cross-linking, the slower the degradation, preventing access by enzymes and decreasing hydrogel swelling, and increasing the rate of hydrolytically or enzymatically sensitive bonds [33]. Further, environmental parameters such as pH exploited can be, e.g., acid-labile hydrogels decay more rapidly in inflammatory microenvironments. By these modifications, it is possible to program a controlled degradation of hydrogel polysaccharides to apply them in the fields of drug delivery, tissue engineering, and greener materials [34].

2.2.3 Mechanical Properties

To ensure a successful tissue engineering activity, bioprinted hydrogel materials have to be strong enough to withstand the stresses of bioprinting and maintain their structure after printing and degrade slowly over time to assist the regeneration of the tissue. The bioink similarly replicates the mechanical characteristics of the target tissue to enable healthy integration and functionality. Natural polysaccharide hydrogels have superior biocompatibility and biodegradability, but their natural low mechanical properties can make them incapable of withstanding bioprinting pressures or the long-term stability of load-bearing tissues. To resolve this problem, scientists usually enhance polysaccharide-based bioinks with synthetic polymers or nanomaterials [35, 36]. The modification provides the material with greater structural stability to support mechanically strained tissues such as cartilage, bone, or cardiovascular systems [15]. The other major problem that demands custom cross-linking strategies or composite blends in order to generate tissue-like functionality is the biodegradation-mechanical strength trade-offs.

2.2.4 Swelling Behavior and Water Retention

Hydrogel polymers made of polysaccharide have outstanding ability to absorb water because of cross-linked polymeric networks. These hydrogel systems contain ionic groups that greatly increase liquid uptake capacity of these hydrogels which is important in facilitating effective transport of nutrients and cellular interactions inside the gel matrix [37]. Kinetics of drug release of hydrogel-based delivery systems are also effectively modulated by this property. It has also been observed that a blend of covalent and physical cross-linking approaches had a beneficial effect on the swelling character and reduced the elasticity index, making this type of hydrogel more flexible where it was used in the biomedical environment [38]. Various factors affect the swelling characteristics of polysaccharide hydrogels, such as density of cross-linking, ionic strength, synthesis, and type of chemical bonds [39]. The swelling ratio, defined as the weight ratio of a swollen hydrogel to its dry state, serves as a key indicator of hydrogel performance [17]. Higher cross-linking densities typically result in lower swelling ratio values due to restricted network expansion, whereas loosely cross-linked hydrogels exhibit greater swelling capacity. Additionally, the balance between hydrophilic and hydrophobic functional groups in the polymer structure directly affects water absorption, hydrogels with more hydrophilic moieties demonstrate superior swelling properties [40].

3 3D Printing Techniques for Polysaccharide Hydrogels

Three-dimensional bioprinting has revolutionized tissue engineering by enabling the precise fabrication of biomimetic scaffolds that can recreate optimal cellular microenvironments. This advanced manufacturing technique utilizes computer-aided design (CAD) to guide the layer-by-layer deposition of bioinks, allowing for unprecedented control over scaffold architecture. The technology's precision enables the creation of complex, multilayered structures and patient-specific implants that closely match anatomical requirements. In addition to conventional fabrication technique, 3D bioprinting has considerable benefits in speed and accuracy of tissue-engineered constructs.

3.1 Overview of 3D Printing Technologies

Modern 3D bioprinting has three main approaches to the fabrication of tissue biomimicked, autonomous self-assembly, and approaches involving microtissues.

- The biomimetic strategy is a direct replication of the natural structure of target tissues which entails detailed information about the cellular assembly, ECM composition, and biochemical signal transactions. The rationale behind this approach is manufacturing artificial constructs that are similar to native tissue structures both on a micro and macroscopic scale [41].
- Self-assembly method is based on the inherent abilities of cells to reorganize and recreate functional tissue structures in the absence of an external directive. This is done through cellular processes of ECM production and tissue morphogenesis, which gives rise to biologically natural structures that have greater integration capacity when implanted [42].
- Bioprinting allows this strategy by creating simple cellular building blocks which can further be autonomously assembled. The microtissue approach builds upon these ideas, forming functional microtissue units that may be further assembled into bigger, more complicated structures. In this hybrid approach, it is recognized that the whole organs consist of different functional subunits, and uses either a designed assembly or spontaneous cellular reorganization mechanisms to provide hierarchical tissue assembly [43]. This integrative method constitutes a major development toward the development of advanced, physiologically relevant tissue constructs to regenerative applications.

The 3D bioprinting is composed of three key steps which jointly define the quality and functionality of the end tissue construct:

- This preliminary stage determines the outline of success in the fabrication. The first stage is the digital modeling, in which the CAD process is informed with the advanced medical imaging (CT, MRI) that gives anatomical data [44]. These scans of very high resolution are translated into very accurate digital templates that define the architecture of the construct, both in micro- and macro-scale.

- The fundamental fabrication step includes three coordinated processes, namely, bioink formulation and optimization, printer choice and setup depending on the complexity of its structure, and accurate deposition of bioinks [45, 46]. This step involves the close calibration of many parameters such as the properties of the materials, cellular viability, and printing conditions. Ready bioinks are placed in dedicated cartridges, and printhead choice is especially essential to obtaining fine architectural features [47].
- The last stage converts printed constructs into functional tissues by: structural stabilization using crosslinking approaches, prolonged tissue maturation in regulated culture environment, and cellular proliferation and ECM deposition [14].

3.1.1 Extrusion-Based Printing

Extrusion-based bioprinting methods, such as Direct Ink Writing (DIW) and Fused Deposition Modeling (FDM), are extensively applied in the production of polysaccharide hydrogel structures in tissue engineering and regenerative medicine. The extrusion printing is based on the controlled deposition of bioinks, which deposits through a nozzle to form 3D structures in layers. DIW especially works well with polysaccharide-based hydrogels because of their shear-thinning characteristics that enable easy extrusion with shape fidelity after deposition. Modified polysaccharide composites, capable of being gelated thermally, have been modified to allow FDM to be used in bioprinting [48]. One of the major strengths of these methods is that it can be used in a wide range of polysaccharide formulations, with low and high viscosity. With a varying extrusion pressure, nozzle diameter, and printing speed, DIW can pattern cell-laden hydrogels to high precision, including at high cell densities [49]. FDM also provides further regulation of structural integrity, through the temperature-dependent gelling of some polysaccharides. Nevertheless, issues arise such as the threat of shear-induced cell injury during extrusion, and to optimize ink rheology to avoid nozzle clogging and to maintain structural stability [50]. DIW and FDM are both superior in scaffolding tissue constructs of large scale and complexity despite moderate resolution. Crosslinking plans are typically used after printing to increase strength of mechanical properties. The extrusion-based strategies are further evolving with current studies aiming at enhancing cell viability, resolution, and cell functional performance in regenerative medicine applications [51].

3.1.2 Inkjet-Based Printing

The concept of inkjet bioprinting has become a flexible method to producing polysaccharide hydrogel scaffold constructs in tissue engineering and regenerative medicines. This non-contact printing method is based on thermal, piezoelectric, or electrohydrodynamic droplet deposition of bioink droplets in a layer-based mode [4]. The unique benefits of inkjet printing are high resolution, high fabrication rates, and high cell viability [52–61, 61] because the droplets are formed with low shear stresses. The technique can be especially efficient with low-viscosity bioinks. Through crosslinking strategies of deterministic spatial control of cell placement and biochemical signals into more complex 3D hydrogel structures are achieved through incorporating controlled crosslinking strategies [62]. The viscosity constraints however prevent use of numerous native polysaccharide formulations, and may necessitate chemical modification or co-polymerization with other polymers to attain appropriate printability [63]. Precise Inkjet bioprinting of polysaccharide hydrogels has shown potential in their use in high precision applications, including skin tissue engineering, vascular networks, and drug screening platforms. The recent developments of multi-material inkjet systems have also increased the ability of this technique to produce complex, heterogenous, tissue constructs with custom mechanical and biological properties. The biocompatibility of polysaccharides and the resolution offered by inkjet printing makes this method especially useful to making personalized regenerative therapies and in vitro tissue models [64].

3.1.3 Stereolithography

Stereolithography (SLA) has been a potent bioprinting method of polysaccharide hydrogels, shown to take advantage of specific light-mediated crosslinking in building complex 3D tissue scaffolds with remarkable resolution. In this technique, UV or visible light is used to selectively polymerize photo-reactive solutions of polysaccharides in layers on a liquid resin vat, allowing the formation of complex microarchitectures suitable to recapitulate natural tissue microenvironments [65]. The benefits of SLA to polysaccharide systems are minimal shear stress SLA, high shape fidelity, and control over mechanical properties by varying crosslinking [66]. It also has its own distinct challenges, which include chemical modification of natural polysaccharides to be photopolymerized, and possible cytotoxicity of UV that must be highly optimized concentrations of photoinitiators and exposure parameters [67]. Although the precision-oriented focus of SLA at present makes it relatively limited in scalability to large tissue constructs, SLA continues to be advantageous in tissue engineering and biomedicine applications that need microscale precision, such as vascular systems, neural scaffolds, and organ-on-a-chip models, where current developments in visible-light systems and hybrid materials are progressively broadening its applications in polysaccharide-based tissue engineering [68, 69].

3.1.4 Other Methods

Laser-assisted bioprinting provides a nozzle-free method of patterning polysaccharide hydrogels with high precision, where laser-induced forward transfer is used to deposit cell-laden bioink droplets on a substrate [70]. The method is especially beneficial when viscous polysaccharide solution has to be processed and would otherwise cause blockage in traditional extrusion systems, where a pulsed laser pulses are used to create vapor bubbles to force bioink off a donor slide without mechanical force. Although laser-assisted bioprinting has demonstrated remarkable resolutions capable of fabricating complex hydrogel microarchitectures, the use with polysaccharides must be optimized to ensure laser parameter (wavelength, pulse length, energy) will reduce the thermal effects that may cause damage to sensitive biopolymers and encapsulated cells. The fact that the technology can be used at high cell densities and can print a variety of materials simultaneously makes it useful in the engineering of heterogeneous polysaccharide-based tissues, although limitations in scalability and cost currently hamper its use in tissue engineering applications that necessitate large-scale construct generation [5]. By choosing LAB in polysaccharide hydrogel bioprinting, scientists should determine a trade-off between architectural specificity and the possible thermal consequences related to the integrity of hydrogel and cell survival, especially in the case of fragile polysaccharide gels. It is by a thorough evaluation of various vital factors that the most effective 3D bioprinting technology is selected in applications to tissue engineering, which requires the bioink rheology, cell loading capacity, printing resolution, fabrication rate, and economic viability to be carefully balanced against the structural and functional demands of the predestined tissue.

3.2 Challenges in Printing with Hydrogels

The concept of "bioink" and "bio paper" was first introduced in 2003 in the context of organ printing technology. Bioink, a crucial element in 3D bioprinting, consists of a combination of living cells, biological factors, and biocompatible materials that together form functional three-dimensional structures [71]. Originally, this concept envisioned using cell aggregates or tissue spheroids as the printing material, with hydrogel serving as the supportive "bio paper" substrate. In its initial formulation, bioink referred specifically to biological components that could be precisely patterned within or deposited onto three-dimensional hydrogel matrices to create tissue-like constructs [72].

3.2.1 Printability and Viscosity Control

Bioprintability characterizes a biomaterial's capacity to maintain structural fidelity while being precisely deposited into three-dimensional architectures during the printing process, forming the foundation of 3D bioprinting technology that enables the computer-guided, layer-by-layer fabrication of living tissue constructs. This advanced manufacturing approach employs various techniques including extrusion printing, inkjet printing, laser-assisted printing, and light-mediated photopolymerization, all of which facilitate the creation of cell-laden bioink structures with increasing precision. The growing biomedical potential of these bioprinting methods has stimulated substantial research interest, particularly in developing novel bioink formulations and optimizing printing parameters to better replicate native tissue complexity and functionality [73]. The clinical performance of polysaccharides in biomedical applications depends on several critical parameters, including structural integrity, porosity, swelling capacity, degradation rate, mechanical strength, and drug release profile, all of which are significantly influenced by the hydrogel preparation method [74–76].

The successful 3D bioprinting of hydrogels critically depends on their rheological properties, which govern both the printing process and final scaffold quality. Rheology, the study of material deformation under applied forces, reveals how bioinks transition through distinct phases during extrusion-based printing: from a static gel state through a high-shear flowing state within the nozzle, followed by structural recovery upon deposition. These dynamic changes are characterized by key parameters including viscosity, viscoelastic moduli (G' and G''), yield stress, and elastic recovery, all of which collectively determine extrudability, shape fidelity, and structural integrity [13]. While optimal viscosity enhances printing resolution and cell encapsulation efficiency, excessive viscosity increases shear stress that compromises cell viability. These properties are primarily influenced by polymer molecular weight and concentration, with non-Newtonian shear-thinning behavior being particularly desirable as it enables smooth extrusion while maintaining post-printing shape retention. Such a harsh bioprinting environment places severe demands on bioink formulations, with a significant requirement being the balance between rheological performance to print (e.g., suitable shear-thinning to micro-extrude) and biological performance [77]. The complexity of this interaction between material properties and processing conditions has a final effect on the precision, resolution and overall biological performance of the tissue constructs produced in bioprinting, and rheological characterization is the key step to improving the application of bioprinting in tissue engineering and regenerative medicine.

3.2.2 Structural Integrity and Resolution

Structural integrity and printing resolution are the two essential properties of 3D-printed polysaccharide hydrogel as they govern the functional performance of the hydrogel in tissue engineering, drug delivery, and regenerative medicine. Structural integrity involves the ability of the hydrogel to retain dimensional, mechanical, and architectural integrity following printing and is further improved by optimized polysaccharide concentrations and crosslinking methods. Although greater polymer content raises mechanical strength, it can reduce extrudability, and trade-offs between yield stress parameters and mechanical strength are required to avoid structural collapse during deposition. Subsequent processes such as instantaneous crosslinking, secondary curing (UV/thermal), and control of adhesion between layers (regulated by moisture content, viscosity, and curing kinetics) also determine the stability of the final construct. Printing resolution, i.e. the precision of fabricated features and layer definition, is controlled by synergistic interactions between the chosen printing technology, bioink rheology, process characteristics, and material intrinsic properties. The interaction of these structural and precision characteristics determines eventually the applicability of the hydrogel in a particular application, where optimum performance is found in achieving adequate mechanical stability and high feature resolution.

3.3 Printing Process Optimization for Polysaccharide Hydrogels

In order to 3D print polysaccharide hydrogel successfully, the entire process will have to be optimized in three key stages, namely, pre-printing formulation, printing parameters, and post-printing stabilization. Pre-printing optimization aims at optimizing the rheological and chemical nature of the bioink, such as polymer concentration, ratio of crosslinkers, additives that stabilize it and accurate control of the viscosity and shear-thinning dynamics to achieve optimal flow behavior. The choice of parameters during printing should be technology-specific: in extrusion-based techniques (DIW/FDM), the most important aspects are the nozzle diameter (trading between resolution and pressure needs), print rate (trading off between layer adhesion and speed), extrusion pressure (balanced with ink viscosity), the layer height (usually $\leq$ 80% of nozzle diameter) and temperature control (trading off between material flow and solidification). In the case of photopolymerization methods (DLP/SLA), important factors include the concentration of photoinitiator (0.5–2%), exposure time of UV radiation (tradeoff between strength and resolution), and wavelength (polymer-dependent). Post-printing stabilization is equally vital, employing immediate crosslinking, secondary curing processes, and controlled drying conditions (often in humidified chambers) to preserve shape fidelity, enhance mechanical properties, and prevent structural defects such as cracking. This multi-phase optimization approach enables the production of polysaccharide hydrogel constructs with precisely controlled architectures and functional properties for advanced applications.

4 Crosslinking and Functionalization of Polysaccharide Hydrogels

4.1 Crosslinking Methods for Strength and Stability

4.1.1 Physical Versus Chemical Crosslinking

Physically crosslinked hydrogels derive their structural stability through reversible secondary interactions, including electrostatic forces, hydrogen bonding, and hydrophobic associations, each imparting distinct functional characteristics [78]. Ionically crosslinked systems form when oppositely charged polyelectrolytes interact, as demonstrated by divalent cation-mediated gelation of anionic polysaccharides like alginate and gellan gum, where polymer chain length and charge density govern the resultant hydrogel's mechanical properties and biocompatibility [79]. Thermosensitive hydrogels exploit sol–gel transitions through lower critical solution temperature/upper critical solution temperature polymer systems, achieving rapid gelation via temperature-induced hydrophobic interactions, while pH-responsive chitosan-based formulations utilize amine deprotonation often augmented by polyol additives like β-glycerophosphate [80]. Supramolecular hydrogel architectures employ host–guest chemistry, particularly through cyclodextrin inclusion complexes with hydrophobic moieties on linear polymers, as exemplified by β-cyclodextrin-adamantane modified hyaluronic acid systems, where modular design of complementary host and guest components enables precise tuning of material properties [81]. These physically crosslinked networks offer unique advantages including cytocompatibility, stimuli-responsiveness, and dynamic reversibility, making them particularly valuable for biomedical applications requiring in situ gelation or environmentally triggered property modulation.

Chemically crosslinked polysaccharide hydrogels derive their structural integrity through covalent bond formation between functional groups on polymer chains, offering superior mechanical stability and tunable properties compared to physical networks. Michael addition reactions facilitate hydrogel formation by coupling electrophilic and nucleophilic moieties on modified polysaccharides (chondroitin sulfate, dextran, chitosan), enabling precise control over elasticity and degradation kinetics for tissue engineering applications [8]. Bio-orthogonal click chemistry approaches, particularly azide-alkyne cycloaddition and Diels–Alder reactions, provide efficient crosslinking with high specificity and cytocompatibility, allowing simultaneous cell encapsulation and mechanical reinforcement for cartilage regeneration [82]. Schiff base formation between amine and aldehyde groups creates injectable systems with adjustable gelation rates, demonstrated in oxidized hyaluronic acid/chondroitin sulfate networks that balance degradability with structural integrity [83].

Enzymatic crosslinking using horseradish peroxidase/H_2O_2 to couple tyramine-modified polysaccharides (chitosan, dextran) produces biomimetic extracellular matrices with programmable stiffness and biodegradation profiles that support chondrogenic differentiation through enhanced glycosaminoglycan and collagen production [84]. These covalent strategies collectively enable the design of polysaccharide hydrogels with precisely engineered mechanical properties, degradation rates, and bioactivity for advanced therapeutic applications.

4.1.2 UV Crosslinking and Thermally Induced Crosslinking

Polysaccharide hydrogels are engineered through distinct crosslinking approaches, with photo-initiated and temperature-dependent methods offering complementary advantages [85]. Photo-crosslinked systems utilize light-activated polymerization of acrylate/methacrylate-modified polysaccharides (chitosan, alginate, chondroitin sulfate) through radical-mediated reactions at physiological conditions. This technique provides spatiotemporal control over gelation by modulating photoinitiator concentration, light exposure parameters, and macromer composition, while maintaining biocompatibility through low-viscosity precursors [86].

In contrast, thermally induced crosslinked hydrogels rely on reversible macromolecular rearrangements, where heating above the helix–coil transition temperature disrupts secondary structures, followed by cooling-induced reformation of stronger junction zones [10]. The presence of specific cations (K^+, Na^+) further stabilizes these networks through enhanced helix aggregation. These orthogonal crosslinking strategies enable precise tuning of hydrogel properties, photo-crosslinking offers spatial patterning capability for cell encapsulation, while thermal methods facilitate injectable delivery with subsequent in situ stabilization, collectively expanding the therapeutic potential of polysaccharide-based biomaterials.

4.2 Functionalization Techniques to Enhance Performance

4.2.1 Chemical Modification

Chemical modification is a powerful strategy to tailor the structural and functional properties of polysaccharide hydrogels. By altering molecular weight, chemical composition, and molecular architecture, these modifications significantly influence their bioactivity, mechanical properties, and physicochemical behavior. Polysaccharides possess abundant reactive functional groups (–OH, –COOH, and $-NH_2$), enabling selective functionalization to achieve desired hydrogel characteristics [87, 88]. Common modification techniques include sulfation, acetylation, phosphorylation, carboxymethylation, selenization, selective oxidation, "click" chemistry, polymer grafting, and bioinspired modifications such as mussel-inspired chemistry. These approaches enhance hydrogel performance, making them suitable for diverse biomedical and industrial applications.

- Sulfation of monosaccharides enhances the immunomodulatory properties of natural polysaccharides, with the incorporation position of sulfate groups affecting their functionality. Common methods for sulfation include sulfur chlorosulfonic acid-pyridine, concentrated sulfuric acid, and trioxide-pyridine, though challenges such as degradation and low solubility persist [89].
- Polysaccharide acetylation involves converting hydroxyl groups into acetyl groups, altering their spatial arrangement, and consequently modifying their physicochemical and biological properties. Common methods include acetic anhydride-pyridine and alkaline acetylation techniques [11].
- Phosphorylation enhances the properties of natural polysaccharides, improving water solubility, biological activity, and reducing viscosity. Common reagents for this modification include H_3PO_4, phosphates, phosphorus oxychloride, and phosphorus pentoxide [52].
- Carboxymethylation enhances the water solubility and bioactivity of natural polysaccharides, such as cellulose, overcoming limitations in food industry applications. This modification presents better prospects for using carboxymethylated polysaccharides in drugs, nutrition, and food production [53].
- Selenization introduces selenium into polysaccharides, combining their natural properties with the benefits of selenium. Common reagents include nitric acid-sodium selenite, favored for its high efficiency and straightforward reaction conditions [54].
- Selective oxidation of polysaccharides is crucial for developing new food packaging and high-bioactivity products, using reagents like periodate and (2,2,6,6-Tetramethylpiperidine-1-oxyl radical) TEMPO. While some oxidation methods can lead to overreaction and depolymerization, high-pressure CO_2 offers improved yields and purification despite lower oxidation yields under milder conditions [12].
- Since its introduction in 2001, "Click" chemistry has become a vital tool for designing novel molecular structures to enhance bioactivity and facilitate drug development [55]. Techniques like copper catalyzed azide–alkyne cycloaddition (CuAAC), metal-free [3 + 2] cycloaddition, Diels–Alder reaction, Thiol-Michael addition, and thiol-ene reaction, and oxime click reactions are particularly promising for modifying polysaccharides to achieve tailored structures and multifunctionalities [56].
- Polymer-grafted polysaccharides combine the benefits of synthetic polymers and natural polysaccharides, enhancing properties like solubility and bioactivity. The main methods for attachment include "graft-from" and "graft-to," with "graft-from" being preferred for its chemical versatility despite some challenges in grafting density [57].

4.2.2 Composite Formation with Other Materials

The integration of polysaccharide hydrogels with reinforcing materials through 3D printing has opened new possibilities in tissue engineering, drug delivery, and biomedical devices. By forming composite structures, these hydrogels gain enhanced mechanical strength, bioactivity, and functional versatility.

- Combining different hydrogels or synthetic polymers improves mechanical properties (strength, flexibility), degradation resistance, and biological activity. For example, blending alginate with gelatin enhances structural integrity while maintaining biocompatibility [58].
- Incorporating nanoparticles or microparticles modifies hydrogel properties for specialized applications: bioglass nanoparticles in alginate-gelatin hydrogels promote cell proliferation and mineralization, making them ideal for bone regeneration [59]. Similarly, graphene oxide enhances the elastic modulus and electrical conductivity of methacrylated chitosan hydrogels, useful in neural and cardiac tissue engineering [18].
- Embedding micro- or nanofibers improves mechanical stability and biological performance: polylactic acid nanofibers reinforce alginate hydrogels, enhancing their suitability for cartilage repair due to improved load-bearing capacity and cell adhesion [60].
- Directionally structured fillers (e.g., clay nanoparticles like Laponite) enhance self-supporting ability and stiffness: PEGDA/alginate/gelatin hydrogels with Laponite exhibit higher Young's modulus, enabling precise 3D printing of complex scaffolds [61].
- Dynamic, stimuli-responsive systems enable advanced functionalities: Cell-laden hydrogels support tissue growth by providing a bioactive microenvironment [61]. Drug-eluting composites allow controlled release in response to pH, temperature, or enzymatic triggers (Table 1).

5 Applications of Polysaccharide Hydrogel 3D Printing

5.1 Tissue Engineering and Regenerative Medicine

Tissue engineering is an interdisciplinary field that combines biomaterials, cell biology, and medical technologies to develop artificial tissues capable of replacing damaged ones. A key challenge in this field is selecting the ideal scaffold material, which must support critical cellular functions such as proliferation, adhesion, and growth, while also delivering bioactive molecules like growth factors to enhance regeneration. Polysaccharides have become a promising candidate because it is biocompatible, bioactive, tunable, and likened to the human ECM in structure. They also show good mechanical strength that makes them withstand physiological stresses. A critical aspect of scaffolds lies in a balanced degradation rate such that they degrade slowly as new tissue grows, and ultimately, replaced by natural ECM. 3D printing technology has further enhanced the potential of polysaccharide-based scaffolds whereby they can be controlled to have the proper architecture, porosity,

Table 1 Compositions of various polysaccharides for fabrication via additive manufacturing across different applications

Sr. No.	Materials	Printing technique	Applications	Refs.
1	Hyaluronic acid/polylactide	FDM	Orthopedic implant	[90]
2	Alginate	DIW	Regenerate the jaw bone	[91]
3	Alginate/gelatin methacryloyl	FDM	Hydrogel fibers	[92]
4	Hyaluronic acid/alginate	DIW	Articular cartilage engineering	[93]
5	Agarose/acrylamide	In-situ polymerization	Human ear or nose	[94]
6	Alginate-GO hydrogels	SLA	Multifunctional devices for aqueous and marine environments	[95]
7	Methacrylate/carrageenan hydrogels	DLP	Engineering cell-loaded tissue scaffolds	[96]
8	Pullulan/ hydroxypropylmethylcellulose solutions	Dual-extrusion	Oral films	[97]
9	Norbornene-functionalized pullulan hydrogels	DLP	3D-printed objects	[98]
10	Starch/gellan gum	Bioprinter	Scaffolds for wound healing	[99]
11	Pectin/sodium alginate/olive leaf extracts	DIW	Wound healing patches	[100]
12	Protein/xanthan gum	DIW	Food printing	[101]
13	Hybrid hydrogel of chitosan and polyethylene glycol diacrylate	SLA	Repair of complex tissue geometries	[102]
14	Chitosan-poly(ε-caprolactone)	FDM	Drug delivery implants	[103]
15	Gellan gum/lignin hydrogel	Bioprinter	Cartilage regeneration	[104]

and mechanical properties. This progress has facilitated their effective use in skin, cartilage, neural, cardiac, vascular and bone tissue engineering and has thus become a multi-purpose regenerative medicine tool.

5.1.1 Scaffold Design and Cell Encapsulation

The necessity to replace bones has led researchers to develop superior biomaterials that can manage to replicate the characteristics of natural bone. Hydrogels of chitosan, reinforced with calcium phosphate (CaP) particles, are currently a promising option because of their biocompatibility, osteoconductivity, and modulable mechanical strength. Shear-thinning chitosan-CaP composite inks were used in robocasting to maximize scaffold fabrication. These inks are platelet-shaped dicalcium phosphate dihydrate particles suspended in an acidic chitosan suspension to create a stable but printable suspension. Rheological characteristics of the inks such as shear thinning tendency and yield stress were well studied to make the inks optimally printable. They were extruded into a simple water/ethanol coagulation bath to initiate the conversion of the chitosan-CaP suspension into a mineralized physical hydrogel. The 3D printed scaffolds resulting were porous chitosan scaffolds (Fig. 1a) with incorporation of poorly crystalline apatite, which is similar to the structural composition of natural bone. Major conclusions were that ink composition can play a crucial role in determining printability, whereby greater chitosan concentrations (e.g., 0.19 mol L^{-1}) enhanced structure formation during deposition. This approach demonstrates the potential of 3D-printed polysaccharide scaffolds in bone tissue engineering, offering a customizable and biomimetic solution for regenerative medicine [6].

Starch was combined with a gel to create a starch-based gel ink, which demonstrated excellent printability. By optimizing the infill density, 3D-printed scaffolds with a 50% infill density and a highly ordered microstructure were successfully produced (Fig. 1b). The incorporation of calcium carbonate nanoparticles–gluconic delta-lactone ($CaCO_3$ NPs-GDL) significantly influenced the scaffolds' swelling behavior, in vitro digestion, water stability, and pore distribution. When 0.075 g of $CaCO_3$ NPs was added to the starch-based gel, the resulting 3D-printed scaffold with 50% infill density proved most suitable for cell-based meat cultivation, exhibiting pore sizes of 80–120 μm and a compression modulus of 246.76 Pa. After 7 days of proliferation, C2C12 mouse skeletal myoblasts showed an approximately 2.81-fold increase in cell numbers. The fusion index and maturation index of the C2C12 cells on the scaffolds reached $57.00 \pm 0.45\%$ and $34.56 \pm 0.56\%$, respectively. The starch-based gel scaffolds displayed strong water stability and in vitro degradability. Additionally, C2C12 cells successfully proliferated and differentiated on the scaffolds, ultimately contributing to cell-based meat production. This study developed a starch-based composite gel scaffold for advancing cell-based meat manufacturing [105].

5.1.2 3D Printed Tissues and Organs

Biomimetic materials have attracted a lot of attention in recent years in relation to bone tissue engineering. None of these, however, favourable immunogenicity and mechanical properties, tend to restrict their performance in vivo. Collagen is a promising material as a major constituent of the bone ECM: it is biodegradable, plastic, and low in immunogenicity. However, pure collagen per se does not give the best outcome and needs to be modified or added with other bioactive materials to more closely mimic the bone tissue microenvironment. Consequently, researchers

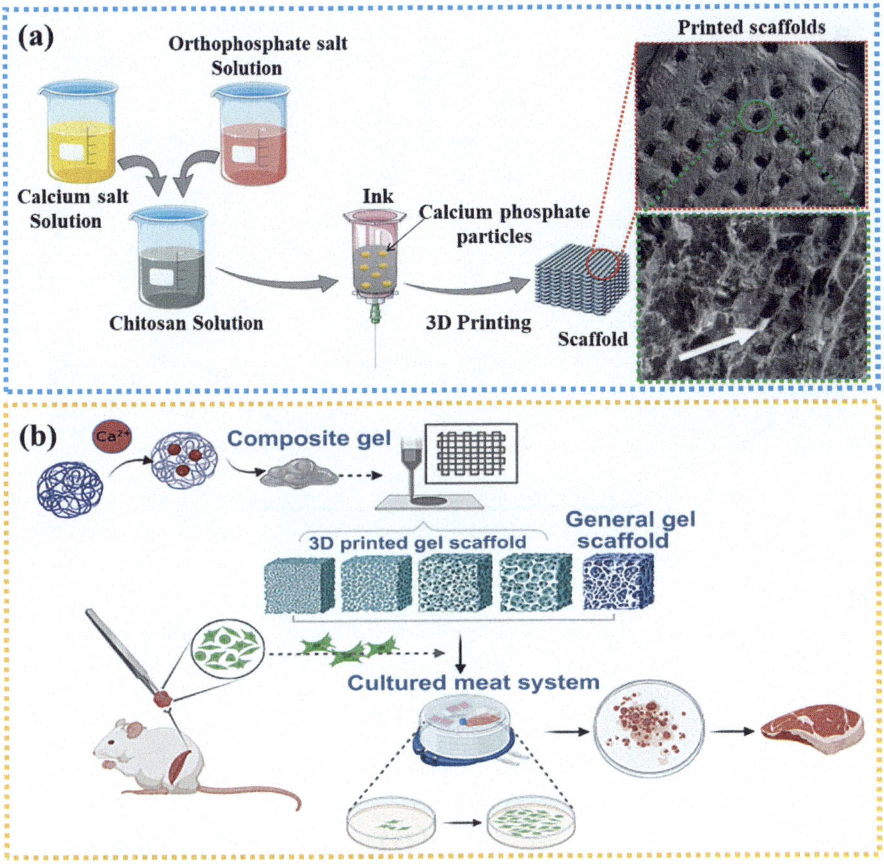

Fig. 1 **a** Diagram and scanning electron micrographs showing the structure of a 3D-printed chitosan-calcium phosphate composite scaffold. Springer Nature copyright 2019, reproduced from ref [6]. **b** 3D printed starch hydrogel matrix developed for cultured meat applications. ACS copyright 2024, reproduced from ref. [105]

often enhance collagen by integrating it with reinforcing scaffolds or functional additives to improve its mechanical and biological properties. Studies have demonstrated that collagen-based matrices exhibit biological activity, promoting cell migration and proliferation within scaffolds. Additionally, modified collagen can be incorporated into metallic scaffolds, such as titanium, to enhance mechanical strength while preserving its beneficial properties. These advancements highlight the potential of collagen-based biomaterials in developing functional bone tissue engineering solutions (Fig. 2a) [16].

Polymer hydrogels are widely studied as scaffolds for cell-loading and tissue engineering due to their extracellular-matrix-like structure. However, many hydrogels with favorable processing properties suffer from poor mechanical strength and

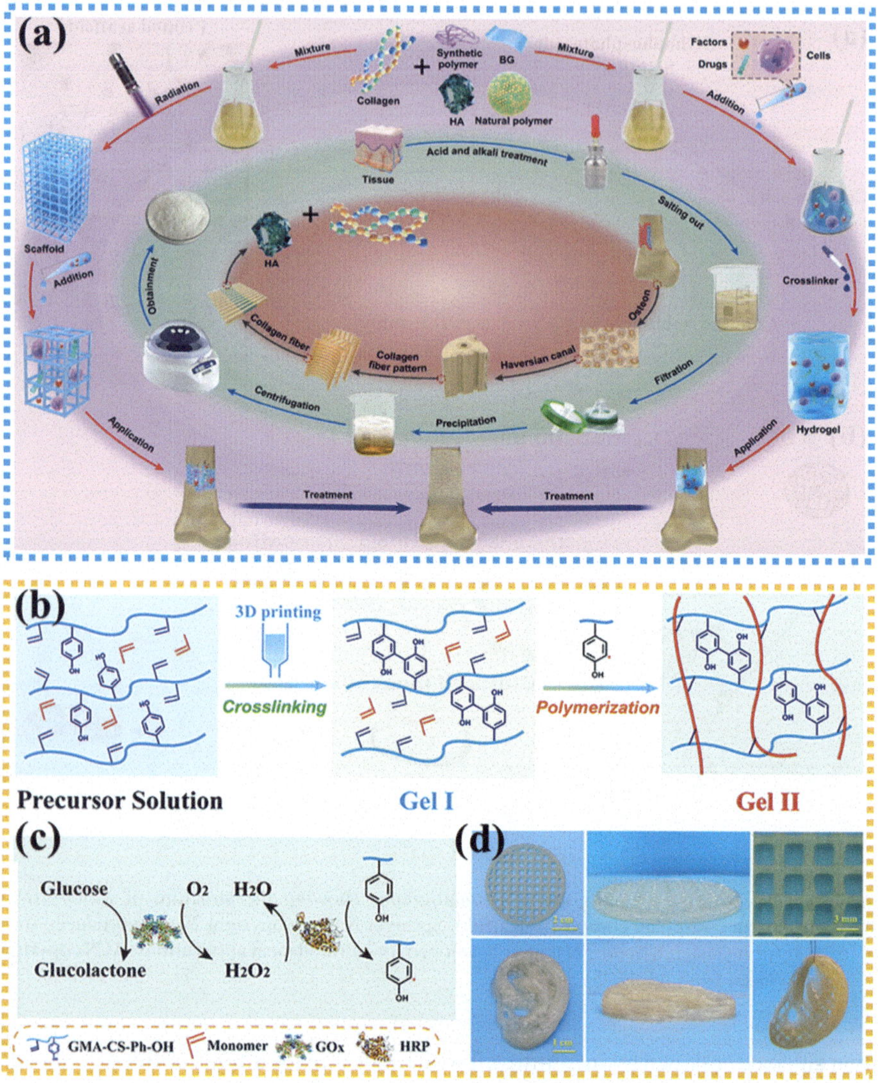

Fig. 2 **a** (Central circle) Collagen mineralization. (Intermediate circle) Collagen extraction procedure. (Peripheral circle) Utilization of collagen in hydrogels and scaffolds for bone tissue engineering applications. Elsevier copyright 2021, reproduced from ref [16]. **b** Schematic representation of Gel I and Gel II hydrogel formation; **c** Enzymatic reaction mechanism of the HRP@GOx system; **d** Multi-layered 3D-printed structures showing both button and ear geometries. Reproduced from ref. [106]

fragility. A promising solution to this challenge is the use of post-gelation reinforcement strategies, which allow tunable mechanical properties after hydrogel formation (Fig. 2b). A novel approach to fabricate composite hydrogels with adjustable strength using an HRP@GOx dual-enzyme system was applied (Fig. 2c). This system simultaneously initiates the rapid crosslinking of tyrosine-grafted chondroitin sulfate and the gradual polymerization of monomers to form a reinforced composite hydrogel. The resulting composite hydrogel combines the advantages of enzymatically crosslinked networks and polymerized hydrogels, enabling precise control over mechanical properties while maintaining favorable processing characteristics. These dual-enzyme-regulated hydrogels exhibit versatile potential as bioactive scaffolds for organoid culture and other biomedical applications, where tunable mechanical properties and biocompatibility are essential (Fig. 2d) [106].

5.2 Drug Delivery Systems

5.2.1 Controlled Release and Targeted Delivery

Despite advances in cancer therapeutics, off-target toxicity remains a major challenge in treatment. Research into innovative drug delivery systems has expanded, utilizing diverse materials and fabrication techniques to improve therapeutic efficacy. Among these, gellan gum, a natural anionic exopolysaccharide produced by *Sphingomonas elodea*, has emerged as a promising candidate due to its unique physicochemical properties (Fig. 3a). This biocompatible and non-toxic polymer forms gels in the presence of cations, making it suitable for controlled drug release across various physiological conditions and anatomical sites. Gellan gum can be processed through multiple physical and chemical gelation methods to create different formulations, such as hydrogels, nanohydrogels, beads, films, and patches, each with tunable mechanical and physicochemical properties (Fig. 3b). These formulations have demonstrated enhanced drug bioavailability, improved solubility, and sustained release kinetics while maintaining biocompatibility and minimizing adverse effects. Engineered gellan gum has the potential to actively deliver therapeutic agents, including antibiotics, antiallergens, anti-inflammatories, and antifungals, with particular emphasis on its application in anticancer drug delivery [22].

5.2.2 Polysaccharide-Based Hydrogel Drug Carriers

Hydrogels made of polysaccharides have been generating a lot of interest as drug carriers because they can control the release of therapeutic agents by entraping them in a polymer network. An example of these is the use of alginate hydrogels to encapsulate insulin in the treatment of diabetes, whereby the drug is released over long durations. Equally, tumor-targeting antibodies have been conjugated to chitosan-based hydrogels to achieve targeted delivery of chemotherapeutic drugs only to the

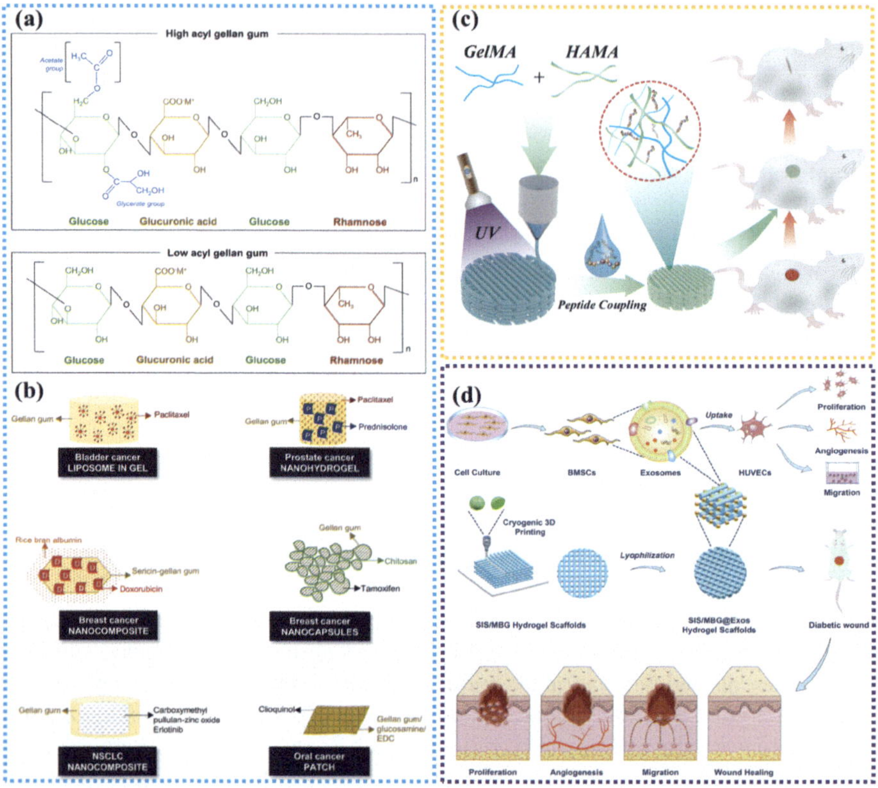

Fig. 3 **a** Chemical structures of high-acyl gellan gum, low-acyl gellan gum, and acetylated gellan gum. **b** Versatile applications of gellan gum in multiple processing techniques for anticancer drug delivery. ACS copyright 2021, reproduced from ref [22]. **c** Illustrating the manufacturing process of 3D-printed peptide patches for cutaneous wound treatment. Elsevier copyright 2022, reproduced from ref [23]. **d** Diagram showing the preparation of cryogenically 3D-printed SIS/MBG@Exos hydrogel scaffolds and their therapeutic evaluation in diabetic wound healing. Elsevier copyright 2021, reproduced from ref. [107]

cancer cells without causing side effects to the entire system. Chitosan hydrogels have special strengths in ocular drug delivery. They are administered to the eye where they go through in situ gelation to create a local drug reservoir that releases the anti-inflammatory drugs or antibiotics slowly. The method makes drugs bioavailable and more patient-compliant, as compared to the traditional therapies. Hyaluronic acid hydrogels have also been promising ocular applicants and especially in the management of glaucoma. These injectable hydrogel depots create a gel depot in the eye to release the drug slowly and to minimize the high dosage levels of conventional eye drops.

5.3 *Other Biomedical Applications*

5.3.1 Wound Healing and Dressing Materials

Severe trauma causes chronic wounds that pose a significant clinical challenge because the healing process is compromised, and they are at risk of infections. A bioactive 3D-printed polysaccharide-based scaffold that promoted wound regeneration was developed to resolve this problem. Using the technology of 3D bioprinting, these scaffolds are capable of being tailored to the exact geometry of the wound to make sure it fits and functions optimally. Gelatin methacryloyl (GelMA) and hyaluronic acid methacryloyl (HAMA) were used as the scaffold matrix. The pro-angiogenic peptide QHREDGS was next chemically grafted into the GelMA/HAMA network to increase vascularization and allow a controlled release and angiogenic activity. The experimental outcomes revealed that the 3D-printed polysaccharide scaffolds enhanced significantly cell viability, blood vessel formation, and tissue regeneration in the laboratory and animal experiments (Fig. 3c) [23].

A novel strategy to fabricate a polysaccharide-based scaffold in 3D by extrusion-based cryogenic 3D printing technology. The scaffold consists of decellularized small intestinal submucosa (SIS), mesoporous bioactive glass (MBG), and therapeutic exosomes to form a bioactive wound dressing (SIS/MBG@Exos) which facilitates exosome release in a controlled manner. This engineered scaffold has an ideal 3D porous architecture, high biocompatibility, and good hemostatic characteristics. In vitro experiments show that the scaffold stimulates the proliferation of human umbilical vein endothelial cells, their migration, and angiogenic activity. In-vivo diabetic wound studies indicate that the scaffold can dramatically enhance the rate of healing through the increase of blood flow in the area, and induction of intense angiogenesis. Additionally, the scaffold promotes the formation of granulation tissue, organized collagen deposition, and functional neovascularization, all critical for effective wound repair (Fig. 3d) [107].

5.3.2 Diagnostic and Therapeutic Tools

A novel hydrogel incorporating copper oxide nanoparticles has been developed to address these limitations through multifunctional therapeutic mechanisms. This implantable system demonstrates controlled, sustained release of CuO nanoparticles during hydrogel biodegradation. The released nanoparticles exhibit dual therapeutic functions: (1) as a source of Cu^{2+} ions that generate intracellular reactive oxygen species (ROS) and (2) as photothermal conversion agents. The system demonstrates synergistic effects where photothermally generated heat enhances Fenton-like reaction efficiency. Furthermore, the platform induces ferroptosis through Cu^{2+}-mediated glutathione depletion via GPX4 inactivation. The hydrogel scaffold's unique properties enable:

- Controlled nanoparticle release kinetics

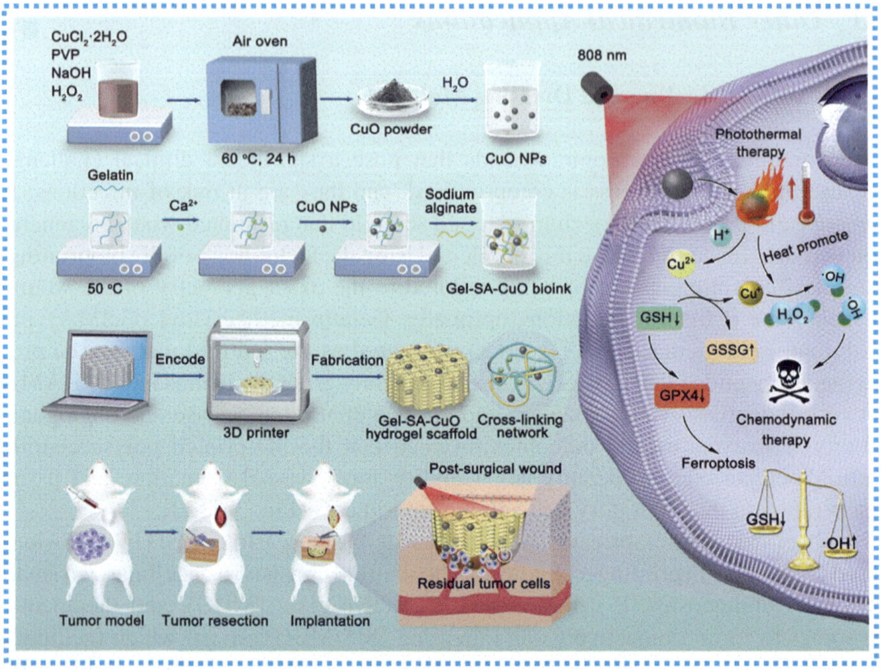

Fig. 4 Diagrammatic representation of 3D-bioprinted Gel-SA-CuO hydrogel matrices for preventing post-surgical cancer recurrence. Reproduced from ref. [108]

- Dual-mode therapeutic action (chemodynamic and photothermal)
- Ferroptosis induction capability
- Localized treatment with reduced systemic effects

This multifunctional implant represents a significant advancement in postoperative cancer care, combining multiple therapeutic modalities in a single biodegradable platform (Fig. 4). The demonstrated efficacy in preventing local recurrence suggests promising clinical potential for improving outcomes in hepatocellular carcinoma treatment following surgical resection [108].

5.4 Industrial and Environmental Applications

5.4.1 Filtration and Environmental Remediation

Conventional filtration systems using traditional adsorbents often suffer from limited capacity and efficiency, particularly in removing persistent contaminants such as crystal violet dye. An innovative approach combining 3D printing technology with bioinspired design to develop a high-performance filtration system. The novel filter

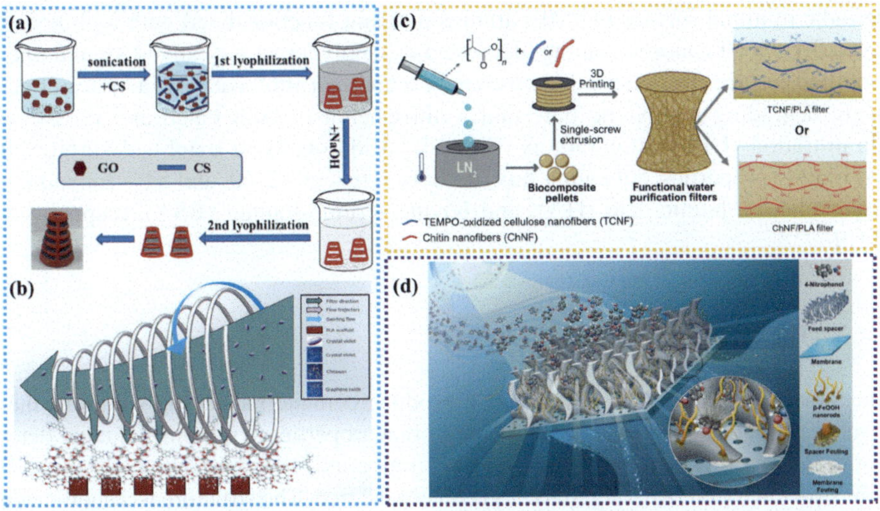

Fig. 5 **a** Illustrating the two-stage fabrication process for PLA-GS composite scaffolds: initial production of 3D porous sponge structures followed by dual freeze-drying treatment to achieve the final architecture. **b** Schematic representation showing the bioinspired structural design concept for enhanced crystal violet removal. Elsevier copyright 2018, reproduced from ref. [109]. **c** Summary of the experimental process for preparing biocomposites, 3D-printable filaments, 3D-printed filters, and a visual depiction of the distribution of nanofibers on the filter surface. Elsevier copyright 2023, reproduced from ref [19]. **d** Innovative photocatalytic 3D-printed feed spacers designed for wastewater treatment applications. Elsevier copyright 2022, reproduced from ref. [110]

incorporates a chitosan-based composite material fabricated through a combination of 3D printing and freeze-drying techniques (Fig. 5a). The filtration system features a biomimetic fish-mouth structure inspired by natural filtration mechanisms, created using polylactic acid (PLA) framework infused with a graphene oxide/chitosan (GO/CS) sponge (Fig. 5b). The system operates on principles of vertical cross-step filtration and swirling flow dynamics [109].

The development of eco-friendly, high-performance filtration systems remains a critical need for sustainable water purification. An innovative approach utilizing 3D printing technology to fabricate fully biodegradable water filters from polysaccharide-reinforced biocomposites. Through FDM, manufactured porous filtration structures using polylactic acid (PLA) matrices enhanced with TEMPO-oxidized cellulose nanofibers (TCNF) or chitin nanofibers (ChNF). The resulting filters were printed in optimized cylindrical and hourglass configurations featuring precisely controlled, multiscale porosity designed to maximize contaminant contact time while maintaining high flow rates (Fig. 5c) [19].

An innovative 3D-printed photocatalytic spacer designed to enhance membrane-based water and wastewater treatment. Unlike conventional spacers that solely provide structural support, this multifunctional component integrates pollutant degradation and membrane cleaning capabilities. The spacer was fabricated using a triply periodic minimal surface (TPMS) architecture and functionalized with a photocatalytic β-FeOOH nanorod coating, deposited on a polydopamine-polyethyleneimine-modified surface. The spacer demonstrated exceptional photocatalytic activity, effectively degrading methylene blue and 4-nitrophenol in both batch and crossflow ultrafiltration (UF) configurations (Fig. 5d). Additionally, it exhibited significant antifouling properties, achieving flux recovery ratios of 92, 60, and 54% for sodium alginate (SA), humic acid (HA), and bovine serum albumin (BSA), respectively [110].

5.4.2 Packaging and Agricultural Use

The storage and transportation of fresh produce demand packaging solutions that combine effective cushioning with antimicrobial properties to minimize quality deterioration. To meet such requirements, a new compound composite aerogel with shell-core structure (CNGA/C-AgNPs) has been engineered. The structure is a shell made of carboxymethyl nanocellulose, glycerin, and derivatives of acrylamide, and the core includes chitosan-embedded silver nanoparticles (AgNPs). This new packaging material exhibits great functionalities: (1) Excellent cushioning capability; (2) An average compression resilience rate of more than 90%. (3) Prolonged antimicrobial action on ubiquitous foodborne pathogens (*E. coli* and *S. aureus*) by the regulated release of AgNPs. (4) 100%. biodegradability to make the environment sustainable. Shell-core is to be used to offer mechanical protection and to ensure efficient antimicrobial activity to meet different packaging needs (Fig. 6a). Such a flexible design method allows optimization in various produce types and distribution contexts. It is a major breakthrough in the intelligent food packaging technology, as it will be a mixture of basic protection features and environmental friendliness [20].

Polyvinyl alcohol is also highly compatible with polysaccharides and is useful in making advanced packings of food materials. PVA formed strong cross-linked

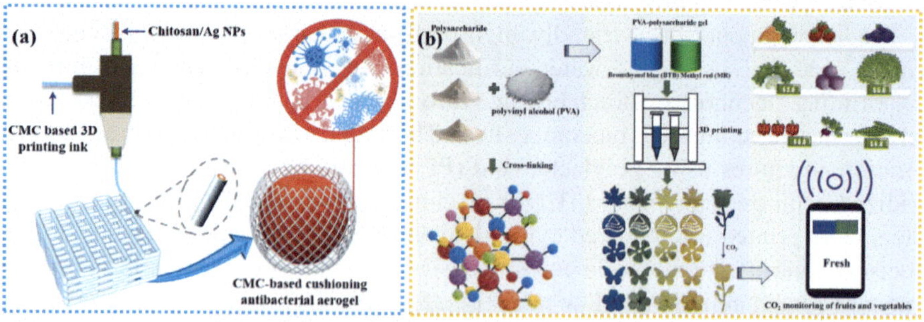

Fig. 6 **a** Illustrating the fabrication procedure of 3D-printed carboxymethyl nanocellulose-based aerogels with dual functions of cushioning and antibacterial properties for food packaging. Adapted from [20]. **b** 3D-printed polysaccharide-based gels and the creation of CO_2 monitoring labels for fruits and vegetables. Elsevier copyright 2024, reproduced from ref. [111]

networks with sodium alginate, and CMC, significantly enhancing material processability, with 7% PVA improving sodium alginate's structural precision by 20.83-fold while also optimizing the flow properties of potato starch, pectin, konjac glucomannan, and xanthan gum composites. The resulting films exhibited smooth surfaces with good structural integrity, though combinations showed minor cracks and wrinkles due to formulation imbalances. This combination proved optimal, balancing hydrophilicity, mechanical strength, and flexibility while enabling smart indicator functionality through bromothymol blue (BTB) and methyl red (MR) that effectively monitored CO_2 levels in tomato packaging during storage (Fig. 6b). This advances intelligent packaging solutions by combining tunable material properties with sensing capabilities for real-time food quality monitoring, offering a versatile platform for developing functional packaging tailored to specific preservation needs without compromising performance or sustainability [111].

The growing global population and limitations of conventional fertilizers necessitate the development of sustainable, slow-release agrochemicals. Agar–urea composites as a potential solution, where agar serves as a biodegradable matrix for urea as a nitrogen source. 3D printed structure successful urea entrapment within the agar matrix, and water sorption studies indicated reduced moisture uptake due to urea, water interactions disrupting agar's hydrogen-bonding network. Nitrogen release studies showed that while conventional urea exhibited near-complete release within 48 h, agar–urea formulations achieved slower, more controlled release rates, of urea, suggesting their potential to mitigate the rapid nutrient loss associated with traditional fertilizers (Fig. 7). This composite development is environmentally friendly, slow-release fertilizers that improve nitrogen-use efficiency in agricultural applications [21].

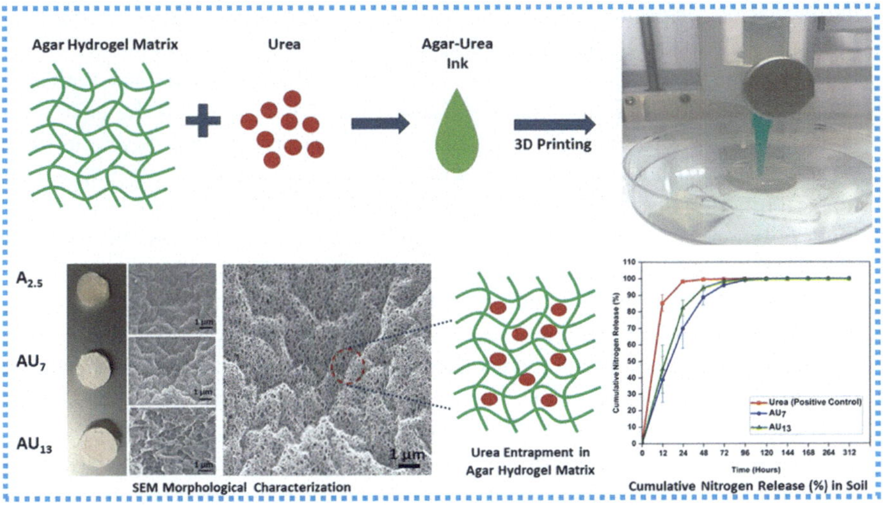

Fig. 7 3D-Printable agar–urea hydrogels: an effective approach for nitrogen delivery in agricultural practices. Adapted from [21]

6 Recent Advances in Polysaccharide Hydrogel 3D Printing

6.1 Innovations in Material Properties and Processing Techniques

Polysaccharide hydrogels have gained significant attention in biomedical, soft robotics, food, and sustainable material applications due to their biocompatibility, tunable properties, and environmental benefits. Recent advancements in 3D printing have further enhanced their functionality through innovations in material design and processing techniques.

Traditional polysaccharide hydrogels often suffer from limited mechanical strength, but recent developments have addressed these challenges through:

- Advanced crosslinking methods or chemical modification to enhance structural integrity.
- Reinforced composites for improved toughness, shape-morphing ability, and extrudability.
- Smart stimuli-responsive hydrogels that adapt to physical, chemical, and biological changes.

To optimize printability and performance, cutting-edge 3D printing strategies include:

- FDM and DIW are widely used for 3D printing polysaccharides. Recent advancements have focused on optimizing bioink rheology, particularly viscosity and shear-thinning properties, which are essential for achieving high-resolution and mechanically stable prints. The integration of hybrid bioinks has further enhanced the properties of printed structures, enabling greater customization of material properties for specific applications [112].
- It is possible to print fine features with polysaccharide-based inks with IJP. Nanoparticle additives are other innovations that have enhanced the characteristics of printed structures. The accuracy of this approach can also be used to create polysaccharide-based sensors with functionalized surfaces, including ones that are functionally coated with enzymes or antibodies [113, 114].
- Photopolymerization techniques (SLA and DLP) have advanced through the development of polysaccharide-based bioinks that can be rapidly cured while maintaining biocompatibility and bioactivity. The methods are especially applicable to high-resolution prints that are essential when medical equipment and biosensors need accuracy. Recent research has centered on enhancing printing resolution on employing customizable light sources and resin specialties, which resulted in more accurate methodology of fabrication [113].
- Electrohydrodynamic jet (E-jet) printing: This method can be used to print nanoscale, and microscale structures with bioink made of polysaccharides and an unparalleled feature resolution. The high-precision deposition of the technique

and recent advances in the optimization of electrostatic forces and voltages enable the printing of objects with great accuracy in exquisitely fine detail [115].
- 3D printing of polysaccharides, aided by solvents, results in improved printability and mechanical strength by controlling solvent interactions. Advances in bioink rheological properties have overcome the earlier limitations of printing of complex polysaccharide architecture [116].

6.2 Hybrid Systems and Multi-material Printing

The modules of 3D printing have distinct abilities of fabricating delicate tissue structures. Researchers should consider combined techniques of printing to produce the best tissue structures.

- Marko Milojevic et al. [117] have established a novel technique of making hybrid scaffolds with both thermoplastics and hydrogels. These composite structures enable tunable structural and chemical properties, which makes them especially useful in applications to tissue engineering and regenerative medicine [117].
- Hybrid 3D printing combines several manufacturing methods in order to use structural and functional materials during one printing. Such a technique allows the manufacture of versatile products by inserting polysaccharides into hybrid material systems [118].
- Proteo-saccharides (combining protein with polysaccharides), or bioactive hydrogels incorporating growth factors, antimicrobial agents, or cell-adhesive motifs for enhanced functionality, scaffold's physiological signaling, and mechanical strength.
- Polysaccharides combined with other polysaccharides to form composite polysaccharide hydrogels, as polysaccharide hybrids, such as alginate–nanocellulose, alginate–methylcellulose, and hyaluronan–agarose.

6.3 Integration with Other Technologies (E.G., Stem Cells, Growth Factors)

The incorporation of polysaccharides with advanced biological components including stem cells and growth factors is transforming regenerative medicine. These hybrid systems enhance tissue regeneration and wound healing through synergistic biological interactions.

- Alginate-based hydrogels combined with mesenchymal stem cells promote chondrogenic differentiation under mechanical and biochemical stimulation, supporting cartilage repair [63].

- Alginate hydrogels enable controlled delivery of BMP-2 and mesenchymal stem cells, significantly improving osteogenesis and bone formation in in-vivo models [119].
- Conjugation of growth factors (e.g., vascular endothelial growth factor or fibroblast growth factor) to hyaluronic acid matrices accelerates angiogenesis and wound healing.
- Chitosan functionalized with epidermal growth factor or platelet-derived growth factor enhances epithelialization and granulation tissue formation in chronic wounds.
- Chitosan-coated nanoparticles deliver vascular endothelial growth factor to stem cells with high precision, improving vascularization and tissue recovery in ischemic injuries [120].

7 Challenges and Limitations

7.1 Mechanical and Structural Limitations

The widespread adoption of 3D-printed polysaccharides in industrial and practical applications faces several mechanical and structural limitations:

- Alginate hydrogels exhibit low stiffness and are prone to mechanical failure under stress. Chitosan hydrogels suffer from brittleness and tend to crack when subjected to mechanical loads.
- Most polysaccharide-based materials are designed as rigid or semi-rigid structures, restricting their flexibility and ability to maintain shape under dynamic conditions.
- Achieving uniform mechanical properties in 3D-printed polysaccharides remains challenging. Variations in extrusion speed, temperature, and humidity during printing often lead to layer-by-layer inconsistencies, compromising the overall strength and stability of the final product.
- Poor layer adhesion and resolution cause defects such as delamination, significantly reducing the structural integrity of printed objects.
- Polysaccharides naturally degrade in biological environments, which is problematic for applications requiring long-term stability. Rapid degradation often results in premature loss of structural integrity.

7.2 Scalability and Commercialization

The transition from lab-scale prototypes to large-scale production of polysaccharide-based 3D-printed products faces several critical challenges that hinder commercialization:

- The refinement and purification of polysaccharides from natural sources remain cost-prohibitive. Obtaining high-quality materials in quantities sufficient for large-scale production is economically unfeasible compared to synthetic alternatives.
- Consistent availability of polysaccharides in large quantities is challenging due to variations in source materials, harvesting conditions, and seasonal production fluctuations.
- Polysaccharide-based bioinks often demonstrate batch-to-batch variability in viscosity, gelation rates, and degradation profiles, compromising product consistency and quality control.
- Current 3D printing technologies for polysaccharides are primarily optimized for small-scale prototyping. Scaling these processes while maintaining consistent product quality presents significant technical challenges.
- Extended curing times and crosslinking requirements for bioinks to achieve desired mechanical properties reduce production rates and increase manufacturing costs.

7.3 Regulatory and Ethical Considerations in Biomedical Applications

Significant regulatory and ethical hurdles must be addressed before widespread clinical adoption of 3D-printed polysaccharide-based technologies can occur. These challenges primarily concern patient safety, treatment efficacy, and compliance with healthcare regulations.

- The use of 3D-printed polysaccharide scaffolds and implants in human trials presents several ethical concerns that require careful attention:

  Patients must receive comprehensive information about potential risks and benefits before participating in trials involving these experimental technologies.
  While these innovations could revolutionize personalized medicine, there are valid concerns that they may initially only be accessible to privileged populations, potentially exacerbating existing healthcare disparities.
  Special ethical considerations must be made for vulnerable groups (e.g., children, elderly patients, and immunocompromised individuals), particularly regarding uncertain long-term effects and potential risks.

- Biomedical and food packaging applications face stringent regulatory standards. The absence of established frameworks for 3D-printed polysaccharide products further complicates commercialization efforts.

8 Future Directions and Prospects

The integration of polysaccharides with 3D printing technology has created transformative opportunities across biomedical, environmental, and industrial sectors. Future developments in this field will focus on several key directions:

- Smart Responsive Hydrogels: Research will advance stimulus-responsive polysaccharide-based hydrogels that can adapt to environmental changes such as pH, temperature, or biochemical signals.
- Advanced Bioprinting Applications: Polysaccharide hydrogels will remain fundamental to tissue engineering, particularly as 3D bioprinting evolves to fabricate complex, multilayered tissues and organ structures. This will involve precise spatial arrangement of cells, growth factors, and hydrogel matrices to replicate native tissue architecture.
- Hybrid Material Development: A promising avenue involves creating composite materials by combining polysaccharides with other biomaterials, nanomaterials, and synthetic polymers. Such hybridization will enhance mechanical properties while enabling additional functionalities through nanoparticle incorporation.
- Sustainable Manufacturing: Future innovations will emphasize green chemistry approaches, prioritizing renewable feedstocks, biodegradable components, and low-toxicity processes to reduce the environmental footprint of 3D printing technologies.

These advancements will position polysaccharide-based materials as versatile platforms for next-generation applications in regenerative medicine, environmental remediation, and beyond.

9 Conclusion

Polysaccharide-based hydrogels integrated with 3D printing technology represent a transformative platform for diverse biomedical applications, including drug delivery systems, tissue engineering, and personalized medicine. These materials combine the inherent advantages of polysaccharides, biocompatibility, biodegradability, and hydrophilicity, with the precision and customization capabilities of additive manufacturing. This synergy enables the fabrication of patient-specific drug carriers, implants, and tissue scaffolds capable of controlled and targeted therapeutic delivery. The molecular-level control afforded by 3D printing facilitates groundbreaking advances in personalized medicine, from tailored drug delivery platforms to complex regenerative scaffolds. Future developments will focus on three key areas: (1) intelligent hydrogels responsive to physiological stimuli, (2) hybrid material systems for enhanced mechanical performance, and (3) incorporation of bioactive molecules to improve therapeutic efficacy. While challenges persist, including optimization of mechanical properties, printability refinement, drug release consistency, and regulatory considerations, rapid advancements in materials science and fabrication technologies are steadily addressing these limitations. The unique combination of customizability, biodegradability, and bioactivity positions 3D-printed polysaccharide hydrogels as a cornerstone technology for next-generation biomedical solutions. As research progresses, these advanced materials will revolutionize therapeutic approaches, enabling more precise, effective, and patient-specific treatments that push the boundaries of modern healthcare and regenerative medicine.

References

1. Rodell CB, Mealy JE, Burdick JA. Supramolecular guest—host interactions for the preparation of biomedical materials. Bioconjug Chem. 2015;26(12):2279–89.
2. Bonde S, Chandarana C, Prajapati P, Vashi V. A comprehensive review on recent progress in chitosan composite gels for biomedical uses. Int J Biol Macromol. 2024:132723.
3. Hassan M, Dave K, Chandrawati R, Dehghani F, Gomes VG. 3D printing of biopolymer nanocomposites for tissue engineering: nanomaterials, processing and structure-function relation. Eur Polymer J. 2019;121:109340.
4. Park JA, Lee Y, Jung S. Inkjet-based bioprinting for tissue engineering. Organoid. 2023;3.
5. Fernandez AM. Laser bioprinting for single cell biological studies. Universidad Politécnica de Madrid; 2021.
6. Ramirez Caballero SS, Saiz E, Montembault A, Tadier S, Maire E, David L, et al. 3-D printing of chitosan-calcium phosphate inks: rheology, interactions and characterization. J Mater Sci Mater Med. 2018;30(1):6.
7. Zhang Q, Zhu E, Li T, Zhang L, Wang Z. High-value utilization of cellulose: intriguing and important effects of hydrogen bonding interactions—a mini-review. Biomacromolecules. 2024;25(10):6296–318.
8. Gao Y, Peng K, Mitragotri S. Covalently crosslinked hydrogels via step-growth reactions: crosslinking chemistries, polymers, and clinical impact. Adv Mater. 2021;33(25):2006362.
9. Rial-Hermida MI, Rey-Rico A, Blanco-Fernandez B, Carballo-Pedrares N, Byrne EM, Mano JF. Recent progress on polysaccharide-based hydrogels for controlled delivery of therapeutic biomolecules. ACS Biomater Sci Eng. 2021;7(9):4102–27.
10. Gieroba B, Kalisz G, Krysa M, Khalavka M, Przekora A. Application of vibrational spectroscopic techniques in the study of the natural polysaccharides and their cross-linking process. Int J Mol Sci. 2023;24(3):2630.
11. Li H, Wang Y, Zhao P, Guo L, Huang L, Li X, et al. Naturally and chemically acetylated polysaccharides: structural characteristics, synthesis, activities, and applications in the delivery system: a review. Carbohyd Polym. 2023;313:120746.
12. Isogai A. TEMPO-catalyzed oxidation of polysaccharides. Polym J. 2022;54(4):387–402.
13. Bercea M. Rheology as a tool for fine-tuning the properties of printable bioinspired gels. Molecules. 2023;28(6):2766.
14. Unal AZ, West JL. Synthetic ECM: bioactive synthetic hydrogels for 3D tissue engineering. Bioconjug Chem. 2020;31(10):2253–71.
15. Han L, Liu K, Wang M, Wang K, Fang L, Chen H, et al. Mussel-inspired adhesive and conductive hydrogel with long-lasting moisture and extreme temperature tolerance. Adv Func Mater. 2018;28(3):1704195.
16. Li Y, Liu Y, Li R, Bai H, Zhu Z, Zhu L, et al. Collagen-based biomaterials for bone tissue engineering. Mater Des. 2021;210:110049.
17. Wang R, Cheng C, Wang H, Wang D. Swollen hydrogel nanotechnology: advanced applications of the rudimentary swelling properties of hydrogels. Chem Phys Mater. 2024.

18. Jiang L, Chen D, Wang Z, Zhang Z, Xia Y, Xue H, et al. Preparation of an electrically conductive graphene oxide/chitosan scaffold for cardiac tissue engineering. Appl Biochem Biotechnol. 2019;188(4):952–64.
19. Fijot N, Aguilar-Sanchez A, Ruiz-Caldas M-X, Redlinger-Pohn J, Mautner A, Mathew AP. 3D printed polylactic acid (PLA) filters reinforced with polysaccharide nanofibers for metal ions capture and microplastics separation from water. Chem Eng J. 2023;457:141153.
20. Zhou W, Fang J, Tang S, Wu Z, Wang X. 3D-printed nanocellulose-based cushioning—antibacterial dual-function food packaging aerogel. Molecules. 2021;26(12):3543.
21. Dissanayake W, Najaf Zadeh H, Nazmi AR, Stevens C, Huber T, Abhayawardhana PL. Exploring the potential of 3D-printable agar—urea hydrogels as an efficient method of delivering nitrogen in agricultural applications. Polysaccharides. 2024;5(1):49–66.
22. Villarreal-Otalvaro C, Coburn JM. Fabrication methods and form factors of gellan gum-based materials for drug delivery and anti-cancer applications. ACS Biomater Sci Eng. 2023;9(7):3832–42.
23. Guan G, Qizhuang L, Liu S, Jiang Z, Zhou C, Liao W. 3D-bioprinted peptide coupling patches for wound healing. Mater Today Bio. 2022;13:100188.
24. Hamza M, Tariq M, Sabahat S, Ahmad A, Muhammad N, Rahim A. Biological methods for fabricating nanomaterial-based metal-organic frameworks. In: Nanomaterial-based metal organic frameworks for single atom catalysis. Elsevier; 2023. pp. 75–106.
25. Bean SR, Zhu L, Smith BM, Wilson JD, Ioerger BP, Tilley M. Chapter 6—starch and protein chemistry and functional properties*. In: Taylor JRN, Duodu KG, editors. Sorghum and millets (second edition). AACC International Press; 2019. pp. 131–70.
26. Aljohani W, Ullah MW, Zhang X, Yang G. Bioprinting and its applications in tissue engineering and regenerative medicine. Int J Biol Macromol. 2018;107(Pt A):261–75.
27. Moussian B. Chitin: structure, chemistry and biology. In: Targeting chitin-containing organisms. 2019:5–18.
28. Sahariah P, Hjalmarsdottir MA, Masson M. Antimicrobial properties of chitosan and chitosan derivatives. Marine Glycobiology: CRC Press; 2016. p. 365–88.
29. Yadav M, Ahmadi Y. Alginates: source, chemistry, and properties. Alginates: Apple Academic Press; 2019. p. 1–24.
30. Makarova AO, Derkach SR, Khair T, Kazantseva MA, Zuev YF, Zueva OS. Ion-induced polysaccharide gelation: peculiarities of alginate egg-box association with different divalent cations. Polymers. 2023;15(5):1243.
31. Chaudhary S, Jain VP, Jaiswar G. The composition of polysaccharides: monosaccharides and binding, group decorating, polysaccharides chains. In: Innovation in nano-polysaccharides for ecosustainability. Elsevier; 2022. pp. 83–118.
32. Mukherjee C, Varghese D, Krishna J, Boominathan T, Rakeshkumar R, Dineshkumar S, et al. Recent advances in biodegradable polymers—properties, applications and future prospects. Eur Polymer J. 2023;192:112068.
33. Jamil N, Saad Ali HM, Yasir M, Hamza M, Sagheer M, Ahmed T, et al. Biosynthesized metallic and bimetallic nanoparticles as effective biocides for plant protection: plausible mechanisms and challenges. J Chem. 2024;2024(1):3328223.
34. Kharkar PM, Kiick KL, Kloxin AM. Designing degradable hydrogels for orthogonal control of cell microenvironments. Chem Soc Rev. 2013;42(17):7335–72.
35. Kanwal Q, Ahmed M, Hamza M, Ahmad M, Yousaf N, Javaid A, et al. Curcumin nanoparticles: physicochemical fabrication, characterization, antioxidant, enzyme inhibition, molecular docking and simulation studies. 2023;13(32):22268–80.
36. Hamza M, Kanwal Q, Ali Z, Zargar S, Alshammari AH, Wani TA, et al. Exploring the molecular biology of zinc-doped hydroxyapatite nanocomposites as fillers for dental materials: a self-defensive approach targeting bacterial DNA. Mater Res Express. 2024;11(9):095401.
37. Fu J, Yap JX, Leo C, Chang CK, Show P-L. Polysaccharide hydrogels for controlling the nutrient release. Sep Purif Rev. 2024;53(3):276–88.
38. Tan J, Luo Y, Guo Y, Zhou Y, Liao X, Li D, et al. Development of alginate-based hydrogels: crosslinking strategies and biomedical applications. Int J Biol Macromol. 2023;239:124275.

39. Hamza M, Ahmad A, Tariq M, Riaz S, Rahim AJN-bmoffsac. Metal—organic frameworks (an overview). 2023:1–38.
40. Mredha MTI, Pathak SK, Cui J, Jeon I. Hydrogels with superior mechanical properties from the synergistic effect in hydrophobic—hydrophilic copolymers. Chem Eng J. 2019;362:325–38.
41. Huang G, Li F, Zhao X, Ma Y, Li Y, Lin M, et al. Functional and biomimetic materials for engineering of the three-dimensional cell microenvironment. Chem Rev. 2017;117(20):12764–850.
42. Athanasiou KA, Eswaramoorthy R, Hadidi P, Hu JC. Self-organization and the self-assembling process in tissue engineering. Annu Rev Biomed Eng. 2013;15(1):115–36.
43. Schon B, Hooper G, Woodfield T. Modular tissue assembly strategies for biofabrication of engineered cartilage. Ann Biomed Eng. 2017;45:100–14.
44. Gorski F. Computer aided design of 3D printable anatomically shaped medical devices: methodologies and applications. 2025.
45. Freeman S, Calabro S, Williams R, Jin S, Ye K. Bioink formulation and machine learning-empowered bioprinting optimization. Front Bioeng Biotechnol. 2022;10:913579.
46. Schwab A, Levato R, D'Este M, Piluso S, Eglin D, Malda J. Printability and shape fidelity of bioinks in 3D bioprinting. Chem Rev. 2020;120(19):11028–55.
47. Wang J, Cui Z, Maniruzzaman M. Bioprinting: a focus on improving bioink printability and cell performance based on different process parameters. Int J Pharm. 2023;640:123020.
48. Iftekar SF, Aabid A, Amir A, Baig M. Advancements and limitations in 3D printing materials and technologies: a critical review. Polymers. 2023;15(11):2519.
49. Gillispie G, Prim P, Copus J, Fisher J, Mikos AG, Yoo JJ, et al. Assessment methodologies for extrusion-based bioink printability. Biofabrication. 2020;12(2):022003.
50. Marnot A, Koube K, Jang S, Thadhani N, Kacher J, Brettmann B. Material extrusion additive manufacturing of high particle loaded suspensions: a review of materials, processes and challenges. Virtual Phys Prototyp. 2023;18(1):e2279149.
51. Krishna DV, Sankar MR. Persuasive factors on the bioink printability and cell viability in the extrusion-based 3D bioprinting for tissue regeneration applications. Eng Regen. 2023;4(4):396–410.
52. Gao Y, Liu Z, Dai S, Zhao J, Guo Y, Cai X, et al. Modification, structure identification, and biological activities of phosphorylated polysaccharides: a review. Starch-Starke. 2024;76(7–8):2300223.
53. Chakka VP, Zhou T. Carboxymethylation of polysaccharides: Synthesis and bioactivities. Int J Biol Macromol. 2020;165:2425–31.
54. Malinowska E, Klimaszewska M, Straczek T, Schneider K, Kapusta C, Podsadni P, et al. Selenized polysaccharides—biosynthesis and structural analysis. Carbohyd Polym. 2018;198:407–17.
55. Dar UA, Ali A. Click chemistry a promising tool to develop polysaccharide-based hydrogels. Polysaccharides-based hydrogels: Elsevier; 2024. p. 91–117.
56. Meng X, Edgar KJ. "Click" reactions in polysaccharide modification. Prog Polym Sci. 2016;53:52–85.
57. Chen T, Sun C, Tian X, Jiang X, Zhang M. Natural polysaccharide: modification and application. Paper Biomater. 2021;6(2).
58. Nguyen T-P, Lee B-T. Fabrication of oxidized alginate-gelatin-BCP hydrogels and evaluation of the microstructure, material properties and biocompatibility for bone tissue regeneration. J Biomater Appl. 2012;27(3):311–21.
59. Mostajeran H, Baheiraei N, Bagheri H. Effects of cerium-doped bioactive glass incorporation on an alginate/gelatin scaffold for bone tissue engineering: in vitro characterizations. Int J Biol Macromol. 2024;255:128094.
60. Gunes OC, Kara A, Baysan G, Bugra Husemoglu R, Akokay P, Ziylan Albayrak A, et al. Fabrication of 3D Printed poly(lactic acid) strut and wet-electrospun cellulose nano fiber reinforced chitosan-collagen hydrogel composite scaffolds for meniscus tissue engineering. J Biomater Appl. 2022;37(4):683–97.

61. Guo Z, Zhang S, Guo Y, Xia J, Wu X, Hu H, et al. Bioinspired coacervate-based bioinks for construction of multiscale tissue engineering scaffolds. Nano Res. 2024;17(9):8209–19.
62. Valot L, Martinez J, Mehdi A, Subra G. Chemical insights into bioinks for 3D printing. Chem Soc Rev. 2019;48(15):4049–86.
63. Zamboulis A, Michailidou G, Koumentakou I, Bikiaris DN. Polysaccharide 3D printing for drug delivery applications. Pharmaceutics. 2022;14(1):145.
64. Safarzadeh Kozani P, Safarzadeh Kozani P, Hamidi M, Valentine Okoro O, Eskandani M, Jaymand M. Polysaccharide-based hydrogels: Properties, advantages, challenges, and optimization methods for applications in regenerative medicine. Int J Polym Mater Polym Biomater. 2022;71(17):1319–33.
65. Bauman L. Photopolymerization based 3D printing of thermoresponsive hydrogel precursors. University of Waterloo; 2024.
66. Zanon M. Photocurable hydrogels for tissue engineering applications. 2023.
67. Generalova AN, Demina PA, Akasov RA, Khaydukov E. Photopolymerization in 3D printing of tissue-engineered constructs for regenerative medicine. Russ Chem Rev. 2023;92(2).
68. Xie M, Su J, Zhou S, Li J, Zhang K. Application of hydrogels as three-dimensional bioprinting ink for tissue engineering. Gels. 2023;9(2):88.
69. Fatimi A, Okoro OV, Podstawczyk D, Siminska-Stanny J, Shavandi A. Natural hydrogel-based bioinks for 3D bioprinting in tissue engineering: a review. Gels. 2022;8(3):179.
70. Zennifer A, Subramanian A, Sethuraman S. Design considerations of bioinks for laser bioprinting technique towards tissue regenerative applications. Bioprinting. 2022;27:e00205.
71. Williams D, Thayer P, Martinez H, Gatenholm E, Khademhosseini A. A perspective on the physical, mechanical and biological specifications of bioinks and the development of functional tissues in 3D bioprinting. Bioprinting. 2018;9:19–36.
72. Decante G, Costa JB, Silva-Correia J, Collins MN, Reis RL, Oliveira JM. Engineering bioinks for 3D bioprinting. Biofabrication. 2021;13(3):032001.
73. Derakhshanfar S, Mbeleck R, Xu K, Zhang X, Zhong W, Xing M. 3D bioprinting for biomedical devices and tissue engineering: a review of recent trends and advances. Bioactive Mater. 2018;3(2):144–56.
74. Sharma S, Bhende M, Goel A. A review: polysaccharide-based hydrogels and their biomedical applications. Polym Bull. 2024;81(10):8573–94.
75. Benalaya I, Alves G, Lopes J, Silva LR. A review of natural polysaccharides: sources, characteristics, properties, food, and pharmaceutical applications. Int J Mol Sci. 2024;25(2):1322.
76. Hamza M, Kanwal Q, Hussain MI, Khan K, Asghar A, Liu Z, et al. Recent progress in 3D printed piezoelectric materials for biomedical applications. Mater Sci Eng R Rep. 2025;164:100962.
77. Kumar MS, Varma P, Kandasubramanian B. From lab to life: advances in in-situ bioprinting and bioink technology. Biomed Mater. 2024;20(1):012004.
78. Yang J, Chen Y, Zhao L, Zhang J, Luo H. Constructions and properties of physically cross-linked hydrogels based on natural polymers. Polym Rev. 2023;63(3):574–612.
79. Bustamante-Torres M, Romero-Fierro D, Arcentales-Vera B, Palomino K, Magafia H, Bucio E. Hydrogels classification according to the physical or chemical interactions and as stimuli-sensitive materials. Gels. 2021;7(4):182.
80. Torgal S, Subramani G, Manian R. Comprehensive insights into chitosan hydrogels: from crosslinking and characterization to immunomodulation, microbiome interactions and biomedical uses. Biomass Convers Biorefinery. 2024:1–36.
81. Anthony CY. Polymer-nanoparticle materials for prophylaxis: from passive immunity to wildfire prevention. Stanford University; 2020.
82. Battigelli A, Almeida B, Shukla A. Recent advances in bioorthogonal click chemistry for biomedical applications. Bioconjug Chem. 2022;33(2):263–71.
83. Cai C, Li W, Zhang X, Cheng B, Chen S, Zhang Y. Natural polymer—based hydrogel dressings for wound healing. Adv Wound Care. 14(6):295–322.
84. Martinez-López AL, Carvajal-Millan E, Lizardi-Mendoza J, López-Franco YL, Rascén-Chu A, SalasMufioz E, et al. The peroxidase/H_20_2 system as a free radical-generating

agent for gelling maize bran arabinoxylans: rheological and structural properties. Molecules. 2011;16(10):8410–8.
85. Li Z, Lin Z. Recent advances in polysaccharide-based hydrogels for synthesis and applications. Aggregate. 2021;2(2):e21.
86. Azeredo HMC, Waldron KW. Crosslinking in polysaccharide and protein films and coatings for food contact—a review. Trends Food Sci Technol. 2016;52:109–22.
87. Niu H, Chen X, Luo T, Chen H, Fu X. Relationships between the behavior of three different sources of pectin at the oil-water interface and the stability of the emulsion. Food Hydrocoll. 2022;128:107566.
88. Sun X, Pei F, Fang Y. The effects of hydrocolloids on the thermomechanical, viscoelastic and microstructural properties of whole wheat flour dough. Food Chem. 2022;370:130976.
89. Huang L, Shen M, Morris GA, Xie J. Sulfated polysaccharides: immunomodulation and signaling mechanisms. Trends Food Sci Technol. 2019;92:1–11.
90. Sylvester CB, Pugazenthi A, Grande-Allen KJ, Ghanta RK. Cell-laden bioactive poly(ethylene glycol) hydrogels for studying mesenchymal stem cell behavior in myocardial infarct-stiffness microenvironments. Cardiovasc Eng Technol. 2021;12(2):183–99.
91. Senatov FS, Niaza KV, Zadorozhnyy MY, Maksimkin A, Kaloshkin S, Estrin Y. Mechanical properties and shape memory effect of 3D-printed PLA-based porous scaffolds. J Mech Behav Biomed Mater. 2016;57:139–48.
92. Sun H, Hu C, Zhou C, Wu L, Sun J, Zhou X, et al. 3D printing of calcium phosphate scaffolds with controlled release of antibacterial functions for jaw bone repair. Mater Des. 2020;189:108540.
93. Mirani B, Pagan E, Shojaei S, Dabiri SMH, Savoji H, Mehrali M, et al. Facile method for fabrication of meter-long multifunctional hydrogel fibers with controllable biophysical and biochemical features. ACS Appl Mater Interfaces. 2020;12(8):9080–9.
94. Antich C, de Vicente J, Jiménez G, Chocarro C, Carrillo E, Montañez E, et al. Bio-inspired hydrogel composed of hyaluronic acid and alginate as a potential bioink for 3D bioprinting of articular cartilage engineering constructs. Acta Biomater. 2020;106:114–23.
95. Guo J, Zhang R, Zhang L, Cao X. 4D printing of robust hydrogels consisted of agarose nanofibers and polyacrylamide. ACS Macro Lett. 2018;7(4):442–6.
96. Valentin TM, Landauer AK, Morales LC, DuBois EM, Shukla S, Liu M, et al. Alginate-graphene oxide hydrogels with enhanced ionic tunability and chemomechanical stability for light-directed 3D printing. Carbon. 2019;143:447–56.
97. Kumari S, Mondal P, Chatterjee K. Digital light processing-based 3D bioprinting of k-carrageenan hydrogels for engineering cell-loaded tissue scaffolds. Carbohyd Polym. 2022;290:119508.
98. Elbadawi M, Nikjoo D, Gustafsson T, Gaisford S, Basit AW. Pressure-assisted microsyringe 3D printing of oral films based on pullulan and hydroxypropyl methylcellulose. Int J Pharm. 2021;595:120197.
99. , Zhaoxuan F, Jinsong L, Dasen Z, Hui S, Jiagi L, Wengin B. A novel photocurable pullulan-based bioink for digital light processing 3D printing. Int J Bioprinting. 2022;9(2):657.
100. Joseph A, Vijayan AS, Xavier J, KB M, Karthikeyan A, Gopinath N, et al. 3D printed arrowroot starchgellan scaffolds for wound healing applications. Int J Biol Macromol. 2024;264:130604.
101. Patitucci F, Motta MF, Dattilo M, Malivindi R, Leonetti AE, Pezzi G, et al. 3D-printed alginate/pectin-based patches loaded with olive leaf extracts for wound healing applications: development, characterization and in vitro evaluation of biological properties. Pharmaceutics. 2024;16(1):99.
102. An Z, Liu Z, Mo H, Hu L, Li H, Xu D, et al. Preparation of Pickering emulsion gel stabilized by tea residue protein/xanthan gum particles and its application in 3D printing. J Food Eng. 2023;343:111378.
103. Morris VB, Nimbalkar S, Younesi M, McClellan P, Akkus O. Mechanical properties, cytocompatibility and manufacturability of chitosan: PEGDA hybrid-gel scaffolds by stereolithography. Ann Biomed Eng. 2017;45:286–96.

104. Yang Y, WuH, Fu Q, Xie X, Song Y, Xu M, et al. 3D-printed polycaprolactone-chitosan based drug delivery implants for personalized administration. Mater Des. 2022;214:110394.
105. Bonifacio MA, Cometa S, Cochis A, Scalzone A, Gentile P, Scalia AC, et al. A bioprintable gellan gum/lignin hydrogel: a smart and sustainable route for cartilage regeneration. Int J Biol Macromol. 2022;216:336–46.
106. Wang J, Dai S, Xiang N, Zhang L, Zhong W, Shao P, et al. Cell-based meat scaffold based on a 3D printed starch-based gel. J Agric Food Chem. 2024;72(34):19143–54.
107. Shen S, Shen J, Shen H, Wu C, Chen P, Wang Q. Dual-enzyme crosslinking and post-polymerization for printing of polysaccharide-polymer hydrogel. Front Chem. 2020;8–2020.
108. Hu Y, Wu B, Xiong Y, Tao R, Panayi AC, Chen L, et al. Cryogenic 3D printed hydrogel scaffolds loading exosomes accelerate diabetic wound healing. Chem Eng J. 2021;426:130634.
109. Dang W, Chen W-C, Ju E, Xu Y, Li K, Wang H, et al. 3D printed hydrogel scaffolds combining glutathione depletion-induced ferroptosis and photothermia-augmented chemodynamic therapy for efficiently inhibiting postoperative tumor recurrence. J Nanobiotechnology. 2022;20(1):266.
110. Zhou G, Wang KP, Liu HW, Wang L, Xiao XF, Dou DD, et al. Three-dimensional polylactic acid@graphene oxide/chitosan sponge bionic filter: highly efficient adsorption of crystal violet dye. Int J Biol Macromol. 2018;113:792–803.
111. Sreedhar N, Kumar M, Al Jitan S, Thomas N, Palmisano G, Arafat HA. 3D printed photocatalytic feed spacers functionalized with B-FeOOH nanorods inducing pollutant degradation and membrane cleaning capabilities in water treatment. Appl Catal B. 2022;300:120318.
112. Tang T, Zhang M, Mujumdar AS. Film forming properties of 3D printed polysaccharide based gels and development of CO_2 monitoring labels. Food Packag Shelf Life. 2024;46:101401.
113. Hasanin MS, Saied HE, Kamel S. Polysaccharides based biosensors for medical applications: prospective and future aspects. Starch-Starke. 2023;75(11–12):2300037.
114. Chiesal, De Maria C, Ceccarini MR, Mussolin L, Coletta R, Morabito A, et al. 3D printing silk-based bioresorbable piezoelectric self-adhesive holey structures for in vivo monitoring on soft tissues. ACS Appl Mater Interfaces. 2022;14(17):19253–64.
115. Hamza M, Qadeer A, Alsaiari M, Alsayari S, Kanwal Q. Rahim AJN-SE. Recent Adv Enzym Immobil Nanomater. 2022;126:1–66.
116. Shokrani H, Shokrani A, Seidi F, Mashayekhi M, Kar S, Nedeljkovic D, et al. Polysaccharide-based biomaterials in a journey from 3D to 4D printing. Bioeng Transl Med. 2023;8(4):e10503.
117. Khan SA, Ahmad H, Zhu G, Pang H, Zhang Y. Three-dimensional printing of hydrogels for flexible sensors: a review. Gels. 2024;10(3):187.
118. Milojevic M, Harih G, Vihar B, Vajda J, Gradisnik L, Zidari€é T, et al. Hybrid 3D printing of advanced hydrogel-based wound dressings with tailorable properties. Pharmaceutics. 2021;13(4):564.
119. Wang YL, Han L, Zhang XL, Cao L, Hu K, Li LH, et al. 3D bioprinting of an electroactive and selfhealing polysaccharide hydrogels. J Tissue Eng Regen Med. 2022;16(1):76–85.
120. Guo Y, Liu Y, Zhang Z, Zhang X, Jin X, Zhang R, et al. Biopolymer based fibrous aggregate materials for diagnosis and treatment: design, manufacturing, and applications. Adv Mater. 2025:2414877.

3D Engineered Polysaccharides for Environmental Remediation

3D Polysaccharide Hydrogels for Soil and Water Remediation

Zaccheus Shehu, Kalu Michael Kalu, Michael Emmanuel, Wilson Lamayi Danbature, Ibrahim Muhammad Nazifi, Abigail Mbozi, Bakari Chaka, Koffi Enoc Elom Amessinou, Pachris Muamba Kapanga, Raphael Muamba Tshimanga, Merolyne Natuhwera, and George William Atwoki Nyakairu

Abstract The challenges associated with water and soil contamination is becoming an issue of global concern, these challenges are attributed to concentrated urban, industrial, and agricultural activities necessitating for sustainable and innovative remediation solutions. This article provides an all-inclusive review of three-dimensional polysaccharide-based hydrogels (PBH), as a novel remediation class of biomaterial matrix. PBH are derived from polymers which are biodegradable,

Z. Shehu (✉) · K. M. Kalu · M. Emmanuel · W. L. Danbature
Department of Chemical Sciences, Gombe State University, Tudun-Wada, Gombe, Gombe State, Nigeria
e-mail: zaccheusshehu@gsu.edu.ng

I. M. Nazifi
Department of Science Laboratory Technology, School of Science, Federal Polytechnic Kaltungo, Kaltungo, Gombe State, Nigeria

A. Mbozi
School of Natural and Applied Sciences, Mulungushi University, Kabwe, Zambia

B. Chaka
Department of Mathematics and Physical Sciences, Maasai Mara University, Narok, Kenya

K. E. E. Amessinou
Resources Valorization, Environment and Sustainable Development Research Team (RVESD), Department of Mines, Mines School of Rabat, Agdal, Rabat, Morocco

Research Institute On Mines and Environment (RIME), University of Québec in Abitibi Témiscamingue (UQAT), Rouyn-Noranda, QC, Canada

Geology and Sustainable Mining Institute, Mohammed VI Polytechnic University, Ben Guerir, Morocco

P. M. Kapanga · R. M. Tshimanga
Congo Basin Water Resources Research Center (CRREBaC), University of Kinshasa (UNIKIN), Kinshasa, Democratic Republic of the Congo

Regional School of Water (ERE), University of Kinshasa (UNIKIN), Kinshasa, Democratic Republic of the Congo

M. Natuhwera · G. W. A. Nyakairu
Department of Chemistry, Makerere University, Kampala, Uganda

M. H. Abu Elella et al. (eds.), *3D Engineered Polysaccharides*,
https://doi.org/10.1007/978-3-032-26154-0_3

abundant, and renewable, some of which are alginate, chitosan, and cellulose which provides an alternative to those used conventionally which are in comparison, not ecofriendly, costly, and whose method of treatment is detrimental. A critical review was provided by this article on the PBH synthesis methods, which includes the chemical and physical crosslinking methods, which have been found to be critical in streamlining their physicochemical characteristics such as mechanical stability, swelling capacity, and porosity. The various applications of PBH were also presented in this study, with a major focus on water remediation, specifically on heavy metal ions, antibiotics, organic dyes removal, and other notable emerging pollutants, the roles played by PBHs in treatment of soil such as in serving as carriers in agrochemical release which are regulated, conditioning of soil for retention of water, and materials for the containment of pollutants fixed in soil. Although polysaccharide hydrogels demonstrate significant promise, there exist profound issues regarding their mechanical strength in various real-world environments, their long term-stability, and production scale-up. A focus on the development of smart and advanced multifunctional hydrogel composites that incorporates stimuli-responsive behaviours with enhanced durability, creating avenues for the next generation, sustainable technologies that adhere to the principles of a circular economy are the future research efforts.

Keywords Hydrogels · Soil remediation · Contamination · Water remediation · 3-D polysaccharides

List of Abbreviations

TGA	Thermogravimetric analysis
ECH	Epichlorohydrin
IPNs	Interpenetrating polymer networks
DMA	Dynamic Mechanical Analysis
GO	Graphene oxide
HPC	Hydroxypropyl cellulose
PBH	Polysaccharide-based hydrogels
XRD	X-ray diffraction
DNs	Double networks
MBA	Methylene bis(acrylamide)
FTIR	Fourier-transform infrared spectroscopy
UV–Vis	Ultraviolet–visible spectroscopy
NMR	Nuclear magnetic resonance
CNTs	Carbon nanotubes
TEMPO	2,2,6,6-Tetramethyl-1-piperidinyloxy
CMC	Carboxymethyl cellulose
DSC	Differential scanning calorimetry

1 Introduction

The escalating challenge of soil and water contamination, driven by industrial and agricultural activities, demands innovative and sustainable remediation technologies. Three-dimensional polysaccharide-based hydrogels (PBH) have risen as a leading biomaterial class to meet this need. Derived from abundant, renewable, and biodegradable sources like chitosan, alginate, and cellulose, these hydrogels offer an environmentally friendly alternative to those used conventionally, often costly and detrimental, treatment methods. Their unique, interconnected porous structure provides a vast surface area for interaction with contaminants, making them highly effective filtration and absorption matrices [1–5].

In water remediation, the inherent chemical functionalities of these polysaccharides, such as NH_2 (amine) and –COO– (carboxyl groups), act as powerful chelating agents, enabling the efficient elimination of heavy metal chemical species, organic dyes, including other pollutants considered as emerging.

For soil applications, these hydrogels are invaluable as conditioners, significantly fostering the retention of water and thus, serving as carriers for monitored/controlled release of agrochemicals, ultimately improving agricultural efficiency while minimizing environmental runoff. As research progresses, the focus is shifting towards developing advanced multifunctional hydrogel composites that promise even greater durability and stimuli-responsive behaviours, creating a blueprint for the next generation of smart, sustainable environmental solutions that complements principles of circular economy [6–8].

2 3D Hydrogels

These hydrogels are grounded on the natural polysaccharides, namely are, chitosan, cellulose, and alginate, with potential in the immobilization of contaminants and improving soil characteristics [9]. Recently, PBH have been shown to be very effective in removing or eliminating heavy metals and organic pollutants from aqueous contaminated system. For instance, it has been stated that hydrogels of chitosan possess high/elevated adsorption capacity for Pb^{2+} and Cd^{2+} ions and were potential in water treatment applications. Moreover, nanoparticles contained within these hydrogels has been reported to foster the adsorption ability and mechanical robustness of hydrogels [10].

In soil, polysaccharide hydrogels could enhance water accumulation, and nutrient release, especially being applied in arid zones. Research by Agbna and Zaidi [11] was able to show that the application of alginate hydrogel in soil remediation was significantly marked by an improved retention of water, particularly in sandy soils, this thus led to increase in growth and demanded less frequent watering. In addition, these hydrogels may act as vehicles for slow-release fertilizers, which reduce nutrient leaching and improve soil fertility [8, 12].

The environmental remediation applications can be finely optimized by their chemical composition and crosslinking strategies of 3D polysaccharide hydrogel. It should be noted that the 3D polysaccharide hydrogels show a versatile feature in environmental remediation as they could be designed for specific applications by modifying the chemical structure or crosslinking type. As studies develop, these hydrogels are expected to make relatively significant impact on the sustainable management of soil and water.

3 Polysaccharides: Sustainable and Functional Biopolymers for Hydrogel Synthesis

In the production of hydrogels, an important consideration is made on the choice of base polymer which informs its application in the environment. It has been reported that synthetic polymer can be functionally optimized, nut since they are petrochemical-based, they are weakly biodegradable and pose pollution issues [13]. It is as a result of this that a lot of efforts are been made to dwell on the employment of natural origin biopolymers – polysaccharides, to form the major units of hydrogels employed in remediation. Polysaccharides on the other hand, offer a lot of advantages that align with the goals of circular economy and green chemistry. They have been recorded as the most plentiful natural substances on earth, which can be gotten from recyclable raw systems like the algae (alginate), animals (chitin), plants (cellulose, starch), and also microorganisms (dextran), as such, they are not regarded expensive, they are sustainable and readily available [13–15]. In their functionality, these polysaccharides are profoundly versatile whose polymer framework is composed majorly of active functionalities like –OH, –COOH, and $–NH_2$, offering two major roles; they first offer active sites for the adsorption of pollutants via processes like ion exchange, chelation, and H-bonding, thus imparting the resulting hydrogels high intrinsic sorption capacity [16–18]. Secondly, these systems are also susceptible to chemical modification so that the researchers can append new functionality or crosslink polymer chains such that they particularly design the properties of the hydrogel for example, selectivity, swelling characteristics, and mechanical resistance—to be specific to a particular pollutant [16].

The application of these natural biopolymers is a paradigm shift in remediation thinking. It is a shift from the resolution of pollution issues produced by petroleum-based, synthetic industries using more energy-consuming technologies and toward the use of natural, frequently waste-material sources (for example, chitin from waste in the shellfish industry) as the solution [13, 19]. This process implements the concepts of circular economy, converting byproducts into high-value materials to restore nature. A range of common polysaccharides applied to synthesize hydrogel is given in Table 1.

Table 1 Overview of common polysaccharides for environmental hydrogel synthesis

Polysaccharide matrix	Natural source (s)	Monomeric form (s)	Functionality	Remediation properties
Chitosan/ Chitin	Fungi, Crustacean shells	Glucosamine/*N*-acetyl glucosamine	Amine ($-NH_2$), Hydroxyl (−OH)	Excellent chelating agent for heavy metals, cationic nature, and antimicrobial properties [13]
Starch	Corn, rice, potatoes (plants)	Amylopectin and Amylose	Hydroxyl (−OH)	Can be easily modified through the process of grafting aimed at promoting swelling, also inexpensive, and highly biodegradable [20]
Cellulose	Bacteria, plants (wood, cotton)	D-glucopyranose	Hydroxyl (−OH)	Abundant, possess high mechanical strength, and can be modified to incorporate carbonyl functionalities (for example, carboxymethylation) [21]
Dextran	Bacteria (*Leuconostoc mesenteroides*)	D-glucose	Hydroxyl (−OH)	Possess characteristic flexible polymer backbone, highly soluble in water, and can be modified to be anionic or cationic [22]
Alginate	Brown seaweed	Mannuronic and Guluronic acid	Carboxyl (−COOH), Hydroxyl (−OH)	They are anionic, capable of forming gels through ionic crosslinking with divalent cations, and possess high affinity for heavy metals [13]

(continued)

Table 1 (continued)

Polysaccharide matrix	Natural source (s)	Monomeric form (s)	Functionality	Remediation properties
Carrageenan	Red seaweed	3,6-anhydro-D-galactose and D-galactose	Hydroxyl (−OH), Sulfate (−SO_3H)	The sulfate groups provide strong binding sites for cationic species; they are anionic and capable of forming thermo-reversible gels [23]
Guar Gum	Guar beans (*Cyamopsis tetragonolobus*)	Galactose and mannose	Hydroxyl (−OH)	Fast swelling properties, high viscosity in solution, can be easily functionalized but are non-ionic [24]

4 Method of Hydrogel Preparation and Physicochemical Properties

The synthesis of a hydrophilic, water-insoluble, and functional 3D hydrogel network, from soluble polysaccharide chains, is referred to as crosslinking. Selection of the crosslinking method is critical, because the choice of method will have a direct influence over the final physical and chemical characteristics of the hydrogel, like swelling, strength, and stability. The relationship among crosslinking method, hydrogel characteristics, and final function is an important concept in hydrogel engineering; one application requires certain properties, and those properties are only developed with careful consideration of the synthetic procedure. Crosslinking methods can be classified in one of two ways: physical crosslinking, which is formed by reversible, non-covalent interactions and chemical crosslinking, which forms permanent covalent bonds [25].

4.1 Synthesis Strategies for PBH: Cellulose, Chitosan, and Additive Manufacturing

Hydrogen is captured physically (in terms of physical entrapment with reversible bonding) or chemically (as strong bonding) through the process of crosslinking. Functional groups (for example, hydroxyl—OH, carboxyl—COOH, amine—NH_2) confer significant features (for example, hydrophilicity, reactivity) that make it valuable for use in various fields (for example, drug carrier, and water purification) [26]. Hydrogels also respond to, for example, temperature and pH as reported. They can then be employed as a basis to produce or develop smart materials for a myriad of sophisticated applications [27].

It is of interest for dispersible hydrogel particles to selectively adsorb pollutants using functionalized network. The preparation of a kind of hydrogels for heavy metal/organic pollutants absorption by modifying certain functional groups to make these polymers have attractiveness to the pollutants is still a challenging topic. As an instance, the negative charge of the carboxyl group can cause the formation of ionic bonds with the positive charge of metal ions, and carboxylated hydrogels ($-COO^-$) has been reported as effective adsorption materials for heavy metal ions such as Pb and Cd [2].

Moreover, amine groups in hydrogels can absorb anionic dyes and other negatively charged pollutants from the water matrix by electrostatic attraction [28]. It has been shown that hybrid hydrogels, composed of both synthetic and natural polymers, can further enhance their structural and functional performances. For instance, chitosan-polyacrylamide hybrid hydrogel has demonstrated excellent potential in the removal of dyes and heavy metals from water. Due to the cationic and biocompatibility nature of chitosan, and the strength of polyacrylamide, this hydrogel utilizes the best of both material properties [29].

4.1.1 Physical Crosslinking

Because of its stiff backbone, cellulose, which is the most prevalent polysaccharide biomass resource known in nature, possess an intriguing one-dimensional shape [18, 30] coupled to this, cellulose has a great incorporation of –OH (hydroxyl) groups that runs throughout its backbone, which forms the intermolecular and intramolecular hydrogen bonding between cellulose backbones and low dispersibility in common solvents [31]. Since hydrogels are made using a two-staged process that involves dissolution and crosslinking, the primary challenge in creating cellulose-based hydrogels is identifying suitable solvents to dissolve cellulose. Two strategies are utilized to dissolve cellulose: (i) direct dissolution of raw cellulose using ionic liquids or base/urea aqueous solution [32]; (ii) cellulose modification for enhanced solubility in regular solvents. Note that cellulose derivatives are available in the market (for example, carboxymethyl cellulose [CMC] and hydroxypropyl cellulose [HPC]) and they are able to dissolve in a regular solvent. These strategies have great potential for the synthesis of hydrogels that are cellulose-based with tunable performances.

A good number of physically crosslinked cellulose-based hydrogels incorporated with different chemical species of cellulose bearing functionalities have been explored [33]. A very significant class of the nanocellulose which possess high degree of crystalline phases is the cellulose nanocrystal, which are obtainable by the process called acid hydrolysis of the unprocessed cellulose. This nanocrystal tends to exhibit a whisker-like or 1D rod like kind of morphology, with its diameter lying in the range of 3–50 nm, whose length is int range of 50–500 nm [34]. It has also been reported that the cellulose nanocrystal possesses slight degree of dispersibility in aqueous solutions that can be attributed to its surface induced negatively charged $-SO_3H$

component that is formed in the process of hydrolyzing in the presence of H_2SO_4 [35]. The now sulfonated cellulose nanocrystal is converted into a birefringent gel at concentration that is greater than 12.6%, attributable to alterations of electrostatic repulsion and its polydispersed rod-like kind of morphology [34].

The uncharged cellulose nanocrystal however, can be extracted through the process of hydrolysis using HCl, but then they are not capable of dispersing in an aqueous solution. To ensure that dispersibility is enhanced for the cellulose nanocrystal, the process of surface oxidation is often adopted to extract the carboxylic acid components by employing TEMPO (2,2,6,6-tetramethyl-1-piperidinyloxy) [36]. The now cellulose nanocrystal is able to gelate at reduced pH which can be attributed to the attractive H-bonding within the –COOH groups, thus forming a percolating network via the interacting fibers of the cellulose nanocrystal [37].

Chitosan on the other hand, are capable of exhibiting self-cross-linkage in the absence of organic solvent or external crosslinking agents, which could be considered harmful for medical applications. A large number of hydrogels can be produced with an increase in the pH, greater than 6.3 with strong bases such as NaOH or NH_4OH, this is due to the microcrystalline physical crosslinking initiated by hydrogen bonds and Van der Waals interactions during the process of neutralization [38] couple to this, chitosan′chitin without the process of derivatization are capable of dissolving in aqueous solution of base/urea. On the other hand, elevated-strength hydrogels that is composed of nanofibrous matrix have been produced through freezing–thawing process [39–41].

4.1.2 Chemical Crosslinking

Although adding such cellulose above might improve the mechanical properties, in general the cellulose/derivatives are just physically trapped in the systems, the weak physical interactions could not obviously boost the performance of the hydrogels. It is priority to achieve such goal where chemical celluloses are incorporated into the hydrous monomer networks to further improve the features of cellulose [42–44]. In general, there are two methods to capture celluloses into hydrogels using chemical approaches:

(i) Cellulose as the crosslinker itself in (ii) a covalent network, involves the connection to other processes (polymer chains). Thus, this serves as a versatile system for the preparation of high-strength covalently crosslinked hydrogels because of the diverse crosslinking methods and crosslinking chemistries. It is the simplest covalent incorporation of cellulose into hydrogel networks with radical-initiated by the redox polymerization [45] and also by high-energy irradiations (for example, UV, γ-ray, [46] and electron-beam radiation [47, 48]). The active glycosidic radical by active sulfate anion radicals can bring about a hydrogen abstraction from the −OH group in cellulose, generating further radicals, and these radicals can activate the vinyl monomers growth.

By contrast, the celluloses act as multifunctional crosslinkers during the polymerization [43, 49]. For example, CMC grafted copolymerization of methylene bis(acrylamide) (MBA) and AA/AM/2-acrylamido-2-methyl-1-propanesulfonic acid crosslinkers has been prepared and evaluated as superabsorbent [43]. High-energy radiation does not involve any added chemicals, random recombination of the polymers and crosslinkers builds up a network since the H_2O molecule—initiated active radicals caused by the high-energy radiation results in polymerization and crosslinking [50]. Furthermore, e-beam-radiation could make the process free of radioactive source, which is hazardous and toxic for human bodies [50].

In contrast to physical crosslinking, adding chemical crosslinking can endow chitosan hydrogel with favorable mechanical properties and prepare chitosan gels with uniform structure, which is relatively stable for long-term application. Epichlorohydrin (ECH) is selective to hydroxyl groups and crosslinking of chitosan hydrogels using ECH have been reported. Moreover, other crosslinkers such as genipin and ethylene glycol diglycidyl ether are more prone to be attacked from active chitosan amino groups, generating 3-D network [51]. Additionally, chitosan retains lots of amine groups in the backbone upon deacetylation, which can in situ react extremely well with aldehyde (–CHO), forming reversible imine bonds by a Schiff base condensation reaction at room temperature [52]. The mostly frequently used crosslinkers for chitosan hydrogels are the poly (ethylene glycol) dialdehyde and formaldehyde [53, 54]. A schematic representation of physical and chemical crosslinking of hydrogels have been presented in Fig. 1.

4.1.3 Additive Manufacturing

Tailoring additive manufacturing of 3D polysaccharide hydrogels to soil and water remediation opens up exciting possibilities for environmental engineering. Laminated Object Manufacturing (LOM) [56] can be used to fabricate layered hydrogel sheets embedded with bioactive agents like enzymes. These sheets can be deployed in contaminated soils to gradually release remediation agents. For example, a 2023 study in *Environmental Technology & Innovation* explored laminated hydrogel mats made from alginate and cellulose for heavy metal immobilization in mining-impacted soils [42].

Selective Laser Sintering (SLS) [57] described in Fig. 2 and Inkjet Printing [58] offer precision in designing hydrogel structures for targeted water purification. SLS enables the creation of porous hydrogel matrices from powdered polysaccharides like chitosan, which can adsorb pollutants such as dyes or heavy metals. Inkjet printing allows for spatial deposition of functional inks—like those containing activated carbon or metal-binding ligands—onto substrates, forming reactive membranes [59].

Direct Ink Wetting (GIW) [61], Fused Deposition Modeling (FDM) [62], and Stereolithography (SLA) [63] are particularly useful for creating customized hydrogel devices for in-situ remediation. GIW is ideal for printing shear-thinning polysaccharide inks like xanthan gum, which can be infused with microbial consortia for biodegradation of organic pollutants. FDM has been adapted to extrude thermoplastic-polysaccharide blends for soil stabilization structures. SLA enables

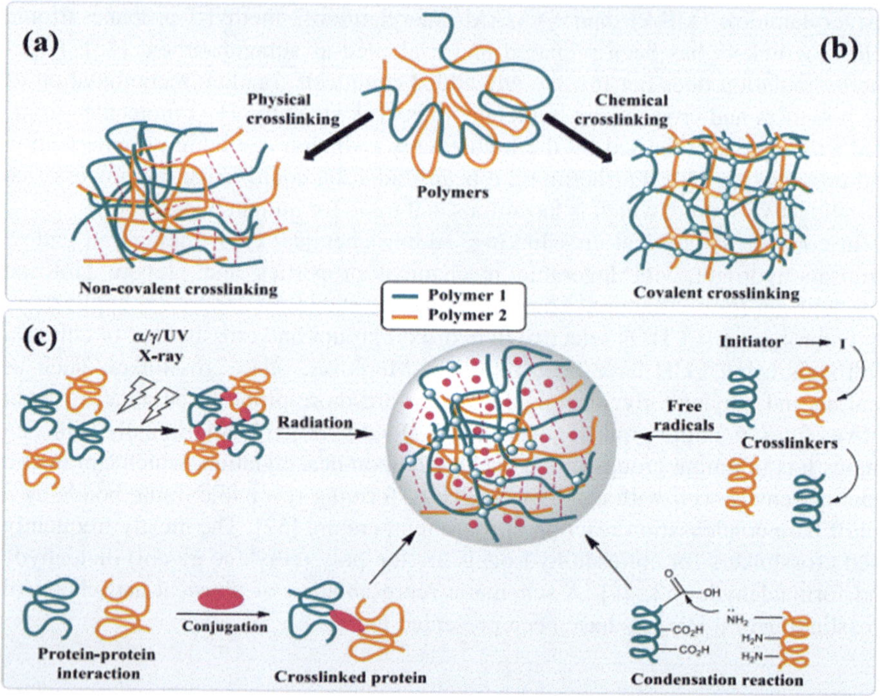

Fig. 1 A schematic illustration of methods of crosslinking: physical crosslinking (**a**) and chemical crosslinking (**b**). Mechanism of hydrogel synthesis via chemical crosslinking (**c**). Adapted from Siqueira et al. [55]

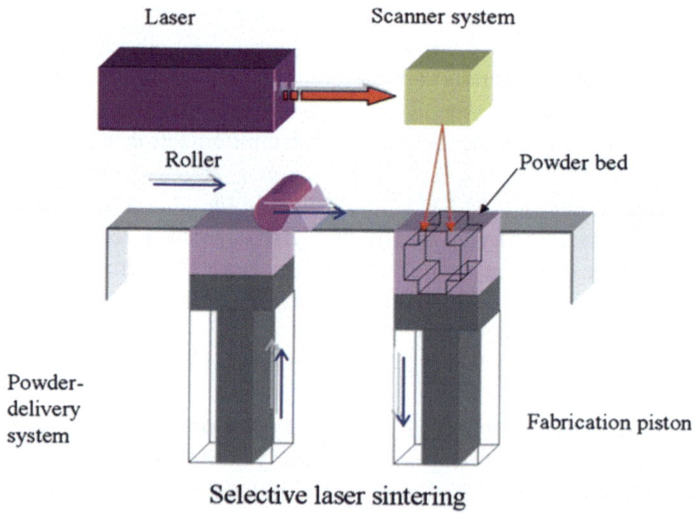

Fig. 2 A schematic description of selective laser sintering. Adapted from Zhang et al. [60]

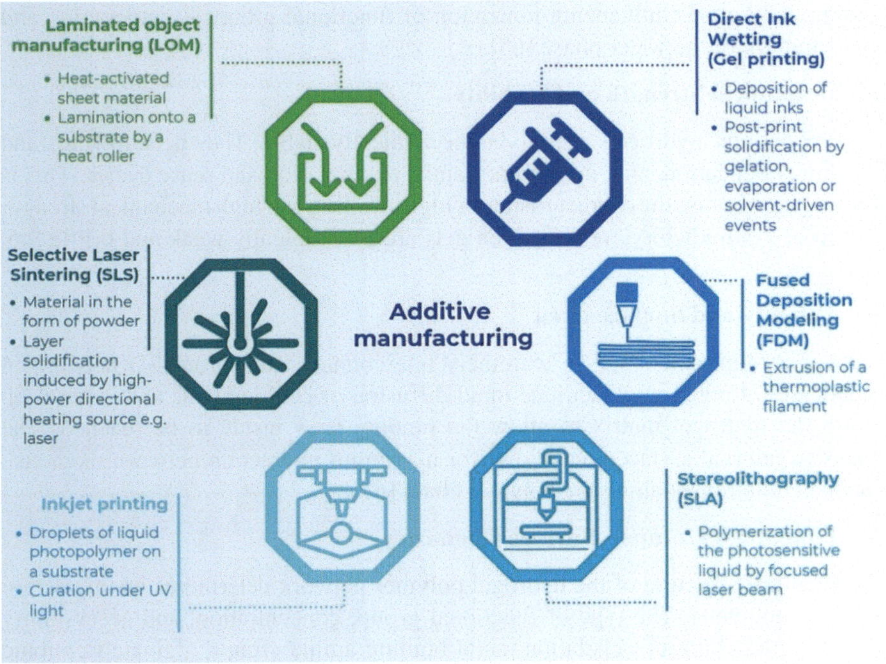

Fig. 3 A schematic showing summary of the different methods of additive materials manufacturing. Adapted from Zhang et al. [60]

high-resolution fabrication of hydrogel filters with embedded photocatalysts for breaking down organic contaminants. A 2025 study in *Water Research* highlighted SLA-fabricated carrageenan hydrogels with titanium dioxide for photodegradation of pharmaceutical residues in wastewater [64]. These methods of additive manufacturing are summarized in Fig. 3.

4.2 *Critical Physicochemical Properties for Environmental Remediation*

The performance of a hydrogel in an everyday setting or environmental application is regulated by a connected system of physicochemical properties, which are themselves direct consequences of the selected polysaccharide and synthesis method.

I. **Swelling Capacity and Kinetics**

This is probably the most basic property of a hydrogel. High swelling capacity is necessary in order to absorb a lot of dirty water and retain water in soil. Swelling is regulated by a balance between the osmotic pressure to pull water in and the retractive elastic force of the crosslinked network [13]. It is highly sensitive to environmental conditions like pH (influencing ionization of functional groups), temperature, and ionic strength of the water phase [65].

II. **Mechanical Strength and Stability**

For applications, hydrogels need to be mechanically stable. They need to withstand handling, application, and, more importantly, regeneration and reuse cycles. This is challenging because the characteristics of high swelling and high mechanical strength tend to be contradictory; very swollen gels are mechanically weak and brittle [66, 67].

III. **Porosity and Surface Area**

There is a requirement for an extremely interconnect and porous 3D network in remediation. Large pores facilitate rapid diffusion of contaminants and water deep within the hydrogel matrix to allow the binding sites inside to be reached. The extensive internal surface area allows for maximum interaction between the active site and contaminant along the polymer chain [68].

IV. **Functional Groups—Surface Chemistry**

The chemical structure of the hydrogel polymer network determines its interaction with contaminants. The type of functional group, concentration, and accessibility (for example, chitosan's chelation metal-binding amine groups, alginate's cationic dye-binding carboxyl groups) determine the adsorbent's selectivity and adsorption capability [69].

V. **Stimuli-Responsiveness**

"Smart" hydrogels can undergo reversible volume phase transition under the influence of external stimuli like temperature, pH, and light [13]. An example would be a pH-sensitive hydrogel which swells and traps a pollutant at a certain pH value and shrinks to release it for recovery at another, enabling the highly sought-after controlled capture-and-release process in recovering resources and regenerating adsorbents [70].

5 Characterization of Hydrogels

5.1 Structural Characterization

I. **NMR**

A number of modes of NMR (NMR-H, NMR-C and pulse field gradient NMR) have been applied in the characterization of hydrogels. The spectral lines of the structure of the polysaccharide changed after preparing the SAP to verify that the cross-linking was successfully completed and the hydrogel was produced [71]. For example, NMR-H was employed to verify the appearance of the curves of a chitosan-SAP (CHCAUR) from chitosan spectral line (CHCA) and (CHUR) [72]. Fang et al. employed NMR-C analysis to verify achieving derivative of chitosan (AEMCS) formation and to obtain chitosan SAPs (CS-SAP and AEMCS-SAP) [73].

II. **FTIR**

As we understand, Fourier transforms infrared spectroscopy (FTIR) is broadly applied for the determination of vibrational energy transitions measurement, polymers chemical structure, and for details on multiple atoms and chemical bonds. Hence FTIR analysis on SAP for characterization is the molecular structure study by observing its absorption spectra [13]. The NaCMC hydrogel formation was confirmed by FT-IR, as observed from the same peaks presented at the spectrum of the raw material, and, the presence of an additional peak at 1751/cm attributed to the electrostatic interaction between –COOH groups due to CMC and Fe^{3+} [49].

III. **XRD**

X-ray diffraction (XRD) technique is commonly employed to distinguish the amorphous and crystalline phases in hydrogel structure and to give the percentage of the crystallinity on the hydrogel network. In addition, the formation of cross-linked network also could be confirmed by XRD [74]. Nie et al. used XRD analysis. Each of these peaks appears to follow the trend found in the structure occurring in the cellulose-based hydrogel beads' structure after the addition of calcium ions with a varied concentration (judging by the increase in the intensity of peaks in the range around 220 with higher concentration of the calcium solution). The optimum concentration of 70% Calcium solution was used for the preparation of beads of good crystallinity [44].

From the X-ray Diffraction study revealed by Nawrotek et al. [75], some notable variations in chitosan hydrogel structure were reported, this was attributed to water conditioning which was carried out before. The systems were composed of chitosan (CH), B-glycerol phosphate (GP), and glutamic acid (GLU).

It was revealed in the diffraction patterns that a largely amorphous structure was observed in the unconditioned hydrogel in it dried form. Subsequently, after conditioning in water for a period of 24 h, the structure of the hydrogel was observed to be much more crystalline and was much more in resemblance to that of the polymer of chitosan. The observed increment in crystallinity was supported by a significant characteristic maximum for CS at $2\theta \approx 20°$ angle in the samples that were conditioned. It was also observed from the analysis variations in diffraction patterns which was dependent on the time of conditioning, resulting in peaks at $2\theta \approx 5°$ and a wide halo around the $2\theta \approx 11°$ angles. Conclusively, the study revealed that the orientations of particles within a hydrogel matrix can be altered as a result of conditioning in water to suit its potential application.

IV. **UV–Vis**

UV–Vis—Ultraviolet and visible absorption spectrometry is regarded as quantitative method that relies on the absorption of light in a molecule in a colorless solution and provides information about the concentration of the substance in the solution. The application of this method is appealing, particularly for research on the drug or the nutrient-release mechanism when employing hydrogels for drug delivery (such as the chloramphenicol) [76] or zero-order-releasing fertilizers [77] or even for the calculation of pollutants adsorption capacity for water treatment, like in the removal of diclofenac sodium from matrix containing pharmaceutical wastewater [78].

V. **Raman**

Raman is a non-invasive analysis of molecular structure [79], in which the hydrogel is illuminated with radiation. Upon excitation, they relax back to their original state by emitting a photon of a different energy, which is then collected by the detector and compared to the originating signal. Such a methodology could be applied to study, for example, the effect of calcium ions on the cellulose structure in SAP beads, which was confirmed by the new peaks at 554, 527, and 481/cm [44].

5.2 Thermal Analysis

Significant details of thermal behaviour of the hydrogels may be discerned through thermogravimetric (TGA) and differential calorimetric (DSC) analysis in decomposition process. They are also useful to study whether they are exothermic (heat releasing) or endothermic nature (heat absorption) and provide information about thermal stability of SAPs. TGA is used to observe the weight or mass change of an SAP at an elevated temperature. This would lead to mass loss (vapor emission, decomposition, and loss of volatile matter) or mass gain (gas fixation, oxidation, etc.). According to TGA curves of several cellulose-based SAPs [80], due to the residual lignin chain bearing on the cellulose, the thermal stability of the prepared SAPs was improved because lignin is more stable in thermal breakdown than cellulose. Moreover, DSC curves indicated that the T_g of the SAPs becomes higher with increasing citric acid. Furthermore, free and bound water of SAPs, like chitosan SAP [81], might be also analyzed by TGA and DSC.

According to Rusu et al. [82], the decomposition behaviour and thermal stabilities of Cs (Chitosan), NMACs (N-maleoyl chitosan), and the HEA hydrogels which are based on NMACs and poly (acrylic acid-co-ethyl acrylate) or P(AA-co-EA) were studied, it was observed that all samples exhibited an initial loss in weight at maximum temperature of 100 °C, this was attributed to removal of inherent water molecules, the chitosan sample exhibited degradatory step whose peak were observed at 249.9 °C. consequently, the thermal stability was elevated when it was functionalized to N-maleoyl chitosan (NMACs), coupled with the main peak of decomposition drifting towards an elevated temperature of 341.3 °C, this phenomena was linked to increase in hydrogen bonding. The HEA (containing acrylic acid and ethyl acrylate) hydrogels on the other hand, which was based on NMACs, was observed to have exhibited a much more complex degradatory stages which were multi-staged. The incorporation due to ethyl acrylate into the matrix of the hydrogel resulted to some degree thermal stability reduction, this was supported by the loss in weight increment coupled with elevated ethyl acrylate ratio.

5.3 Mechanical Analysis

The physical resistance to shear of natural SAPs is lower than of synthetic ones. Nonetheless, the mechanical properties are an essential issue for hydrogels study and they open up the possibility to correlate great structural properties/relations of these systems. For example, the amount of cross-linker agent may affect the mechanical properties of the SAP, as in the case of a Carboxymethyl Cellulose (CMC)—based SAP, which presents an increase in the network rigidity of the SAP when the concentration of the citric acid increases [49]. The corresponding property was also found on the composite SAP of chitosan and alignate, reinforced with ZnO_2 nanoparticles which increased the mechanical performance of the prepared SAPs due to their interactions with polysaccharide chains through H-bonding and electrostatic interactions [83].

5.3.1 Dynamic Mechanical Analysis

In DMA, the viscoelastic behavior of the hydrogel is investigated [84], whereby the modulus of elasticity and the damping are assessed, where an oscillating force (deformation/stress) is applied through the sample depending on buffer or oscillatory frequency. For example, GDSCs of bacterial cellulose-based hydrogel composites prepared by gamma-ray irradiation were analyzed by DMA. The tensile strength of the SAP was enhanced with improving irradiation dose up to 40 kGy and reduced at 50 kGy irradiation. Thus, a dose of 40 kGy was selected as the optimal dose for synthesis, which exhibits a high tensile strength (46.3 MPa) with a strain of 36.6% [85].

6 Environmental Remediation Applications

The distinctive unification of characteristics attributable to 3D polysaccharide hydrogels, such as porous frameworks, abundant functional moieties, elevated water content, and ecological benignancy, hydrogels polysaccharides make them especially useful and versatile materials in the field of environmental remediation. This includes the purification and treatment, as well as the protection and enhancement of contaminated waters and agricultural soils [82].

Some notable hydrogels that have been employed in environmental remediation in soils and water matrices for the elimination of organic dyes and heavy metals, including pesticides have been reported and shown in Table 2.

Table 2 Applications of different hydrogels in the removal of contaminants

Type of hydrogel	Functionality	Removed contaminants	Application	References
Chitosan–polyacrylamide hydrogel	Amine ($-NH_2$), Amide ($-CONH_2$)	Heavy metals, dyes	Dual functionality for broad-spectrum contaminant removal	[29]
Kappa carrageenan hydrogel	Sulfate ($-OSO_3-$)	Heavy metals and organic pollutants	Effective in binding with divalent metal ions and various organics	[86]
Carboxymethyl cellulose hydrogel, Polyacrylic acid hydrogel	Carboxyl	Heavy metals (Cd, Pb, Cu, and phosphates)	Removal of heavy metals from wastewater	[2]
Amine-functionalized hydrogel	Amine	Anionic dyes, heavy metals	Electrostatic removal of negatively charged contaminants	[28]
Iron oxide nanocomposite hydrogel	Magnetic groups	Arsenic, chromium	Magnetic separation of heavy metals from drinking water	[1]
Poly (N-isopropyl acrylamide) hydrogel	Amide	Temperature-sensitive pollutants	Temperature-responsive contaminant capture	[87]
Cyclodextrin hydrogel	Hydroxyl	Organic pollutants, pesticides	Encapsulation and removal of organic chemicals	[88]

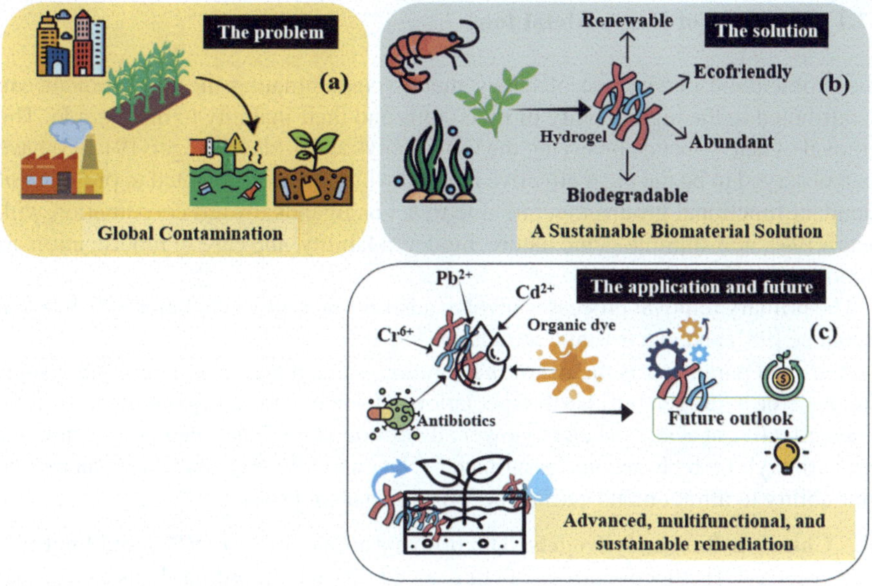

Fig. 4 Biomaterials—a green path to remediation (*Source* This work)

Figure 4 shows an illustration of how water and soils are contaminated via discharges from municipal operations, factories, and agricultural practice and the abundant sources of hydrogels derived from chitosan, alginates owing to their eco-friendliness, abundant, and biodegradable nature, they can be applied as remediation means for soil and water contamination.

6.1 *Water Remediation: Targeting a Spectrum of Aqueous Pollutants*

Hydrogels derived from polysaccharides are used as high-efficiency adsorbents for cleansing water from pollutants. The mechanism involved and efficiency of extraction of pollutants are strongly influenced by the specific chemistry of the hydrogel and the pollutant involved. It is clear that the performance of the hydrogel, in its varied applications, is the surface and bulk chemistry of the hydrogel. The most important factors regarding the hydrogel are its density and the types or characteristic functionality; these factors together define what pollutants can be easily adsorbed and by what methods [89]. This indicates that the greatest potential for progress lies in novel polysaccharide backbone functionalizations [90].

6.1.1 Removal of Heavy Metal Ions

The problematic persistence of heavy metals contaminating the environment can be attributed to the high toxicity of metal ions and their inability to biodegrade. The removal of heavy metals involving the use of polysaccharide hydrogels [91–93] have been observed to be the most effective method which can be attributed to presence of abundant functional groups that are able to act as ligands toward coordination with metals [94], non-immunogenic nature, biodegradability, and ease of modification, as shown in Fig. 5.

The primary removal processes involve adsorption through chelation [96, 97], ion exchange [98], and electrostatic attraction [99, 100].

A critical parameter is the pH of the solution, since it affects not only the surface charge of the hydrogel, but also the speciation of the metal ions. In general, adsorption is less effective at acidic pH due to lower adsorption of the functional groups (amines and carboxyls) which become protonated (NH_3 and COOH), and the reduction of their ability to attract metal cations becomes prevalent [101].

I. **Chitosan-Based Hydrogels**: The primary amine group (–NH_2) and hydroxyl group (–OH) functionalities of chitosan qualify it extraordinarily as a chelating

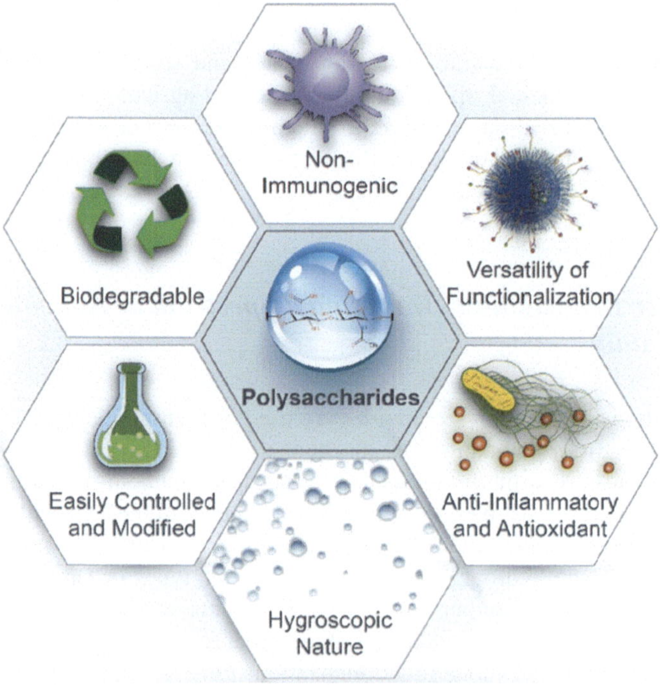

Fig. 5 A schematic representation of the various properties of polysaccharide hydrogel that gave it an edge for application in the removal of heavy metals from solution. Adapted from Zhang et al. [95]

agent for heavy metal removal, this is attributed to its ability to bind with metals like Pb^{2+}, Cu^{2+}, Cd^{2+}, and Cr^{6+}. Modified chitosan hydrogels have been observed to perform particularly well with Cu and Cd, with adsorption capacities of 206 and 178 mg/g respectively [102]. The functionalization of hydrogels with nanoparticles such as Fe_3O_4 not only improves the adsorption process but also allows for efficient magnetic separation of the adsorbent from the treated water, which makes the recovery and regeneration processes even much more efficient [103].

II. **Alginate-Based Hydrogels**: Alginate's capacity to bind metal ions is widely known and is attributed to its high concentration of carboxyl groups. Essentially, it works via ion exchange. The simple calcium alginate beads formed act as very efficient adsorbents. As an example, it has been reported that these beads could adsorb copper, cadmium, and lead in significant quantities, 315.92 mg/g, 232.35 mg/g and 465.22 mg/g respectively. The effectiveness of these alginate hydrogels is shown to be further enhanced through composite materials. As an example, Lee et al. [35] study showed that the incorporation of algae biomass into alginate matrix improved the adsorptive capacity for copper, chromium, and cobalt by 38%, this study was well supported in related investigations [5, 104, 105].

III. **Cellulose-Based Hydrogels**: Although native cellulose does not adsorb well, the introduction of carboxyl groups makes CMC, a cellulose derivative, a very efficient adsorbent. Composite hydrogels of CMC—carboxymethyl cellulose and polyvinyl acetate polymers have been observed to demonstrate remarkable affinity and selectivity, particularly towards ions such as silver (Ag^+) [42]. Cellulose has considerable potential for designing complex hybrid materials. For example, the magnetic hydrogels based on cellulose and chitosan and reinforced with Fe_3O_4 nanoparticles function as strong and recoverable adsorbents and have a great affinity for Fe^{2+}, Cu^{2+}, and Pb^{2+}. The range of adsorption capacity for cellulose-based hydrogels is adjustable by fine tuning the chemical modifications for specific metal ions, spanning from single digit values to more than 2000 mg/g [106, 107].

6.1.2 Removal of Organic Dyes

Synthetic dyes that are gotten from paper, textile, and food industries are representative of a major category of water pollutants that are usually carcinogenic, toxic, and resistant to degradation. Polysaccharide hydrogels aid in the removal of dyes much effectively by adsorption process, this process is facilitated by a combination of hydrogen bonding, electrostatic interactions, and π–π stacking interactions existing between the polymer network and dye molecules [5, 108–110].

I. **Chitosan-Based Hydrogels**: These are cationic polysaccharides and are specifically effective for the adsorption of anionic dyes such as Congo Red and reactive Black 5. The amine groups present in chitosan are protonated at acidic pH, thus resulting to greater electrostatic attraction with negatively charged dye molecular species [111, 112]. Composite hydrogels that are derived from squid pen protein and chitosan, have been recorded to gain adsorption capacities of 151.52 mg/g for reactive blue 4 [113].

II. **Alginate-Based Hydrogels**: In other words, Alginate hydrogels are anionic which makes them a good polysaccharide adsorbent for cationic dyes like the methylene blue and crystal violet. The deprotonated carboxyl groups on the alginate chains are strongly electrostatically bound to positively charged dye molecules at neutral or basic pH. Alginate-based beads exhibited very high adsorption capacity with methyl violet alginate beads having more than 889 mg/g adsorption capacity [114].

III. **Cellulose-Based Hydrogels**: Cellulose undergoes modification to adsorb both forms of dyes. For example, quaternized cellulose supplies permanent positive charges that allow it to be adsorbed by anionic dyes, while sulfated cellulose nanofibers can quickly adsorb cationic dyes, Methylene Blue, for instance [115, 116].

IV. **Functionalized Acrylate-Based Polymer**: To enhance the polymer's adsorption capacity, the base of the polymer is first modified by introducing polar and charged groups that promote electrostatic and hydrogen-bonding interactions with dye molecules [117, 118], while maintaining the material's structural integrity and reusability. For example, as illustrated in Fig. 6, the acrylate-based polymer surface was modified through a sequential chemical functionalization process to enhance its adsorption performance. The procedure begins with a silane coupling reaction, in which silane molecules containing reactive epoxy groups are grafted onto the polymer surface, forming a stable covalent linkage between the silane and the acrylate matrix. This modification introduces epoxy functionalities that serve as reactive sites for subsequent surface activation. An epoxide–amine reaction then occurs between the surface epoxy groups and amine-containing molecules, such as polysaccharides or amine-rich polymers, resulting in the formation of strong covalent bonds and the incorporation of hydrophilic, amine-functionalized layers. This hierarchical modification not only improves the surface chemistry and hydrophilicity of the polymer but also enhances its affinity for dye molecules through electrostatic attraction and hydrogen bonding.

6.1.3 Removal of Antibiotics

The environmental and health impacts of pharmaceutical and agricultural runoff contaminated with antibiotics is very much due to its contribution to the growing issue of developing antibiotic-resistant bacteria. Composite forms of polysaccharide hydrogels have been observed to be effective in the removal of antibiotics. The adsorption mechanisms are complex, involving π–π stacking, electrostatic interactions, and H-bonding.

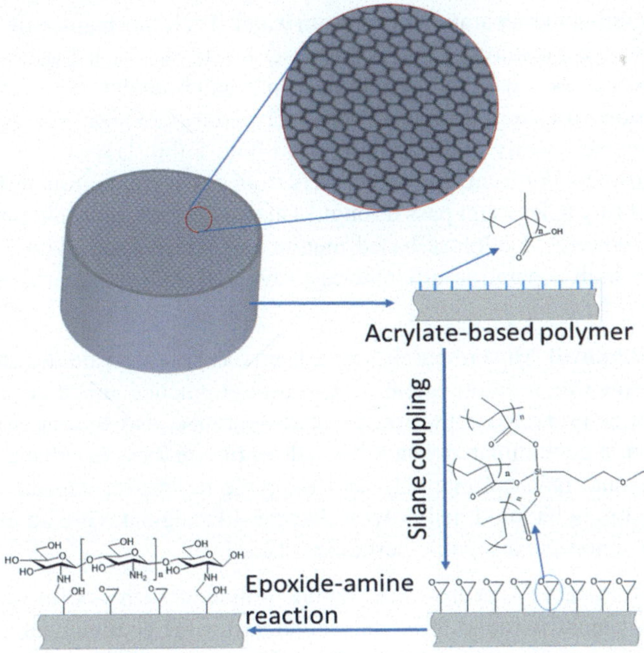

Fig. 6 A schematic representation for the functionalization of acrylate-based polymer to enhance its hydrophilicity. Adapted from Fijol et al. [119]

I. **Chitosan/Graphene Oxide (GO) Composites**: The performance of these hybrid hydrogels is quite remarkable. Graphene oxides has gained popularity as it facilitates large surface areas as well as aromatic rings suitable for π–π interactions. Chitosan has hydroxyl and amine functionalities in abundance which help in providing hydrogen bonds and electrostatic interactions. The combination of these effects has remarkable adsorption capacities. Some of the rGO@ZIF-67@CS hydrogels have been reported to obtain the incredible values of 1685 mg/g and 1890 mg/g for tetracycline and norfloxacin respectively [120, 121].

II. **Magnetic Polysaccharide Hydrogels:** By embedding magnetic nanoparticles into polysaccharide matrices such as chitosan or alginate, it is now possible to develop adsorbents that not only remove waterborne antibiotics such as diclofenac, but that can also be retrieved using an external magnetic field, increasing their utility for volumetric applications [121–123].

6.1.4 Removal of Other Contaminants (Oil/Water Emulsions, Pesticides)

I. **Oil/Water Emulsion Separation**: Wastewater containing oils from industries, as well as water derived from oil spills, is an ecological challenge. Polysaccharide hydrogel-based membranes and coatings provide an effective and energy-efficient method for aiding the separation of oil from water. Their mechanism relies solely on special wettable features [124–126] as described in Fig. 7a–c. Figure 7c shows that the flux increases with increasing mesh size and both the intrusion pressure and separation efficiency decreases with increasing pore size. The hydrogel's surfaces that are heavily hydrophilic tend to hold a thin layer of water which acts as a barrier that prevents the oil droplets from passing or getting through, yet retains the ability to let water pass through. This process is termed as "underwater superoleophobicity". Cellulose-based membranes and polysaccharide aerogels exhibit very high separation efficiencies (>99%) as well as great fluxes [127, 128].

II. **Pesticide Removal**: Most researches have been focused on pesticides and hydrogels concerning their employment in agricultural practice aimed at preventing pollution by control release, the process of absorption is also relevant in pesticide remediation in contaminated water. The adsorption process is enhanced by the partitioning and physical interaction of pesticide molecules into the hydrogel matrix. Hydrophobic modified polysaccharides like alginates are an alternative for capturing non-polar organic pesticides [129, 130].

As illustrated in Table 3, hydrogels of natural polymers such as Chitosan, CMC, K-carrageenan, Lignin is found to be very efficient for the contaminant's removal from water. In addition, the natural hydrogels studied are usually biodegradable with relatively low cost, recyclable properties and are excellent in the treatment of heavy metals and dyes. For instance, chitosan hydrogels, including CMAPP dual network chitosan, possess a highly efficient adsorption for diverse dyes. Kappa carrageenan and lignin hydrogels can also be employed in the treatment of a variety of heavy metals and dyes. However, one big disadvantage of such materials is the mechanical weakness and instability. This problem can be solved using composite hydrogels in combination with other materials. For example, the CMC-clay nanocomposite hydrogel exhibits enhanced stability and the high adsorption capacity for dyes.

In addition, some materials such as carbonaceous nanofiber membranes exhibit excellent properties, including large specific area and fast absorption for heavy metals and dyes. And much more effective, though more costly and more elaborate in fabrication than the natural polymer hydrogels. This cost disparity and manufacturing complexity tends to determine their viability and widespread use for water treatment.

6.1.5 Functional 3D Polysaccharide Hydrogel Filters for Water Purification

A new class of environmentally friendly materials for water filtration is represented by functional 3D polysaccharide hydrogel filters. These filters are made from biodegradable biocomposite materials, as seen in Fig. 8, which mix poly(lactic acid) (PLA) with nanofillers such chitin nanofibrils (ChNF) or 2,2,6,6-tetramethylpiperidine-1-oxyl-oxidized cellulose nanofibrils (TCNF) [137, 138]. The combination is formed

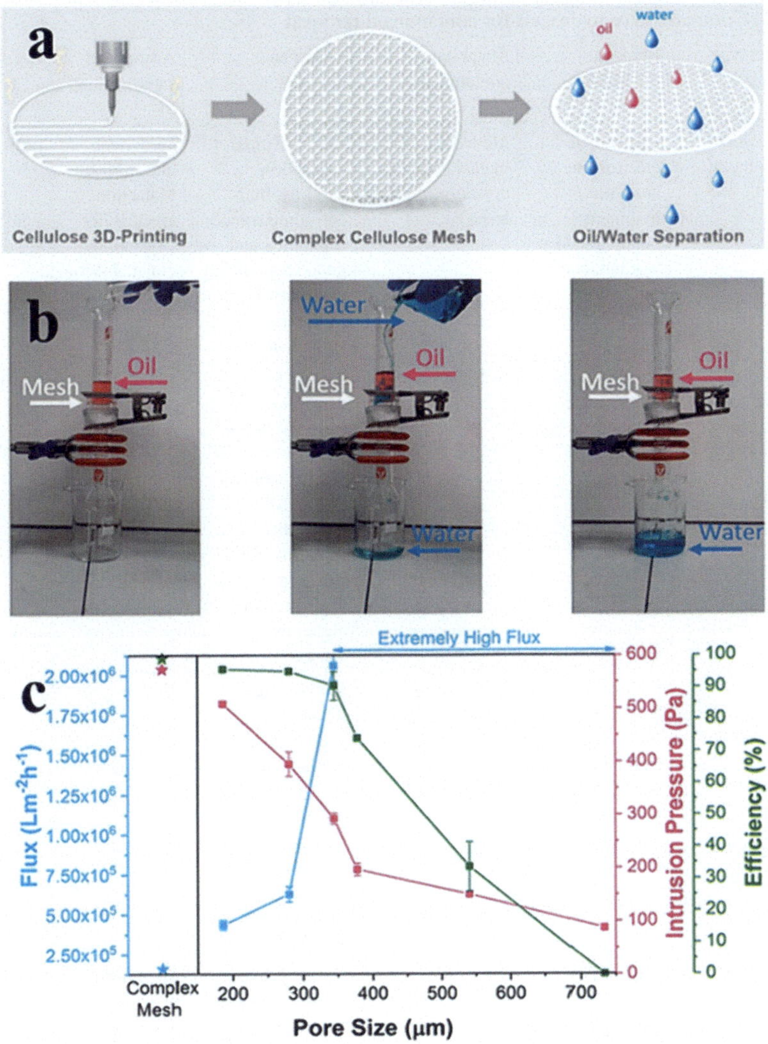

Fig. 7 Mechanism for oil/water separation using 3D printed cellulose mesh (**a**), practical application showing oil/water (**b**), and parameters such as flux, intrusion pressure and efficiency dependence on the pore size (**c**). Adapted from Zhang et al. [60]

into porous three-dimensional structures, extruded into filaments, and processed into pellets of biocomposite. These filters work by adsorbing metal ions like Cu^{2+} and filtering out microplastics according to their size. The PLA matrix's mechanical performance, surface reactivity, and hydrophilicity are all markedly enhanced by the addition of polysaccharide-based nanofillers. As a result, the final filters support circular economy concepts in material design and offer an effective, environmentally

Table 3 Comparison of hydrogels for contaminant removal

Hydrogel type	Removal mechanism	Metal ions/ contaminants	Merits and demerits	Adsorption capacity (mg/g)	References
Dual-network chitosan-based hydrogel	Adsorption via electrostatic attraction, hydrogen bonding, and n–π interactions	Methylene blue, malachite green, crystalline violet, basic fuchsin	Merits: Superior adsorption capacity, high anti-interference capability, and good recyclability Demerits: Limited adsorption for anionic dyes	Methylene blue: 1800; Malachite green: 1797; Crystalline violet: 2495; Basic fuchsin: 2938	[90]
Carboxymethyl cellulose (CMC)-g-poly (AM-co-ITA)/ clay nanocomposite hydrogel	Adsorption via electrostatic interactions and multilayer adsorption	Brilliant green dye	Merits: High adsorption capacity, better stability with clay, easy separation and high reusability Demerits: Smaller swelling ratio at the expense of higher clay content	1513.55	[90, 131]
Novel chitosan–ethylene glycol hydrogel (CEGH)	Ionic hydrogen bond interaction between carbonyl groups of PFOA and protonated amine groups of CEGH	Perfluorooctanoic acid (PFOA)	Merits: High adsorption performance, easy synthesis and environmentally friendly Demerits: Restricted to certain contaminants such as PFOA	1275.9	[132]
Carbonaceous nanofiber membranes	Adsorption through electrostatic interactions, hydrogen bonding, and surface complexation	Pb^{2+}, Cr^{3+}, methylene blue	Merits: Large surface area, rapid adsorption rate, high capacity, regenerative and can be reused for several cycles Demerits: Reduced performance over time and costly to manufacture	Methylene blue: up to 818.6; Pb^{2+}: up to 423.7; Cr^{6+}: up to 221.3	[133]

(continued)

Table 3 (continued)

Hydrogel type	Removal mechanism	Metal ions/ contaminants	Merits and demerits	Adsorption capacity (mg/ g)	References
Kappa carrageenan/ tamarind kernel powder hydrogel	Adsorption	Brilliant green, Rose Bengal	Merits: Biodegradable, inexpensive, good adsorption capacity, pH-sensitive Demerits: Regeneration may be necessary; stability problems in aqueous media	Brilliant green: 840.33; Rose Bengal: 68.06	[134]
Kappa carrageenan/ cellulose hydrogel	Adsorption, electrostatic interactions, and ion exchange	Pb^{2+} ions	Merits: Large adsorption capacity, environmental friendliness, reusability (>79% after eight cycles), and low cost Demerits: At low pH, the adsorption capacity might drop in every use cycle	486 ± 28.5	[86]
Lignin hydrogels	Adsorption, ion exchange, complexation, hydrogen bonding, and electrostatic interactions	Pb^{2+}, Cr^{3+}, and Cu^{2+} ions	Merits: Biodegradable, natural, affordable, renewable, versatile Demerits: Low mechanical strength, poor hydrolysis and chemical resistance. Requires pretreatment. Risk of leaching and low swelling potential	308, 65.72, 541 respectively	[135, 136]

acceptable substitute for eliminating ionic and particle impurities from water [137, 138].

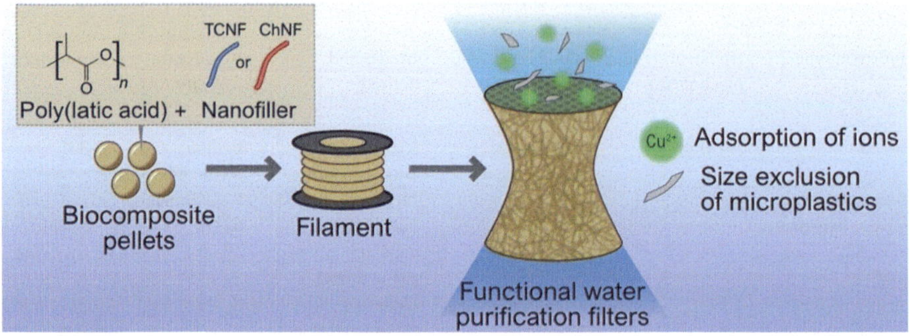

Fig. 8 Schematics of a functional 3D polysaccharide hydrogel filter for water purification. Adapted from Fijol et al. [119]

6.2 *Soil Treatment: Enhancing Agricultural Sustainability and Safety*

Beyond the process of water purification, the applications of polysaccharide hydrogels the treatment of soil have been transformative, where they are able to address issues related to inefficient nutrient use, scarcity of water, and mobility of contaminants. It is in this regard that these hydrogels show exceptional duality in their functional groups, this is observed when they act simultaneously as agents of soil enhancement through nutrient and water delivery, also as protective agents through the immobilization of damaging contaminants. This amazing combination could lead to creating simple, all-in-one solutions to improve the health and safety of poor or damaged farmland [139–144].

6.2.1 Application as Soil Conditioners for Water Retention

Hydrogels based on polysaccharides have attracted a lot of interest as soil conditioners that are intended to enhance water retention in agricultural applications, particularly in arid and drought-prone environments [145, 146]. These hydrogels are efficient water reservoirs in soil matrices because of their hydrophilic nature and highly porous three-dimensional architectures, which allow them to absorb and hold significant amounts of water [145]. When the soil itself is dried up, the hydrogel networks are able to slowly release the absorbed volume of water targeted to the root systems of the plan, offering a continuous water source. This action increases the irrigation times, while optimizing water use, also enhancing the crops drought tolerance. Aside from these notable merits, the contraction and swelling of the hydrogel network can foster the health of soil system by promoting aeration and porosity, while reducing runoff [140]. To improve crop growth in sandy, water-scarce soils, for example, a hydrogel made from locust bean gum and borax showed an amazing water absorption capacity of over 130 g/g in just a few minutes and maintained soil moisture levels for up to 14 days [145]. Also, hydrogel made from renewable biomass has shown to increase the moisture content of soil by 30% [1].

6.2.2 Systems for Controlled Release of Agrochemicals

It has been suggested that the extreme use of poorly managed fertilizers and pesticides can pose detrimental effects on the immediate environment by contaminating the water and soil sources, resulting in the over-enrichment of aquatic bodies [140, 145, 146]. Polysaccharide hydrogels are capable of acting as systems of delivery for the efficient slow release of agrochemicals into the soil when plants require them. This approach is more effective because it minimizes the need for frequent reapplication of pesticides and fertilizers and stops waste from evaporating and running off [141, 142]. So many hydrogels such as sodium carboxymethyl cellulose, isothiocyanate, alginate, and carboxymethyl chitosan [145, 146] were synthesized and applied soil borne pathogens. The results indicated that these hydrogels inhibited pathogenic bacteria, improve soil water content, and promote seed germination.

6.2.3 Immobilization of Soil Contaminants

Polysaccharide hydrogels (PH) can be employed in-situ immobilization for soils that have been contaminated with heavy metals, this process aids in the reduction in mobility and contaminants bioavailability. The introduction of PH in soil causes the functional groups to bind efficiently to the heavy metal ions already dissolved, such as Pb^{2+} and Cd^{2+} ions, the PH is able to have these ions locked in withing the matrix of the hydrogel [142]. Thus, disallowing the uptake of the dissolved heavy metal ions through the roots of plant, thereby facilitating the elimination of such chemical species in the food chain and ensuring the safety of crops and overall health of humans who are also part of the food chain [143]. Furthermore, because these hydrogels are biodegradable, their application won't result in secondary contamination, and the byproducts of their breakdown can improve soil health [146]. Hence, they can be used to clean up polluted soils in existing agricultural systems, promoting both environmental protection and agricultural sustainability [146].

It has been shown from studies, for example, that the employment of a composite modified calcium alginate and biochar exhibited profound cadmium content reduction present in lettuce that was cultivated in contaminated soil. In addition to this, the synchronized immobilization of toxic heavy metals from soils can be achieved by employing advanced hydrogels whilst causing a release in essential nutrients and offering a combination of benefits of fertilization and remediation [144].

7 Limitations, Future Prospective, and Challenges

The field of polysaccharide hydrogels for environmental remediation has enormous potential and is rapidly developing, but in order to advance the progress and going from the pilot-scale research to extensive, real-world use, some resolutions on a number of intrinsic limits and practical issues has to be considered. This will necessitate interdisciplinary advances in chemistry, materials science, and engineering to get past these challenges (Fig. 9).

7.1 *Inherent Limitations and Practical Challenges*

It has been reported often that the inadequate mechanical strength of PBH significantly impacts their efficient utilization. This is more profound when the PH is in a highly swollen state, it has been reported that much of these PH are brittle, soft and possess low toughness and tensile strength [13, 16, 18].

This limitation in mechanical rigidity can potentially result in structural breakdown or degradation in the process of their application, most especially in conditions involving changes in environment such as the flow of water through treatment facilities or under the influence of physical stress derived from the soil matrix. The fragility of the PH makes them uneasy to handle, recover, and also regenerate, thus Impeding their economic viability and ease of reuse [70].

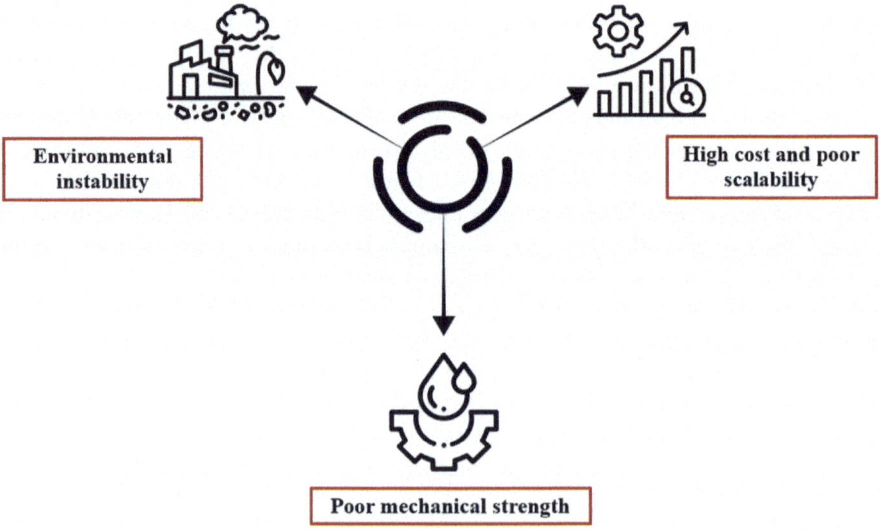

Fig. 9 Schematic diagram showing key barriers for polysaccharide hydrogels (*Source* This work)

This particular issue is very much linked to what could be termed as a mechanical-biodegradability trade-off. The covalent bonds that have been introduced permanently via the process of crosslinking can effectively enhance the robustness of the PH mechanical strength, and this in turn could cause the PH matrix to be more biodegradation-resistant and making it more difficult to be reused or recycled, thus impeding one of its key merits for environmental applications [147].

On the other hand, the reversible and weak physical interactions that enables the process of biodegradation leads often to poor mechanical materials. In multifaceted environmental conditions, hydrogels stability can be very much of a concern, this can be as a result of the potential for physically crosslinked hydrogels (such as those reliant on ionic interactions), been able to breakdown at certain pH values or at higher concentrations of competing ions [7]. Another impending barrier lies in its cost of production and scalability, even though the raw polysaccharides are cheap, the processes involved in purification, the stage-wise chemical modifications, and sophisticated techniques of fabrication needed in producing efficient and high-performance hydrogels can be expensive and much complex to handle, thus impeding their viability for industrial deployment [13, 70].

7.2 *Strategies for Enhancing Performance and Stability*

In order to overcome the limitations associated with polysaccharide hydrogel for efficient remediation, researchers are actively exploring viable means of advanced hybrid and composite hydrogels.

I. **Nanocomposite Hydrogels**: One of such strategies that have been found effective is the incorporations of what is known as "nanofillers" into the matrix of the polysaccharide. Some of these employed nanomaterials such as carbon nanotubes (CNTs), graphene oxide (GO), metal/metal-oxide nanoparticles (such as ZnO, Fe_3O_4), and clay minerals (for example, montmorillonite) can act as agents of reinforcements and multifunctional components. These agents can form very strong bonding with the polymer matrix chains, which profoundly improves the thermal stability, mechanical strength of the hydrogel, it has been observed often to also foster the capacity of its adsorption through the means of additional sites of binding or catalytic activity [55, 148–150].
II. **Interpenetrating and Double-Network Hydrogels**: Just like how natural tissues such as cartilage get their strength, researchers have created special materials called interpenetrating polymer networks (IPNs) and double networks (DNs). These structures use two different polymer networks that are intertwined. By combining a rigid, brittle network with a soft, ductile one, they become exceptionally good at dissipating fracture energy. This results in hydrogels with incredible toughness, stretchability, and fatigue resistance—way better than single-network versions [151, 152].

III. **Advanced Chemical Modification**: There is a growing focus by researchers on the development of new, "green" methods of crosslinking that are void of the employment of toxic chemicals. This process adopts bio-based crosslinkers that are derived from natural sources or compounds, and from the exploration of dynamic covalent chemistries. The sole aim is to produce reversible bonds that are strong enough, which can solve the potential issue related to mechanical biodegradability of the hydrogel [153].

7.3 Future Perspective: Towards Smart, Multifunctional, and Circular Hydrogel Systems

The future outlook of PHs as agents of environmental remediations is shifting beyond the use of materials which functions as passive, simple sponges or filters. The future of polysaccharide hydrogels in environmental remediation is moving beyond materials that serve as simple, passive sponges. The future path is reliant on the development of smart and dynamic systems that are also multifunctional.

I. **Smart, Stimuli-Responsive Systems**: Synthesis of "smart" hydrogels that are responsive to particular external or environmental stimuli (specific analytes, temperature, pH, light) is perhaps one of the most promising areas for future investigation. They would enable on-demand contaminant capture and release, and remediation would become a resource recovery process. Taking for example, hydrogel can be synthesized easily in a manner it can specifically recover valuable heavy metals selectively from effluents released by industries and then release them as concentrates upon a simple pH shift, thus allowing for water treatment and metal recycling applications [13, 154].

II. **Multifunctional Platforms**: The addition of a range of functionalities within a single hydrogel system offers promising potential. This has been posited to involve hydrogels that can integrate adsorption with photocatalysis to in synergy breakdown and capture persistent organic pollutants, or self-healing hydrogels that can autonomously heal damage and thereby promoting or extending their lifespan of operation as well as resulting in the reduction of waste. In the case of soil application, it has also been posited that these systems can be capable of enhancing supply-controlled nutrients release, storage of water, and immobilize pollutants [155–157].

III. **Principles of Circular Economy Focus**: One of the most basic changes in outlook is developing hydrogels that are in synergy with principles of circular economy. It is more than considering biodegradability and performance to include the whole material life cycle. Emerging investigations are expected to focus on the matrix of hydrogels that are not just only renewable in nature but can also be highly reused, readily regenerable at minimal energy inputs, and eventually recyclable or compostable in a safe manner after their shelf life [6, 157, 158].

8 Conclusion

Three-dimensional hydrogels from naturally-derived material such as chitosan, alginate, and chitosan represent a perfect, eco-friendly solution to numerous environmental issues. These hydrogels are a "green" approach to conventionally very energy-demanding and ecologically unfriendly methods. They can be engineered to be extremely efficient at purifying water from a broad range of dyes, heavy metals, and antibiotics contaminants. They also have dual merits: they enhance soil health by improving water retention and can even safeguard the food supply chain by immobilizing pollutants in the soil.

But these materials do come with a few problems. They themselves are not usually very robust, which means that they can only be made so durable and recyclable. Getting them to perform in high-stress real-world scenarios and producing them at low cost is also an enormous challenge.

In the future, attention is being directed toward the development of "smarter" materials. Attention is going toward the establishment of the next generation systems that, in addition to adsorbing contaminants, will be able to sequester and release target compounds on command, and even have self-healing or catalytic properties. With the combination of the natural properties of polysaccharides and modern material science, scientists are developing strong, efficient, and truly eco-friendly solutions to soil as well as water cleanups. It is continuous research that contributes significant role in preserving our surroundings and provision of clean resources in the future.

References

1. Zhang Y, Li Z. Heavy metals removal using hydrogel-supported nanosized hydrous ferric oxide: synthesis, characterization, and mechanism. Sci Total Environ. 2017;580:776–86.
2. Perumal S, Atchudan R, Immanuel Edison TNJ, Babu RS, Karpagavinayagam P, Vedhi C. A short review on recent advances of hydrogel-based adsorbents for heavy metal ions. Metals. 2021;11:864.
3. Serra IR, Fradique R, Vallejo MCS, Correia TR, Miguel SP, Correia IJ. Production and characterization of chitosan/gelatin/β-TCP scaffolds for improved bone tissue regeneration. Mater Sci Eng C. 2015;55:592–604.
4. Iglesias-Mejuto A, García-González CA. 3D-printed alginate-hydroxyapatite aerogel scaffolds for bone tissue engineering. Mater Sci Eng C. 2021;131:112525.
5. Wang B, Wan Y, Zheng Y, Lee X, Liu T, Yu Z, et al. Alginate-based composites for environmental applications: a critical review. Crit Rev Environ Sci Technol. 2019;49(4):318–56.
6. Niemczyk-Soczynska B, Sajkiewicz PŁ. Hydrogel-based systems as smart food packaging: a review. Polymers. 2025;17:1005.
7. El-Husseiny HM, Mady EA, Hamabe L, Abugomaa A, Shimada K, Yoshida T, et al. Smart/stimuli-responsive hydrogels: Cutting-edge platforms for tissue engineering and other biomedical applications. Mater Today Bio. 2022;13:100186.
8. Mandal M, Singh Lodhi R, Chourasia S, Das S, Das P. A review on sustainable slow-release N, P, K fertilizer hydrogels for smart agriculture. ChemPlusChem. 2025;90:e202400643.
9. Zhang W, Xu Y, Mu X, Li S, Liu X, Lei Z. Research progress of polysaccharide-based natural polymer hydrogels in water purification. Gels. 2023;9:249.

10. Singh BV, Girase IP, Sharma M, Vidhya CS, Tiwari AK, Baral K, et al. Nanoparticle-enhanced approaches for sustainable agriculture and innovations in food science. Int J Environ Clim Change. 2024;14(1):293–313.
11. Agbna GHD, Zaidi SJ. Hydrogel performance in boosting plant resilience to water stress—a review. Gels. 2025;11:276.
12. Ungureanu E, Mikhailidi A, Tofanica BM, Fortună ME, Rotaru R, Ungureanu OC, et al. Sustainable gels from polysaccharides in agriculture. Polysaccharides. 2025;6:37.
13. Berradi A, Aziz F, El Achaby M, Ouazzani N, Mandi L. A comprehensive review of polysaccharide-based hydrogels as promising biomaterials. Polymers. 2023;15:2908.
14. Kalia A, Sharma S, Semor N, Babele PK, Sagar S, Bhatia RK, et al. Recent advancements in hydrocarbon bioremediation and future challenges: a review. 3 Biotech. 2022;12:135.
15. El-Sherbiny IM, Yacoub MH. Hydrogel scaffolds for tissue engineering: progress and challenges. Glob Cardiol Sci Pract. 2013;2013(3):38.
16. Zhao C, Liu G, Tan Q, Gao M, Chen G, Huang X, et al. Polysaccharide-based biopolymer hydrogels for heavy metal detection and adsorption. J Adv Res. 2023;44:53–70.
17. Ghiorghita CA, Dinu MV, Lazar MM, Dragan ES. Polysaccharide-based composite hydrogels as sustainable materials for removal of pollutants from wastewater. Molecules. 2022;27:8574.
18. Rumon MMH, Akib AA, Sarkar SD, Khan MAR, Uddin MM, Nasrin D et al. Polysaccharide-based hydrogels for advanced biomedical engineering applications. ACS Polym Au. 2024.
19. Drishya PK, Reddy MV, Mohanakrishna G, Sarkar O, Isha Rohit MV, et al. Advances in microbial and plant-based biopolymers: synthesis and applications in next-generation materials. Macromol. 2025;5(2):21.
20. Kolya H, Kang CW. A new approach for agricultural water management using pillows made from covid-19 waste face masks and filled with a hydrogel polymer: preliminary studies. Agriculture (Switzerland). 2023;13(1):152.
21. Wang LY, Wang MJ. Removal of heavy metal ions by poly(vinyl alcohol) and carboxymethyl cellulose composite hydrogels prepared by a Freeze-Thaw method. ACS Sustain Chem Eng. 2016;4(5):2830–7.
22. Ye G, Li G, Wang C, Ling B, Yang R, Huang S. Extraction and characterization of dextran from *Leuconostoc pseudomesenteroides* YB-2 isolated from mango juice. Carbohydr Polym. 2019;207:218–23.
23. Álvarez-Viñas M, González-Ballesteros N, Torres MD, López-Hortas L, Vanini C, Domingo G, et al. Efficient extraction of carrageenans from *Chondrus crispus* for the green synthesis of gold nanoparticles and formulation of printable hydrogels. Int J Biol Macromol. 2022;206:553–66.
24. Mandal S, Hwang S, Shi SQ. Guar gum, a low-cost sustainable biopolymer, for wastewater treatment: a review. 2022
25. Hussein DJ, Satar M, Al-Hamzawi AK, Shabbani H, Mutar MA, Othman MR. Photopolymerization of hydrogel using three types of cross-linking agents: effects of water content, mechanical and thermal properties. S Afr J Chem Eng. 2025;53:400–9.
26. Caló E, Khutoryanskiy VV. Biomedical applications of hydrogels: a review of patents and commercial products. Eur Polym J. 2015;65:252–67.
27. Hoare TR, Kohane DS. Hydrogels in drug delivery: progress and challenges. Polymer. 2008;49:1993–2007.
28. Ferfera-Harrar H, Benhalima T, Sadi A. Development of functional chitosan-based superabsorbent hydrogel nanocomposites for adsorptive removal of Basic Red 46 textile dye. Polym Bull. 2022;79(8):6141–72.
29. Wang Z, Zhang G, Li Y. Preparation of chitosan/polyacrylamide/graphene oxide composite membranes and study of their methylene blue adsorption properties. Materials. 2020;13(19):1–18.
30. Preibisch I, Niemeyer P, Yusufoglu Y, Gurikov P, Milow B, Smirnova I. Polysaccharide-based aerogel bead production via jet cutting method. Materials. 2018;11(8):1287.
31. Sofiah AGN, Pasupuleti J, Samykano M, Kadirgama K, Koh SP, Tiong SK, et al. Harnessing nature's ingenuity: a comprehensive exploration of nanocellulose from production to cutting-edge applications in engineering and sciences. Polymers. 2023;15:3044.

32. Shen X, Shamshina JL, Berton P, Gurau G, Rogers RD. Hydrogels based on cellulose and chitin: Fabrication, properties, and applications. Green Chem. 2015;18:53–75.
33. Souza AL, Souza VGL, Jesus M, Mata F, de Oliveira TV, Soares NFF. Modification of cellulose nanocrystals using polydopamine for the modulation of biodegradable packaging, polymeric films: a mini review. Sustainability (Switzerland). 2025; 17:5633.
34. Thomas B, Raj MC, Joy J, Moores A, Drisko GL, Sanchez C. Nanocellulose, a versatile green platform: from biosources to materials and their applications. Chem Rev. 2018;118(24):11575–625.
35. Lee JH, Park SH, Kim SH. Surface alkylation of cellulose nanocrystals to enhance their compatibility with polylactide. Polymers (Basel). 2020;12(1):1–16.
36. Shang Z, An X, Seta FT, Ma M, Shen M, Dai L, et al. Improving dispersion stability of hydrochloric acid hydrolyzed cellulose nano-crystals. Carbohydr Polym. 2019;222:115037.
37. Veloso SRS, Azevedo AG, Teixeira PF, Fernandes CBP. Cellulose nanocrystal (CNC) gels: a review. Gels. 2023;9:574.
38. Domard A. A perspective on 30 years research on chitin and chitosan. Carbohydr Polym. 2011;84(2):696–703.
39. Duan J, Liang X, Cao Y, Wang S, Zhang L. High strength chitosan hydrogels with biocompatibility via new avenue based on constructing nanofibrous architecture. Macromolecules. 2015;48(8):2706–14.
40. Wang Z, Nie J, Qin W, Hu Q, Tang BZ. Gelation process visualized by aggregation-induced emission fluorogens. Nat Commun. 2016;7:12033.
41. Zhuikova Y, Zhuikov V, Varlamov V. Biocomposite materials based on poly(3-hydroxybutyrate) and chitosan: a review. Polymers. 2022;14:5549.
42. Rana AK, Gupta VK, Hart P, Thakur VK. Cellulose-alginate hydrogels and their nanocomposites for water remediation and biomedical applications. Environ Res. 2024;243:5549.
43. Ma J, Li X, Bao Y. Advances in cellulose-based superabsorbent hydrogels. RSC Adv. 2015;5(73):59745–57.
44. Nie G, Zang Y, Yue W, Wang M, Baride A, Sigdel A, et al. Cellulose-based hydrogel beads: Preparation and characterization. Carbohydr Polym Technol Appl. 2021;2:100074.
45. Bhattacharya A, Misra BN. Grafting: a versatile means to modify polymers: techniques, factors and applications. Prog Polym Sci (Oxford). 2004;29:767–814.
46. Li Z, Lin Z. Recent advances in polysaccharide-based hydrogels for synthesis and applications. Aggregate. 2021;2(2):e21.
47. Jiang F, Zhang W, Guo S, Zhang X. Preparation and properties of pH-responsive intelligent medical dressing based on bacterial cellulose/polyacrylic acid. Mater Sci Appl. 2022;13(03):107–231.
48. Xia J, Liu Z, Chen Y, Wang Z, Cao Y. Fabrication of thermo-sensitive lignocellulose hydrogels with switchable hydrophilicity and hydrophobicity through an SIPN strategy. RSC Adv. 2019;9(51):29600–8.
49. Zhou Y, Fu S, Zhang L, Zhan H. Superabsorbent nanocomposite hydrogels made of carboxylated cellulose nanofibrils and CMC-g-p(AA-co-AM). Carbohydr Polym. 2013;97(2):429–35.
50. Mohd Amin MCI, Ahmad N, Halib N, Ahmad I. Synthesis and characterization of thermo- and pH-responsive bacterial cellulose/acrylic acid hydrogels for drug delivery. Carbohydr Polym. 2012;88(2):465–73.
51. Muzzarelli RAA. Genipin-crosslinked chitosan hydrogels as biomedical and pharmaceutical aids. Carbohydr Polym. 2009;77:1–9.
52. Thirupathi K, Raorane CJ, Ramkumar V, Ulagesan S, Santhamoorthy M, Raj V, et al. Update on chitosan-based hydrogels: preparation, characterization, and its antimicrobial and antibiofilm applications. Gels. 2023;9:35.
53. Guo Y, Qiao D, Zhao S, Liu P, Xie F, Zhang B. Biofunctional chitosan–biopolymer composites for biomedical applications. Mater Sci Eng R Reports. 2024;159:100775.
54. Cheung RCF, Ng TB, Wong JH, Chan WY. Chitosan: an update on potential biomedical and pharmaceutical applications. Mar Drugs. 2015;13:5156–86.

55. de Siqueira EC, de França JAA, de Souza RFM, da Leoterio DMS, Cordeiro JN, Doboszewski B. Mechanisms of the chemical crosslinking to obtain the hydrogels: Synthesis, conditions of crosslinking and biopharmaceutical applications. Res Soc Dev. 2023;12(8):e18312943072.
56. Kumar S. Laminated object manufacturing. In: A concise encyclopedia of additive manufacturing. Cham: Springer; 2025. p. 245–249.
57. Singh S, Kaur D, Singh M, Balu R, Mehta A, Vasudev H. Challenges and issues in manufacturing of components using polymer-based selective laser sintering (SLS): a review. Int J Interact Des Manuf. 2024.
58. Wu Y, Zhang Y, Yan M, Hu G, Li Z, He W, Wang X, Abulimit A, Li R. Research progress on the application of inkjet printing technology combined with hydrogels. J Appl Mater Today. 2024;36:102036.
59. Cheng C, Williamson EJ, Chiu GTC, Han B. Engineering biomaterials by inkjet printing of hydrogels with functional particulates. Med-X. 2024;2:9.
60. Zhang H, Xue Y, Jiang C, Liu D, Zhang L, Lang G, Mao T, Effrem DB, Iimaa T, Surenjav U, Liu M. 3-Dimensional printing of polysaccharides for water treatment: a review. Int J Biol Macromol. 2024;265(Pt 2): 131117. https://doi.org/10.1016/j.ijbiomac.2024.131117.
61. Rahman MU, Liu Z, Liu D, Wei J. Microscale studying and machine learning of polymer wetting behavior in direct ink writing under different gravitational intensity. Phys Fluids. 2025;37(8):60–8.
62. Ali S, Deiab I, Pervaiz S. State-of-the-art review on fused deposition modeling (FDM) for 3D printing of polymer blends and composites: innovations, challenges, and applications. Int J Adv Manuf Technol. 2024;135:5085–113.
63. Celikci N, Ziba CA, Dolaz M, Tümer M. Comparison of composite resins containing UV light-sensitive chitosan derivatives in stereolithography (SLA)-3D printers. Int J Biol Macromol. 2024
64. Hernandez Monroy L, Tavares JR, Dumont MJ. Application of photocatalytic hydrogels used for the degradation of dyes, pharmaceuticals and other contaminants in water. Adsorption. 2025;31(3):52.
65. Iliasov L, Shibaev A, Panova I, Kushchev P, Philippova O, Yaroslavov A. Weakly cross-linked anionic copolymers: kinetics of swelling and water-retaining properties of hydrogels. Polymers (Basel). 2023;15(15):3244.
66. Nanda D, Behera D, Pattnaik SS, Behera AK. Advances in natural polymer-based hydrogels: synthesis, applications, and future directions in biomedical and environmental fields. Discov Polym. 2025;2(1).
67. Sandu T, Chiriac AL, Zaharia A, Iordache TV, Sarbu A. New trends in preparation and use of hydrogels for water treatment. Gels. 2025;11.
68. Zhang A, Wong JKU, Xia Y, Bilek M, Yeo G, Akhavan B. Surface biofunctionalised porous materials: advances, challenges, and future prospects. Prog Mater Sci. 2025;154.
69. Seida Y, Tokuyama H. Hydrogel adsorbents for the removal of hazardous pollutants—requirements and available functions as adsorbent. Gels. 2022;8:224.
70. Calderón Moreno JM, Chelu M, Popa M. Eco-friendly conductive hydrogels: towards green wearable electronics. Gels. 2025;11.
71. Li W, Wang S, Li Y, Ma C, Huang Z, Wang C, et al. One-step hydrothermal synthesis of fluorescent nanocrystalline cellulose/carbon dot hydrogels. Carbohydr Polym. 2017;175:7–17.
72. Narayanan A, Kartik R, Sangeetha E, Dhamodharan R. Super water absorbing polymeric gel from chitosan, citric acid and urea: synthesis and mechanism of water absorption. Carbohydr Polym. 2018;191:152–60.
73. Fang S, Wang G, Li P, Xing R, Liu S, Qin Y, et al. Synthesis of chitosan derivative graft acrylic acid superabsorbent polymers and its application as water retaining agent. Int J Biol Macromol. 2018;115:754–61.
74. Karvinen J, Kellomäki M. Characterization of self-healing hydrogels for biomedical applications. Eur Polym J. 2022;181.

75. Nawrotek K, Modrzejewska Z, Douglas TEL, Zarzycki R, Douglas T. Progress on chemistry and application of chitin and its. 2012;XVII. https://www.researchgate.net/publication/259331541
76. Rihawy MS, Alzier A, Allaf AW. Investigation of chloramphenicol release from PVA/CMC/HEA hydrogel using ion beam analysis, UV and FTIR techniques. Appl Radiat Isotop. 2019;153:108806.
77. Zhang Y, Wu F, Liu L, Yao J. Synthesis and urea sustained-release behavior of an eco-friendly superabsorbent based on flax yarn wastes. Carbohydr Polym. 2013;91(1):277–83.
78. Feng Z, Odelius K, Hakkarainen M. Tunable chitosan hydrogels for adsorption: property control by biobased modifiers. Carbohydr Polym. 2018;196:135–45.
79. Raman CV. Plate FRS. A new radiation. Indian J Phys. 1928;2.
80. Lacoste C, Lopez-Cuesta JM, Bergeret A. Development of a biobased superabsorbant polymer from recycled cellulose for diapers applications. 2019.
81. Aijaz MO, Haider S, Al-Mubaddel FS, Khan R, Haider A, Alghyamah AA, et al. Thermal, swelling and stability kinetics of chitosan based semi-interpenetrating network hydrogels. Fibers Polym. 2017;18(4):611–8.
82. Rusu AG, Popa MI, Lisa G, Vereştiuc L. Thermal behavior of hydrophobically modified hydrogels using TGA/FTIR/MS analysis technique. Thermochim Acta. 2015;613(1):28–40.
83. Zhang M, Qiao X, Han W, Jiang T, Liu F, Zhao X. Alginate-chitosan oligosaccharide-ZnO composite hydrogel for accelerating wound healing. Carbohydr Polym. 2021;266.
84. Zeng M, Feng Z, Huang Y, Liu J, Ren J, Xu Q, et al. Chemical structure and remarkably enhanced mechanical properties of chitosan-graft-poly(acrylic acid)/polyacrylamide double-network hydrogels. Polym Bull. 2017;74(1):55–74.
85. Mulijani S, Erizal E, Irawadi TT, Katresna TC. Composite copolymer acrylamide/bacterial cellulose hydrogel synthesis and characterization by the application of gamma irradiation. Adv Mater Res. 2014;974:91–6.
86. Kalaiselvi K, Mohandoss S, Ahmad N, Khan MR, Manoharan RK. Adsorption of Pb^{2+} ions from aqueous solution onto porous Kappa-Carrageenan/cellulose hydrogels: isotherm and kinetics study. Sustainability (Switzerland). 2023;15(12).
87. Saito T, Nishiyama Y, Putaux JL, Vignon M, Isogai A. Homogeneous suspensions of individualized microfibrils from TEMPO-catalyzed oxidation of native cellulose. Biomacromolecules. 2006;7(6):1687–91.
88. Cova TF, Murtinho D, Aguado R, Pais AACC, Valente AJM. Cyclodextrin polymers and cyclodextrin-containing polysaccharides for water remediation. Polysaccharides. 2021;2:16–38.
89. López-Maldonado EA, Abdellaoui Y, Abu Elella MH, Abdallah HM, Pandey M, Anthony ET, Ghimici L, Álvarez-Torrellas S, Pinos-Vélez V, Oladoja NA. Innovative biopolyelectrolytes-based technologies for wastewater treatment. Int J Biol Macromol. 2024;273(2).
90. Alsaka L, Alsaka L, Altaee A, Zaidi SJ, Zhou J, Kazwini T. A review of hydrogel application in wastewater purification. Separations. 2025;12:51.
91. Abu Elella MH, Abdallah HM, AboBakr Ali E, Makhado E, Abd El-Ghany NA. Recent developments in conductive polysaccharide adsorbent formulations for environmental remediation: a review. Int J Biol Macromol. 2025;304(2):140915.
92. Liu S, Cai Y, Du M, Liu J, Zhang X, Yuan H, Xu Z, Chen X, Kan X, Zhang S, Zhong L. Research on adsorption of heavy metal ions by polysaccharide hydrogels. Chem Europe. 2024;9(41):e202401752.
93. Levy-Ontman O, Dayan S, Paz-Tal O, Wolfson A. Adsorption of heavy metal ions from multi-ion solutions using polysaccharide hydrogels. Polysaccharides. 2025;6(2):42.
94. Iddrisu L, Danso F, Cheong KL, Fang Z, Zhong S. Polysaccharides as protective agents against heavy metal toxicity. Foods. 2024;13:853.
95. Zhang D, Xiao J, Guo Q, Yang J. 3D-printed highly porous and reusable chitosan monoliths for Cu(II) removal. J Mater Sci. 2019;54:6728–41. https://doi.org/10.1007/s10853-019-03332-y.

96. Abd El-Ghany NA, Elella MH, Abdallah HM, Mostafa MS, Samy M. Recent advances in various starch formulation for wastewater purification via adsorption technique: a review. J Polym Environ. 2023;31(7):2792–825.
97. Grewal MK, Fatona A, Su Y, Ungureanu J, Moran-Mirabal JM. Tetrahydrazide-EDTA cross-linked cellulose hydrogels for water treatment by heavy metal chelation. ACS Appl Mater Interf. 2025;17(27):39075–88.
98. Ling Felicia WX, Rovina K, Supri S, Matanjun P, Mohd Amin SF, Abdul Rahman MN. Next-generation sodium alginate hydrogels for heavy metal ion removal: properties, dynamic adsorption–desorption mechanisms, and sustainable application potential. Polym Bull. 2025.
99. Bashir A, Malik LA, Ahad S, Manzoor T, Bhat MA, Dar GN, et al. Removal of heavy metal ions from aqueous system by ion-exchange and biosorption methods. Environ Chem Lett. 2019;17:729–54.
100. Ma M, Ke X, Wang T, Li J, Ye H. A novel double-network hydrogel made from electrolytic manganese slag and polyacrylic acid-polyacrylamide for removal of heavy metals in wastewater. J Hazard Mater. 2024;462:132722.
101. Cruz-Lopes LP, Macena M, Esteves B, Guiné RPF. Ideal pH for the adsorption of metal ions Cr^{6+}, Ni^{2+}, Pb^{2+} in aqueous solution with different adsorbent materials. Open Agric. 2021;6(1):115–23.
102. Shankar S, Joshi S, Srivastava RK. A review on heavy metal biosorption utilizing modified chitosan. Environ Monitor Assess. 2023;195:1350.
103. Dong L, Shan C, Liu Y, Sun H, Yao B, Gong G, et al. Characterization and mechanistic study of heavy metal adsorption by facile synthesized magnetic xanthate-modified chitosan/polyacrylic acid hydrogels. Int J Environ Res Public Health. 2022;19(17):11123.
104. Benettayeb A, Ghosh S, Usman M, Seihoub FZ, Sohoo I, Chia CH, et al. Some well-known alginate and chitosan modifications used in adsorption: a review. Water (Switzerland). 2022;14:1353.
105. da Costa TB, da Silva TL, Costa CSD, da Silva MGC, Vieira MGA. Chromium adsorption using Sargassum filipendula algae waste from alginate extraction: Batch and fixed-bed column studies. Chem Eng J Adv. 2022;11:100341.
106. Persano F, Malitesta C, Mazzotta E. Cellulose-based hydrogels for wastewater treatment: a focus on metal ions removal. Polymers. 2024;16:1292.
107. Chen X, Huang Z, Luo SY, Zong MH, Lou WY. Multi-functional magnetic hydrogels based on *Millettia speciosa* Champ residue cellulose and chitosan: highly efficient and reusable adsorbent for Congo red and Cu^{2+} removal. Chem Eng J. 2021;423:130198.
108. Adeyemo AA, Adeoye IO, Bello OS. Adsorption of dyes using different types of clay: a review. Appl Water Sci. 2017;7:543–68.
109. Khanam A, Nadeem F, Praveena SM, Rashid U. Removal of dyes using alginated, calcinized and hybrid materials-a comprehensive review. IJCBS. 2017;12. www.iscientific.org/Journal.html
110. Motshabi BR, Makhado E, Malatji N, Modibane KD, Hato MJ, Botlhoko OJ, et al. Synthesis of locust bean gum/titanium dioxide hydrogel nanocomposites for efficient removal of methylene blue from aqueous solution. J Polym Environ. 2024;32(11):6114–30.
111. Vakili M, Zwain HM, Mojiri A, Wang W, Gholami F, Gholami Z et al. Effective adsorption of reactive black 5 onto hybrid hexadecylamine impregnated chitosan-powdered activated carbon beads. Water (Switzerland). 2020:12(8).
112. Qian L, Zhang H. Controlled freezing and freeze drying: a versatile route for porous and micro-/nano-structured materials. J Chem Technol Biotechnol. 2011;2011:172–84.
113. Nakasu PYS, Martinez MA, Melanie S, Shmool TA, Hallett JP. Chitosan-based biocomposite hydrogels with squid pen protein for anionic dyes adsorption. ACS Mater Lett. 2025;7(3):1012–8.
114. Al Samman MM, Sánchez J. Chitosan- and alginate-based hydrogels for the adsorption of anionic and cationic dyes from water. Polymers (Basel). 2022;14(8):1498.
115. Zhang Z, Lu Y, Gao S, Wu S. Sustainable and efficient wastewater treatment using cellulose-based hydrogels: a review of heavy metal, dye, and micropollutant removal applications. Separations. 2025;12(3):72.

116. Liu L, Gao ZY, Su XP, Chen X, Jiang L, Yao JM. Adsorption removal of dyes from single and binary solutions using a cellulose-based bioadsorbent. ACS Sustain Chem Eng. 2015;3(3):432–42.
117. Kolya H, Kang CW. Recent advances in polymer nanocomposites for the adsorptive removal of toxic azo dyes from water. Discov Water. 2025;5(1):28.
118. Bouazizi N, Vieillard J, Samir B, Le Derf F. Advances in amine-surface functionalization of inorganic adsorbents for water treatment and antimicrobial activities: a review. Polymers. 2022;14:378.
119. Fijoł N, Aguilar-Sánchez A, Mathew AP. 3D-printable biopolymer-based materials for water treatment: a review. Chem Eng J. 2022;430(Pt 3): 132964. https://doi.org/10.1016/j.cej.2021.132964.
120. da Silva Alves DC, Healy B, Yu T, Breslin CB. Graphene-based materials immobilized within chitosan: Applications as adsorbents for the removal of aquatic pollutants. Materials. 2021;14:3655.
121. Yang H, Wang S, Liu Y, Hu Y, Shen W. ZIF-67 grows in chitosan-rGO hydrogel beads for efficient 2 adsorption of tetracycline and norfloxacin. https://ssrn.com/abstract=4541763
122. Facchi DP, Cazetta AL, Canesin EA, Almeida VC, Bonafé EG, Kipper MJ, et al. New magnetic chitosan/alginate/Fe_3O_4@SiO_2 hydrogel composites applied for removal of Pb(II) ions from aqueous systems. Chem Eng J. 2018;337:595–608.
123. Murcia-Salvador A, Rodríguez-López MI, Pellicer JA, Gómez-Morte T, Auñón-Calles D, Yáñez-Gascón MJ, et al. Development of chitosan polysaccharide-based magnetic gel for direct red 83:1 removal from water. Gels. 2024;10(8):496.
124. Abu-Thabit NY, Abu Elella MH, Azad AK, Ratemi E, Hakeem AS. Superwetting metal mesh membranes for oil/water separation: a comprehensive review. Sep Purif Technol. 2025;363(1):1383–5866.
125. Abu Elella MH, Abu-Thabit NY, Uwaezuoke OJ, Azad AK. Superwetting cotton textiles for separation of oil/water mixtures. Cellulose. 2023;30:7427–62.
126. Azad AK, Abu-Thabit NY, Abu Elella MH. Sustainable and durable superwetting textiles coated with polydopamine for oil/water separation. RSC Sustain. 2025;3:2501–33.
127. Zakuwan SZ, Ahmad I, Tahrim NA, Mohamed F. Functional hydrophilic membrane for oil–water separation based on modified bio-based chitosan–gelatin. Polymers (Basel). 2021;13(7):1176.
128. de Medeiros ADM, da Silva Junior CJG, de Amorim JDP, Durval IJB, de Santana Costa AF, Sarubbo LA. Oily wastewater treatment: methods, challenges, and trends. Processes. 2022;10:743.
129. Pooresmaeil M, Namazi H. Application of polysaccharide-based hydrogels for water treatments. In: Hydrogels based on natural polymers. Elseiver, 2019. pp 411–455
130. Cosgrove S, Jefferson B, Jarvis P. Pesticide removal from drinking water sources by adsorption: a review. Environ Technol Rev. 2019;8:1–24.
131. Aljeboree AM, Alkaim AF. Studying removal of anionic dye by prepared highly adsorbent surface hydrogel nanocomposite as an applicable for aqueous solution. Sci Rep. 2024;14(1).
132. Long L, Hu X, Yan J, Zeng Y, Zhang J, Xue Y. Novel chitosan–ethylene glycol hydrogel for the removal of aqueous perfluorooctanoic acid. J Environ Sci (China). 2019;84:21–8.
133. Liang HW, Cao X, Zhang WJ, Lin HT, Zhou F, Chen LF, et al. Robust and highly efficient free-standing carbonaceous nanofiber membranes for water purification. Adv Funct Mater. 2011;21(20):3851–8.
134. Vaid V, Jindal R. An efficient pH-responsive kappa-carrageenan/tamarind kernel powder hydrogel for the removal of brilliant green and rose bengal from aqueous solution. J Appl Polym Sci. 2022;139(21):52218.
135. Thakur S, Govender PP, Mamo MA, Tamulevicius S, Mishra YK, Thakur VK. Progress in lignin hydrogels and nanocomposites for water purification: future perspectives. Vacuum. 2017;146:342–55.
136. Zhang L, Lu H, Yu J, Fan Y, Yang Y, Ma J, et al. Synthesis of lignocellulose-based composite hydrogel as a novel biosorbent for Cu^{2+} removal. Cellulose. 2018;25(12):7315–28.

137. Fijoł N, Aguilar-Sánchez A, Ruiz-Caldas MX, Redlinger-Pohn J, Mautner A, Mathew AP. 3D printed polylactic acid (PLA) filters reinforced with polysaccharide nanofibers for metal ions capture and microplastics separation from water. Chem Eng J. 2023;457.
138. Agbakoba VC, Mokhena TC, Ferg EE, Hlangothi SP, John MJ. PLA bio-nanocomposites reinforced with cellulose nanofibrils (CNFs) for 3D printing applications. Cellulose. 2023;30(18):11537–59.
139. Arem S, Patil S, Archana K. Occupational health hazards of workers use of hydrogels in agriculture. AgriGate—An International Multidisciplinary e-Magazine. www.agrigatemagazine.com
140. Guilherme MR, Aouada FA, Fajardo AR, Martins AF, Paulino AT, Davi MFT, et al. Superabsorbent hydrogels based on polysaccharides for application in agriculture as soil conditioner and nutrient carrier: a review. Eur Polym J. 2015;72:365–85.
141. Pathak VM, Verma VK, Rawat BS, Kaur B, Babu N, Sharma A, et al. Current status of pesticide effects on environment, human health and it's eco-friendly management as bioremediation: a comprehensive review. Front Microbiol. 2022;13:962619.
142. Enzo M. The impacts of agricultural chemicals on soil and water quality. J Environ Hazards Short Commun. 2024.
143. Sewar Y, Hussain S, Ali D, Khan S, Yahya Y, Sewar A et al. A review of soil contamination and remediation on basis of case studies. Int J Agric Biol Sci. https://www.researchgate.net/publication/359512312
144. Vasilachi IC, Stoleru V, Gavrilescu M. Analysis of heavy metal impacts on cereal crop growth and development in contaminated soils. Agriculture (Switzerland). 2023;13:1983.
145. Rodrigo L, Munaweera I. Employing sustainable agriculture practices using eco-friendly and advanced hydrogels. RSC Adv. 2025;15(26):21212–28. https://doi.org/10.1039/d5ra03035h.
146. Azeem MK, Islam A, Khan RU, Rasool A, Qureshi MAUR, Rizwan M, Sher F, Rasheed T. Eco-friendly three-dimensional hydrogels for sustainable agricultural applications: current and future scenarios. Polym Adv Technol. 2023. https://doi.org/10.1002/pat.6122.
147. Sun S, Wang Y, Zhang Y, Wu L, Wang X, Wang G, et al. Development and application of a modified biochar-calcium alginate composite (MB-CA) for in situ remediation of cadmium-contaminated soil. Gels. 2025;11(5):375.
148. Zhang X, Zhao Y, Wang S, Jing X. Cross-linked polymers based on B-O bonds: synthesis, structure and properties. Mater Chem Front. 2021;5:5534–48.
149. Pires J, de Paula CD, Souza VGL, Fernando AL, Coelhoso I. Understanding the barrier and mechanical behavior of different nanofillers in chitosan films for food packaging. Polymers. 2021;13:1–29.
150. Yadav V, Pal D, Poonia AK. Nanofillers as a potential key for shaping the future of the industries. Hybrid Adv. 2024;7:100340.
151. Grigoraș CG, Simion AI, Drob C. Hydrogels based on chitosan and nanoparticles and their suitability for dyes adsorption from aqueous media: assessment of the last-decade progresses. Gels. 2024;10:211.
152. Fuchs S, Shariati K, Ma M. Specialty tough hydrogels and their biomedical applications. Adv Healthc Mater. 2020;9:211.
153. Lu P, Ruan D, Huang M, Tian M, Zhu K, Gan Z, et al. Harnessing the potential of hydrogels for advanced therapeutic applications: current achievements and future directions. Signal Transduc Target Therap. 2024;9:166.
154. Garcia-Garcia A, Muñana-González S, Lanceros-Mendez S, Ruiz-Rubio L, Alvarez LP, Vilas-Vilela JL. Biodegradable natural hydrogels for tissue engineering, controlled release, and soil remediation. Polymers. 2024;16:2599.
155. Rando G, Scalone E, Sfameni S, Plutino MR. Functional bio-based polymeric hydrogels for wastewater treatment: from remediation to sensing applications. Gels. 2024;10:498.
156. Revete A, Aparicio A, Cisterna BA, Revete J, Luis L, Ibarra E, et al. Advancements in the use of hydrogels for regenerative medicine: properties and biomedical applications. Int J Biomater. 2022;2022:3606765.

157. Rial-Hermida MI, Rey-Rico A, Blanco-Fernandez B, Carballo-Pedrares N, Byrne EM, Mano JF. Recent progress on polysaccharide-based hydrogels for controlled delivery of therapeutic biomolecules. ACS Biomater Sci Eng. 2021;7:4102–27.
158. Elgamal AM, Abu Elella MH, Saad GR, Abd El-Ghany NA. Synthesis, characterization and swelling behavior of high-performance antimicrobial biocompatible copolymer based on carboxymethyl xanthan. Mater Today Commun. 2022;33: 104209. https://doi.org/10.1016/j.mtcomm.2022.104209.

Polysaccharide-Based Aerogels for Soil and Water Remediation

Nahed A. Abd El-Ghany

Abstract Porous polysaccharides are employed in a wide range of applications and have recently attracted a lot of attention due to their porosity, availability, and exceptional properties like sustainability and biocompatibility. Because of their unique capacity to form porous structures, studies on the use of polysaccharides such as cellulose, chitosan, alginate, and starch as aerogels have recently increased. Polysaccharide-based aerogels are promising for water remediation due to their high porosity, large surface area, and biodegradability; they are eco-friendly and can be functionalized to remove various pollutants. They can effectively adsorb toxic heavy metals such as copper, mercury, and lead and are used to remove coloring dyes from industrial wastewater. Modified polysaccharide aerogels, such as those with hydrophobic properties, are efficient for oil-water separation. This chapter presents recent advancements in polysaccharide-based porous aerogels, including their synthesis and applications in water and soil remediation. The comparison of the supercritical CO_2 and freeze-drying processes indicates which drying method is best for understanding the aerogel formation of porous polysaccharides. The drying step in the aerogel synthesis process is particularly important. The limitations and potential for future commercialization of porous aerogels are also discussed, along with the recent uses of polysaccharide aerogels in soil and water remediation.

Keywords Polysaccharides · Aerogels · Metals removal · Dye removal · Oil removal · Soil treatments

List of Abbreviations

Abbreviation	Full name
CO_2	Carbon dioxide
BC	Bacterial cellulose
GO	Graphene oxide

N. A. Abd El-Ghany (✉)
Department of Chemistry, Faculty of Science, Cairo University, Giza, Egypt
e-mail: abdelghanyn@sci.cu.edu.eg

M. H. Abu Elella et al. (eds.), *3D Engineered Polysaccharides*,
https://doi.org/10.1007/978-3-032-26154-0_4

CNC	Cellulose nano-crystals
RCAs	Regenerated cellulose aerogel (RCAs)
K++	Potassium ions
Co++	Cobalt ions
Ca++	Calcium ions
Fe+++	Ferric ions
CR	Crystal red
Pb++	Lead ions
Cd++	Cadmium ions
Cu++	Cupper ions
MO	Methyl orange
NC	Nanostructured cellulose
MB	Methylene blue
TEMPO (catalyst)	2,2,6,6-tetramethyl-1-piperidinyloxy
MOF	Metal Organic Framework
CS	Chitosan
SA	Sodium alginate
CMC	Sodium carboxymethyl cellulose
IPD	Intraparticle diffusion
CV	Crystal violet
RB	Rhodamine B
Cr(VI)	Hexavalent chromium
((L)CNFs)	Ligno cellulose nanofibers

1 Introduction

Every day, environmental contamination rises and causes serious, irreparable harm to the planet. It comes in various forms, including air, water, soil, noise, light, etc., and is caused by factories releasing hazardous waste such as plastics, heavy metals, dyes, burning fossil fuels, acid rain, oil spills, and industrial toxins. Of these, water and soil pollution are the most common, given the significant health risks to humans [1–4].

Globally, the amount of wastewater produced and its overall pollutant load is rising due to factors like population growth, fast urbanization, and economic development. Both surface and ground water are contaminated by the use of chemical pesticides and fertilizers, as well as by irrigation that uses untreated wastewater. Waste from numerous industries is still dumped straight into rivers. There is a significant underestimation of the potential sustainability and cost-effectiveness of safely reusing wastewater as a source of water, fertilizers, power, and other recoverable resources [2, 5].

Soil can get contaminated by a variety of things, including chemicals, improper disposal of industrial waste, long-term pipe breakdowns, and product spills of all kinds. Contaminants commonly found in soil include pesticides, heavy metals, oils, paints, solvents, and other synthetic materials. In soil remediation projects, contaminants such as pesticides, heavy metals, oils, paints, solvents, and other synthetic materials are commonly found. Humans and animals can be exposed to soil toxins in a number of ways, including through the intake of soil, inspiration of dust and pollutants, skin absorption, or eating of food grown in contaminated soil [6].

Water and soil contamination is a public health concern that can lead to issues including bacterial imbalances in the environment, damage to aquatic life, and obstruction of water purification plants. Furthermore, water's dispersibility and penetration make it easier for different pollutants to spread throughout other natural sources, which adds another remediation problem [7].

Therefore, one important way to address this issue is to develop materials for the efficient treatment of wastewater and soil. Aerogels have gained a lot of interest as an adsorption medium for the removal of numerous pollutants that pose a hazard to human health and the environment among the many high-performance applications [7–9].

Aerogels are solid, three-dimensional materials made of a combination of inorganic and organic precursors that have a high porosity and surface area. Professor Kistler created the first aerogel in 1931, and because of their adaptable chemical structure and wide range of possible uses, they have become more significant in recent decades [10]. Aerogels are three-dimensional porous nanomaterials in which colloidal nanoparticles or polymer molecules combind to each other to form a nanoporous, nano-lamella framework. The pores of these materials are filled with a gaseous dispersal medium.

Researchers are very attentive with the generation of polysaccharide-based aerogels because of the need to be ecologically friendly [9, 11, 12], it has become popular to use as many polysaccharide constituents as possible in aerogel formulations [2]. The various techniques for creating polysaccharide aerogels and their development for soil and water remediation are the main topics of this chapter.

Aerogels are generally classified as either hybrid, inorganic, or organic [7, 13]. They can be further divided into three categories according to the drying method: supercritically dried (aerogels), ambient pressure-dried (xerogels), and freeze-dried (cryogels. Aerogels can be hydrophilic, hydrophobic, amphiphilic, oleophilic, or oleophobic, depending on their polarity and surface functioning [14]. Figure 1 shows the classification of aerogels according to their appearance, origin, microstructures, chemical structure, and methods of synthesis [15].

Aerogels based on polysaccharides are unique functional materials that offer all the benefits of conventional aerogels such as low toxicity [16], biocompatibility, high repeatability, and a lot of potential for use in soil and water remediation. The significance of polysaccharide-based aerogels is found in the fact that the majority of them contain many function groups as amino, carboxylic, and hydroxy groups, which facilitate the customization of aerogels characteristics [17].

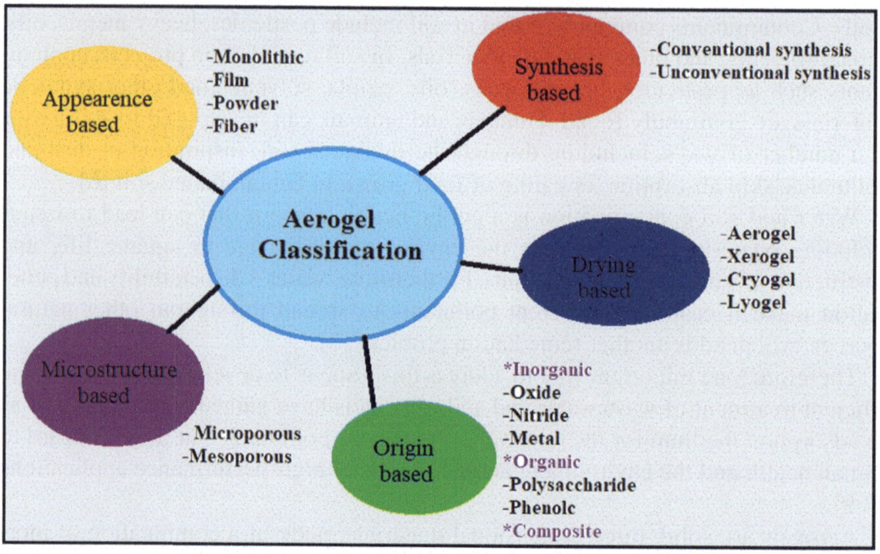

Fig. 1 Aerogel classification based on origin, shape, synthesis, and chemistry and microstructures

2 Synthesis of Polysaccharides-Based Aerogels (General Methods)

As seen in Fig. 2, Bio-aerogels are synthesized differently, with the polymer dissolving first, followed by gelation and drying. The type of initial raw polysaccharide components determines which aerogel preparation method is used. The polysaccharide is completely dissolved in the solvent at the start of this process by applying heat for a predetermined amount of time. The solvent determines the proper heating temperature normally [18].

Gelation is the process that produces the network structure of the aerogel it enhances the aerogel's hydrophobicity, solubility, and biological activity. This can be achieved by filling a mold with the solution and allowing it to settle for 24–72 h at room temperature [19]. Heating polysaccharides in polar solvents such as methanol or ethanol, ionic liquids, or alkaline/urea solutions causes them to gel [20].

The porous structure of the aerogels usually attained during the drying process after the solvent is extracted. The solvent is usually extracted when the pressure and temperature are suitable for supercritical CO_2 drying. This prevents the linked aerogel structure from shrinking and ensures that the liquid doesn't evaporate directly [8]. Compared to physical crosslinking, chemical crosslinking produces a more stable structure and gel at a faster rate. The presence of many functional groups, including hydroxyl, phosphate, sulfate, and carboxyl groups, on the cell wall of polysaccharides allows them to crosslink or entangle with other precursors [21].

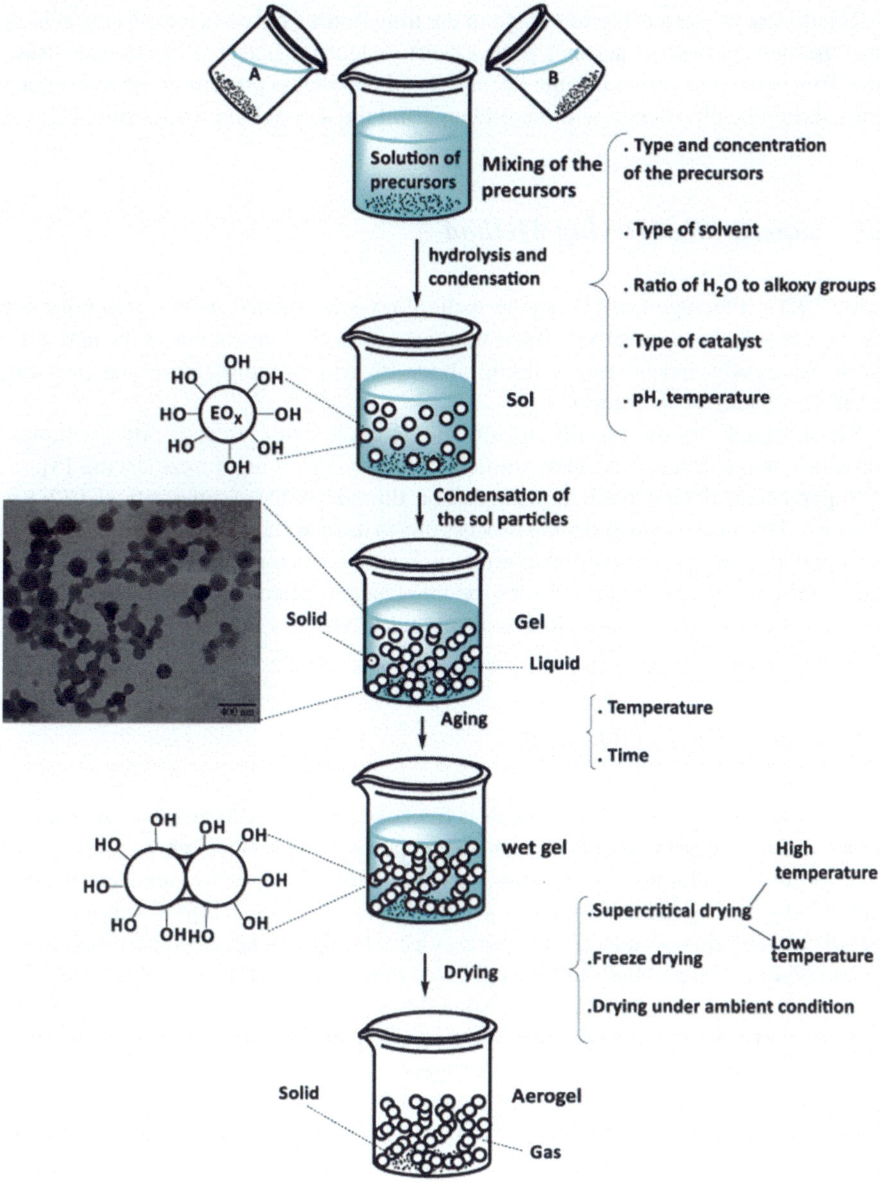

Fig. 2 Schematic presentation of aerogel preparation [18]

The drying process is the last stage in creating porous polysaccharide aerogels. It turns the hydrogel into an aerogel, reduces surface tension, and creates a porous structure. This is the most crucial stage because a solvent that is present in the hydrogel's pores should be eliminated without endangering the aerogel microstructure [22].

2.1 Supercritical Drying Method

In the 1970s, the supercritical drying technology was created. It is a technique for drying wet gels that uses supercritical liquids to draw the solvent out of the aerogel's pores. Squeezing and heating a liquid above its critical temperature and pressure results in supercritical conditions.

Supercritical drying usually results in aerogels with greater pore volumes, porosities, and surface areas than ambient pressure drying and freeze drying [8].

Supercritical drying method with carbon dioxide (CO_2) (supercritical CO_2) is considered the most popular drying process due to its many advantages, which include its abundance of reserves, affordability, atmospheric origin, environmental safety and sustainability, ease of transformation into a supercritical state, and non-toxicity, non-flammability, and non-explosive properties [23].

2.2 Freeze-Drying Method

Another method for drying the wet gels is freeze drying. By reducing the temperature below its freezing point, the solvent inside the pores is frozen throughout this process [24]. At this temperature, the pressure is usually reduced below the sublimation pressure by introducing vacuum into the system. Without creating a liquid–vapor barrier, the solvent sublimates and exits the pores. Once the solvent has been eliminated, the system is squeezed and the sample is allowed to return to ambient temperature. This process produces aerogels, which are sometimes referred to as cryogels. Because of its great safety and inexpensive cost, the freeze-drying technique is also greatly used in the drying process of formatting aerogels. The efficiency of the freeze-drying procedure is determined by the temperature of crystallization of the solvent at its point of freezing. When the solvent is being in the liquid-solid state, its freezing point must be higher than when it is in the liquid-liquid case. The freeze-drying procedure accomplishes two goals: (1) stabilizing the gel body and (2) encouraging solvent exchange for a low coefficient of expansion and high sublimation pressure. Moreover, the porosity of the aerogel is influenced by the freezing rate, the freezing temperature, and the direction of ice crystal growth [25, 26]. According to Xi et al., temperature and ingredient concentration have a major impact on the ice crystal forming process [27].

2.3 *Ambient Pressure Drying Method*

The technique of ambient pressure drying involves evaporating the pore liquid at room temperature to 200 °C while applying ambient pressures to dry wet gels [28]. The wet gel may shrink considerably and lose its porous structure if it is dried in the open without matrix strengthening. The primary cause of this shrinking is the nanopores, which are the most significant characteristic of aerogels. As the solvent exits the gel network's pores, the vapor-liquid interface's radius of curvature decreases, causing compress on the gel's surface. Excessive temperature or compress increases or decreases are crucial because they will cause the liquid to begin boiling [29].

Comparison between supercritical drying and freezing drying process of polysaccharide-based aerogel was shown in Table 1 and Fig. 3.

3 Characteristics of Polysaccharide-Based Aerogels

Aerogels made of polysaccharides have drawn increased attention because of their many beneficial qualities, including high porosity, biocompatibility, biodegradability, and simplicity of customization [39].

3.1 *Cellulose-Based Aerogels*

The most abundant and widespread polysaccharide in nature is cellulose, which is often derived by microbes and plants. Staudinger in 1932 characterized cellulose as a biopolymer composed of β-1,4-glucosidic bonds and β-D-glucose units [40, 41]. The glucose unit has many polar hydroxyl groups that can form strong hydrogen bonds, increasing the molecular chain's rigidity, and numerous cellulosic molecular chains can then construct the fibrous structure [16]. This provides the fundamentals for manufacturing the cellulosic aerogels with better physical and chemical features and superior properties (Fig. 4). There are three types of cellulose aerogels: derived, regenerated, and natural [42]. The characteristics and concentration of raw cellulose determine the characteristics of aerogels based on cellulose; the majority exhibit good repeatability, biocompatibility, and degradability.

1. The two primary types of natural cellulose aerogels are bacterial (BC) and nanocellulose (NC). Extra light nanocomposite aerogel was prepared from the cellulose (BC) and graphene oxide (GO). Zhang et al. and Weng et al., described the drying method of preparation of cellulose aerogel derived from raw materials like purified cellulose fibers [43, 44].
2. Regenerated cellulose aerogel which is made by dissolving, regenerating, exchanging solvents, and drying. The cellulose crystal structure is destroyed and

Table 1 Comparison between supercritical and freezing drying techniques of polysaccharide-based-aerogels

	Supercritical drying	Freezing drying	References
Mechanism	• Easy to change into a supercritical condition [31] • Is cheap, plentiful in reserves, non-toxic, and non-explosive • It is a safe environmental solvent and derived from the atmosphere [32]	• Used to lower the gel's temperature below the solvent's crystallization point [33] • Lowers the pressure to cause the solvent to sublimate and leave the matrix • Prior to freeze-drying, an appropriate solvent will be chosen for solvent exchange [34]	
Structural comparison	• Produces aerogels with homogenous fiber structure, high surface areas, and high pore volumes and porosities	• Produces aerogels with small surface areas and pore volumes	[35]
	• Can produce uniform, fine, and narrow pores with reduced pore collapse	• During grain growth, a dense fiber structure is produced	[31]
	• The cost is significant since it necessitates special conditions: High pressure, much solvents, and long time to swap solvents	• Offers the benefits of cost effectiveness, ease of use, and environmental friendliness	[36]
Equipment	• Using the autoclave as the main piece of apparatus. The high working pressure and significant operating and maintenance costs make the autoclave system extremely complex	• The freeze drying adopts conventional atmospheric pressure equipment; thus, the equipment is relatively simple and low-price	[37]
Energy consumption	• Supercritical drying requires high-pressure specialized equipment; it uses more power than freeze drying	• Consumes lower power than supercritical drying	[38]
Mechanical properties	• The produced polysaccharide-based aerogels have a low compressive strength	• High compressive strength of the fabricated aerogels	[8, 38]

rearranged throughout the regeneration and solvent exchange processes. Ahsan Zaman et al. [45] constructed regenerated cellulose aerogels using cotton, filter paper, bamboo fiber, and wheat straw as raw materials. The regenerated cellulose aerogel is inexpensive, easy to prepare, and has promising potential for use in soil and water remediation.

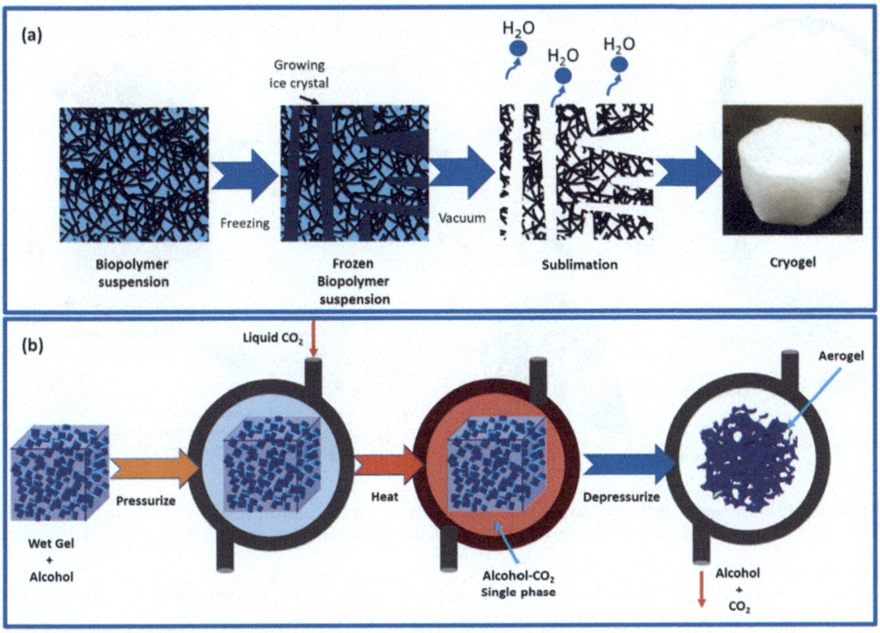

Fig. 3 **a** Principles of freeze-drying process. **b** A two-component CO_2-alcohol process is used to illustrate the supercritical drying process [30]

3. Cellulose and nanocellulose derived aerogels are the two primary categories of cellulose-derived aerogels. Certain cellulose derivatives' solubility in water and traditional organic solvents is used by cellulose derivative aerogels [30, 46].

3.2 *Chitosan-Based Aerogel*

After cellulose, the second most prevalent polysaccharide resource on Earth is chitosan.

Chitosan is mostly generated from the exoskeletons of insects, fungi, and crustaceans. It is a long-chain, insoluble carbohydrate polymer made up of repeated n-acetylglucosamine residues [47–49] Chitosan has excellent potential for use in wastewater treatment because of its special alkaline and cationic qualities, This enable it to activate both the sites of anionic organic pollutants and heavy metal ions [50]. Moreover, chitosan has numerous chemically active groups that are amenable to chemical modification (crosslinking, grafting, etc.) or physically (powdering, nanoparticles, gel, etc.) to enhance its mechanical qualities, adsorption capabilities, thermal stability, and other attributes [49, 51].

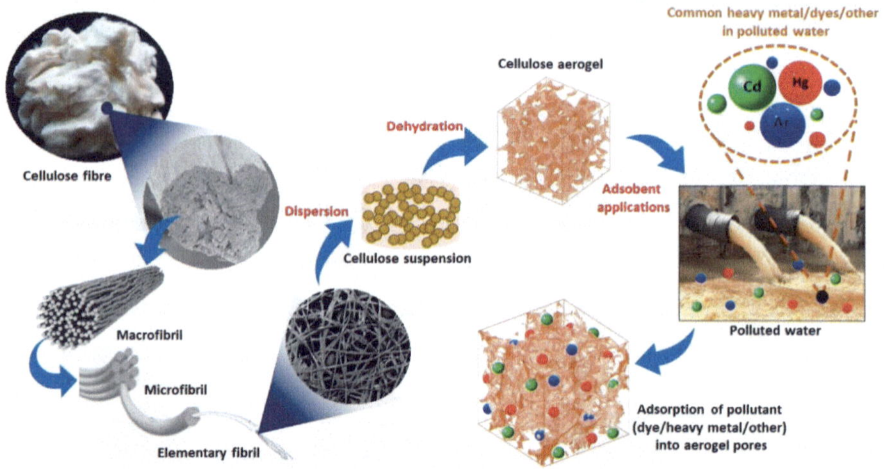

Fig. 4 Cellulose-based aerogels adsorbent in wastewater remediation [30]

A chitosan aerogel a nanonetwork can be generated by chemically or physically crosslinking the chitosan function groups (amino and hydroxyl groups). This sustainable aerogel is biodegradable and biocompatible, has a high degree of porosity, and a large surface area. They are great sorbents for extracting organic solvents and oils from water, as well as ionic contaminants through complexation, chelation, and electrostatic interaction. The chitosan amino and hydroxyl groups help ionic pollutants be adsorbed [52]. This has led to a variety of applications for chitosan aerogel, including oily wastewater treatments, medical applications, and catalysis [53–56].

3.3 Sodium Alginate Aerogels

Aerogels from sodium alginate have garnered a lot of interest because of their special qualities and practicality. They have been used in contemporary scientific domains like material chemistry, environmental research, and electric energy storage. Sodium alginate aerogel's porous structure and superior physicochemical qualities offer a wide range of possible applications [57, 58]. It offers a good alternative for environmental cleanup and has a remarkable elimination effect on organic contaminants and heavy metal ions in wastewater. The sodium alginate aerogel is prepared by a distinctive drying process after changing the pH or adding a metal cationic crosslinking agent to create a gel [59, 60]. Among other qualities, sodium alginate composite aerogel has excellent resistance to compression, salt, and underwater superoleophobicity [61]. It is a highly effective adsorbent with several benefits and prospective uses in the practical separation of greasy wastewater in demanding industrial settings [62].

3.4 Starch Based Aerogels

In higher plants, starch serves as the primary storage carbohydrate. Blennow et al. [63]. Starch comprises two biopolymers: amylose, which is generally linear and composed of 1,4-α-D-glucose units and amylopectin, which is composed of 1,4-α-D-glucose units with a lot of branches constituted of α-1,6-glycosidic linkages. When starch chains reach water, their branching bulk makes it easy to swell into a three-dimensional hydrogel [64–67]. Due to the heterogeneous structure's inability to withstand the pressure during the freezing process, pure amylose was unable to produce aerogels [68]. Amylose can, however, be added appropriately to the aerogel to improve its specific surface area while decreasing its density. For instance, the mechanical characteristics of aerogels may be influenced by the amount of amylose present in the starch [65]. Even with starch aerogel's many benefits, there are still several disadvantages, like its low hardness. However, starting materials can be changed during the aerogel preparation process to improve the mechanical properties. Trisodium citrate cross-linked starch aerogels were created by Negar Abhari et al. [69] in order to increase hardness and decrease adhesiveness. Three processes are used to prepare low density mesoporous starch-derived aerogels with high surface area: hydrogel production, solvent exchange with ethanol, and supercritical CO_2. The source, concentration, level of gelatinization, and retrogradation of the starch all have a significant impact on the properties of aerogel [70]. Applications for starch-derived aerogels include tissue engineering [21], thermal insulators [71], carriers for active chemicals [72], templates for creating new materials [73, 74], and adsorbents [75].

3.5 Carrageenan Based Aerogels

Carrageenan (CR) is composed of two recurring and alternating components: 3-bind β-d-galactopyranoses and 4-bind α-d-galactopyranoses. Carrageenan can be classified into three primary types based on the degree and location of sulfation: κ-carrageenan, ι-carrageenan, and λ-carrageenan [76]. These three types of carrageenan differ from one another due to the binding ions that form helical structures [77]. CR has some unique qualities that demonstrate its applicability as additives in a variety of industries. They are useful in a variety of applications due to their many characteristics, they are easy soluble, stable in water, biocompatible, non-toxic, have gelling ability, binding capability and high viscosity [78]. CR is classified as a superabsorbent hydrogel and can be used in a variety of industries, including food, medicine, cosmetics, hygienics, and the adsorption of metal ions, colors, and pollutants [79]. Three procedures were used to create the κ-carrageenan aerogel matrixes. The initial stage involved dissolving the polymer in ionic liquids, water, or potassium salts, which were used as crosslinking agents. Hydrogels and ionogels were produced at the conclusion of this stage. In the second step, a solvent that was highly soluble in CO_2, like ethanol, was used to replace the water or IL that was present in the gel structure's pores. In order to get the final porous materials, the third stage involved extracting the ethanol using $scCO_2$ assisted drying [39].

4 Application of Polysaccharide-Based Aerogels in Water Treatment

4.1 Removal of Dyes

Dyes can be eliminated from wastewater using a variety of methods, including chemical processes [41, 80–82] like Advanced Oxidation Processes (AOP) and photocatalysis [83], biological processes and physical processes including membrane separation [84, 85], coagulation, flocculation, and adsorption [86, 87]. The application of various biological and chemical methods for dye removal is constrained by low productivity, slow kinetics, limited scalability, and the risk of generating hazardous intermediates. The most often used method for removing dyes is adsorption since it is non-toxic and makes it simple to recover the solute from the sorbent [88].

An effective bio-adsorbent aerogel was fabricated from nanocellulose (NC). Lignocellulose nanofibers (LCNFs) were extracted from barley straw and exposed to freeze-drying to obtain potent and environmentaly green aerogels. The aerogel was used in wastewater treatment for releasing of different pollutants. Methylene blue (MB) dye and other organic dye contaminants may be efficiently removed from wastewater by the (L)CNF-made aerogels [46]. They are distinguished from cellulose aerogels by having a higher crystallinity degree, a bigger shape ratio, a lower rate of shrinkage, and a higher Young's modulus. MB's removal capacities by the aerogels were higher than 90% (at 25 °C for 129 min), ascribed to internal porous structure and the attraction force between NC and the MB molecules. Additionally, the equilibrium data supported the Langmuir isothermal model, and the maximum hypothetical adsorption capacity for MB was determined to be 4.88 mg/g [89].

A bio-based aerogel produced through the freeze-drying process of TEMPO-oxidized nanofibers of cellulose, starch, acrylamide and calcium carbonate made nanoparticles ($CaCO_3$@STA/PAM/TOCN). With its regular, non-collapsing, three-dimensional structure and high pore size, the produced aerogel has a superior ability to adsorb macromolecules and significantly increases its water absorption and retention. The organic dyes Congo red (CR) and MB are primarily adsorbed by the $CaCO_3$@STA/PAM/TOCN aerogel via hydrogen bonding and electrostatic interactions. The Langmuir adsorption isotherm and the pseudo-second-order kinetic model were fitted the adsorption data. After 6–8 h, both dyes were equivalent at a pH of 7–9 at 25 °C, and the MB and CR's adsorption capabilities peaked at neutral or weakly basic. As the pH increased from 3 to 7, the adsorption capabilities similarly increased. According to the Langmuir curve the adsorption capacities for CR and MB at equilibrium were 277.76 and 101.01 mg/g, respectively. After four cycles, the aerogel's adsorption effectiveness ranges from 60 to 80% [90].

The freezing-drying method was used to generate the Metal Organic Framework-5/Cellulose (MOF-5/Cellulose) composite aerogel, which was used to absorb the acid blue (anionic) dye. 76.58% of the dye was eliminated at neutral pH = 7.8 and at 45 °C for 180 min while 96% of the dye was removed in less than 60 min at room temperature in an acidic environment (pH 3). Additionally, the adsorbent material demonstrated reusability, offering a sustainable dye releasing method. The adsorbent MOF-5 cellulose aerogel showed reliable performance across a number of consecutive cycles. During the absorption process, the produced MOF-5/cellulose aerogel composite followed pseudo-second-order kinetics and displayed behavior of Langmuir isotherm [91].

MOF-199@CNCA/CMCS hybrid aerogel spheres were made by Zhangu et al. [92] to extract MB dye from water. Ionic cross-linking and coordination bonding between MOF-199 and carboxylated cellulose nanocrystals (CNCA) and carboxymethyl chitosan (CMCS) hydrogel were used to create this aerogel (Fig. 5). Following a freeze-drying process, the aerogel spheres were produced. The maximum adsorption capacity of any aerogel yet reported against MB was demonstrated by these spheres, with a value of 1112.2 mg/g. This was ascribed to multilayer adsorption via physisorption (increased porosity on the surface of microspheres) and chemisorption (p-p interaction and electrostatic contact between the MOF-199 and MB molecules) [92].

Methyl orange (MO) was removed using chitosan (CS) and chitosan–graphene oxide (CS–GO) aerogels adsorbent made by supercritical gel drying, which was carried out at 200 bar/35 °C under driving force 100 ppm and adsorbent dose 8 g/L. It was noted that the two adsorbents' performance differences were only noticeable at high concentrations: for CS–GO and CS, the maximum equilibrium capacity was 59 mg/g and 28 mg/g, respectively, when using a 100 ppm MO solution. With a CS–GO concentration of 8 g/L, the adsorption efficiencies rise from 24 to 85%. The system behavior is reliably described by the Freundlich isotherm. The multi-linear intraparticle diffusion kinetic model demonstrates that many mass transfer mechanisms govern the diffusion of molecules via nanostructured porous devices [93].

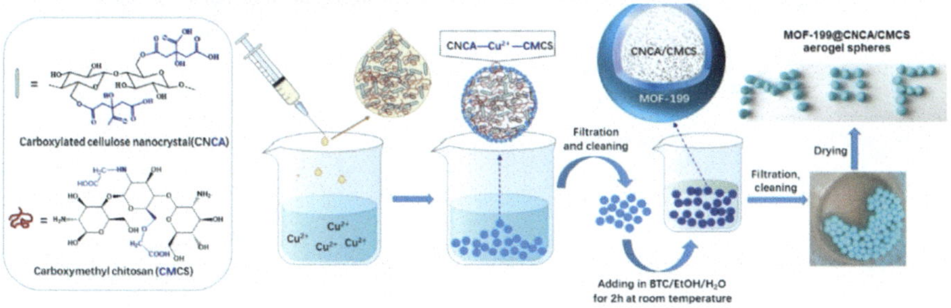

Fig. 5 Schematic representation of the preparation process of MOF-199@CNCA/CMCS aerogel sphere [92]

Reusable, highly compressible, and regenerated cellulose/chitosan composite aerogel (RC/CSCA) with a stable porous structure was made utilizing the freeze-drying approach, with glutaraldehyde (GA) as the crosslinking agent and N-methylmorpholine-N-oxide (NMMO) as the cellulose solvent. The GA is employed to stop shrinkage and maintain RC/CSCA's capacity for recovery. The RC/CSCA aerogel can effectively and selectively remove hazardous anionic dyes from wastewater due to its remarkable adsorption capacity and recyclability (Fig. 6). RC/CSCA maximum adsorption capacity for methyl orange (MO) was 742.68 mg/g with efficiency of 95.63%, respectively [94].

Zeolite 13X powder (ZX) was added to a chitosan (CS) aerogel (1:1 by weight) followed by freezing-drying process to create CS–ZX composite microporous aerogels. The CS–ZX composite adsorped both cationic (MB) and anionic (IC) dyes, with supper absorption efficiency equal 221 mg/g and 108 mg/g, respectively. Furthermore, equilibrium conditions are reached in less than two hours, indicating that the adsorption kinetics are fairly quick. Both the Freundlich and Langmuir models can adequately describe the adsorption behavior of all systems. A 1 M NaCl aqueous solution retains more than 60% of the MB removal capacity across three adsorption–desorption cycles, while a 0.1 M NaOH aqueous solution restores the IC removal ability [95].

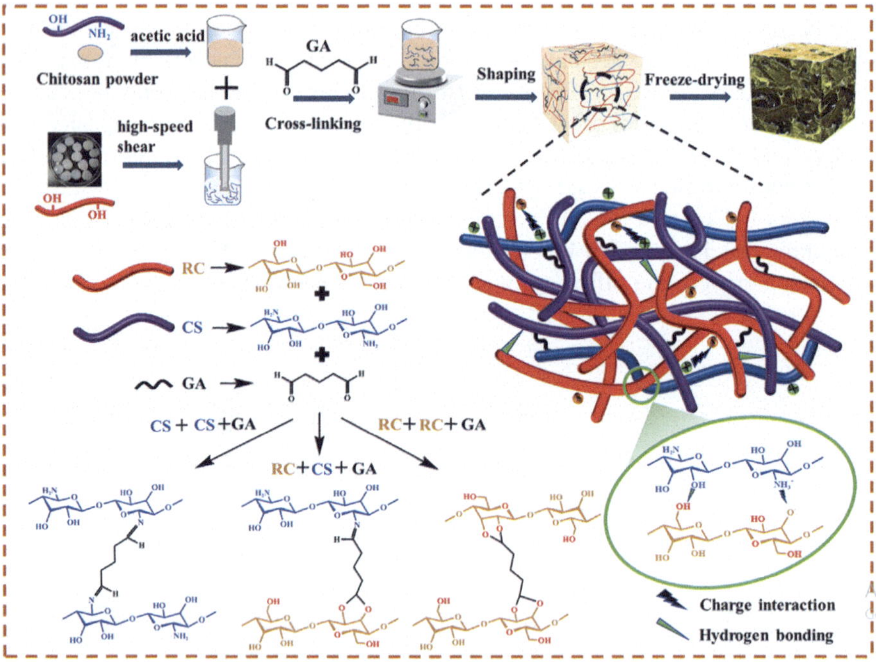

Fig. 6 Schematic representation of preparation and crosslinking of RC/CSCA [94]

The chitosan-carbon aerogel was made using an easy and sustainable method. The composite with increased surface area and porous structure was obtained by freeze-drying technique. The composite's active functional groups are responsible for its exceptional adsorption kinetics and significant potential for eliminating methylene Blue (MB) dye. The adsorption capacity of the chitosan-carbon aerogel increased from 38 to 86.8 mg/g when the pH was raised from 2 to 7. The maximal MB adsorption capacity increased from 22 to 632 mg/g as the initial MB concentration increased from 25 to 2000 mg/L. The chitosan-carbon aerogel had a saturation adsorption efficiency of 668 mg/g. The removal efficiency decreased from 89 to 73.1% after five reuse cycles [96].

Porous starch composite aerogel (PSA) was obtained from potato starch and modified Ca-montmorillonite (OMMT). The freeze-drying procedure was used to create the porous starch aerogel. Composites with varying starch-to-organoclay ratios were employed for removal of MO dye from wastewater. The amount of organoclay in the PSA matrix improved the removal of methyl orange dyes; at 45 °C, PSA's maximal adsorption capacity was approximately 344.7 mg/g. According to the fitting result, chemisorption was the rate-controlling step in MO adsorption, which is based on a monolayer adsorption model. The negatively charged MO dye and the positively charged adsorbent interacted electrostatically to facilitate adsorption. Furthermore, during five cycles of reusability, the removal rate maintained a value of 84.7%, demonstrating exceptional recyclability performance [97].

A potential and sustainable absorbent aerogel were prepared from starch impregnated with different concentrations of silver-graphene oxide (AgGO) nanoparticles to remove organic pollutants as methylene blue dye. The porous aerogels generated via Freezing dry method; the adsorption process revealed that starch aerogels can adsorb up to 78% of methylene blue. The aerogel's adsorption capacity was further increased by up to 87% with the addition of graphene oxide and silver nanoparticles. Furthermore, the composite material has outstanding stability and is effectively reusable [98].

According to Yun Chen et al., dialdehyde starch nanocrystals (DASNCs) and graphene oxide (GO) nanosheets mix to create extremely porous, compressible, and mechanically resilient aerogels that can be employed as effective adsorbents and supercapacitor electrodes. High mechanical characteristics resulted from the aerogel skeleton being more robust after DASNCs connected to the reduced graphene (RGO). According to the results, the highly porous structure and high specific surface of DASNCs-RGO hybrid aerogels (SRGO) allowed them to have excellent adsorption capacities toward rhodamine B (RB) and crystal violet (CV). The highest adsorption capacities for RB and CV were 539 mg g^{-1} and 318 mg g^{-1}, respectively [99].

Recently, allyl sulfonic acid sodium salt (AS) and acrylic acid (AA) were grafted onto carboxymethyl starch sodium (CMS) chains using UV-induced radical polymerization to form a CAA/SAC starch aerogel. This produced a starch-based aerogel was low in density, high in porosity, and had better swelling properties. The aerogel was sulfonated with strong sulfuric acid, which added $-SO_3H$ groups (Fig. 7). Cationic dyes can be adsorbed selectively thanks to these groups and the negative charges on CMS, AA, and AS. The methylene blue dye has a significantly lower adsorption

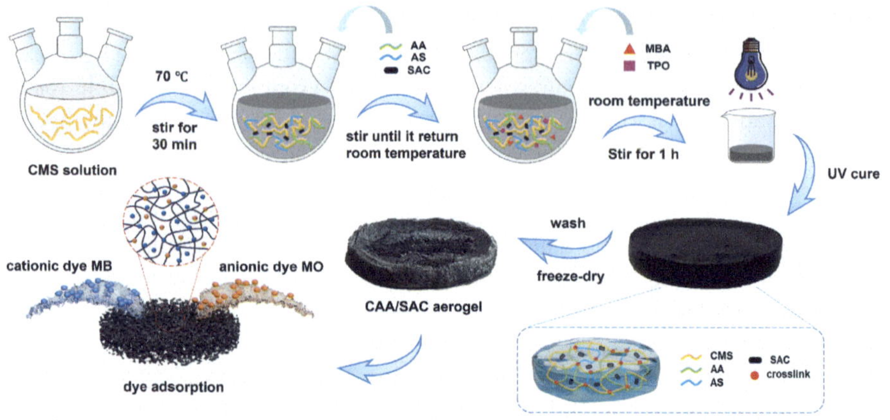

Fig. 7 Preparation process and mechanism CAA/SAC starch-based aerogel [100]

capacity (136.43 mg/g) on CAA/SAC aerogel than methyl orange (69.31 mg/g). The enhanced electrostatic interaction between the dye molecules and the adsorbent function groups allowed for the achievement of the maximal adsorption capacity after five hours of adsorption at 35 °C. The pseudo-second-order kinetics and Langmuir isotherm were followed during the adsorption process [100].

Chetan et al. reported that a simple chemical crosslinking was used to prepare soy protein isolate/sodium alginate/polyethyleneimine aerogel (ASP) for the adsorption of (MB) dye from fluids. The freeze-drying approach was used to create the porous aerogel. MB binds to the ASP aerogel through π–π interactions, electrostatic forces, and hydrogen bonds. The maximum adsorption capacity of the dye by the aerogels was (106.3 mg/g) achieved at adsorbent concentration of 1.5 (mg/L), pH (5.5), and temperature of 40 °C. The adsorption kinetics and isotherms, which exhibit a substantial adaptation with the pseudo-second order and Langmuir models, point to a monolayer chemisorption process including the function groups of ASP aerogel and MB. Additionally, the aerogel demonstrated remarkable reusability after the fourth regenerative cycle, retaining more than 80% of its initial adsorption efficiency [101].

Hexavalent chromium (Cr(VI), Congo red (CR) and Methylene blue (MB) dyes were removed from wastewater using new adsorbent aerogels made from alginate, polyethyleneimine, polydopamine, and nanocellulose fiber as a reinforcing agent. For CR, Cr(VI), and MB, the aerogel's adsorption capabilities were 1078.06, 2827.38, and 596.96 mg/g, respectively. Furthermore, it was determined that the Langmuir model offers the most realistic depiction of the adsorption isotherm, while pseudo-second-order kinetics offers the most accurate means of understanding the adsorption kinetic process. Furthermore, even after five adsorption-desorption cycles, SPCP aerogel maintained its remarkable adsorption capacity [102].

Sodium alginate/graphene oxide (SA–GO) aerogel was prepared by gelation followed by supercritical CO_2 drying method and used for tartrazine elimination from water at room temperature. The SA and its SA–GO aerogel showed a uniform pores dispertion, with compressive strengths up to 25%, volumetric shrinkages, and a Polydispersity Index (PDI) less than 10%. SA–GO aerogel's showed maximal tartrazine adsorption capacity of 420.36 mg/g three times greater than that of pure SA aerogel 129.64 mg/g. The tartrazine removal efficiency was characterized by the adsorption isotherms, which also showed that the Langmuir model fit the absorption process and that it was controlled by the mass transfer mechanism. In comparison to SA aerogels, SA–GO beads removed up to four times as much tartrazine from the aerogels [103].

Hydrophobic aerogel synthesized from sodium alginate, sodium carboxymethyl cellulose (SA/CMC), methyl trimethoxysilane and silica nanoparticles as hydrophobic modifiers. Following a variety of modifications, a three-dimensional aerogel with a low density, high porous structure, and a high adsorption capability was produced. This aerogel was then utilized to adsorb organic liquids that contained chloroform and methylene blue, among other substances. The findings showed that the superhydrophobic SA/CMC aerogel has good adsorption qualities, suitable compressive strength, and outstanding water stability. The equilibrium adsorption capacity and dye removal percentage of the SA/CMC aerogel are (7.22 mg/g) and (78.53%), respectively. The primary contributing factor is the electrostatic attraction between MB and aerogel. A pseudo-second order model fits the adsorption mechanism. Additionally, following washing, it is expected that this aerogel will be recycled [104].

4.2 *Removal of Metal Ions*

Heavy metals can be extracted from water using a variety of techniques, The most widely used is the adsorption method [105].

Chitin and chitosan macromolecules served as the basis for the construction of aerogel with high strength and a 3 macro-meso-microporous structure (HPS-aerogel) (Fig. 8). The macro-meso-microporous structure gave HPS-aerogel its exceptional compression strength (1472 kPa at 80% strain), high porosity, and distinguished mechanical properties. It exhibited potential removal capacity for Cu(II) and Congo red (CR) from water. Cu(II) and CR had adsorption capacities of 59.21 mg/g and 2074 mg/g at 303 K, respectively, and the adsorption process fit pseudo-second-order and Langmuir models with voluntary and endothermic nature. Crucially, HPS-aerogel illustrated excellent reusability and stability together with an efficient adsorption capability. We recognized chelation, electrostatic attraction, the schiff base and hydrogen bonding as the primary adsorption mechanisms for CR and Cu(II), respectively [106].

Taesoon Jang and colleagues utilized the freeze-drying method to produce aerogel beads using κ-carrageenan (CG), polyvinyl alcohol (PVA), and sodium alginate (SA). Kinetic and isotherm studies were used to evaluate the SA/PVA/CG aerogel beads' adsorption capacity and mechanism for the adsorption of heavy metals Cu(II), Pb(II), (MB), and acid blue 25 (AB). The adsorption Kinetic and isotherm data indicated that

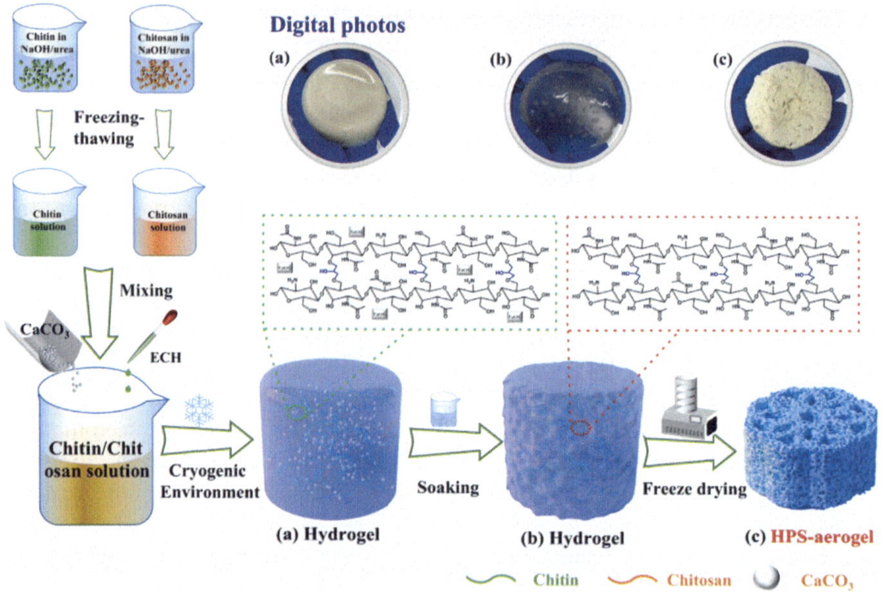

Fig. 8 Digital photos and preparation method of HPS-aerogel [106]

chemical adsorption was the primary rate-governing step. According to the Langmuir model, the aerogel beads showed higher adsorption rates towards the metal ions than the dye, of 85.17, 265.98, and 1324.30 mg/g, respectively.

[107].

Intelligent aerogels made by acrylic acid grafted cellulose nanocrystal (AA-g-CNC) for adsorption of Pb(II), Cd(II), and Cu(II) adsorbents. The resulting hydrogels were freeze-dried in liquid nitrogen. The AA-g-CNC aerogels that were produced had a high zeta potential, several functional groups, and a highly porous honeycomb structure—all characteristics that are necessary for an efficient adsorbent. The maximal removal capabilities of AA-g-CNC aerogels for Pb(II), Cd(II), and Cu(II) were 1026.0, 898.28, and 872.14 mg/g, respectively. Their adsorption fit well with pseudo-second-order and Freundlich isotherm models. The adsorption mechanism depends on the the electrostatic interaction of the metal ions with the carboxylate and sulfonate groups. Four adsorption-desorption cycles resulted in a modest decrease in the sorption capabilities of Pb^{2+}, Cu(II), and Cd(II) [108].

Cu(II) ions were selectively removed from wastewater by cellulose aerogel made from coupling of Brushite and polyethyleneimine (PEI)-modified nanofibrillated cellulose (NFC).

According to the adsorption kinetic and isotherm studies, the removal of Cu(II) was a monolayer chemisorption process, which was in line with the pseudo-second-order model and Langmuir model. A remarkable adsorption capacity of 332.66 mg/g for Cu(II) was demonstrated by this compressible aerogel. Furthermore, the proposed aerogel might be easily employed as a continuous flow filter for the successive purification of industrial wastewater. After being eluted by 0.1 M EDTA-2Na, the Cu(II)-loaded aerogel was regenerable, and even after eight consecutive cycles, its shape ability, structure, and selective adsorption qualities were barely impacted [109].

TEMPO-oxidized cellulose nanofibril (TO-CNF) crosslinked Trimethylolpropane-tris-(2-methyl-1-aziridine) propionate (TMPTAP) and polyethyleneimine (PEI) were gathered to create a functionalized nanocellulose aerogel adsorbent for an efficient and reversible recycle trapping of Cu(II). The resultant TOCNF/TMPTAP/PEI aerogel showed a 3D multi-wall perforated cellular structure with a large number of amino and oxygen-containing groups. The TO-CNF/TMPTAP/PEI aerogel's special design allowed for a maximum Cu(II) ion adsorption capacity of 485.44 mg/g, as indicated by the Langmuir isotherm model. Additionally, the TO-CNF/TMPTAP/PEI aerogels can be utilized for more than four cycles without seeing a reduction in their adsorption capabilities. These composite aerogels are ideal materials for the adsorption of Cu(II) ions in empirical wastewater treatment applications because of their exceptional recyclability and reversibility, unique chemical and structural constancy, and simplicity of separation [110].

Chitosan CO_2-responsive aerogel adsorbent was created by physicochemically crosslinking chitosan and CO_2-responsive poly(acrylic acid-2-(dimethylamino)ethyl methacrylate) for Cu(II) adsorption (CPA). The adsorption capacity of the CO_2-responsive chitosan aerogel was superior to those of the chitosan aerogel. Most significantly, the desorption rate of metal ions was greater than 80%, and the Cu(II) could be desorbed by CO_2 bubbling following its adsorption by CO_2-responsive chitosan aerogel. Aerogel's ability to adsorb Cu(II) was ascribed to complexation and chelation. Repulsion interaction of the protonated $-NH_2$ and –N– groups and Cu(II) ions was the primary mechanism of Cu(II) desorption from the porous aerogels. The adsorption capacity of CPA for metal ions maintained 70% of its initial adsorption capacity after six cycles, and the desorption rate increased to 75% [111].

Biobased aerogels (FBAs) were created using inexpensive cellulose fibers (CF), chitosan (CS), and citric acid (CA) as easy and developable method. In addition to the hydrogen bonds and attraction forces that formed between CF and CS functional groups, the CA crosslinker gave the FBAs exceptional mechanical properties. The materials' surface was more diverse with the addition of CS and CA, which led to extremely height dye and heavy metal adsorption levels (619 mg/g) for copper and (206 mg/g) for methylene blue, respectively. FBAs can be further modified in a simple manner by employing methyltrimethoxysilane to provide aerogel its oleophilic and hydrophobic qualities. With an efficiency of above 96%, the produced FBAs demonstrated quick performance in the separation of water and oil/organic solvents. Furthermore, the performance of the FBA sorbents would not be significantly affected by regeneration and reuse for several cycles [112].

An aerogel constituted from chitosan (CS) and cellulose nano fibers (NFC) with a directed microchannel shape was created via a directive freeze-drying approach and examined for its potential adsorption of Pb (II). As a result of its highly directed microchannel structure, the produced aerogel requires just 5 minutes to obtain 85% of equilibrium adsorption efficiency of Pb (II). The maximal adsorption ability for Pb (II) as determined by the Langmuir isotherm curve is 252.6 mg/g. Moreover, after five adsorption-desorption cycles, the aerogel retains a high removal effectiveness (> 85%) [113].

For effective simple removal of Cr(VI) and Cr(III), flexible nanofibrillated cellulose (NFC)/chitosan (CS) aerogel was prepared by freezing-dry and thermal cross-linking method. After six hours of shaking at 25 °C, the adsorption equilibrium was reached. Using the Langmuir model, the maximal adsorption abilities of Cr(VI) and total Cr were measured to be 197.33 and 134.12 mg/g, respectively. After five adsorption-desorption cycles, the aerogel was easily regenerated and exhibited a high regeneration efficiency of 80.25%. Significantly, 752 mL of wastewater with a Cr(VI) concentration of 49.71 mg/g was purified using a constructed (NFC/CS) aerogel filter coupled with a peristaltic pump. The primary forces behind the adsorption process were complexation, reduction, and electrostatic interactions [114].

Sulfhydryl-modified chitosan was transformed into an aerogel by freeze-drying method that is capable of efficiently adsorbing organic dyes and heavy metals. At 25 °C, the maximal adsorption magnitudes for methyl orange (MO), methylene blue (MB), rhodamine B (RhB), and Sudan I (SI) were 57.65, 51.62, 58.65, and 48.37 mg L^{-1}, respectively, while the adsorption capacities of Cu(II), Pb(II), and Cd(II) ions were 81.15, 38.87, and 38.15 mg/g, respectively. The findings demonstrated that the ideal adsorption pH value for Pb(II) was 5.0 and for Cu(II) and Cd(II) was 6.0. Furthermore, the removal rate of organic dyes can reach 99% and the adsorption equilibrium was achieved after 150, 210, and 210 min for MO, MB, and RhB, respectively. The adsorption kinetic data of metal was fitted with The PFO reaction and PSO models [115].

Heavy metals such as Pb(II), Co(II), Cu(II), Cr(II), and Cd(II) ions in industrial wastewater can be removed using a cost-effective and environmentally friendly hybrid aerogel from sodium alginate-meso-2,3-dimercaptosuccinic acid (SA-DMSA), which was generated using an effortless chemical grafting and vacuum freeze-drying process. The aerogel SA-DMSA was able to extract more than 90% of the metal ions from the related solution with powerful and high selectivity to Pb(II) and Cd(II). The adsorption capacities for Pb(II) and Cd(II) were 116.4 and 91.2 mg/g, respectively. A simple one-step acid wash can regenerate the SA-DMSA adsorbent, and even after repeated use, the adsorption performance remained consistent easy [62].

To remove heavy metal ions like Cd(II), Pb(II), Cu(II), Cr(II), and Co(II) from wastewater, a hybrid aerogel made of calcium alginate and disodium ethylenediaminetetraacetate dihydrate (Alg-EDTA) was created via grafting and freeze-drying techniques. The Alg-EDTA adsorbent exhibits great selectivity for Cd(II) and a high affinity for heavy metal adsorption over 85%. An easy acid-washing procedure was used to regenerate the adsorbent, and even after repeated use, the maximal adsorption capacity for Cd(II) was 177.3 mg/g [116].

In order to create corn starch aerogels for the possible adsorption of Cd (II) and Zn (II) ions from aqueous systems, various concentrations of the crosslinking agent (trisodium citrate) (1,1, 1,1.5, and 1,2) were used. Because of the potential balance during this period and the large specific surface area accessible for Cd (II) adsorption, an effective adsorption capacity was reported after 12 h for the starting concentration of 100 ppm. Furthermore, the starch aerogel's hydroxyl groups had a 30% adsorption efficiency in removing Cd(II) ions. However, the Zn(II) ion was not absorbed by starch aerogels [117].

Alginate/starch aerogel was created as adsorbents for Cu(II) ions from aqueous media. The hydrogel was made using the interior and exterior gelation methods and dried with CO_2 supercritical drying to create aerogels, and with freeze-drying to create cryogels. Regardless of the gelation technique used, pregel mixes containing 3% w/w starch and only 25% w/w alginate produced intriguing materials with meso- and macro-pore structure. The pseudo-second order kinetic model for aerogels and cryogels was well-fitted to the adsorption data regarding Cu(II) removal, demonstrating that aerogels had the maximum adsorption capacity (40 mg/g) and removal efficiency (about 92%). The materials showed five adsorption/desorption cycles of exceptional reusability [118].

Du et al. created a shapeable sodium alginate aerogel for separating of Cd, Cu, and Pb ions from water, that included modified L-cysteine/UiO-67 [51]. These materials were able to quickly approach the equilibrium of adsorption and remove large amounts of metal ions 661.2, 296.2, and 296.2 mg/g for Pb(II), Cd(II), and Cu(II), respectively. The enhanced ion adsorption ability is attributed to the chelation process and ion exchange between the aerogel function groups and metal ions [119].

Hyperbranched polymer aerogel was formed from crosslinking of Lota-carrageenan with Polyamidoamine as potential adsorption to metal ions. The obtained hydrogels were converted to aerogels by using freeze-drying technique. The maximum adsorption efficiency of the aerogel as fitted from Langmuir model for the metal ions Cr(VI), Mn(II), Co(II), Cu(II), or Cd(II) and blue dye from liquids was 105, 144.9, 133, 158.7, 128, and 140.8 mg/g, respectively. The mechanism of adsorption may be a chemisorption reaction. The adsorption process of the various metal ions and the acid dye on the aerogel was fitted with a pseudo-second-order representation instead of a pseudo-first-order model [78] (Table 2).

4.3 Removal of Oil from Water

For separation of oil and water, aerogels are appropriate materials, they have a higher sorption capacity and large surface area compared to textiles, metal meshes, and sponges, making them more efficient for oil removal [146]. Aerogels are also highly reusable, lightweight, and can be cost-effective, especially when made from waste materials. Although sponges have a great capacity for absorption, they may become less effective after repeated use. Additionally, some sponges can need more chemical treatments, which could pollute the environment [147]. Further, metal meshes are reusable and long-lasting, their absorption capacity may be reduced, and they may be impacted by environmental chemical changes [148]. Despite synthetic fabrics are efficient, hydrophobic, and resistant to tearing, they usually have a lesser absorption capacity than aerogels.

[149].

Table 2 Comparison between polysaccharide, organic and inorganic based aerogels in water remediation

Adsorbent	Pollutant	Adsorption mechanism	Adsorption performance (mg/g)	Structure and porosity	Ref.
Cellulose filament/ bisacrylamide	Cu(II)	Physical interaction	51.3	Large pores size with uniform distribution	[120]
Cellulose nanofibrils (CNFs)/ carboxymethylcellulose	Cu(II)	Chelation, ion exchange and electrostatic interaction	452.49	Highly porous structure	[121]
Sodium alginate-derived carbon aerogel	MB	Physical interaction	70.42	Structure is extremely porous and has a specific surface area of 470 m^2/g	[122]
Nanocellulose	Pb(II)	Chemisorption/ complexation	20, 40 and 115	Combined nano-whiskers; irregular surface, porous and flurry	[123]
Cellulose nanofiber/PVA/ acrylic acid	Cu(II) Pb(II)	Complexation and ionic interactions	30 131.5	Highly porous 3D stable structure	[124]
Cellulose/chitosan	Cu(II) Cd(II) Pb(II)	Physical interaction	181.92 163.85 170.19	Symmetrical 3D-orientate porous structure	[125]
Cellulose nanofiber grafted with siloxane derivative	Cu(II)	Chelation	45.6	3D cellulosic structure with high porosity and low density	[126]
Cellulose (nanocrystals) grafted with acrylic acid	Cu(II) Pb(II) Cd(II)	Chemical adsorption and complexation	872 898.8 1026	Macroporous honeycomb structure with high porosity	[108]
β-cyclodextrin, chitosan, and HMTA	Cr(VI)	Hydrogen bonding and ionic interaction	333.8	Bead-like composite adsorbent with high porosity	[127]
Cellulose/MOF (metal–organic frame work) composites	Pb(II) Cu(II)	Complexation and coordination	81.34–89.47 31.20–39.36	Surface contains plenty of micron-sized macropores frame	[128]

(continued)

Table 2 (continued)

Adsorbent	Pollutant	Adsorption mechanism	Adsorption performance (mg/g)	Structure and porosity	Ref.
Phenyl-bridged silica aerogels	Pb(II)	Chelation	704 99.95%	Network that resembles a homogeneous cage and has a high strength, low bulk density, and high specific surface area	[129]
Chitosan grafted with poly(acrylic acid-2-(dimethylamino)ethyl methacrylate)	Cu(II)	Complexation and chelation	660	Low density, irregular, and porous structure	[111]
Alginate/chitosan/melamine	Pb(II)	Complex formation by aromatic ring nitrogen and amino groups	1331.6	Heterogeneous pores in three D network structures	[130]
Chitosan/cellulose	Cr(VI)	Protonated amine groups and Cr(VI) oxyanions interact electrostatically	210.6–211.4	Extremely light (20–30 mg/cm3) and extremely porous (more than 97.5%)	[131]
Modified chitosan/polydopamine and glutaraldehyde	Cr(VI) Pb(II)	Chelation and electrostatic interaction	374.4 441.2	Large holes and an ideal three-dimensional network structure	[132]
Chitosan/citrus flake and bentonite	Cu(II)	Chelation and charge interaction	861.58	A three-dimensional porous structure that is more compact, dense, and ductile.	[133]
Chitosan/thiourea	Pb(II), Ag(I)	Interaction between ions with the –NH and –S groups		High surface area and low density	[134]
Sodium alginate–streptomycin sulfate composite	Pb(II) Cu(II)	Ion exchange and chelation interaction	280 160.2	High surface energy and specific surface area	[135]

(continued)

Table 2 (continued)

Adsorbent	Pollutant	Adsorption mechanism	Adsorption performance (mg/g)	Structure and porosity	Ref.
Sodium alginate aerogel	Cd(II) Cu(II) Pb(II)	Complexation and ion exchange	661.2 296.2 326.4	Metal-organic frameworks (MOFs) possessing advantageous stability, flexible designability, adjustable pore size, and high specific surface area	[119]
3D composite aerogel of graphene oxide/ι-carrageenan	MB	Electrostatic interaction and π–π interaction	245.3	Three-dimensional composite aerogel of graphene oxide and ι-carrageenan of high porosity	[136]
Silica aerogel/ polyacrylonitrile/ polyvinylidene fluoride nanofiber	BR18	Physical adsorption and electrostatic interaction	182	Porous morphology of nanofibers	[137]
Graphene oxide polyvinyl alcohol aerogel	NR, CV, MG, MB, and RhB	Electrostatic and interfacial interaction	900	Compressible, dimension stable and solvent resistive	[138]
Copper ferrite/rGO	RhB	Electrostatic interactions and π-cation interaction	87.8% removal efficiency	3D hierarchical interconnected porous structure	[139]
Chitosan and cellulose nanofibrils aerogel	AB93	Physical adsorption and electrostatic interaction	1428.7	Highly connected porous structure and large specific surface area	[140]
Carboxymethyl cellulose-supported graphene oxide aerogel	RB	Electrostatic interaction	312.5	3D interconnected porous structure	[141]

(continued)

Table 2 (continued)

Adsorbent	Pollutant	Adsorption mechanism	Adsorption performance (mg/g)	Structure and porosity	Ref.
Fe_3O_4@polydopamine/carboxymethyl chitosan aerogel	MB	Electrostatic, hydrogen bonding, hydrophobic and π–π stacking interaction	217.4	Abundant and interconnected pores	[142]
Carboxymethyl starch sodium/sulfonated activated carbon	MB MO	Physical adsorption and electrostatic interaction	136.43 69.31	Structure with a high specific surface area, high porosity, and low density	[100]
Cellulose nanofibrils/(3-mercaptopropyl) trimethoxysilane compressible aerogels	MO	Electrostatic interactions	186.7	Compressible aerogel with porous structure.	[143]
GO/alkali lignin	MB	Electrostatic interaction	1185.98	Three-dimensional porous structure.	[144]
Carbon aerogels	Mb, Phenols	Electrostatic interaction	194.07–1030.05	Interconnected coral-like microstructure of carbon	[145]

Note CR (Congo red), MG (Malachite green), AB93 (Acid Blue 93), CV (Crystal violet); MO (Methyl orange), BR18 (Basic red 18), RB (Rhodamine B)

Recently, polysaccharide-based aerogels have garnered a lot of interest as highly porous materials for separation of oil and water. They can be roughly classified into two groups according to their wettability: superhydrophobic/superoleophobic (those that eliminate water) and superhydrophobic/superlipophilic (those that absorb oil) [150–153]. The main properties of aerogels utilized in oil-water isolation are as follows:

1. **Mechanical characteristics**

In order to attain good mechanical qualities in oil-water separation, aerogel typically refers to two indicative indications of bending and compression resistance. For effective oil-water separation following serial recycling, aerogels need to possess a specific mechanical strength and high flexibility [61].

2. **Surface wettability**

In order to separate the emulsions and produce the desired separation outcomes, aerogel needs to possess superwetting capabilities. Thus, controlling the aerogel's hydrophobic/lipophilic and hydrophilic/oleophobic properties is essential [154].

3. **Thermal stability**

Aerogels must possess strong resistance to thermal degradation. It can extract the oil or organic compounds from the aerogel using the distillation method, but it cannot decompose at a higher temperature during recycling and reuse. The aerogel's structure and performance won't be harmed once the distillation process is finished [155].

4. **Environmental adaptability**

Since oil spills cause marine contamination, effectively and efficiently separating water and oil in the marine environment has gained international attention. After being submerged in an environment, Aerogels based on polysaccharides may disintegrate, swell, and deform, and their initial moisture content may progressively disappear. Consequently, environmental adaptation properties have emerged as critical needs for polysaccharide-based aerogels [156]. Table 3 represents the characteristics of some Polysaccharides based aerogels in oil/ water separation.

5 Polysaccharides-Based Aerogel for Soil Treatment

Soil treatment is the process of controlling the acidity and basicity of soil. It can also be defined as the process of removing pollutants from the soil to protect the environment and public health. It balances the acidity and baseness of the soil while revitalizing and purifying it. Human health risks from direct contact with polluted soil, vapors emitted by contaminants, and secondary contamination of water sources are the primary concerns underlying soil remediation. Therefore, soil treatment methods, organic additions, pH adjusters, and microbes can all be used to treat the polluted soil.

Table 3 Properties of some polysaccharide-based aerogels for oil removal

Polysaccharide	Pollutant	Contact angle (°)	Porosity (%)	Maximum oil adsorption (gg^{-1}), Removal efficiency (%)	Drying technique	Structure and main results	References
Chitosan/cellulose nanofiber	Oil and organic solvent	148	> 98	253 g g^{-1}	Freeze-drying	The composite aerogel has ultralow density high temperature-invariant and anisotropic properties with low thermal conductivity	[157]
Cellulose/sodium carboxymethyl cellulose / polydimethylsiloxane	n-hexane	145.8	96	186.03 g g^{-1}	Freeze-drying	3 D network structure with superhydrophobic properties	[158]
Sodium alginate/ cellulose	Toluene, hexane and chloroform	148	96.47	99%	Freeze-drying	The aerogel has lamellar microstructures and exhibited super-elasticity with the minimal energy dissipation and minimal plastic deformation	[159]
Regenerated cellulose	Diesel oil and ethanol	125	98.05	32 g g^{-1}	Freeze-drying	The aerogel exhibited excellent sorption capacity and hydrophopicity	[160]

(continued)

Table 3 (continued)

Polysaccharide	Pollutant	Contact angle (°)	Porosity (%)	Maximum oil adsorption (gg^{-1}), Removal efficiency (%)	Drying technique	Structure and main results	References
Cellulose nanofiber/ sodium alginate/ calcium carbonate	Diesel oil	135	99.47	52–126 $g\ g^{-1}$	Freeze-drying	The aerogel exhibited microstructure, with hydrophobicity, good elasticity and reusability.	[161]
Sodium alginate/ nanofibrillated cellulose aerogel	Soybean oil, pump oil, toluene and n-hexane	160.2	98.86	99.65%	Freeze-drying	3 D interconnected network structure, high mechanical strength and good flexibility.	[162]
Cellulose aerogel from waste paper	Crude oil, marine diesel oil, and biodiesel	–	96–98	29.67 $g\ g^{-1}$	Freeze-drying	The aerogel samples are resistant to mechanical stresses and are hydrophobic without being brittle	[163]
Cellulose/chitosan composite aerogels	Oils and organic solvents	160	98	> 99%	Freeze-drying	Reinforced 3D interconnected network structure, high mechanical performance and excellent underwater superoleophobicity	[164]

(continued)

Table 3 (continued)

Polysaccharide	Pollutant	Contact angle (°)	Porosity (%)	Maximum oil adsorption (gg^{-1}), Removal efficiency (%)	Drying technique	Structure and main results	References
Chitosan and carboxymethyl cellulose/ citric acid	Olive oil, corn oil, engine oil and organic solvents	137.7	> 96	43.87–80.28, 90.7%	Freeze-drying	The composite aerogel exhibits light weight, low density, high porosity, hydrophilic and water retention properties, and satisfactory underwater oleophobicity	[165]
Chitosan, polyvinyl alcohol graphene oxide/ reduced graphene oxide	Diesel oil	122–132	96	18.36–23.05	Freeze-drying	3-D porous network with interconnected layers, exhibit a honeycomb-like structure with a larger number of pores and an increased roughness	[165]

(continued)

Table 3 (continued)

Polysaccharide	Pollutant	Contact angle (°)	Porosity (%)	Maximum oil adsorption (gg^{-1}), Removal efficiency (%)	Drying technique	Structure and main results	References
Cellulose nanofiber / chitosan aerogel	Diesel oil, pump oil, mineral oil, benzene and n-hexane	147	91.5	47.61, 94%	Freeze-drying	Low density, high porosity, significant hydrophobicity and robust cycling performance. Exhibits better mechanical strength and distinction separation properties for oils and emulsions	[166]
Alginate–halloysite nanocomposite aerogels	Corn oil, pump oil and n-hexane	145.6	99	99.6–98.8%	Freeze-drying	Stable three-dimensional porous microstructure with uniformly dispersed halloysite nanotubes with underwater oleophobicity	[167]

Because petroleum leaks produce heavy oil pollutants with a high molecular weight and high viscosity, they can cause long-term harm to human health as well as significant damage to the soil ecosystem. Recently, high temperature pyrolysis and freeze-drying techniques were used to create biochar (BC) aerogel from plentiful sisal leaves, which was subsequently used to clean up heavy oil-contaminated soil. After 60 days, the 600-aerogel, 700-aerogel, and 800-aerogel treatments' respective degradation efficiencies were 80.69%, 86.04%, and 86.62%. Because of their potential to increase soil fertility and quality, immobilize and convert contaminants (both organic and heavy metals), and store carbon, biochars have garnered a lot of interest [168]. More active functional groups and active adsorption sites for Biochar can be provided by polysaccharide modification, allowing polysaccharide-modified BC to remediate environmental contaminants in a number of ways (Fig. 9).

Furthermore, the inclusion of polysaccharides substantially magnified the functional groups on BC and supplied a large number of functional groups that could be utilized for complexation with metal ions. Distinct polysaccharide modifications showed distinct benefits. For example, the CS-modified BC enhanced the soil's ability to buffer pH, whereas the SA-modified BC greatly aided in the immobilization of heavy metals. Additionally, the composite could improve the soil's ability to buffer against pH variations and drastically lower the amount of accessible HMs in the soil [169, 170]. Because of its special qualities, including high adsorption capacity, large specific surface area, and significant porosity, biochar has drawn a lot of interest as a soil remediation agent in recent years [171–173].

One popular method in soil treatments is to immobilize cadmium in the soil to lower its bioaccessibility. Cadmium release, however, can still result from variations in the soil pH, redox potential, and activity of microbs. $MgFe_2O_4$/chitosan carbon

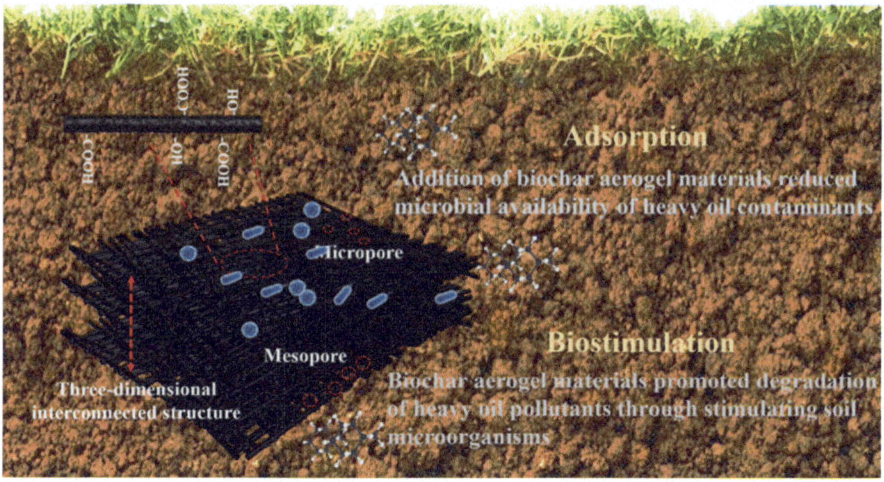

Fig. 9 The mechanism of biochar aerogel materials for the remediation of heavy oil-contaminated soil [168]

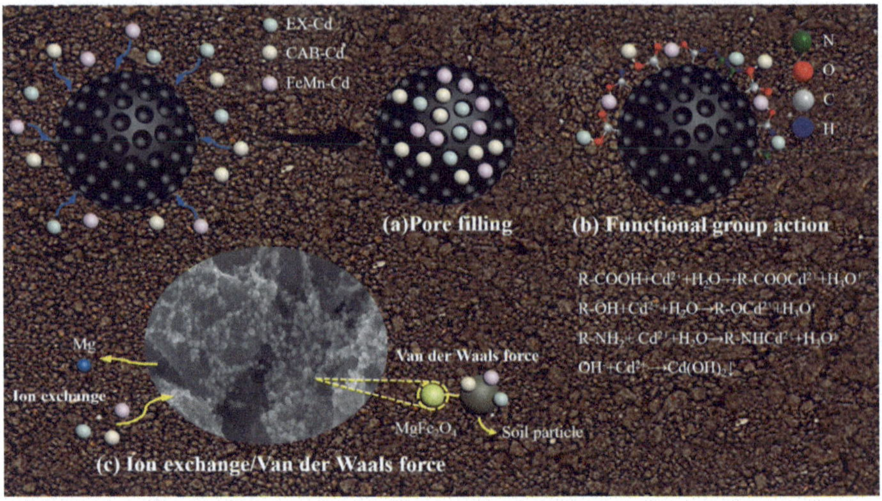

Fig. 10 Possible immobilization mechanisms of MCSCA in soil remediation [174]

aerogel beads (MCSCA) were synthesized for in-situ remediation to eliminate Cd from the soil (Fig. 10). After three remediation cycles, the results showed that 1%, 2%, and 3% MCSCA edits, respectively, eliminated 55.46%, 72.27%, and 75.70% of the total cadmium. This corresponds to the elimination of 76.511%, 87.391%, and 90.282% of available cadmium. Even after three cycles, the regenerated MCSCA retained a greater treatment efficiency, eliminating bioavailable cadmium percent of 62.720%, 75.201%, and 80.965%, using MCSCA with recovery rates of 94.40%, 92.30%, and 91.40% [174].

6 Conclusion, Challenges and Perspectives

Due to their low toxicity, biocompatibility, renewability, and sustainability, polysaccharide-based aerogels have been considered the most sustainable aerogel material. The process entails the extraction of raw materials (polysaccharide), their modification, hydrogel creation, and dehydration (using the freeze-drying and supercritical drying methods). There are many potentials uses for polysaccharide-based aerogels in the realms of soil treatment and water remediation. Future research will focus on improving the implementation of the aerogels made from polysaccharides to satisfy the demands of modern society. Aerogels based on polysaccharides can also use their unique features to replace conventional materials in order to serve humanity as environmental degradation increases. Several suggestions can be made in light of the literature mentioned above, which could offer precise guidance for the advancement of polysaccharide-based aerogels in the future:

1. **Enhancing performance through the utilization of natural agricultural waste as raw materials**

The agricultural waste has a specific physical form, a huge product, a wide origin, and a low price. Getting some of this structure at the micro- or even nano-scale should improve the performance of polysaccharide-based aerogels.

2. **Increase the hydrophobicity**

The hydrophobicity of polysaccharide-based aerogel composites presents certain difficulties because polysaccharides are hydrophilic substances. Researchers must focus on increasing the hydrophobicity of polysaccharide-based aerogels since weather, water, and microorganisms will cause problems that will lead to their destruction and degradation. Some chemical techniques to boost hydrophobicity have been discovered in recent years, but the majority of these techniques will negatively impact the environment. In order to extend the shelf life of aerogels, we must achieve a balance between hydrophilia and hydrophobia. Poor split-up efficiency and low material exploitation brought on by oil pollution blocking the aerogel's surface can be more effectively addressed by the underwater superoleophobic aerogel. Consequently, one of the biggest challenges in the future will be coming up with a better technique for making hydrophobic polysaccharide-based aerogels.

3. **Expansion of the production scale**

Although aerogel preparation is easy in the lab, industrialization is difficult. Operating for large-scale production entails some risk, and setting up large-scale supercritical drying equipment is still difficult. Therefore, in order to achieve mass production in industry and applications in a wide range of fields, it is essential to learn how to build and optimize equipment and adjust operational parameters to reach large capacity and continuous output.

4. **Equipment and methods of aerogel preparation**

Aerogel preparation techniques, tools, procedures, and process conditions could produce polysaccharide aerogels with exceptional properties. The creation of a regenerating adsorbent may provide a more cost-effective, time-efficient, and energy-efficient alternative.

5. **Crosslinking of Polysaccharides during aerogel preparation**

Selecting appropriate chemical substances for crosslinking of polysaccharides is crucial since is not a good environmental option.

6. **Drying process**

Drying is the most important step in aerogel production, and two popular techniques are freeze drying and sc-CO_2. The aerogel's shape, porosity, and structural integrity are all impacted by the drying process. Although the sc-CO_2 drying approach can produce a robust, porous structure, it necessitates extremely high temperatures and pressures, making it a costly, risky, and energy-intensive procedure. In general, these circumstances are not ideal for industrial use and water treatment. Additionally, the network structure is readily disrupted during the lengthy freeze-drying process. Overcoming these limitations requires improving process conditions.

References

1. Bediako JK, El Ouardi Y, Mouele ESM, Mensah B, Repo E. Polyelectrolyte and polyelectrolyte complex-incorporated adsorbents in water and wastewater remediation—a review of recent advances. Chemosphere. 2023;325:138418.
2. Liu X, Sathishkumar K, Zhang H, Saxena KK, Zhang F, Naraginiti S, et al. Frontiers in environmental cleanup: recent advances in remediation of emerging pollutants from soil and water. J Hazard Mater Adv. 2024:100461.
3. Elella MHA, Abdallah HM, Ali EA, Makhado E, Abd El-Ghany NA. Recent developments in conductive polysaccharide adsorbent formulations for environmental remediation: a review. Int J Biol Macromol. 2025:140915.
4. Attia NF, Policicchio A, Zakria AM, Bonaventura CP, Bartucci S, Alkahlawy A, et al. Rational fabrication of ultramicroporous flexible porous carbon fabrics for efficient capture of CO_2 and CO_2/N_2 and CO_2/H_2 separation. Surf Interfaces. 2025;56:105584.
5. Pandey M, Pradhan N, Elella MHA, Makhado E, Samy M, Botlhoko OJ, Allouss D. Frontiers in polymer/carbon dots nanocomposites: recent advances in synthesis and diverse applications. Surf Interfaces. 2025:106101.
6. Stanciu I. Soil treatment technologies through bioremediation. Bioremediation Global Environ Conserv. 2023:25.
7. Romero-Montero A, Valencia-Bermúdez JL, Rosas-Meléndez SA, Núñez-Tapia I, Piña-Barba MC, Leyva-Gómez G, Del Prado-Audelo ML. Biopolymeric fibrous aerogels: the sustainable alternative for water remediation. Polymers. 2023;15(2):262.
8. Muhammad A, Lee D, Shin Y, Park J. Recent progress in polysaccharide aerogels: their synthesis, application, and future outlook. Polymers. 2021;13(8):1347.
9. Mekonnen BT, Ding W, Liu H, Guo S, Pang X, Ding Z, Seid MH. Preparation of aerogel and its application progress in coatings: a mini overview. J Leather Sci Eng. 2021;3:1–16.
10. Fu Y, Guo Z. Natural polysaccharide-based aerogels and their applications in oil–water separations: a review. J Mater Chem A. 2022;10(15):8129–58.
11. Gu X, Ling Y. Research progress of aerogel materials in the field of construction. Alex Eng J. 2024;91:620–31.
12. Makhado E, Seleka WM, Mahlaule-Glory L, Selowa K, Elella MHA, Botlhoko OJ, et al. Comparative extraction and characterisation of cellulose nanostructures from sawdust and maize stalk biomass. Int J Biol Macromol. 2025;147842
13. Zubair NA, Abouzari-Lotf E, Nasef MM, Abdullah EC. Aerogel-based materials for adsorbent applications in material domains. E3S Web Conf; 2019.
14. Ganesamoorthy R, Vadivel VK, Kumar R, Kushwaha OS, Mamane H. Aerogels for water treatment: a review. J Clean Prod. 2021;329:129713.
15. Rai N, Chauhan I. Multifunctional aerogels: a comprehensive review on types, synthesis and applications of aerogels. J Sol-Gel Sci Technol. 2023;105(2):324–36.
16. Wang Y, Su Y, Wang W, Fang Y, Riffat SB, Jiang F. The advances of polysaccharide-based aerogels: preparation and potential application. Carbohydr Polym. 2019;226:115242.
17. Şahin İ, Özbakır Y, Inönü Z, Ulker Z, Erkey C. Kinetics of supercritical drying of gels. Gels. 2017;4(1):3.
18. Maleki H, Durães L, García-González CA, Del Gaudio P, Portugal A, Mahmoudi M. Synthesis and biomedical applications of aerogels: possibilities and challenges. Adv Colloid Interf Sci. 2016;236:1–27.

19. Salgado M, Santos F, Rodríguez-Rojo S, Reis RL, Duarte ARC, Cocero MJ. Development of barley and yeast β-glucan aerogels for drug delivery by supercritical fluids. J CO_2 Util. 2017;22:262–9.
20. Rubina MS, Elmanovich IV, Shulenina AV, Peters GS, Svetogorov RD, Egorov AA, et al. Chitosan aerogel containing silver nanoparticles: from metal-chitosan powder to porous material. Polym Test. 2020;86:106481.
21. Zhu F. Starch based aerogels: production, properties and applications. Trends Food Sci Technol. 2019;89:1–10.
22. Takeshita S, Sadeghpour A, Sivaraman D, Zhao S, Malfait WJ. Solvents, CO_2 and biopolymers: structure formation in chitosan aerogel. Carbohydr Polym. 2020;247:116680.
23. Darpentigny C, Nonglaton G, Bras J, Jean B. Highly absorbent cellulose nanofibrils aerogels prepared by supercritical drying. Carbohydr Polym. 2020;229:115560.
24. Dorcheh AS, Abbasi M. Silica aerogel; synthesis, properties and characterization. J Mater Process Technol. 2008;199(1–3):10–26.
25. Tomic N, Djekic I, Hofland G, Smigic N, Udovicki B, Rajkovic A. Comparison of supercritical CO_2-drying, freeze-drying and frying on sensory properties of beetroot. Foods. 2020;9(9):1201.
26. Dilamian M, Noroozi B. Rice straw agri-waste for water pollutant adsorption: relevant mesoporous super hydrophobic cellulose aerogel. Carbohydr Polym. 2021;251:117016.
27. Xi X-T, Luo X-Q, Xia Y, Yi L-F, Wang Y, Song D-Y, et al. Ice crystal growth mechanism and structure-activity relationships of graphene oxide/poly (vinyl alcohol) aerogels. Chin J Polym Sci. 2022;40(7):772–80.
28. Aegerter MA, Leventis N, Koebel MM. Advances in sol-gel derived materials and technologies. Aerogels Handbook. New York, NY, USA: Springer; 2011.
29. Wei G, Zhang J, Usuelli M, Zhang X, Liu B, Mezzenga R. Biomass vs inorganic and plastic-based aerogels: structural design, functional tailoring, resource-efficient applications and sustainability analysis. Prog Mater Sci. 2022;125:100915.
30. Ahmad A, Kamaruddin MA, HPS AK, Yahya EB, Muhammad S, Rizal S, et al. Recent advances in nanocellulose aerogels for efficient heavy metal and dye removal. Gels. 2023;9(5):416.
31. Ciftci D, Ubeyitogullari A, Huerta RR, Ciftci ON, Flores RA, Saldaña MD. Lupin hull cellulose nanofiber aerogel preparation by supercritical CO_2 and freeze drying. J Supercrit Fluids. 2017;127:137–45.
32. Menshutina N, Tsygankov P, Khudeev I, Lebedev A. Intensification methods of supercritical drying for aerogels production. Dry Technol. 2022;40(7):1278–91.
33. Zhang X, Yu Y, Jiang Z, Wang H. The effect of freezing speed and hydrogel concentration on the microstructure and compressive performance of bamboo-based cellulose aerogel. J Wood Sci. 2015;61:595–601.
34. Hirashima H. Preparation of TiO_2 aerogels-like materials under ambient pressure. Aerogels handbook. 2011:145–53.
35. Jin H, Nishiyama Y, Wada M, Kuga S. Nanofibrillar cellulose aerogels. Colloids Surf A Physicochem Eng Asp. 2004;240(1–3):63–7.
36. García-González CA, Alnaief M, Smirnova I. Polysaccharide-based aerogels—promising biodegradable carriers for drug delivery systems. Carbohydr Polym. 2011;86(4):1425–38.
37. Wang Y, Xiang F, Wang W, Wang W, Su Y, Jiang F, et al. Sound absorption characteristics of KGM-based aerogel. Int J Low-Carbon Technol. 2020;15(3):450–7.
38. Wang Z, Zhu W, Huang R, Zhang Y, Jia C, Zhao H, et al. Fabrication and characterization of cellulose nanofiber aerogels prepared via two different drying techniques. Polymers. 2020;12(11):2583.
39. Agostinho DA, Paninho AI, Cordeiro T, Nunes AV, Fonseca IM, Pereira C, et al. Properties of κ-carrageenan aerogels prepared by using different dissolution media and its application as drug delivery systems. Mater Chem Phys. 2020;253:123290.
40. Benalaya I, Alves G, Lopes J, Silva LR. A review of natural polysaccharides: sources, characteristics, properties, food, and pharmaceutical applications. Int J Mol Sci. 2024;25(2):1322.

41. Elmehbad NY, Mohamed NA, El-Ghany NAA. Synthesis of novel antimicrobial and food-preserving hydrogel nanocomposite films based on carboxymethylcellulose. Starch-Stärke. 2024;76(7–8):2300258.
42. Wan C, Jiao Y, Wei S, Zhang L, Wu Y, Li J. Functional nanocomposites from sustainable regenerated cellulose aerogels: a review. Chem Eng J. 2019;359:459–75.
43. Cherian J, Paulose J, Vysakh P. Harnessing nature's hidden material: nano-cellulose. Mater Today Proc. 2018;5(5):12609–14.
44. Mai T, Wang P, Ma M. Promising cellulose-based aerogel composites: preparation methods and advanced applications. Interface Sci Technol. 2024;38:289–327.
45. Zaman A, Huang F, Jiang M, Wei W, Zhou Z. Preparation, properties, and applications of natural cellulosic aerogels: a review. Energy Built Environ. 2020;1(1):60–76.
46. Chen Y, Zhang L, Yang Y, Pang B, Xu W, Duan G, et al. Recent progress on nanocellulose aerogels: preparation, modification, composite fabrication, applications. Adv Mater. 2021;33(11):2005569.
47. Iber BT, Kasan NA, Torsabo D, Omuwa JW. A review of various sources of chitin and chitosan in nature. J Renew Mater. 2022;10(4):1097.
48. Mohamed NA, Abd El-Ghany NA. Evaluation of in vitro antimicrobial, antioxidant and anticancer activity of a new carboxymethylchitosan-based cyanoguanidine copolymer. Cell Chem Technol. 2024;58:481–93.
49. El-Naggar M, Gomha SM, Abd El-Ghany NA, Abolibda TZ, Alruwaili AH, Zaki ME, et al. Green synthesis and molecular docking studies of bisthiazole and bisthiadiazole derivatives using p-aminobenzoic acid chitosan/CuONP composites. Green Chem Let Rev. 2025;18(1):2511197.
50. Basem A, Jasim DJ, Majdi HS, Mohammed RM, Ahmed M, Al-Rubaye AH. Adsorption of heavy metals from wastewater by chitosan. A review. Res Eng. 2024:102404.
51. Elmehbad NY, Mohamed NA, Abd El-Ghany NA, Abdel-Aziz MM. Reinforcement of the antimicrobial activity and biofilm inhibition of novel chitosan-based hydrogels utilizing zinc oxide nanoparticles. Int J Biol Macromol. 2023;246:125582.
52. Paul J, Qamar A, Ahankari SS, Thomas S, Dufresne A. Chitosan-based aerogels: a new paradigm of advanced green materials for remediation of contaminated water. Carbohydr Polym. 2024:122198.
53. Elmehbad NY, Mohamed NA, Abd El-Ghany NA. Evaluation of the antimicrobial and anti-biofilm activity of novel salicylhydrazido chitosan derivatives impregnated with titanium dioxide nanoparticles. Int J Biol Macromol. 2022;205:719–30.
54. Gomha SM, El-Ghany NAA, Ebaid MS, Abolibda TZ, Zaki MEA, Alhilal M, et al. Eco-friendly synthesis of thiazole derivatives using recyclable cross-linked chitosan hydrogel biocatalyst under ultrasonic irradiation as anti-hepatocarcinogenic agents. Catalysts. 2024;14(12):840.
55. Hussein AM, Gomha SM, El-Ghany NAA, Zaki ME, Farag B, Al-Hussain SA, et al. Green biocatalyst for ultrasound-assisted thiazole derivatives: synthesis, antibacterial evaluation, and docking analysis. ACS Omega. 2024;9(12):13666–79.
56. Cuong HN, Kien ND, Hoa NV. Chitosan-based aerogel microspheres for wastewater treatment. Vietnam J Chem. 2024;62(5):615–23.
57. Elella MA, Abdel-Aziz MM, Abd El-Ghany NA. Synthesis of a high-performance antimicrobial o-quaternized alginate—a promising potential antimicrobial agent. Cellul Chem Technol Synth. 2021;55:75–86.
58. Elmehbad NY, Mohamed NA, Abd El-Ghany NA, Abdel-Aziz MM. Green synthesis of nano-silver/sodium alginate/carboxymethyl xanthan gum hydrogel and evaluation of its anti-inflammatory and anti-helicobacter pylori activity. Cellul Chem Technol. 2022;56(9–10):983–95.
59. Zhao J, Cao L, Dong Y. Composite aerogel comprised of sodium alginate and bentonite via supercritical CO_2 drying: an efficient adsorbent for lysozyme. Gels. 2022;8(6):359.
60. Elella MHA, Kamel AM, López-Maldonado EA, Uzondu SW, Abdallah HM. A review of recent progress in alginate-based nanocomposite materials for tissue engineering applications. Int J Biol Macromol. 2025;139840

61. Wang C, He G, Cao J, Fan L, Cai W, Yin Y. Underwater superoleophobic and salt-tolerant sodium alginate/N-succinyl chitosan composite aerogel for highly efficient oil–water separation. ACS Appl Polym Mater. 2020;2(3):1124–33.
62. Wang Z, Wu S, Zhang Y, Miao L, Zhang Y, Wu A. Preparation of modified sodium alginate aerogel and its application in removing lead and cadmium ions in wastewater. Int J Biol Macromol. 2020;157:687–94.
63. Blennow A, Nielsen TH, Baunsgaard L, Mikkelsen R, Engelsen SB. Starch phosphorylation: a new front line in starch research. Trends Plant Sci. 2002;7(10):445–50.
64. Abd El-Ghany NA, Elella MHA. Food packaging films generated from 2D biopolymer-based nanocomposites: an overview. In: Food packaging: safety, management and quality; 2022.
65. Abd El-Ghany NA, Mahmoud ZM. Synthesis, characterization and swelling behavior of high-performance antimicrobial amphoteric hydrogels from corn starch. Polym Bull. 2021;78:6161–82.
66. Abd El-Ghany NA, Aziz MSA, Abdel-Aziz MM, Mahmoud Z. Antimicrobial and swelling behaviors of novel biodegradable corn starch grafted/poly (4-acrylamidobenzoic acid) copolymers. Int J Biol Macromol. 2019;134:912–20.
67. Abd El-Ghany NA, Aziz MSA, Abdel-Aziz MM, Mahmoud ZM. Reinforcement of antimicrobial activity and swelling ability of starch-g-poly 4-acrylamidobenzoic acid using chitosan nanoparticles. Cellul Chem Technol. 2023;57:803–13.
68. Druel L, Bardl R, Vorwerg W, Budtova T. Starch aerogels: A member of the family of thermal superinsulating materials. Biomacromolecules. 2017;18(12):4232–9.
69. Abhari N, Madadlou A, Dini A. Structure of starch aerogel as affected by crosslinking and feasibility assessment of the aerogel for an anti-fungal volatile release. Food Chem. 2017;221:147–52.
70. Dogenski M, Navarro-Díaz HJ, de Oliveira JV, Ferreira SRS. Properties of starch-based aerogels incorporated with agar or microcrystalline cellulose. Food Hydrocoll. 2020;108:106033.
71. Han JH, Lee J, Kim SK, Kang D, Park HB, Shim JK. Impact of the amylose/amylopectin ratio of starch-based foams on foaming behavior, mechanical properties, and thermal insulation performance. ACS Sustain Chem Eng. 2023;11(7):2968–77.
72. Caldonazo A, Almeida SL, Bonetti AF, Lazo REL, Mengarda M, Murakami FS. Pharmaceutical applications of starch nanoparticles: a scoping review. Int J Biol Macromol. 2021;181:697–704.
73. Abd El-Ghany NA, Elella MHA, Abdallah HM, Mostafa MS, Samy M. Recent advances in various starch formulation for wastewater purification via adsorption technique: a review. J Polym Environ. 2023;31(7):2792–825.
74. Abd El-Ghany NA, Mohamed NA, Mahmoud ZM. Synthesis and characterization of novel highly swellable antimicrobial starch hydrogels impregnated with ZnO nanoparticles as edible food coating.
75. Costa TB, Matias PM, Sharma M, Murtinho D, Rosa DS, Valente AJ. Recent advances on starch-based adsorbents for heavy metal and emerging pollutant remediation. Polymers. 2024;17(1):15.
76. Kalsi G, Hazarika U, Baruah LD, Bordoloi PL, Gogoi M. Comprehensive review of carrageenan's multifaceted role in health and food systems. Discov Food. 2025;5(1):115.
77. Cao Y, Li S, Fang Y, Nishinari K, Phillips GO, Lerbret A, Assifaoui A. Specific binding of trivalent metal ions to λ-carrageenan. Int J Biol Macromol. 2018;109:350–6.
78. Abdellatif FHH, Abdellatif MM. Bio-based i-carrageenan aerogels as efficient adsorbents for heavy metal ions and acid dye from aqueous solution. Cellulose. 2020;27(1):441–53.
79. Sharma G, Khosla A, Kumar A, Kaushal N, Sharma S, Naushad M, et al. A comprehensive review on the removal of noxious pollutants using carrageenan based advanced adsorbents. Chemosphere. 2022;289:133100.
80. Tang KHD, Darwish NM, Alkahtani AM, AbdelGawwad MR, Karácsony P. Biological removal of dyes from wastewater: a review of its efficiency and advances. Tropical Aquatic and Soil Pollution. 2022;2(1):59–75.

81. Donkadokula NY, Kola AK, Naz I, Saroj D. A review on advanced physico-chemical and biological textile dye wastewater treatment techniques. Rev Environ Sci Biotechnol. 2020;19:543–60.
82. Elgamal AM, Abd El-Ghany NA, Saad GR. Highly reactive adsorbent based on carboxymethyl xanthan gum-g-poly (4-vinylpyridine) copolymer for the potential removal of Acid Orange 10 dye and Cr(VI) ions for water treatment. J Appl Polym Sci. 2022;139(47):e53179.
83. Cheng S, Zhao S, Xing B, Liu Y, Zhang C, Xia H. Preparation of magnetic adsorbent-photocatalyst composites for dye removal by synergistic effect of adsorption and photocatalysis. J Clean Prod. 2022;348:131301.
84. Bal G, Thakur A. Distinct approaches of removal of dyes from wastewater: a review. Mater Today Proc. 2022;50:1575–9.
85. Solayman H, Hossen MA, Abd Aziz A, Yahya NY, Leong KH, Sim LC, et al. Performance evaluation of dye wastewater treatment technologies: a review. J Environ Chem Eng. 2023;11(3):109610.
86. Abd Malek NN, Jawad AH, Ismail K, Razuan R, ALOthman ZA. Fly ash modified magnetic chitosan-polyvinyl alcohol blend for reactive orange 16 dye removal: adsorption parametric optimization. Int J Biol Macromol. 2021;189:464–76.
87. Harja M, Buema G, Bucur D. Recent advances in removal of Congo Red dye by adsorption using an industrial waste. Sci Rep. 2022;12(1):6087.
88. Agarwala R, Mulky L. Adsorption of dyes from wastewater: a comprehensive review. ChemBioEng Reviews. 2023;10(3):326–35.
89. Pirozzi A, Rincón E, Espinosa E, Donsì F, Serrano L. Nanostructured cellulose-based aerogels: influence of chemical/mechanical cascade processes on quality index for benchmarking dye pollutant adsorbents in wastewater treatment. Gels. 2023;9(12):958.
90. Li P, Yang C, Xu X, Miao C, He T, Jiang B, Wu W. Preparation of bio-based aerogel and its adsorption properties for organic dyes. Gels. 2022;8(11):755.
91. Shiri M, Hosseinzadeh M, Shiri S, Javanshir S. Adsorbent based on MOF-5/cellulose aerogel composite for adsorption of organic dyes from wastewater. Sci Rep. 2024;14(1):15623.
92. Zhang Z, Hu J, Tian X, Guo F, Wang C, Zhang J, Jiang M. Facile in-situ growth of metal–organic framework layer on carboxylated nanocellulose/chitosan aerogel spheres and their high-efficient adsorption and catalytic performance. Appl Surf Sci. 2022;599:153974.
93. Zanotti A, Baldino L, Cardea S, Reverchon E. Methyl Orange adsorption using chitosan-based composite aerogels produced by supercritical gel drying. Polysaccharides. 2025;6(1):1.
94. He S, Li J, Cao X, Xie F, Yang H, Wang C, et al. Regenerated cellulose/chitosan composite aerogel with highly efficient adsorption for anionic dyes. Int J Biol Macromol. 2023;244:125067.
95. Luo J, Yu D, Hristovski KD, Fu K, Shen Y, Westerhoff P, Crittenden JC. Critical review of advances in engineering nanomaterial adsorbents for metal removal and recovery from water: mechanism identification and engineering design. Environ Sci Technol. 2021;55(8):4287–304.
96. El-Newehy MH, Aldalbahi A, Thamer BM, Moydeen AM. Chitosan-carbon aerogel for efficient removal of methylene blue dye: synthesis, characterization, adsorption mechanism and antimicrobial therapies. ChemistrySelect. 2024;9(37):e202401818.
97. Jemai R, Djebbi MA, Boubakri S, Ben Rhaiem H, Ben Haj Amara A. Effective removal of methyl orange dyes using an adsorbent prepared from porous starch aerogel and organoclay. Colorants. 2023;2(2):209–29.
98. Radhi A, Iffah WF. Physicochemical properties of starch aerogel incorporated with silver nanoparticles and graphene oxide for methylene blue removal. In: AIP Conference Proceedings. AIP Publishing LLC.; 2025.
99. Chen Y, Dai G, Gao Q. Starch nanoparticles–graphene aerogels with high supercapacitor performance and efficient adsorption. ACS Sustain Chem Eng. 2019;7(16):14064–73.
100. Shao S, Xu Z, Xia S, Gao G. Mechanistic and adsorption kinetics study of sulfonated activated carbon enhanced starch-based aerogel for high efficient adsorption of cationic and anionic dyes. Int J Biol Macromol. 2025;309:142555.

101. Revadekar CC, Godiya CB, Park BJ. Novel soy protein isolate/sodium alginate-based functional aerogel for efficient uptake of organic dye from effluents. J Environ Manag. 2024;352:120011.
102. Li J, Lei L, Liu Z, Qiu J, Wang H. Sodium alginate/polyethyleneimine/polydopamine@ cellulose nanofiber composite aerogel as a novel adsorbent for Cr(VI) and dyes removal. Polym Eng Sci. 2023;63(10):3492–506.
103. Lentz L, Mayer DA, Dogenski M, Ferreira SRS. Hybrid aerogels of sodium alginate/graphene oxide as efficient adsorbents for wastewater treatment. Mater Chem Phys. 2022;283:125981.
104. Li H, Huang J, Shen S, Meng C, Wang H, Fu J. Superhydrophobic sodium alginate/cellulose aerogel for dye adsorption and oil–water separation. Cellulose. 2023;30(11):7157–75.
105. Elgamal AM, Abd El-Ghany NA, Saad GR. Synthesis and characterization of hydrogel-based magnetite nanocomposite adsorbents for the potential removal of Acid Orange 10 dye and Cr (VI) ions from aqueous solution. Int J Biol Macromol. 2023;227:27–44.
106. Kuang J, Cai T, Dai J, Yao L, Liu F, Liu Y, et al. High strength chitin/chitosan-based aerogel with 3D hierarchically macro-meso-microporous structure for high-efficiency adsorption of Cu(II) ions and Congo red. Int J Biol Macromol. 2023;230:123238.
107. Jang T, Yoon S, Choi J-H, Kim N, Park J-A. Simultaneous removal of heavy metals and dyes on sodium alginate/polyvinyl alcohol/κ-carrageenan aerogel beads. Gels. 2025;11(3):211.
108. Liu X, Xu M, An B, Wu Z, Yang R, Ma C, et al. A facile hydrothermal method–fabricated robust and ultralight weight cellulose nanocrystal-based hydro/aerogels for metal ion removal. Environ Sci Pollut Res. 2019;26:25583–95.
109. Wang Q, Zuo W, Tian Y, Kong L, Cai G, Zhang H, et al. Flexible brushite/nanofibrillated cellulose aerogels for efficient and selective removal of copper (II). Chem Eng J. 2022;450:138262.
110. Mo L, Pang H, Tan Y, Zhang S, Li J. 3D multi-wall perforated nanocellulose-based polyethylenimine aerogels for ultrahigh efficient and reversible removal of Cu(II) ions from water. Chem Eng J. 2019;378:122157.
111. Fan S, Chen J, Fan C, Chen G, Liu S, Zhou H, et al. Fabrication of a CO_2-responsive chitosan aerogel as an effective adsorbent for the adsorption and desorption of heavy metal ions. J Hazard Mater. 2021;416:126225.
112. Varamesh A, Abraham BD, Wang H, Berton P, Zhao H, Gourlay K, et al. Multifunctional fully biobased aerogels for water remediation: applications for dye and heavy metal adsorption and oil/water separation. J Hazard Mater. 2023;457:131824.
113. Li Y, Guo C, Shi R, Zhang H, Gong L, Dai L. Chitosan/nanofibrillated cellulose aerogel with highly oriented microchannel structure for rapid removal of Pb(II) ions from aqueous solution. Carbohydr Polym. 2019;223:115048.
114. Wang Q, Zuo W, Tian Y, Kong L, Cai G, Zhang H, et al. An ultralight and flexible nanofibrillated cellulose/chitosan aerogel for efficient chromium removal: adsorption-reduction process and mechanism. Chemosphere. 2023;329:138622.
115. Zhang J, Wang Y, Liang D, Xiao Z, Xie Y, Li J. Sulfhydryl-modified chitosan aerogel for the adsorption of heavy metal ions and organic dyes. Ind Eng Chem Res. 2020;59(32):14531–6.
116. Wang M, Wang Z, Zhou X, Li S. Efficient removal of heavy metal ions in wastewater by using a novel alginate-EDTA hybrid aerogel. Appl Sci. 2019;9(3):547.
117. Camani PH, Midhun Dominic C, Parra DF, Maltez HF, Rosa DS. Divalent metal ion removal from simulated water using sustainable starch aerogels.
118. Lencina MS, Piqueras CM, Vega DA, Villar MA, Del Barrio MC. Environmentally friendly starch/alginate aerogels for copper adsorption from aqueous media. A microstructural and kinetic study. J Environ Sci Health A. 2023;58(4):369–81.
119. Du M, Cao Y, Luo X, Yang W, Lin W, Wang Y, et al. Shapeable sodium alginate aerogel beads incorporated with L-cysteine-modified defective UiO-67 for heavy metal ions removal. Chem Eng J. 2023;475:146289.
120. Liu J, Chen T-W, Yang Y-L, Bai Z-C, Xia L-R, Wang M, et al. Removal of heavy metal ions and anionic dyes from aqueous solutions using amide-functionalized cellulose-based adsorbents. Carbohydr Polym. 2020;230:115619.

121. Mo L, Tan Y, Shen Y, Zhang S. Highly compressible nanocellulose aerogels with a cellular structure for high-performance adsorption of Cu(II). Chemosphere. 2022;291:132887.
122. Raeisi Kheirabadi N, Salman Tabrizi N, Sangpour P. Synthesis of sodium alginate-derived carbon aerogel for adsorptive removal of methylene blue. Iran J Chem Chem Eng. 2020;39(5):157–68.
123. Vadakkekara GJ, Thomas S, Nair CR. Maleic acid modified cellulose for scavenging lead from water. Int J Biol Macromol. 2019;129:293–304.
124. She J, Tian C, Wu Y, Li X, Luo S, Qing Y, Jiang Z. Cellulose nanofibrils aerogel cross-linked by poly (vinyl alcohol) and acrylic acid for efficient and recycled adsorption with heavy metal ions. J Nanosci Nanotechnol. 2018;18(6):4167–75.
125. Rong N, Chen C, Ouyang K, Zhang K, Wang X, Xu Z. Adsorption characteristics of directional cellulose nanofiber/chitosan/montmorillonite aerogel as adsorbent for wastewater treatment. Sep Purif Technol. 2021;274:119120.
126. Ji Y, Wen Y, Wang Z, Zhang S, Guo M. Eco-friendly fabrication of a cost-effective cellulose nanofiber-based aerogel for multifunctional applications in Cu(II) and organic pollutants removal. J Clean Prod. 2020;255:120276.
127. Lian Wang X, Mei Guo D, Da An Q, Yi Xiao Z, Ru Zhai S, High-efficacy adsorption of Cr(VI) and anionic dyes onto β-cyclodextrin/chitosan/hexamethylenetetramine aerogel beads with task-specific, integrated components. Int. J. Biol. Macromol. 2019;128:268–278.
128. Lei C, Gao J, Ren W, Xie Y, Abdalkarim SYH, Wang S, et al. Fabrication of metal-organic frameworks@ cellulose aerogels composite materials for removal of heavy metal ions in water. Carbohydr Polym. 2019;205:35–41.
129. Wang Q, Ransing AA, Mahadik D, Meti P, Choi H, Kim T, Park H-H. Phenyl-bridged silica aerogels with enhanced textural and mechanical properties for heavy metal removal. J Water Process Eng. 2025;71:107240.
130. Gao C, Wang X-L, An Q-D, Xiao Z-Y, Zhai S-R. Synergistic preparation of modified alginate aerogel with melamine/chitosan for efficiently selective adsorption of lead ions. Carbohydr Polym. 2021;256:117564.
131. Tan LN, Nguyen NCT, Trinh AMH, Do NH, Le KA, Le PK. Eco-friendly synthesis of durable aerogel composites from chitosan and pineapple leaf-based cellulose for Cr(VI) removal. Sep Purif Technol. 2023;304:122415.
132. Guo D-M, An Q-D, Xiao Z-Y, Zhai S-R, Yang D-J. Efficient removal of Pb(II), Cr(VI) and organic dyes by polydopamine modified chitosan aerogels. Carbohydr Polym. 2018;202:306–14.
133. Nie J, Feng D, Shang J, Nasen B, Jiang T, Liu Y, Hou S. Green composite aerogel based on citrus peel/chitosan/bentonite for sustainable removal Cu(II) from water matrices. Sci Rep. 2023;13(1):15443.
134. Ghiorghita C-A, Lazar MM, Platon I-V, Humelnicu D, Doroftei F, Dinu MV. Feather-weight cryostructured thiourea-chitosan aerogels for highly efficient removal of heavy metal ions and bacterial pathogens. Int J Biol Macromol. 2023;235:123910.
135. Zhu C, Feng Z, Meng Y, Wang M, Wang Z. Effective removal of lead and copper ions from water using a novel sodium alginate–streptomycin sulfate composite aerogel. New J Chem. 2023;47(14):6739–48.
136. Song G, Shi Y, Li A, Wang H, Ding G. Facile preparation of three-dimensional graphene oxide/ ι-carrageenan composite aerogel and its efficient ability for selective adsorption of methylene blue. J Mater Sci. 2021;56(26):14866–79.
137. Mahmoodi NM, Mokhtari-Shourijeh Z, Langari S, Naeimi A, Hayati B, Jalili M, Seifpanahi-Shabani K. Silica aerogel/polyacrylonitrile/polyvinylidene fluoride nanofiber and its ability for treatment of colored wastewater. J Mol Struct. 2021;1227:129418.
138. Xiao J, Zhang J, Lv W, Song Y, Zheng Q. Multifunctional graphene/poly (vinyl alcohol) aerogels: in situ hydrothermal preparation and applications in broad-spectrum adsorption for dyes and oils. Carbon. 2017;123:354–63.
139. Yao T, Qi Y, Mei Y, Yang Y, Aleisa R, Tong X, Wu J. One-step preparation of reduced graphene oxide aerogel loaded with mesoporous copper ferrite nanocubes: a highly efficient catalyst in microwave-assisted Fenton reaction. J Hazard Mater. 2019;378:120712.

140. Esmaeili Z, Izadyar S, Hamzeh Y, Abdulkhani A. Preparation and characterization of highly porous cellulose nanofibrils/chitosan aerogel for acid blue 93 adsorption: kinetics, isotherms, and thermodynamics analysis. J Chem Eng Data. 2021;66(2):1068–80.
141. Lv Y, Xing B, Zheng M, Yi G, Huang G, Zhang C, et al. Hydrothermal synthesis of ultralight coal-based graphene oxide aerogel for efficient removal of dyes from aqueous solutions. Nano. 2018;8(9):670.
142. Lei C, Wen F, Chen J, Chen W, Huang Y, Wang B. Mussel-inspired synthesis of magnetic carboxymethyl chitosan aerogel for removal cationic and anionic dyes from aqueous solution. Polymer. 2021;213:123316.
143. Grishkewich N, Li Y, Liu K, Tam KC. Synthesis and characterization of modified cellulose nanofibril organosilica aerogels for the removal of anionic dye. J Polym Res. 2022;29(7):261.
144. Wu Z, Huang W, Shan X, Li Z. Preparation of a porous graphene oxide/alkali lignin aerogel composite and its adsorption properties for methylene blue. Int J Biol Macromol. 2020;143:325–33.
145. Li K, Zhou M, Liang L, Jiang L, Wang W. Ultrahigh-surface-area activated carbon aerogels derived from glucose for high-performance organic pollutants adsorption. J Colloid Interface Sci. 2019;546:333–43.
146. Segneanu A-E, Herea D-D, Buema G, Bradu IA, Cepan M, Grozescu I. Advanced aerogels for water remediation: unraveling their potential in fats, oils, and grease sorption—a comprehensive review. Gels. 2025;11(4):268.
147. Shiu RF, Lee CL, Hsieh PY, Chen CS, Kang YY, Chin WC, Tai NH. Superhydrophobic graphene-based sponge as a novel sorbent for crude oil removal under various environmental conditions. Chemosphere. 2018;
148. Zhu M, Liu Y, Chen M, Xu Z, Li L, Zhou Y. Metal mesh-based special wettability materials for oil-water separation: a review of the recent development. J Pet Sci Eng. 2021;205:108889.
149. Dong C, Hu Y, Zhu Y, Wang J, Jia X, Chen J, Li J. Fabrication of textile waste fibers aerogels with excellent oil/organic solvent adsorption and thermal properties. Gels. 2022;8(10):684.
150. Azad AK, Abu-Thabit NY, Elella MHA. Sustainable and durable superwetting textiles coated with polydopamine for oil/water separation. RSC Sustain. 2025;
151. Xin Y, Qi B, Wu X, Yang C, Li B. Different types of membrane materials for oil-water separation: status and challenges. Colloid Interface Sci Commun. 2024;59:100772.
152. Sun Y, Deng C, Dong H. Oil/water separation with hydrophobic/oleophilic silica-shelled cellulose aerogels. Geoenergy Science and Engineering. 2024;240:213043.
153. Li J, Liu Q, He J, Zhao Y, Mu L, Liu X, et al. A review of superwetting aerogel-based oil-water separation materials. Mater Today Sustain. 2024;26:100741.
154. Maggay IV, Chang Y, Venault A, Dizon GV, Wu C-J. Functionalized porous filtration media for gravity-driven filtration: reviewing a new emerging approach for oil and water emulsions separation. Sep Purif Technol. 2021;259:117983.
155. Boccia AC, Neagu M, Pulvirenti A. Bio-based aerogels for the removal of heavy metal ions and oils from water: novel solutions for environmental remediation. Gels. 2023;10(1):32.
156. Li Y, Zhang H, Fan M, Zheng P, Zhuang J, Chen L. A robust salt-tolerant superoleophobic alginate/graphene oxide aerogel for efficient oil/water separation in marine environments. Sci Rep. 2017;7(1):46379.
157. Zhang M, Jiang S, Han F, Li M, Wang N, Liu L. Anisotropic cellulose nanofiber/chitosan aerogel with thermal management and oil absorption properties. Carbohydr Polym. 2021;264:118033.
158. Chen J, Chen J, Zheng L, Zheng H. Cellulose-based aerogels for efficient dye sorption and oil-water separation in textile wastewater treatment. Int J Biol Macromol. 2025;143612
159. Yang J, Xia Y, Xu P, Chen B. Super-elastic and highly hydrophobic/superoleophilic sodium alginate/cellulose aerogel for oil/water separation. Cellulose. 2018;25:3533–44.
160. Paulauskiene T, Uebe J, Ziogas M. Cellulose aerogel composites as oil sorbents and their regeneration. PeerJ. 2021;9:e11795.
161. Ye R, Long J, Peng D, Wang Y, Zhang G, Xiao G, et al. Oil/water separation using elastic bioaerogels derived from bagasse: role of fabrication steps. J Hazard Mater. 2022;438:129529.

162. Li Y, Zhang H, Fan M, Zhuang J, Chen L. A robust salt-tolerant superoleophobic aerogel inspired by seaweed for efficient oil–water separation in marine environments. Phys Chem Chem Phys. 2016;18(36):25394–400.
163. Paulauskiene T, Uebe J, Karasu AU, Anne O. Investigation of cellulose-based aerogels for oil spill removal. Water Air Soil Pollut. 2020;231(8):424.
164. Zhang H, Li Y, Shi R, Chen L, Fan M. A robust salt-tolerant superoleophobic chitosan/nanofibrillated cellulose aerogel for highly efficient oil/water separation. Carbohydr Polym. 2018;200:611–5.
165. Lu H, Jiang X, Wang J, Hu R. Degradable composite aerogel with excellent water-absorption for trace water removal in oil and oil-in-water emulsion filtration. Front Mater. 2022;9:1093164.
166. Zhang C, Chen G, Lang D, Liu G, Wu R, Wang W, et al. Superhydrophobic cellulose-nanofiber aerogels from waste cotton stalks for superior oil–water and emulsion separation. Cellulose. 2024;31(14):8519–38.
167. Bhagyaraj S, Krupa I. Alginate–halloysite nanocomposite aerogel: preparation, structure, and oil/water separation applications. Biomolecules. 2020;10(12):1632.
168. Lv Y, Bao J, Dang Y, Liu D, Li T, Li S, et al. Biochar aerogel enhanced remediation performances for heavy oil-contaminated soil through biostimulation strategy. J Hazard Mater. 2023;443:130209.
169. Turan V, Khan SA, Iqbal M, Ramzani PMA, Fatima M. Promoting the productivity and quality of brinjal aligned with heavy metals immobilization in a wastewater irrigated heavy metal polluted soil with biochar and chitosan. Ecotoxicol Environ Saf. 2018;161:409–19.
170. He X, Nkoh JN, Shi R-y, Xu R-k. Application of chitosan-and alginate-modified biochars in promoting the resistance to paddy soil acidification and immobilization of soil cadmium. Environ Pollut. 2022;313:120175.
171. El-Naggar A, Lee M-H, Hur J, Lee YH, Igalavithana AD, Shaheen SM, et al. Biochar-induced metal immobilization and soil biogeochemical process: an integrated mechanistic approach. Sci Total Environ. 2020;698:134112.
172. Yuan P, Wang J, Pan Y, Shen B, Wu C. Review of biochar for the management of contaminated soil: preparation, application and prospect. Sci Total Environ. 2019;659:473–90.
173. Tang W, Zanli BLGL, Jing F, Hu T, Chen J. Low temperature pyrolytic biochar is a preferred choice for sulfonamide-Cu (II) contaminated soil remediation in tropical climate region. Sci Total Environ. 2023;876:162792.
174. Li Y, Ma X, Du Y, Chen S, Wu P, Zhang S. Cd-contaminated soil remediation using $MgFe_2O_4$/chitosan carbon aerogel beads. J Environ Chem Eng. 2024;12(1):111918.

3D-Printed Polysaccharides for Interfacial Solar Evaporation and Water Desalination

Nedal Abu-Thabit and Ismail Awan

Abstract Traditional water treatment techniques have disadvantages such as high operational costs, low separation efficiency, large footprints, and possible secondary pollution. At the same time, solar energy has become an alternative source to fossil fuels and crude oil which address the sustainability and renewability issues. This chapter discusses the application of the use of 3D printed polysaccharide for water purification by using solar water evaporation and solar desalination. The chapter provide a brief overview of various technologies for the 3D printing including their scope and limitations focusing on common polysaccharides. The chapter discusses various types of polysaccharides including chitin, chitosan, cellulose, alginates, etc., for fabricating aerogels, hydrogels, and monoliths. The chapter discusses the fabrication methods, performance of each type of 3D printed material, and their limitation. The chapter shows the current advantages and future research direction in this field.

Keywords Additive manufacturing · Cellulose · Chitosan · Alginate · Hydrogel · Aerogel · Photothermal · Drinking water · Desalination · Solar

Abbreviations

Abbreviation	Full Form
3DP-BHE	3D-Printed Bionic Hydrogel Evaporator
ABS	Acrylonitrile Butadiene Styrene
AlgMA	Methacrylated Alginate
AM	Additive Manufacturing
BJT	Binder Jetting
CA	Cellulose Acetate

N. Abu-Thabit (✉)
Chemical Engineering Department, Jubail Industrial College, Jubail Industrial City, Saudi Arabia

I. Awan
Centre of Advanced Materials, Manufacturing and Design, University of Auckland, Auckland, New Zealand

M. H. Abu Elella et al. (eds.), *3D Engineered Polysaccharides*,
https://doi.org/10.1007/978-3-032-26154-0_5

CACH	Cellulose/Alginate/Carbon Black Hydrogel
CAD	Computer-Aided Design
CB	Carbon Black
CHE	Conventional Hydrogel Evaporator
ChitoMA	Methacrylated Chitosan
CMC	Carboxymethyl Cellulose
CMCs	Ceramic Matrix Composites
CNC	Cellulose Nanocrystals
CNF	Cellulose Nanofibers
CNT	Carbon Nanotube
DED	Direct Energy Deposition
DIW	Direct Ink Writing
DLP	Digital Light Processing
DMLS	Direct Metal Laser Sintering
DOD	Drop-on-Demand
FDM	Fused Deposition Modelling
FG	Flash Graphene
G'	Storage Modulus
G''	Loss Modulus
GelMA	Methacrylated Gelatin
HAMA	Hyaluronic Acid Methacrylate
HPMC	Hydroxypropyl Methylcellulose
IJP	Inkjet Printing
LAP	Lithium Phenyl-2,4,6-trimethylbenzoylphosphinate
MAX phase	Mn + 1AXn phase
MEX	Material Extrusion
MJT	Material Jetting
MMCs	Metal Matrix Composites
MOF	Metal-Organic Framework
NFC	Nanofibrillated Cellulose
PBF	Powder Bed Fusion
PC	Polycarbonate
PCL	Polycaprolactone
PEDOT:PSS	Poly(3,4-ethylenedioxythiophene) Polystyrene Sulfonate
PEEK	Polyether Ether Ketone
PEGDA	Poly(Ethylene Glycol) Diacrylate
PETG	Polyethylene Terephthalate Glycol
PLA	Polylactic Acid
PMCs	Polymer Matrix Composites
SA	Sodium Alginate
SHL	Sheet Lamination
SLA	Stereolithography
SLS	Selective Laser Sintering
STL	Stereolithography (file format)
TCPP	meso-tetra(4-carboxyphenyl)porphyrin

TIPS	Thermally Induced Phase Separation
TPU	Thermoplastic Polyurethane
VPP	Vat Photopolymerization
WHO	World Health Organization
η	Viscosity
σy	Yield Stress

1 Introduction

The fresh water in the earth is about 0.25 percent of the total water in the Earth's water and the remaining presents the salty seawater [1]. Water shortage is an important concern because of the large, polluted water resources caused by the industrial activities around the world. Due to this fact, there is an urgent need to use clean and advanced technologies for water purification and water treatment to address such concern [1, 2]. The current available technologies for water treatment and purification (e.g. reverse osmosis, membrane ultrafiltration, etc.) have limitations including high capital cost investment, large space requirement [3, 4]. There are also limitations for the conventional wastewater treatments (e.g. chemical precipitations, flocculation, ion-exchange, etc.) regardless of their large use at the current time [5–8].

There is a large amount of sludge produced by inorganic flocculant materials which produces another type of secondary pollution [9–11]. In a similar situation, colloidal sulfides are formed as byproducts from chemical precipitation reaction during water treatment. Adsorption are a cheap method used for water treatment, but it needs extensive regeneration and treatment of the sorbent materials which make the process more complicated. The ion-exchange process is used for water treatment, but its application is limited by the cases where there are high concentrations of metals. The previous methods can help to reduce the contaminants to a good extent, but they cannot meet the water drinking standards. Therefore, there is an urgent requirement and needs to search and develop new methods and technologies that are sustainable and less cost for the treatment of industrial wastewater effluents. The problem of membrane fouling is one of the main limitations of membrane technologies [12–14].

The 3D printing technology has emerged as a new strategy for manufacturing industry and converting different design and prototypes into a final product. On of the areas that 3D printing has been applied is the fabrication of 3D printed membranes for oil/water separation, water treatment, and water purification [15, 16]. Fabrication of membranes using thermally induced phase separation (TIPS) is effective but includes using toxic and hazardous chemicals, monomers, and solvents, which causes a secondary pollution [17]. In the case of 3d printing, this problem of using toxic chemicals and solvents is avoided with minimal generation of chemical waste [18–21]. Various types of 3D printing techniques use polymers as the main feed, except in cases such as direct energy deposition form [22–25]. Recent advances in smart functional materials such as smart polymers, stimuli responsive hydrogels, and advanced plastic materials, which have revolutionized the additive manufacturing by allowing the 4D printing for robotics, healthcare, and advanced aerospace applications [26–29]. The high precision of the 3D printing also allows for supporting the fabrication of intelligent devices in diverse fields [30–34].

One of the interesting and most available natural polymers are polysaccharides including cellulose, chitin, chitosan, alginates, etc. Polysaccharides have been employed as sustainable and renewable polymers for 3D printing of scaffolds, monoliths, membranes, hydrogels, and aerogels for water purification and treatment. This is because these natural biopolymers are superhydrophilic, water-dispersible, and has adsorption and catalytic properties [35–37]. Various polysaccharides (cellulose, chitosan, alginates, etc.) have been utilized successfully in various 3d printing methods [38–40]. 3D printed polysaccharides have been applied in different areas in medicine, drug delivery, tissue engineering, and water treatment [35, 41, 42]. This chapter highlights the recent developing in biopolymers derived from natural polysaccharides for 3D printing of hydrogels, aerogels, membranes, and scaffolds for interfacial solar water evaporation and water desalination. The discussion includes fabrication methods, performances, limitations, and the future research trends possible in this area of science.

2 Polysaccharide-Based 3D Printing Technologies

2.1 Introduction to 3D Printing Techniques

Additive manufacturing (AM) is an advanced manufacturing technique used to fabricate 3D objects directly from computer-aided design (CAD) models. AM was developed to fabricate functional end-use parts with complex geometries and with minimal metal waste [43].

2.1.1 Classification of 3D Printing Techniques

As sketched in Fig. 1, the 3D printing methods are organized into different technologies based on material deposition, curing, and binding mechanisms [45, 46]. These include Material Extrusion (MEX), such as Fused Deposition Modelling (FDM) and Direct Ink Writing (DIW); Binder Jetting (BJT); Material Jetting (MJT); Powder Bed Fusion (PBF), such as Selective Laser Sintering (SLS) and Direct Metal Laser Sintering (DMLS); Vat Photopolymerization (VPP), such as Stereolithography (SLA); Sheet Lamination (SHL) and Direct Energy Deposition (DED).

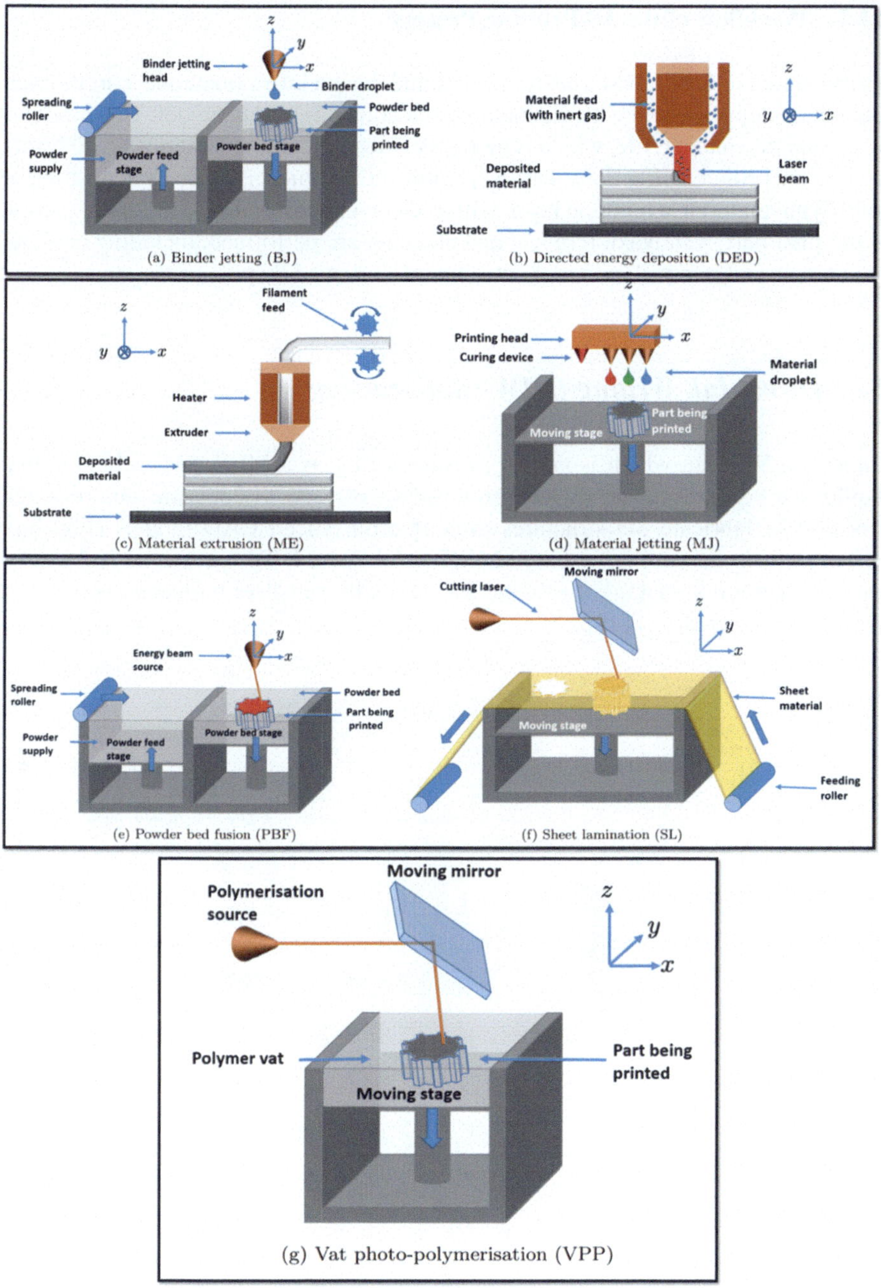

Fig. 1 Schematic illustration of the seven ISO/ASTM 52900 AM 3D printing technologies. Reproduced with permission from the reference [44]

2.1.2 Workflow of the 3D Printing Process

As illustrated in Fig. 2, AM transforms a digital design into a functional part through a series of steps. It starts with the making of a digital 3D CAD model. This model is converted into an STL file, which is then processed (sliced) into a series of 2D layers to generate printer instructions. During printing, a 3D object is made by depositing or solidifying material layer upon layer, with each layer conforming to the cross-section of the final part [43]. Next, removal and cleaning are performed, including cooling, excess material removal, and part detachment, followed by post-processing.

2.2 *Direct Ink Writing (DIW)/Robocasting*

Direct Ink Writing, which is known as robocasting, is a MEX AM technique that employs a compressed-air-driven printhead to extrude viscoelastic inks through a nozzle to fabricate 3D structures directly from sliced CAD models (Fig. 3a). Different from thermoplastic-based MEX techniques, DIW operates at ambient or mild temperatures, making it suitable for thermally sensitive biomaterials, such as natural polysaccharides [45].

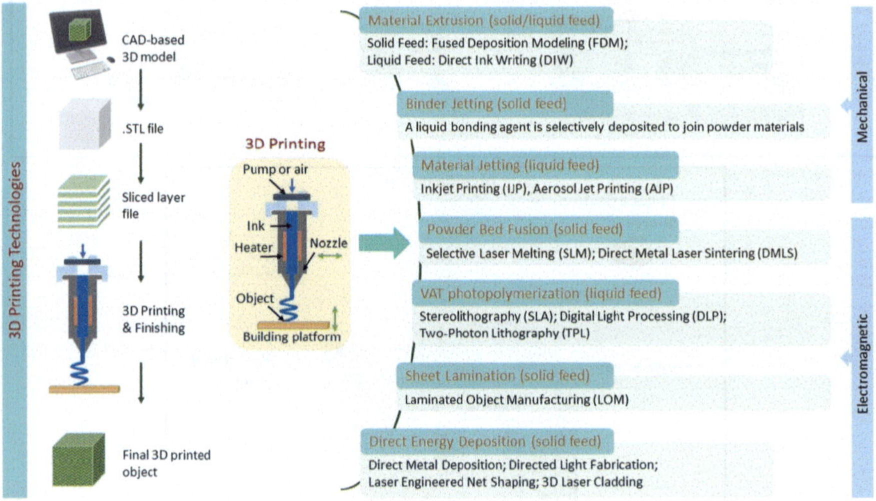

Fig. 2 Outline of the procedure from a CAD file to a functional artifact using 3D printing technologies [47]

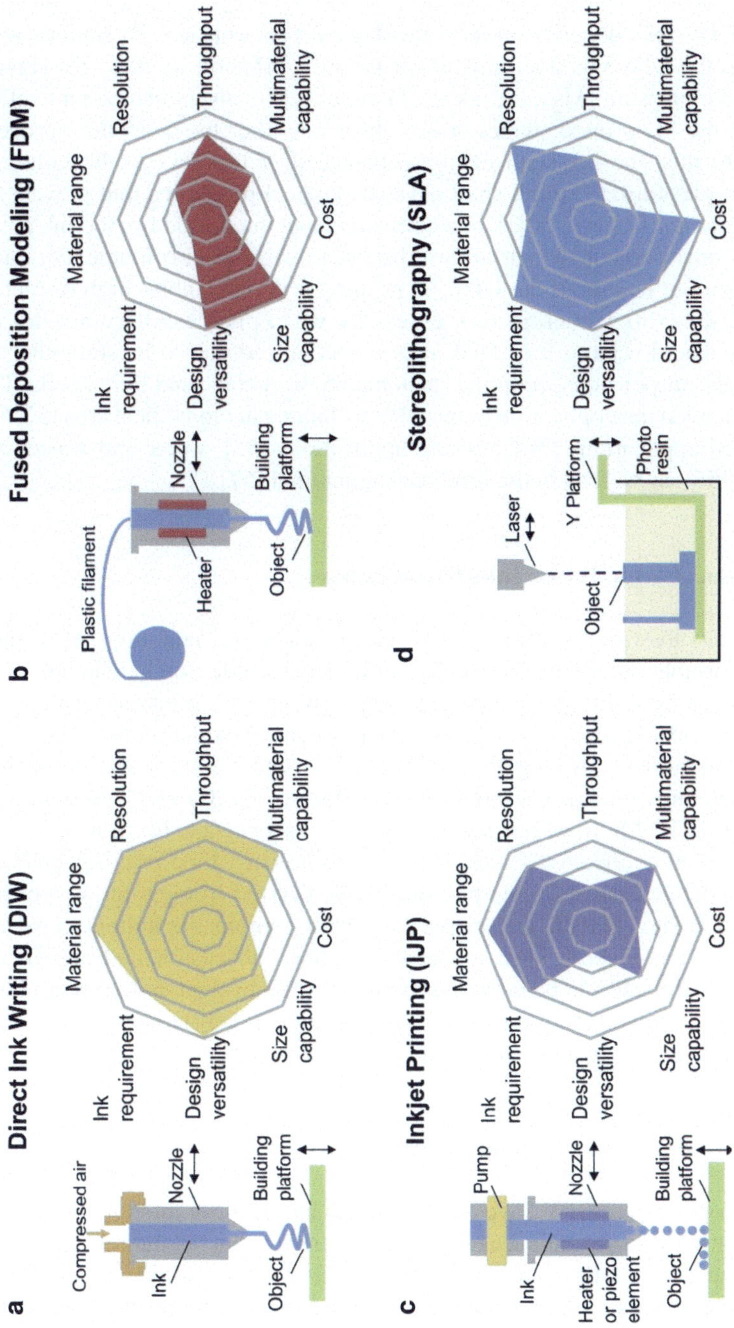

Fig. 3 Schematic principles and features of the main 3D printing techniques used for Polysaccharides 3D printing [48]

2.2.1 DIW Compatible Inks

The use of DIW inks depends on their rheological performance. To achieve good performance, the DIW ink shall have shear thinning property to allow for smooth extrusion under pressure. Also, the ink shall have elastic recovery behaviour to allow for retaining the 3D printed shapes at low shear rate after the deposition process. In addition to this, the ink shall exhibit a balanced yield stress to the failure of the ink in droplet form. The ink shall have rheological properties that prevent the blockage of the nozzles [49, 50]. Upon extrusion from the nozzle tip, the ink enters a zero-shear environment, which allows the ink to solidify into filaments rapidly [51]. An important feature of the DIW 3D printing is that it enables high resolution features (i.e. down to 1 micrometre), especially when microcapillary nozzles are applied [47]. The diverse use of DIW inks is recognized due to its suitability for polymeric gels, suspensions, ceramics, nanomaterials, metals, and living cells. This feature has enabled the application of the DIW technique in biomedical area [52, 53], energy-related applications [54], sensing applications [55], water and wastewater remediation [56], as well as in the food packaging field [57].

2.2.2 Polysaccharide-Based Inks Formulations

Polysaccharides are widely utilized in DIW due to their biodegradability, gel-forming ability, and variable rheological properties. Polymers include starch, which requires gelatinization or the addition of a plasticizer (e.g., glycerol) to induce shear-thinning. Used in biodegradable scaffolds and food-grade structures, with optimized amylose-to-amylopectin ratios improving printability [41, 58]. Other polysaccharides include Cellulose Derivatives (e.g., Carboxymethyl cellulose (CMC) and hydroxypropyl methylcellulose (HPMC)), which enable control of viscosity and film-forming properties for different applications, such as tissue engineering, food printing, and drug delivery [59]. Chitosan, a cationic polysaccharide derived from chitin, is typically printed using acidic solvents and crosslinked with genipin. Its antimicrobial and bioactive properties make it ideal for wound healing, biosensors, and implantable scaffolds [60]. Alginate can be rapidly crosslinked with divalent cations (e.g., Ca^{2+}) during or post-printing. Commonly used in bioprinting of cell-laden hydrogels, due to its high water retention and ionic gelation properties [61].

2.2.3 Recent Functional Enhancements

Recent developments in DIW have incorporated nanocellulose, biopolymer blends, and functional additives to enhance mechanical strength, print resolution, and biological performance [55]. DIW supports bio-functional printing, allowing encapsulation of cells, drugs, or active biomolecules, expanding its role in bioprinting, controlled release systems, and sustainable packaging [35, 62].

2.3 *Fused Deposition Modelling (FDM)*

Fused Deposition Modelling (FDM) is a material extrusion (MEX) 3D printing technique that builds objects layer by layer from a sliced CAD model using thermoplastic filaments. These filaments are heated to a semi-molten state in the extruder and deposited through a nozzle to form the desired structure. Initially developed for rapid prototyping, FDM remains widely used due to its low cost, minimal post-processing, and ability to fabricate complex geometries without the need for tooling [51].

2.3.1 FDM Compatible Materials

However, unlike techniques such as Direct Ink Writing (DIW), FDM is restricted to thermoplastic materials. Common feedstocks include PLA, ABS, Nylon, TPU, and PC. Polysaccharides, such as cellulose, chitosan, starch, and alginate, are renewable, biodegradable, and biocompatible, making them attractive for sustainable additive manufacturing. In natural form, they are unsuitable for FDM; therefore, non-thermoplastic nature polysaccharides require careful filament formulation and process optimization due to [47]:

- **Thermal degradation** that occurs at standard FDM extrusion temperatures, which are typically higher (>180 °C)
- **Poor melt processability** due to their lack of thermoplastic behaviour
- **High hydrophilic** nature that causes higher moisture absorption and dimensional instability

Standard FDM printers can process biopolymer-based filaments with minor modifications, such as:

- **Nozzle Temperature**: Reduced to $\approx$ 140–200 °C to prevent degradation
- **Bed Temperature**: Maintained at 40–60 °C based on matrix polymer
- **Print Speed**: Slower speeds (20–40 mm/s) improve resolution
- **Multi-material Printing**: Allows combining polysaccharides with structural thermoplastics

Although polysaccharides have been used in many research works, there are some limitations like the weak adhesion between the printed layers, low dimensional stability, and the need for controlling the pinting conditions like the outdoor humidity, temperature, etc. These challenges can be overcome by one or combination of the following solutions which are expected to allow the fabrication of polysaccharides filaments suitable for the FDM printing.

- **Blending** with thermoplastics (e.g., PLA, PCL)
- **Chemical modifications**, such as plasticization or esterification
- **Composite formulations**, using compatibilizers and binders

2.3.2 FDM Filament Fabrication Process

Producing FDM-compatible polysaccharide filaments involves [63]:

1. **Dehydration**: To dry the polysaccharide to reduce moisture sensitivity
2. **Compounding**: To mix with thermoplastic matrix and additives (e.g., plasticizers, chain extenders)
3. **Extrusion**: To produce filament (1.75 or 2.85 mm diameter)
4. **Property Calibration**: To improve thermal and tensile properties for FDM compatibility

Moisture control is critical during both storage and printing, often requiring desiccated storage or in-line filament dryers.

2.4 Inkjet Printing (IJP)

Inkjet printing is a non-contact, drop-on-demand additive manufacturing three dimensional printing techniques that deposits small, discrete droplets of low-viscosity inks onto susbtrates with high three-dimensional resolution [47].

2.4.1 IJP Compatible Inks

Inkjet printing uses low viscosity inks (< 10 mPa·s), high surface tension, and rapid evaporation characteristics to ensure droplet formation and to prevent nozzle from blocking [47]. Polysaccharides need modification or dilution to meet the mentioned properties, which results in compromising mechanical strength or bio-functionality. Therefore, formulations need to be adjusted to balance the printability and the functional achievement [56].

2.4.2 Polysaccharides for Inkjet Printing

Alginates due to their mild gelation via ionic crosslinking (e.g., Ca^{2+}). Pre-gelation or in situ gelation is often employed to improve structure fidelity [61], Chitosan [60], Cellulose Derivatives [35, 64], and starch-based materials [65].

2.4.3 Applications and Advantages of IJP

Its ambient temperature processing, material efficiency, and digital control over droplet placement make IJP a compelling platform for bio-fabrication, particularly where low-viscosity, cell-friendly, and solvent-free deposition is essential [35].

Recent developments include multi-material inkjet printing, picolitre droplet engineering, and integration with post-print crosslinking systems to enhance structural fidelity and function in bio-based architectures [35, 64]. Inkjet printing of polysaccharides is widely used in the following applications:

- **Bioprinting**: Precise placement of cells in hydrogel matrices [66].
- **Pharmaceuticals**: Custom dosing, drug-loaded films, and orally disintegrating systems [67].
- **Diagnostics** (e.g., biosensors, microarrays for point-of-care testing) [68].
- **Food engineering**: Functional edible coatings of nutrients or flavours and microstructures [65].
- **Flexible electronics**: Bio-based printable substrates for sensors and circuits [69].

2.5 *Stereolithography (SLA) Printing*

Stereolithography (SLA) is a vat photopolymerization-based additive manufacturing technique that fabricates complex 3D structures with high spatial resolution (0.5 µm) by selectively curing photo-crosslinkable resins using a UV or visible light source [47]. Traditionally used with synthetic photopolymers, recent advances have focused on adapting naturally derived polysaccharides for SLA through chemical functionalization to enable photocuring [35].

2.5.1 SLA Compatible Resins

Polysaccharides such as alginate, chitosan, cellulose, dextran, and hyaluronic acid lack intrinsic photo-reactivity and must be chemically modified, typically via methacrylation (e.g., converting hydroxyl or amine groups to methacrylate moieties), to form photocurable derivatives [47]. The most common example is methacrylated gelatin (GelMA), which has inspired similar approaches for polysaccharide-based bio-resins [70].

The key challenges for polysaccharide-based bio-resins [70]:

- Maintaining biocompatibility and biodegradability post-functionalization
- Achieving adequate crosslinking density for structural reliability
- Optimizing resin viscosity and light penetration for uniform layer curing

Photo-initiator selection (e.g., Irgacure 2959, LAP) and light intensity must be carefully controlled to enable polymerization without degrading bioactivity, particularly for cell-laden systems.

2.5.2 Polysaccharide-Based Photocurable Resins

- **Methacrylated Alginate (AlgMA)**: Supports ionic and photo-crosslinking, used in soft tissue scaffolds, hydrogels, and drug delivery carriers [70].
- **Methacrylated Chitosan (ChitoMA)**: Offers antimicrobial properties and is employed in biosensors, wound healing matrices, and bone tissue engineering. Its cationic nature enables enhanced cell adhesion [70].
- **Cellulose Derivatives** (e.g., **hydroxyethyl cellulose methacrylate**): Used in biodegradable films, cartilage scaffolds, and microfluidic devices due to their tunable mechanical properties and print resolution [71].
- **Hyaluronic Acid Methacrylate (HAMA)**: Commonly used in bioprinting, especially for cartilage and corneal tissue models, due to its inherent cell-friendly properties [70].

2.5.3 Applications of SLA Printing

SLA enables the creation of highly defined, cell-compatible scaffolds and micro-architectures for use in diverse areas including [35, 66]:

- Tissue engineering (e.g., bone, cartilage, skin)
- Bio-fabrication of soft tissues
- Microfluidic and biomedical devices
- Controlled drug delivery platforms

SLA continues to evolve as a promising technique for high-resolution fabrication of photocurable, bio-based structures. Recent trends include the use of visible-light photo-initiators, multi-material SLA systems, and hybrid bio-resins combining polysaccharides with synthetic polymers (e.g., PEGDA) to enhance printability, mechanical strength, and degradation kinetics [35, 66].

3 3D-Printed Polysaccharides for Interfacial Solar Water Evaporation

3.1 3D-Printed Polysaccharide Monoliths

Various types of used polysaccharides for constructing solar water evaporators are listed in Table 1. Natural wood has a well-designed water transport system, including aligned microchannels and pits, which provides an efficient circulation for water and nutrients in trees. Regardless, the relatively slow water flow velocity restricts the direct use of such a system for efficient water evaporation. Inspired by this natural design, researchers used 3D printing to create a wood-mimetic evaporator, employing diverse cellulosic nanostructures (e.g., nanocellulose, cellulose nanocrystals, nanofibers) as precursors for cellulose-based inks. Chen et al. reported the fabrication of a novel 3D-printed solar evaporator with a unique tripodal porous structure, including interconnected large, micro, and submicron pores to facilitate the spontaneous salt rejection [72]. The 3D-printed device was constructed from sustainable cellulose nanofibers (CNF) and carbon black ink, followed by a freeze-drying step to produce a 3D porous hierarchical architecture. The fabricated device achieved excellent performance, including over 97% light absorption, low thermal conductivity (0.075 W m^{-1} K^{-1}), and a high evaporation rate of $\approx$ 5.79 kg m^{-2} h^{-1} under five sun irradiation. The fabricated device achieved stable, salt-free operation for over 100 hours in 15 wt.% NaCl, with the distilled water produced meeting safe drinking standards. Zhu et al. reported the fabrication of a 3D-printed solar evaporator by printing CNF inks into porous scaffolds, followed by coating with a CNF/carbon nanotube (CNT) film [73]. The fabricated 3D-printed scaffolds were able to purify wastewater, achieving an outdoor production rate of 0.63 g g^{-1} day^{-1}. Compared with the previous models, the reported design showed enhanced performance, with a 1.6-fold higher absorption rate and a 2.4-fold higher release rate. In a separate experiment, a cellulose/carbon black ink was used to print a wood-mimetic evaporator that achieved a rate of 1 kg m^{-2} h^{-1} and durable salt rejection over extended periods. Also, at high sodium chloride concentrations (15%), an evaporation efficiency of up to 75% was achieved, indicating excellent desalination performance.

Cellulose is a hydrophilic polymer due to the presence of numerous polar hydroxyl (− OH) groups on its backbone, which form strong hydrogen bonds with water. Cellulose is a semi-crystalline due to the dense packing arrangement of chains that results in highly ordered crystalline domains. This dual nature arises from the equilibrium between the polar hydroxyl groups, which attract in water, and the robust intra- and intermolecular hydrogen bonds among chains and van der Waals forces that establish the stable, crystalline structure [80, 81]. This semi-crystalline nature hinders water penetration through the cellulosic scaffolds. To enable water permeation, cellulose crystallinity has been reduced to a highly amorphous, interpenetrated network through cellulose regeneration techniques [82, 83]. Koh et al. reported the fabrication of 3D-printed hydratable plastics based on highly amorphous regenerated cellulose for solar desalination [74]. The two-step fabrication process involved DIW of cellulose acetate (CA) white ink and CA/carbon black nanoparticles ink as light absorbers, Fig. 4. The next step involved the regeneration of amorphous cellulose via chemical deacetylation of the cellulose acetate scaffold. The 3D-printed bilayer structure, combining a water-transporting amorphous cellulosic substrate and a light-absorbing top layer, provided excellent mechanical robustness and controlled microarchitecture. The device achieved an evaporation rate of $\approx$ 3 kg m^{-2} h^{-1} under one-sun illumination and an unprecedented 7.35 kg m^{-2} h^{-1} under three-sun irradiation, surpassing previous hydrogel and carbon-based systems. Raman spectroscopy and molecular dynamics simulations revealed that the reduced vaporization enthalpy ($\approx$ 1850 J/g vs. 2450 J/g for bulk water) is attributed to weaker cellulose-water hydrogen bonds, which enable efficient rehydration, long-term stability, and a four-order-of-magnitude reduction in water salinity below WHO standards.

Table 1 3D-Printed polysaccharide structures for solar water evaporation

Material/ Composite	Structure type	Fabrication method	Evaporation rate (1 sun illumination)	Key features	Reference
CNF and Carbon Black	Monolith	DIW and Freeze-drying	≈ 5.79 kg m^{-2} h^{-1} (under 5 suns)	Tripodal porous structure; 97% light absorption; spontaneous salt rejection.	[72]
CNF and CNT film	Monolith	DIW and Coating	Outdoor: 0.63 g g^{-1} day^{-1}	Hierarchical porous scaffolds for atmospheric water harvesting.	[73]
Regenerated Cellulose and Carbon Black	Monolith (bilayer)	DIW and Chemical Regeneration	≈ 3.0 kg m^{-2} h^{-1}	Highly amorphous cellulose reduces the vaporization enthalpy and has excellent mechanical robustness.	[74]
CNF/SA/ MXene	Hydrogel	DIW and Ionic Crosslinking	2.60 kg m^{-2} h^{-1}	Biomimetic dual-scale channels; 413 kPa compressive strength; >99% ion rejection.	[75]
Starch/ activated carbon (3DP-BHE)	Hydrogel	DIW and Freeze/thaw crosslinking	2.13 kg m^{-2} h^{-1}	Bionic leaf-trunk structure; stable for 7 days in 10 wt.% brine.	[76]
Cellulose/ alginate/ carbon black (CACH)	Hydrogel	DIW and Chemical Coagulation/ crosslinking	1.33 kg m^{-2} h^{-1}	Hierarchical porous structure; superhydrophilicity; self-cleaning salt rejection.	[77]
PCN-224/ CB@NFC/ SA	Hydrogel	DIW and Ionic Crosslinking	~2.0 kg m^{-2} h^{-1}	Multifunctional: Stable salt rejection & ~97% photocatalytic dye degradation.	[78]
Flash graphene/ CNF	Aerogel (Janus)	DIW, Freeze-drying and Heat Treatment	≈ 3.40 kg m^{-2} h^{-1}	Superior salt resistance in 15 wt.% brine; uses only 0.08 wt.% FG.	[79]

Most CNF aerogels and 3D-printed monoliths become elastic when saturated with water due to the plasticizing and hydrogen bonding effects of water molecules [84]. However, soaking provides poor control over water content, leading to weak hydrogels with easily expelled water [85]. Incorporating hygroscopic salts [86] or MOFs [87, 88], which can controllably absorb moisture from air, is proposed as a strategy

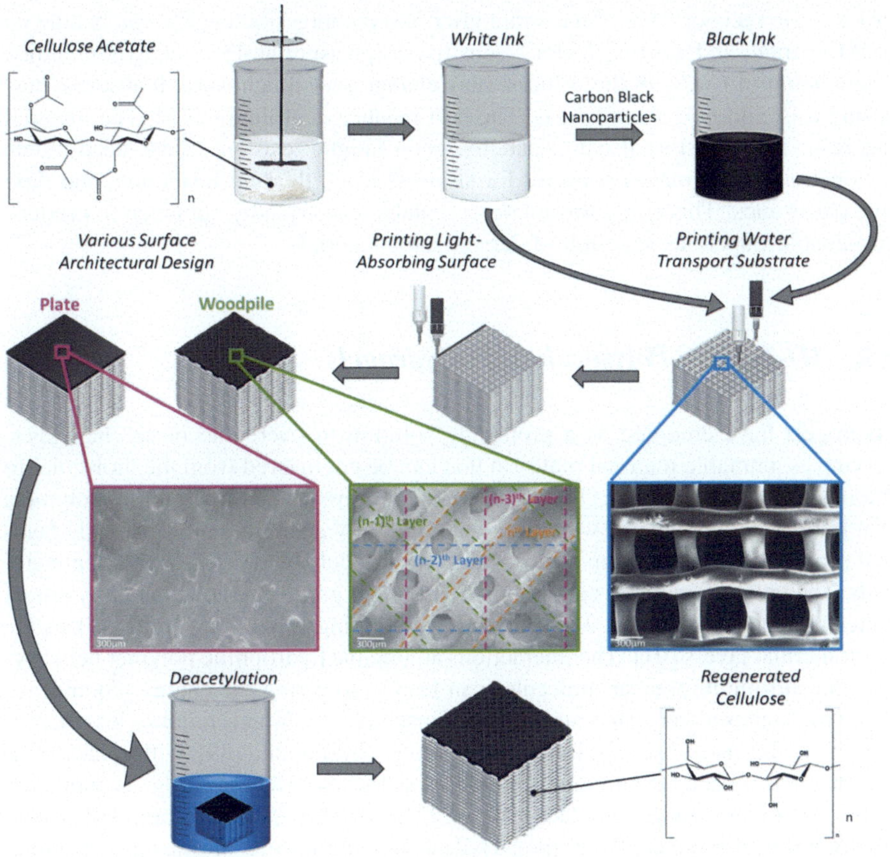

Fig. 4 Schematic of the fabrication process of 3D-printed highly amorphous regenerated cellulose-based solar evaporator with various surface architectural designs, including scanning electron microscopy images of the water transport substrate and the surface morphology of the Plate and Woodpile samples [74]

to create superelastic 3D-printed CNF monoliths. Chen et al. reported the fabrication of the first 3D-printed cellulose nanofibril (CNF) monolith that demonstrates superelasticity, hygroscopicity, and ionic conductivity, thereby overcoming the difficulty of fabricating flexible cellulose structures constrained by robust hydrogen bonding [89]. The innovation involves incorporating hygroscopic calcium chloride ($CaCl_2$) into a hierarchically organized CNF framework created via direct ink writing and freeze-drying. $CaCl_2$ serves as both a cross-linker (via Ca^{2+} COO^- ionic interactions) and a humidity-dependent plasticizer, drawing in moisture from the air to make the material more flexible and conductive. The process of making these materials entails printing TEMPO-oxidized CNF inks (2.0–2.5 wt.%) into porous structures, cross-linking them in $CaCl_2$ baths, and then freeze-drying them to create lightweight aerogels with a density of around 0.09 g cm^{-3}. The monoliths that emerge from this process can recover 91% of the strain after 500 compression cycles, can handle up to 90% strain, and are 0.337 kPa^{-1} sensitive to pressure at 43% relative humidity. The amount of $CaCl_2$ in the solution may change how much water it absorbs, how strong it is, and how flexible it is. The best balance is around 0.5 M. The material has reversible moisture-driven elasticity, good fatigue resistance, and steady ionic conductivity. This makes it useful for strain sensors, thermal insulation, and flexible electronics. This study formulates a scalable eco-friendly approach to produce sustainable, superelastic cellulose-derived metamaterials.

3.2 3D-Printed Polysaccharide Hydrogels

Hydrogels have emerged as a promising solution to overcome these challenges, serving as a tunable material platform that can be engineered from the molecular to the macroscopic scale [90]. These materials are cross-linked hydrophilic polymers that hold large amounts of water and feature highly adaptable physical and chemical properties, making them superior to conventional solar vapor generation materials. The polymer networks are rich in hydrophilic groups, which efficiently attract water molecules into their 3D structures, facilitating rapid water transport to the photothermal layers [91]. The interactions among the hydrophilic polymer networks and the surrounding water molecules can lead to lowering the energy required for water molecules to get evaporated. In some instances, hydrogels requires less than 50 percent of the energy needed for bulk water to get evaporated [89], with some examples that has reached evaporation enthalpies as low as 1187–1426 J/g, as compared to the water vaporization standard value of 2260 J/g [92]. Balancing printability with functional performance. Therefore, Hydrogels exhibit very promising behaviour, making them targetable choice as compared to the traditional solar evaporation materials. Hydrogels have shown that in some cases they have achieved high evaporation rates up to 3.6 kg m^{-2} h^{-1} under one sun irradiation with solar-to-vapor efficiencies exceeding 90% [93]. This level can meet the daily drinking water demands for while operating under natural solar environments [94]. The hierarchical three-dimensional topography of the hydrogel networks enable the construction of multiple transport regimes for water, including internal gaps, tiny channels, and molecular meshes which allow for the uniform supply of water molecules to the evaporation surfaces through capillary forces and osmotic pressure effects [93]. This design guarantees continuous water transport and prevents the accumulation of salt [95].

Guo et al. reported the fabrication of a 3D-printed dual-scale porous hydrogel evaporator that replicates the vascular structure of natural wood to accomplish high-efficiency solar desalination and long-term salt rejection [75]. The innovative aspect of this work is its biomimetic hierarchical channel design, which combines 3D-printed vertically aligned macro-channels with self-assembled microchannels. This design accelerates the movement of water and salt ions via Marangoni convection, preventing salt crystallization at the interface. The hydrogel evaporator is made by DIW, followed by ionic crosslinking. It uses a sustainable composite ink made of cellulose nanofibers for hydrophilicity and mechanical strength, sodium alginate as a rheological and crosslinking modifier, and MXene ($Ti_3C_2T_n$) nanosheets as a broadband photothermal material. To make MXene, Ti_3AlC_2 (MAX phase) is etched with LiF/HCl and then broken down into fragile flakes. The CNF/SA/MXene ink exhibits a significant shear-thinning effect, enabling accurate printing of 3D shapes via a 600 μm nozzle. It also has stable Ca^{2+} induced crosslinking that creates a strong dual-porosity network. The improved hydrogel evaporator has a wide range of solar absorption (≈ 96% from 250 to 2500 nm) and high compressive strength (413 kPa). The system achieves a solar-to-vapor conversion efficiency of approximately 92% and an evaporation rate of 2.60 kg m^{-2} h^{-1} under 1-sun irradiation. Outdoor field experiments produced 3.74–5.54 L m^{-2} day^{-1} of freshwater with over 99% ion rejection. This met the WHO drinking criteria. This research presents a scalable, eco-friendly method for creating programmable 3D-printed solar evaporators from biomass, supporting sustainable desalination and hypersaline wastewater treatment.

Zhang et al. reported the fabrication of a 3D-printed bionic hydrogel evaporator (3DP-BHE) that mimics the tree's leaf trunk system for efficient and practical solar desalination [76]. The fabrication process was carried out using direct ink writing 3D printing. The hydrogels were created through physical crosslinking *via* a freeze/thaw cycle of pregelatinized starch biopolymer. Figure 5a Photothermal absorber layers were created by adding activated carbon material to the pregelatinized starch polymer. The resulting hydrogel exhibits highly interconnected porous networks that mimic a tree's natural system, Fig. 5b, c. The top starch/activated carbon layer acts as a bionic leaf for light absorption and vapor diffusion, and the bottom starch layer functions as a bionic trunk with bimodal pores, large printed wick channels, and smaller hydrogel microchannels, facilitating efficient water transport. During desalination, a concentration gradient forms between these channels, driving salt diffusion and convection that prevent salt crystallization on the surface, Fig. 5d. In contrast, conventional hydrogel evaporators with high tortuosity suffer from salt accumulation that clogs transport channels and reduces evaporation efficiency, Fig. 5e. The fabricated device achieved an evaporation rate of 2.13 kg m^{-2} h^{-1} under one sun, with a solar-to-vapor efficiency of ≈ 91%, while maintaining a steady rate of ≈ 2 kg m^{-2} h^{-1} in a 10 wt.% % NaCl solution for seven days, demonstrating excellent salt resistance and durable performance.

Yuan et al. reported the use of a 3D printing approach to construct a hierarchical porous cellulose/alginate/carbon black hydrogel (CACH) for solar water desalination [77]. The ink was prepared by dispersing the three components in NaOH/Urea/H_2O solution at −12.5 °C, Fig. 6a. The prepared construct was treated and crosslinked in a sequence of solutions to form the final hydrogel. Initially, the printed scaffold was treated in a 10% sodium chloride solution as a coagulation bath to regenerate cellulose. Next, the construct was treated in calcium chloride solution to crosslink the sodium alginate polymer, Fig. 6b. Finally, it was washed with hydrochloric acid solution to remove excess unreacted ions. Under one sun illumination, the CACH evaporator achieved an evaporation rate of 1.33 kg m^{-2} h^{-1} and an impressive solar conversion efficiency of ≈ 91%. This performance is attributed to the superhydrophilicity of the fabricated hydrogel and its hierarchical morphology with micro/nano pores that

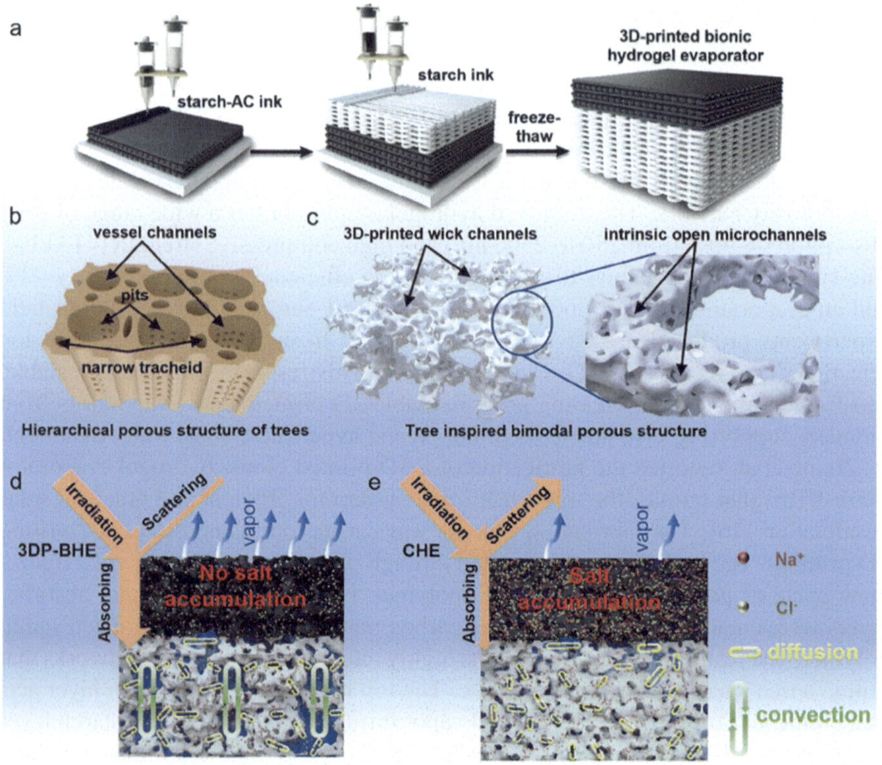

Fig. 5 Schematic the fabrication and salt resistance mechanism of the 3DP-BHE. **a** Prepare the process of the 3DP-BHE. **b** Hierarchical porous structure of trees. **c** Bionic trunk with bimodal porous structure of 3DP-BHE. Salts rejection mechanism of the **d** 3DP-BHE and **e** CHE hydrogel evaporator, [76]

continuously draw water while facilitating salt diffusion back into the bulk solution, preventing crystallization, and enabling self-cleaning salt rejection.

Yang et al. reported the fabrication of a 3D-printed multifunctional hydrogel evaporator (PCN-224/CB@NFC/SA) for efficient solar desalination and photocatalytic degradation of pollutants [78]. The prepared hydrogel precursor was extruded through DIW technology. The fabricated lattice scaffold was ionically cross-linked at 2 wt.% $CaCl_2$ solution for 1 h and then frozen for 24 h to stabilize the hierarchical porous structure. The resulting vertically aligned, interconnected channels enabled rapid capillary water transport and efficient ion back-diffusion, achieving a high evaporation rate of ≈ 2 kg m^{-2} h^{-1} and stable performance in 3.5 wt.% brine for 8 h without salt crystallization, while degrading $\approx 97\%$ of Rhodamine B within 60 min under 1.5 kW m^{-2} illumination. Thus, the incorporation of CB and the porphyrinic MOF PCN-224 enhanced broadband light absorption and enabled

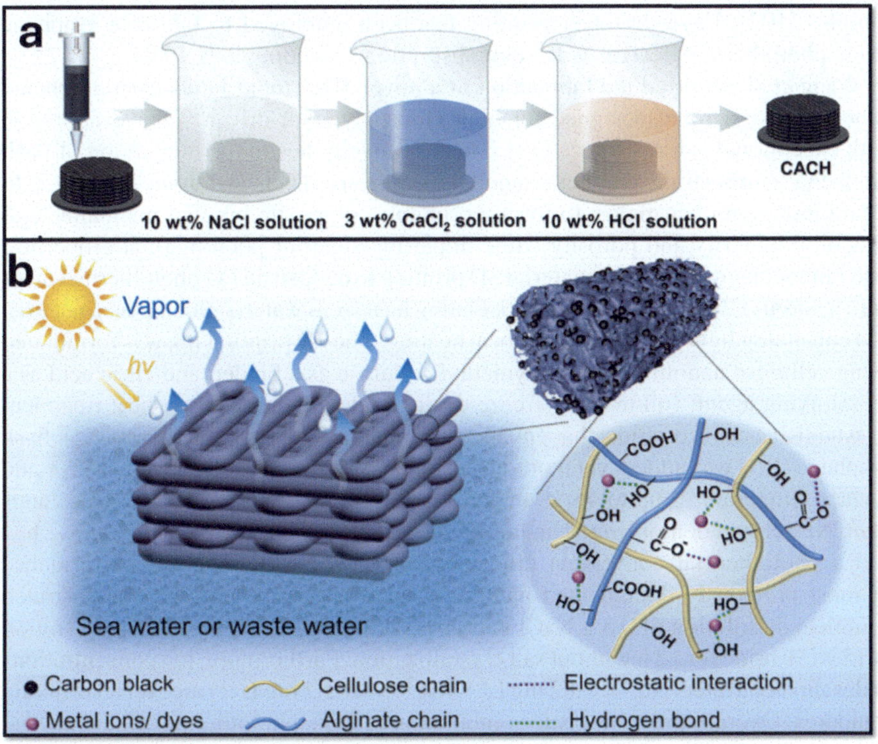

Fig. 6 **a** Scheme for direct ink writing and post-coagulation of 3D-printed CACH, and **b** schematic illustration of the crosslinked porous hydrogel, [77].

visible-light photocatalysis, providing synergistic photothermal heating and pollutant degradation for self-cleaning, high-efficiency solar d3D-Printed Polysaccharide Aerogels.

Aerogels are very promising for making solar vapor because they are very light, do not transfer heat well, and have a lot of pores, which provide them with a lot of internal surface area for evaporation [96]. Their porous architectures trap gas within the polymer network, which keeps them strong and prevents them from collapsing [97]. These characteristics, together with a large surface area and low bulk density, make aerogels great for use as evaporators in interfacial solar desalination [98]. Cellulose-based aerogels further improve sustainability and performance. For example, they can achieve an evaporation rate of 1.61 kg m^{-2} h^{-1} with PEDOT: PSS composites [99], 2.16 kg m^{-2} h^{-1} and $\approx 94\%$ efficiency with cellulose nanofiber/polylactic acid composites [100], and 1.91 kg m^{-2} h^{-1} with vinyltrimethoxysilane-modified bacterial cellulose [101]. Various 3D printing methods have been employed to make aerogel-based solar water evaporation systems. Direct ink writing, which represents one of the most versatile approaches, enabled the fabrication of carbon nitride hybrid aerogels with patterned microstructures that are 2.5 times more active than control samples [102]. Layer-by-layer printing has been employed to fabricate graphene oxide photothermal devices with solar absorption exceeding 97% [94].

Wang et al. reported the fabrication of a novel 3D-printed Janus flash graphene/nanocellulose aerogel designed for high-efficiency solar-driven water evaporation with exceptional salt resistance [79]. The key novelty lies in the integration of flash graphene, a turbostratic graphene nanoflake synthesized via Flash Joule Heating, into a cellulose nanofiber matrix to form a printable composite ink that maintains high structural integrity and porosity while imparting superior photothermal properties. The Janus design uses multi-material 3D printing to deposit an FG photothermal layer onto a porous CNF scaffold, thereby isolating heat from water, reducing thermal loss, and enhancing light absorption and local heating. The 3D printing ink was formulated using cellulose nanofibers, carboxymethyl cellulose as a binder, and citric acid as a crosslinking agent, followed by freeze-drying and dehydrogenation heat treatment to establish ester crosslinks that enhance mechanical stability. Only 0.08 wt.% flash graphene (FG) was added, yet it produced a light absorption efficiency of $\approx 97\%$ and surface temperatures up to 45 °C under one-sun illumination. The optimized Janus FG/CNF aerogel demonstrated an impressive evaporation rate of ≈ 3.40 kg m^{-2} h^{-1} and a photothermal conversion efficiency of 229%, surpassing the performance of most previously documented natural-based solar evaporators. It demonstrated complete dissolution of 0.5 g NaCl within 110 min and maintained stable operation in 15 wt.% brine for 12 h without salt crystallization. Furthermore, ion concentrations in desalinated water ($Na^+ = 2.82$ mg L^{-1}, $Mg^{2+} = 10.87$ mg L^{-1}) met WHO drinking standards. Overall, the synergistic combination of FG's photothermal efficiency and 3D-printed cellulose architecture offers a scalable, low-cost, and sustainable strategy for durable solar desalination and wastewater purification.

4 Conclusions and Future Perspectives

Although there has been a progress in the filed of solar water desalination using polymers and 3D printed natural polymers and polysaccharides. However, as listed in Table 1, there are limited number of research studies currently related to 3D printing of polysaccharides for solar water desalination. For practical applications, there should be number of issues to be considered in the future research which are listed below:

1. **Inefficient energy use**: Future research shall focus on using photothermal materials with the minimum heat loss (e.g. via conduction to bulk water, convection to air, or thermal radiation).
2. **Immature water transport and salt accumulation**: This is crucial because poor water transport can casue rapid salt crystallization on the surface of the absorber, leading to pore blockage, reductionn of light absorption, and degradation of durability in the long term operation.

3. **Vapor escape and condensation inefficiency:** Most of the reported studies did not investigate or focus on the vapor condensation and management. Poor condenser design and heat accumulation is expected to reduce the percentage of the produced water.
4. **Performance under real outdoor conditions:** Most studies report the lab results under one sun illumination. Only very little reports consider the outdoor experiments. The situation in real testing environments shall be considered due to the availability of different conditions such as wind, incident angle, season, humidity, and other factors.
5. **Scalability and commercialization:** The 3D complex structures and architectures may not reproduce the good results or metrics on large scale similar to those achieved on a small bench scale. The presented studies focused on small scale lab prototypes. In real case scenarios, the 3D printing of large prototypes is more challenging due to the unexpected problems, material cost, or different performance parameters.

References

1. Zhang P, Yang M, Lan J, Huang Y, Zhang J, Huang S, et al. Water quality degradation due to heavy metal contamination: health impacts and eco-friendly approaches for heavy metal remediation. Toxics. 2023;11(10):828.
2. Weerasooriya RR, Liyanage LPK, Rathnappriya RHK, Bandara W, Perera T, Gunarathna M, et al. Industrial water conservation by water footprint and sustainable development goals: a review. Environ Dev Sustain. 2021;23(9):12661–709.
3. Ibrahim M, Nawaz MH, Rout PR, Lim JW, Mainali B, Shahid MK. Advances in produced water treatment technologies: an in-depth exploration with an emphasis on membrane-based systems and future perspectives. Water (Basel). 2023;15(16):2980.
4. Moon J, Son S, Kim J, Park K. Critical challenges in high-salinity seawater reverse osmosis systems: technical, energy, and environmental reviews. Desalination. 2025;118811
5. Abu Elella MH, Abu-Thabit NY, Uwaezuoke OJ, Azad AK. Superwetting cotton textiles for separation of oil/water mixtures. Cellulose. 2023;30:7427–62.
6. Abu-Thabit NY, Uwaezuoke OJ, Elella MHA. Superhydrophobic nanohybrid sponges for separation of oil/water mixtures. Chemosphere. 2022;294:133644.
7. Azad AK, Abu-Thabit N, Abu Eella M. Sustainable and durable Superwetting textiles coated with Polydopamine for oil/water separation. RSC Sustainability [Internet] 2025; Available from: https://doi.org/10.1039/D5SU00041F
8. Abu-Thabit NY, Elella MHA, Azad AK, Ratemi E, Hakeem AS. Superwetting metal mesh membranes for oil/water separation: a comprehensive review. Sep Purif Technol. 2025:132016.
9. Yuan S, Zhang S, Tang X. Research progress on flocculation-based technology for the enhancement of sludge dewatering: a review. Sep Sci Technol. 2024;59(10–14):1183–201.
10. Elella MHA, Castro-Riquelme CL, López-Maldonado EA, Abady MM, Mohammed DM, Abu-Thabit NY, et al. Synthesis of NPs by wastewater and bio-sludge. In: Waste-derived nanoparticles. Elsevier; 2024. p. 125–55.
11. Elella MHA, Hassan SS, Abdallah HM, Mostafa MS, Abu-Thabit NY. Metal chalcogenides and their nanocomposites in water purification systems. Met—Chalcogenide Nanocomposites. 2024:59–82.

12. Osman AI, Chen Z, Elgarahy AM, Farghali M, Mohamed IMA, Priya AK, et al. Membrane technology for energy saving: principles, techniques, applications, challenges, and prospects. Adv Energy Sustainability Res. 2024;5(5):2400011.
13. Liu L, Wang Y, Liu Y, Wang J, Zheng C, Zuo W, et al. Insight into key interactions between diverse factors and membrane fouling mitigation in anaerobic membrane bioreactor. Environ Pollut. 2024;123750
14. Foorginezhad S, Zerafat MM, Ismail A, Goh PS. Emerging membrane technologies for sustainable water treatment: a review of recent advances. Environ Sci: Adv. 2025;
15. Diachenko A, Hutsul K, Dontsova T. Prospects for using 3d printing to form ceramic membranes: a brief review. Water Water Purif Technol Sci Tech News. 2024;38(1):14–26.
16. Molavi H, Mirzaei K, Barjasteh M, Rahnamaee SY, Saeedi S, Hassanpouryouzband A, et al. 3d-printed MOF monoliths: fabrication strategies and environmental applications. Nanomicro Lett. 2024;16(1):272.
17. Kim D, Nunes SP. Green solvents for membrane manufacture: recent trends and perspectives. Curr Opin Green Sustain Chem. 2021;28:100427.
18. Liu Z, Jiang Q, Zhang Y, Li T, Zhang HC. Sustainability of 3D printing: a critical review and recommendations. In: International manufacturing science and engineering conference. American Society of Mechanical Engineers; 2016. p. V002T05A004.
19. Javaid M, Haleem A, Singh RP, Suman R, Rab S. Role of additive manufacturing applications towards environmental sustainability. Adv Ind Eng Polym Res. 2021;4(4):312–22.
20. Maines EM, Porwal MK, Ellison CJ, Reineke TM. Sustainable advances in SLA/DLP 3D printing materials and processes. Green Chem. 2021;23(18):6863–97.
21. Thorbole A, Bodhak C, Sahu P, Gupta RK. Green ink revolution: pioneering eco-friendly 3D printing with vegetable oil-derived inks. Polym Eng Sci. 2024;64(12):5879–902.
22. Ligon SC, Liska R, Stampfl J, Gurr M, Mülhaupt R. Polymers for 3D printing and customized additive manufacturing. Chem Rev. 2017;117(15):10212–90.
23. Yuk H, Lu B, Lin S, Qu K, Xu J, Luo J, et al. 3D printing of conducting polymers. Nat Commun. 2020;11(1):1604.
24. Yang E, Miao S, Zhong J, Zhang Z, Mills DK, Zhang LG. Bio-based polymers for 3D printing of bioscaffolds. Polym Rev. 2018;58(4):668–87.
25. Wang Y, Xu Z, Wu D, Bai J. Current status and prospects of polymer powder 3D printing technologies. Materials. 2020;13(10):2406.
26. Suriano R, Bernasconi R, Magagnin L, Levi M. 4D printing of smart stimuli-responsive polymers. J Electrochem Soc. 2019;166(9):B3274–81.
27. Choi I, Jang S, Jung S, Woo S, Kim J, Bak C, et al. A dual stimuli-responsive smart soft carrier using multi-material 4D printing. Mater Horiz. 2023;10(9):3668–79.
28. González-Henríquez CM, Sarabia-Vallejos MA, Rodriguez-Hernandez J. Polymers for additive manufacturing and 4D-printing: materials, methodologies, and biomedical applications. Prog Polym Sci. 2019;94:57–116.
29. Fu P, Li H, Gong J, Fan Z, Smith AT, Shen K, et al. 4D printing of polymers: techniques, materials, and prospects. Prog Polym Sci. 2022;126:101506.
30. Park S, Shou W, Makatura L, Matusik W, Fu KK. 3D printing of polymer composites: materials, processes, and applications. Matter. 2022;5(1):43–76.
31. Ma T, Zhang Y, Ruan K, Guo H, He M, Shi X, et al. Advances in 3D printing for polymer composites: a review. InfoMat. 2024;6(6):e12568.
32. Arefin AME, Khatri NR, Kulkarni N, Egan PF. Polymer 3D printing review: materials, process, and design strategies for medical applications. Polymers (Basel). 2021;13(9):1499.
33. Zhou L, Fu J, He Y. A review of 3D printing technologies for soft polymer materials. Adv Funct Mater. 2020;30(28):2000187.
34. Cardoso PHN, Araújo ES. An approach to 3D printing techniques, polymer materials, and their applications in the production of drug delivery systems. Compounds. 2024;4(1):71–105.
35. Aghajani M, Garshasbi HR, Naghib SM, Mozafari MR. 3D printing of hydrogel polysaccharides for biomedical applications: a review. Biomedicine. 2025;13(3):731.

36. Teixeira MC, Lameirinhas NS, Carvalho JPF, Silvestre AJD, Vilela C, Freire CSR. A guide to polysaccharide-based hydrogel bioinks for 3D bioprinting applications. Int J Mol Sci. 2022;23(12):6564.
37. Abu-Thabit N, Akhtar S, Aldhafeeri TN, Alshammari S, Hakeem AS. Fabrication of high-adhesion hydrophobic filter paper and its application as a device for microscale synthesis. In: Proceedings of the 4th international online conference on materials. Basel, Switzerland: MDPI; 2025.
38. Mandal S, Nagi GK, Corcoran AA, Agrawal R, Dubey M, Hunt RW. Algal polysaccharides for 3D printing: a review. Carbohydr Polym. 2023;300:120267.
39. Shokrani H, Shokrani A, Seidi F, Mashayekhi M, Kar S, Nedeljkovic D, et al. Polysaccharide-based biomaterials in a journey from 3D to 4D printing. Bioeng Transl Med. 2023;8(4):e10503.
40. Tai C, Bouissil S, Gantumur E, Carranza MS, Yoshii A, Sakai S, et al. Use of anionic polysaccharides in the development of 3D bioprinting technology. Appl Sci. 2019;9(13):2596.
41. Mahendiran B, Muthusamy S, Sampath S, Jaisankar SN, Popat KC, Selvakumar R, et al. Recent trends in natural polysaccharide based bioinks for multiscale 3D printing in tissue regeneration: a review. Int J Biol Macromol. 2021;183:564–88.
42. Wang X, Li H, Mu M, Ye R, Zhou L, Guo G. Recent development and advances on polysaccharide composite scaffolds for dental and dentoalveolar tissue regeneration. Polym Rev. 2025;65(1):47–103.
43. Hernandez-Tenorio F, Múnera-Gutiérrez E, Miranda AM, Sáez AA, Marín-Palacio LD, Giraldo-Estrada C. 3D printing of polysaccharide-based formulations: opportunities for innovation. Bioprinting. 2024:e00383.
44. de Pastre MA, Quinsat Y, Lartigue C. Effects of additive manufacturing processes on part defects and properties: a classification review. Int J Interact Des Manuf (IJIDeM). 2022;16(4):1471–96.
45. Kivanani AN, Khalilpourazary S, Mobadersani F. Additive manufacturing for producing microchannel heat sinks. Prog Addit Manuf. 2025;10(1):231–45.
46. Martínez-García A, Monzón M, Paz R. Chapter 12—Standards for additive manufacturing technologies: structure and impact. In: Pou J, Riveiro A, Davim JPBTAM, editors. Handbooks in advanced manufacturing [internet]. Elsevier; 2021. p. 395–408. Available from: https://www.sciencedirect.com/science/article/pii/B9780128184110000136
47. Gusain R, Kumar N, Ray SS. 3D-printed hydrogels and aerogels for water treatment and energy storage applications. ChemistrySelect [Internet]. 2023 8(20):e202300738. Available from: https://doi.org/https://doi.org/10.1002/slct.202300738.
48. Lyu Z, Lim GJH, Koh JJ, Li Y, Ma Y, Ding J, et al. Design and manufacture of 3D-printed batteries. Joule. 2021;5(1):89–114.
49. Li J, Wu C, Chu PK, Gelinsky M. 3D printing of hydrogels: rational design strategies and emerging biomedical applications. Mater Sci Eng: R: Rep [internet] 2020;140:100543. Available from: https://www.sciencedirect.com/science/article/pii/S0927796X20300012
50. Wei P, Cipriani C, Hsieh CM, Kamani K, Rogers S, Pentzer E. Go with the flow: rheological requirements for direct ink write printability. J Appl Phys [Internet]. 2023 134(10):100701. Available from: https://doi.org/10.1063/5.0155896.
51. Zamboulis A, Michailidou G, Koumentakou I, Bikiaris DN. Polysaccharide 3D printing for drug delivery applications. Pharmaceutics. 2022;14
52. Tay RY, Song Y, Yao DR, Gao W. Direct-ink-writing 3D-printed bioelectronics. Materials Today [Internet]. 2023;71:135–51. Available from: https://www.sciencedirect.com/science/article/pii/S1369702123003000
53. Sadeghianmaryan A, Ahmadian N, Wheatley S, Alizadeh Sardroud H, Nasrollah SAS, Naseri E, et al. Advancements in 3D-printable polysaccharides, proteins, and synthetic polymers for wound dressing and skin scaffolding—a review. Int J Biol Macromol [Internet]. 2024;266:131207. Available from: https://www.sciencedirect.com/science/article/pii/S0141813024020129
54. Abas M, Jan Z, Rahman K. Direct ink writing. In: Innovative development in micromanufacturing processes. CRC Press; 2023. p. 188–209.

55. Li C, Feng C, Zhang L, Zhang L, Wang L. Direct ink writing of polymer-based materials—a review. Polym Eng Sci [Internet] 2025 65(2):431–454. Available from: https://doi.org/10.1002/pen.27038.
56. Zhang H, Xue Y, Jiang C, Liu D, Zhang L, Lang G, et al. 3-Dimentional printing of polysaccharides for water-treatment: a review. Int J Biol Macromol [Internet]. 2024;265:131117. Available from: https://www.sciencedirect.com/science/article/pii/S0141813024019226
57. Liu Y, Zhang Y, Cai L, Zeng Q, Wang P. Protein and protein-polysaccharide composites-based 3D printing: the properties, roles and opportunities in future functional foods. Int J Biol Macromol [Internet]. 2024;272:132884. Available from: https://www.sciencedirect.com/science/article/pii/S0141813024036894
58. Ma Y, Morozova SM, Kumacheva E. From nature-sourced polysaccharide particles to advanced functional materials. Adv Mat [Internet]. 2024 36(23):2312707. Available from: doi:https://doi.org/10.1002/adma.202312707.
59. Giri BR, Poudel S, Kim DW. Cellulose and its derivatives for application in 3D printing of pharmaceuticals. J Pharm Investig [Internet]. 2021;51(1):1–22. Available from:. https://doi.org/10.1007/s40005-020-00498-5.
60. Taghizadeh M, Taghizadeh A, Yazdi MK, Zarrintaj P, Stadler FJ, Ramsey JD, et al. Chitosan-based inks for 3D printing and bioprinting. Green Chem [Internet] 2022;24(1):62–101. Available from: doi:https://doi.org/10.1039/D1GC01799C.
61. Piras CC, Smith DK. Multicomponent polysaccharide alginate-based bioinks. J Mater Chem B [Internet]. 2020;8(36):8171–88. Available from: https://doi.org/10.1039/D0TB01005G
62. Paraiso HF, Dib T, Zhu K, Hou Z, Chen S. Polysaccharide-based packaging: storage life and effectiveness in protecting ruminant meat qualities. Food Rev Int [Internet]:1–29. https://doi.org/10.1080/87559129.2025.2487897.
63. Deb D, Jafferson JM. Natural fibers reinforced FDM 3D printing filaments. Mater Today Proc [Internet] 2021;46:1308–1318. Available from: https://www.sciencedirect.com/science/article/pii/S2214785321015170.
64. Dou X, Ren J. The application of dairy products and their derivatives as edible inks in 3D printing technology: a review. Int J Food Sci Technol [Internet]. 2024;59(11):8630–44. Available from:. https://doi.org/10.1111/ijfs.17195.
65. Hakim L, Deshmukh RK, Lee YS, Gaikwad KK. Edible ink for food printing and packaging applications: a review. Sustain Food Technol. 2024;2(4):876–92.
66. Kaith A, Jain N, Kaul S, Nagaich U. Polysaccharide-infused bio-fabrication: advancements in 3D bioprinting for tissue engineering and bone regeneration. Mater Today Commun [Internet] 2024;40:109429. Available from: https://www.sciencedirect.com/science/article/pii/S2352492824014107.
67. Carou-Senra P, Rodríguez-Pombo L, Awad A, Basit AW, Alvarez-Lorenzo C, Goyanes A. Inkjet printing of pharmaceuticals. Adv Mat [Internet]. 2024 36;11:2309164. Available from: doi:https://doi.org/10.1002/adma.202309164.
68. Elviri L. Polysaccharide-based materials for the development of point-of-care devices. In: Handbook of the extracellular matrix. Springer-Nature; 2023.
69. Kacem E, Suk HD, Ponnamma D, Hassan MK, Hawari A, Alshammari BA, et al. Exploring 3D printing and electrospinning technologies for advanced porous membrane fabrication: a review. Adv Mater Technol. 2025:70032.
70. Randhawa A, Dutta SD, Ganguly K, Patel DK, Patil TV, Lim KT. Recent advances in 3D printing of Photocurable polymers: types, mechanism, and tissue engineering application. Macromol Biosci [Internet]. 2023;1(23):2200278. https://doi.org/10.1002/mabi.202200278.
71. Díaz-Montes E. Polysaccharides: sources, characteristics, properties, and their application in biodegradable films. Vol. 3, Polysaccharides. 2022. p. 480 501–501.
72. Chen L, He S, Huang W, Liu D, Bi T, Zhang C, et al. 3D-printed tripodal porous wood-mimetic cellulosic composite evaporator for salt-free water desalination. Compos B Eng [Internet]. 2023;263:110830. Available from: https://www.sciencedirect.com/science/article/pii/S1359836823003335.

73. Zhu P, Yu Z, Sun H, Zheng D, Zheng Y, Qian Y, et al. 3D printed cellulose nanofiber aerogel scaffold with hierarchical porous structures for fast solar-driven atmospheric water harvesting. Adv Mat [Internet]. 2024 36(1):2306653. Available from: doi:https://doi.org/10.1002/adma.202306653.
74. Koh JJ, Lim GJH, Chakraborty S, Zhang Y, Liu S, Zhang X, et al. Robust, 3D-printed hydratable plastics for effective solar desalination. Nano Energy. 2021;79:105436.
75. Guo ZH, Chen L, Liu Q, Lang JQ, Yang SC, Mai T, et al. 3D-printed dual-scale porous hydrogel evaporator with biomimetic architecture for high-efficiency solar desalination and sustainable salt rejection. Chem Eng J. 2025;167800
76. Zhang S, Li M, Jiang C, Zhu D, Zhang Z. Cost-effective 3D-printed bionic hydrogel evaporator for stable solar desalination. Adv Sci. 2024;11(17):2308665.
77. Yuan J, Lei X, Yi C, Jiang H, Liu F, Cheng GJ. 3D-printed hierarchical porous cellulose/alginate/carbon black hydrogel for high-efficiency solar steam generation. Chem Eng J. 2022;430:132765.
78. Yang X, Guo X, Tian Y, Zhou R, Gong Y, Zhang C, et al. 3d PRINTING OF MULTI-FUNCTIONAL COMPOSITE HYDROGEL EVAPORATORS FOR EFFICIENT SOLAR DESALINATION AND PHOTOCATALYTIC DEGRADATion. Available at SSRN 5286481.
79. Wang Z, Yan C, Ji K, Cao J, Chen Y, Gao Y, et al. 3D printing of flash graphene/nano-cellulose aerogel for high-efficiency solar evaporation with superior salt resistance. Chem Eng J. 2025;165496
80. Etale A, Onyianta AJ, Turner SR, Eichhorn SJ. Cellulose: a review of water interactions, applications in composites, and water treatment. Chem Rev. 2023;123(5):2016–48.
81. Chami Khazraji A, Robert S. Interaction effects between cellulose and water in nanocrystalline and amorphous regions: a novel approach using molecular modeling. J Nanomater. 2013;2013(1):409676.
82. Koh JJ, Lim GJH, Zhou X, Zhang X, Ding J, He C. 3D-printed anti-fouling cellulose mesh for highly efficient oil/water separation applications. ACS Appl Mater Interfaces. 2019;11(14):13787–95.
83. Zhou X, Koh JJ, He C. Robust oil-fouling resistance of amorphous cellulose surface underwater: a wetting study and application. Langmuir. 2019;35(4):839–47.
84. Jiang F, Hsieh YL. Amphiphilic superabsorbent cellulose nanofibril aerogels. J Mater Chem A Mater. 2014;2(18):6337–42.
85. Jiang F, Liu H, Li Y, Kuang Y, Xu X, Chen C, et al. Lightweight, mesoporous, and highly absorptive all-nanofiber aerogel for efficient solar steam generation. ACS Appl Mater Interfaces. 2018;10(1):1104–12.
86. Xu J, Li T, Chao J, Wu S, Yan T, Li W, et al. Efficient solar-driven water harvesting from arid air with metal–organic frameworks modified by hygroscopic salt. Angew Chem Int Ed. 2020;59(13):5202–10.
87. Fathieh F, Kalmutzki MJ, Kapustin EA, Waller PJ, Yang J, Yaghi OM. Practical water production from desert air. Sci Adv. 2018;4(6):eaat3198.
88. Kim H, Yang S, Rao SR, Narayanan S, Kapustin EA, Furukawa H, et al. Water harvesting from air with metal-organic frameworks powered by natural sunlight. Science. 2017;356(6336):430–4.
89. Chen Y, Yu Z, Ye Y, Zhang Y, Li G, Jiang F. Superelastic, hygroscopic, and ionic conducting cellulose nanofibril monoliths by 3D printing. ACS Nano. 2021;15(1):1869–79.
90. He N, Wang H, Zhang H, Jiang B, Tang D, Li L. Ionization engineering of hydrogels enables highly efficient salt-impeded solar evaporation and night-time electricity harvesting. Nanomicro Lett. 2023;
91. Yan H, Wang P, Li L, Zhao Z, Xiang Y, Guo H, et al. Development status of solar-driven interfacial steam generation support layer based on polymers and biomaterials: a review. Polymers (Basel). 2024;
92. Zhang J, Zhang L, Zhao J, Qu W, Wang Z. A facile and mild strategy to fabricate an underwater superoleophobic and underoil superhydrophobic mesh with outstanding anti-viscous oil-fouling properties for switchable high viscosity oil/water separation. Green Chem. 2019;21(18):5080–9.

93. Wang Y, Hu J, Yu L, Wu X, Zhang Y, Xu H. Recent strategies for constructing efficient interfacial solar evaporation systems. Nano Res Energy. 2023;
94. Zhang C, Liang HQ, Xu Z, Wang Z. Harnessing solar-driven Photothermal effect toward the water–energy nexus. Adv Sci. 2019;
95. Ji Z, Zhao J, Feng S, Zhu F, Yu W, Ye Y, et al. Insight into the charge-ratio-tuned solar vapor generation of Polyion complex hydrogel/coal powder composites. Polymers (Basel). 2023
96. Xu C, Zhang J, Shahriari-Khalaji M, Gao M, Yu X, Ye C, et al. Fibrous aerogels for solar vapor generation. Front Chem. 2022;10:843070.
97. Wu F, Qiang S, Zhu XD, Jiao W, Liu L, Yu J, et al. Fibrous mxene aerogels with tunable pore structures for high-efficiency desalination of contaminated seawater. Nano Micro Lett. 2023;15(1):71.
98. Wang T, Li M, Xu H, Wang X, Jia M, Hou X, et al. MXene sediment-based poly (vinyl alcohol)/ sodium alginate aerogel evaporator with vertically aligned channels for highly efficient solar steam generation. Nanomicro Lett. 2024;16(1):220.
99. Han S, Ruoko T, Gladisch J, Erlandsson J, Wågberg L, Crispin X, et al. Cellulose-conducting polymer aerogels for efficient solar steam generation. Adv Sustain Syst. 2020;4(7):2000004.
100. Li S, Yang X, He Y, Wang Y, Liao D, Chen Y, et al. Hierarchical porous aero-cryogels for wind energy enhanced solar vapor generation. Cellulose. 2022;29(2):953–66.
101. Xu C, Gao M, Yu X, Zhang J, Cheng Y, Zhu M. Fibrous aerogels with tunable superwettability for high-performance solar-driven interfacial evaporation. Nanomicro Lett. 2023;15(1):64.
102. Tijing L, Dizon JR, Cruz G Jr. 3D-printed absorbers for solar-driven interfacial water evaporation: a mini-review. Adv Sustain Sci, Eng Technol. 2021;3(1):343550.

3D-Printed Polysaccharide Biofilms for Environmental Remediation

Mahmood Ahmed, Ali Abbas Aslam, Ahmad Saeed, and Awais Ali Aslam

Abstract Polysaccharide biofilms 3D-printed have provided a significant level of interest because of their sustainable, biodegradable, and changeable properties. The biofilms have high adsorption capacity, mechanical strength, and controllable porosity, making them useful in the elimination of pollutants like heavy metals, organic, and microplastic substances in aqueous and soil matrices. In this chapter, the synthesis, characterization, and use of 3D-printed polysaccharide biofilms are outlined with a stress on the role of these biofilms in sustainable environmental practices. Major developments, including the strategies of functionalization, scalability, and field deployment, are discussed. In addition, barriers in cost, reproducibility, and regulatory compliance are examined, as well as the potential opportunities to increase their environmental impact.

Keywords 3D-printed biofilms · Environmental remediation · Water purification · Heavy metal removal · Biodegradable

Abbreviations

Alg	Alginate
PCL	Polycaprolactone
CS	Chitosan Subtilis
CEL	Cellulose
PLA	Polylactic acid
CA	Ca Alginate
PHAs	Polyhydroxyalkanoates

M. Ahmed (✉) · A. A. Aslam · A. Saeed
Department of Chemistry, Division of Science and Technology, University of Education, Lahore, Pakistan

A. A. Aslam
Department of Chemical Organic Technology and Petrochemistry, Silesian University of Technology, Gliwice, Poland

M. H. Abu Elella et al. (eds.), *3D Engineered Polysaccharides*,
https://doi.org/10.1007/978-3-032-26154-0_6

CAD	Computer aided design
IP	Inkjet printing
SLS	Selective laser sintering
LOM	Laminated object manufacturing
TCNF	TEMPO-oxidized cellulose nanofibers
CBE	Circular bioeconomy
PBPs	Petroleum-based polymers
LCA	Life cycle analysis
GO	Graphene Oxide
E. Coli	Escherichia Coli
B. Subtilis	Bacillus Subtilis
EPS	Extracellular polymeric substances
PVA	Polyvinyl alcohol
FDM	Fused deposition model
DIW	Direct ink writing
SLA	Stereolithography
SLM	Selective laser melting
BOD	Biochemical oxygen demand
ChNF	Chitin nanofibers
ZLD	Zero liquid discharge
CAGR	Compound annual growth rate
TEA	Thermal electric aerosol

1 Introduction

Water is a vital resource that cannot be ignored, and it is essential for everyday life, agricultural activities, and industrial growth. However, only about 0.26% of the entire hydrosphere of the earth, which is comprised of surface water (e.g., rivers, lakes, mountain streams, etc.) and, to a lesser extent, underground aquifers, is fresh water [1, 2]. Conversely, terrestrial water is mostly salty oceanic water that has a mean salinity of 35,000 mg/L [3]. Nevertheless, freshwater is facing growing pressure because of increased industrialization, which has resulted in widespread contamination in industries such as steel manufacturing, petroleum refining, coal processing, textile dyeing, and electronic manufacturing. To achieve sustainable development, there is an urgent need to introduce measures that make the process of recycling and reusing freshwater easy, while also reducing energy use and pollution rates [4, 5]. Research on wastewater treatment includes reverse osmosis [6], forward osmosis [7], electroosmosis [8], membrane-based treatment [9], enhanced evaporation [10], mechanical vapor compression [11], and solar-driven evaporation [12]. However, the limitations to these technologies are often their susceptibility to fouling, inefficient operation, and high operating costs, which highlight the pressing need to develop novel, cost-effective, and environmentally benign technologies in managing industrial wastewater. 3D printing has become a flexible and potentially valuable technology that allows for creating components with tailored geometries and architectures, improving performance in various applications, including mechanical engineering [13], aerospace [14], medicine [15], tissue engineering [16], and environmental applications [17]. Its flexibility and accuracy have made it a useful tool in research and industry. Recent attention has focused on water-compatible and eco-friendly materials, such as polysaccharides, whose sustainability and versatility in use have made them appealing for 3D printing [18]. Specifically, 3D-printed matrices of physically/chemically modified polysaccharides, achieved through surface grafting, etching, or polymerization, exhibit a variety of desirable properties (Fig. 1), including high biocompatibility [20], super-wettability [21], and catalytic activity [20]. A combination of 3D printing technology with bio-based polysaccharides represents significant potential in developing sustainable programs [19, 22, 23] and a circular bio-economy across various industries [22].

To design functional living systems, 3D-printed polysaccharide biofilms are at the forefront of biotechnological innovation, combining the versatile physicochemical

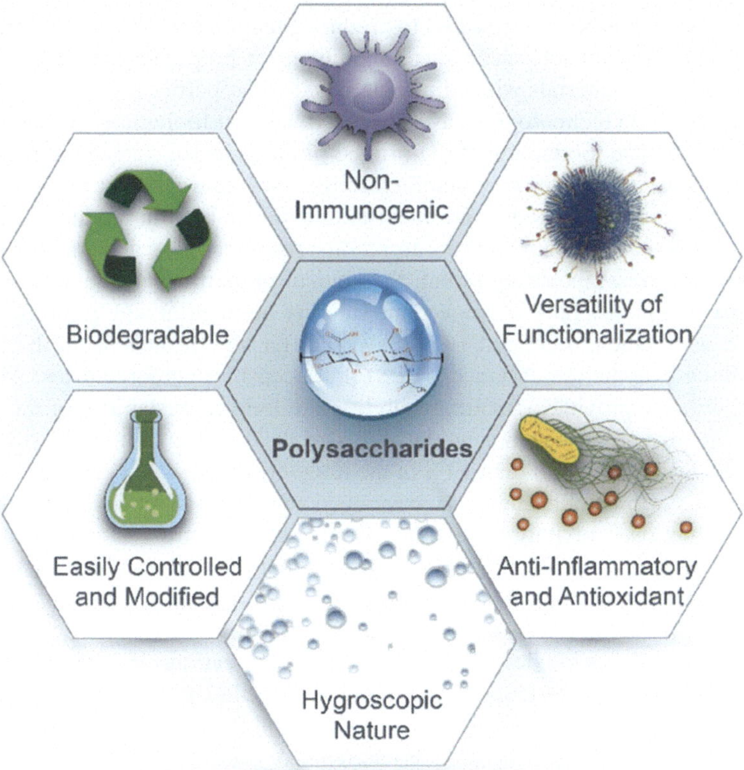

Fig. 1 Properties of 3D polysaccharides. Reproduced with permission from the ref. [19], under creative commons attribution (CC BY) license, 2018

properties of naturally occurring polysaccharides with the precision of additive manufacturing [23]. These biofilms are typically produced using biocompatible polysaccharides, such as alginate. The traditional production protocol consists of mixing microbial cultures with Alg to form a bioink, which is then deposited on the surface of a calcium-rich substrate. The direct cross-linking of Alg under the influence of calcium ions results in well-defined hydrogel networks that firmly fix the bacteria in specific spatial geometries, allowing the development of highly complex 3D structures with mechanical properties that replicate the behavior of natural biofilms and maintain microbial activity over prolonged time [24].

In this field, major advances have been made through the genomic modification of bacterial strains aimed at enhancing the functionality of biofilms within 3D printed structures [25]. For example, in a study, the development involves altering E. coli to produce inducible curli fibers, proteinaceous components commonly found in natural biofilms, which are added to enhance the mechanical strength and environmental stability of printed materials. This genetic control enables the printed microbial systems to maintain their spatial arrangement in the presence of the Alg matrix, even after the degradation of the original Alg matrix, thereby producing a self-sustaining microbial environment [25]. Similarly, using *B. subtilis,* the genetically modified microorganism, has been engineered to produce functional amyloid proteins, giving the biofilm matrix tunable mechanical and chemical properties, and also enabling the generation of materials with adjustable physical properties [23].

Although these technologies have improved, several limitations are likely to be associated with the production of 3D-printed biofilms, particularly in terms of high spatial resolution and microbial viability. Extrusion forces inherent to the extrusion process may compromise bacterial integrity and reduce biofilm productivity. In this respect, the researchers are working to modify parameters, such as those that alleviate cellular stress, including bioink rheological properties, nozzle geometry, and cross-linking kinetics [25]. Moreover, the stability of the print after printing remains an issue, as synthetic matrices, such as Alg, differ significantly from the original extracellular polymeric substances (EPS) produced by microorganisms. Although the Alg initially provides structural stability, the long-term stability of the printed biofilm, in most cases, depends on the ability of the engineered microbes to replace the synthetic scaffold with their EPS in the long term.

The goals of materials chemistry have been achieved with the aid of 3D-printed biofilms composed of polysaccharides. Composite systems integrating materials like CS, cellulose, or synthetic polymers have enhanced both mechanical durability and functional capabilities. New formulations can help produce smart biofilms that contain nanoparticles and stimuli-responsive biomaterials [24, 26].

Research explores the design and biomedical applications of 3D-printed natural polysaccharide films, particularly in regenerative medicine, drug delivery [27–29], and tissue engineering for cartilage, bone, skin, and the vascular system [27]. One such innovation is composite hydrogels based on polysaccharides and proteins, which have the greatest clinical potential. However, as interest in integrating 3D printing with polysaccharide biomaterials continues to grow, few studies have been conducted on the utilization of such materials in environmental settings [28, 29].

The growing amounts of pollution in the air and water require innovative and sustainable solutions because most traditional methods of remediation have resulted in producing more waste and have failed to eliminate the pollutants. The recent advances in the functionalization of materials, large-scale production, and practical implementation indicate that these new methodologies have great potential in their wide application. However, there are a few major issues to consider before full implementation of these technologies takes place. They include the high cost of production, performance consistency, and regulatory factors. Thus, the emphasis of the chapter on biofilms of polysaccharides 3D-printed is quite justified. This chapter highlights their potential as effective, sustainable, and scalable technologies to deal with the current acute environmental issues and minimize the use of conventional remediation strategies.

2 Material Properties and Functionalization of Polysaccharide Biofilms

Biofilms of polysaccharides also have certain material properties such as high-water retention, biocompatibility, biodegradability, and mechanical flexibility. They can be functionalized by chemically modifying them, cross-linking them, or introducing NPs to increase their strength, stability, and antimicrobial activity. These alterations also enhance selective permeability, thus allowing the regulated passage of nutrients and bioactive agents. As a result, these biopolymers functionalized biofilms are useful in environmental remediation.

2.1 Adsorption Properties

3D-printed biofilms are on the verge of becoming the next generation of materials in cleaning up the environment, due to the biocompatibility and chemical versatility of these materials. The materials are improved with the exact structural control offered by additive manufacturing and make use of their functional groups (hydroxyl, carboxyl, and amino) to trap a broad range of contaminants. Their modes of binding adsorption are electrostatic, ionic exchange, hydrogen bonding, and chelation [30, 31]. Most of these include materials based on Alg, which can quickly cross-link with divalent ions, including $Ca^{2+,}$ into porous high-surface-area scaffolds that are ideal to trap contaminants. As an example, direct-ink-writing methods have been used to develop CA scaffolds that can take up to 117.3 mg/g of uranium at pH 2.5, which is significantly better than conventional Alg beads because of their well-organized macroporous structures and available active sites [32, 33]. These hydrogels, which have pore sizes between 10 and 800 μm, enhance diffusion and make it possible to extract 90% of uranium in 30 min in an acidic environment. To further improve such systems, CS is utilized as a source of protonated amino groups in acidic conditions, and this system effectively adsorbs anionic contaminants like methyl orange through the electrostatic forces. The adsorption capacity of Pb on composite systems, including CS and ZIF-67 MOFs in 3D-printed lattice structures, has been shown to have a high capacity of 5.47 mg/g by taking advantage of both binding sites of CS and the microporosity of ZIF-67 [30, 33].

The physical and chemical arrangements of these biofilms have a major influence on their adsorption capacity. Nozzle diameter (100–500 μm) and printing speed (5–15 mm/s) have to balance the supply of high-resolution structures and pollutant-degrading microbiological viability of embedded microbes [34]. An example is the ability of genetically modified biofilms of *Pseudomonas aeruginosa* overproducing Psl exopolysaccharides to entrap microplastics up to 92% of genetically modified biofilm by hydrophobic interactions. Further, the use of cellulose nanocrystals in Alg-matrix can enhance heavy-metal uptake twice because hydroxyl groups are available as free groups [31]. The importance of structural optimization is also substantial, as smooth surfaces of cubic CS hydrogels are 1.2 times better at removing methyl orange compared to rough and semi-spherical ones because of a more effective diffusion process. (Fig. 2) Combinations of ionic Alg with either methacrylate-modified CS as hybrid cross-linking generate a material that is resistant to shear force (up to 0.5 P/A) and is also mechanically intact, with compression moduli in 10–50 kPa [32, 35].

The performance of these biofilms is also affected by the environmental factors. Graphene oxide (GO) enhanced Alg-CS composites exhibit a high adsorption rate in a wide pH range [3–12], whereby the uptake of uranium reached its highest point at 200.1 mg/g at pH 2.8 because of concomitant ion exchange and complexation of oxygen functional groups of GO. When comparing a pure CS-based hydrogel, it is most effective at a pH of around 5–6, where amino groups are the most effective absorbers of dyes. The use of thermoresponsive polymers like poly(N-isopropylacrylamide) enables the release of the pollutant when heated, e.g., 78% of adsorbed methyl orange can be desorbed at 40 °C. However, there is a risk of structural integrity of these materials being compromised with long exposure to UV light, and there is now the use of UV-protective additives to enable them to be used in the field [30, 37].

2.2 *Mechanical Stability and Durability*

Biofilms can be further improved by the incorporation of live microorganisms. The gelation of three-dimensional printed Alg films with genetically modified strains of *E. coli* with the ability to produce curli fibers is found to have about 80% viability after the printing of the gel and has the ability to secrete laccase enzymes continuously, thus decolorizing synthetic dyes. It has also been investigated that the use of *B. subtilis* and *P. putida* has shown a decline in polyethylene microplastic levels by 15% in two weeks [34, 38]. Other embodiments include cyanobacteria containing hydrogels that help in removing 83% of copper ions by biosorption, and scaffold systems inoculated with

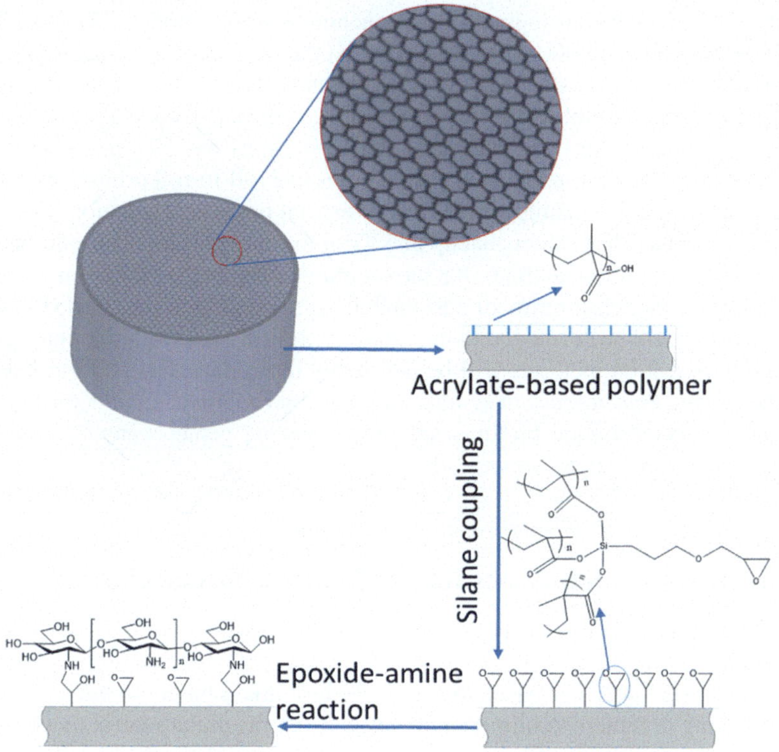

Fig. 2 Schematic diagram of how a complex 3D structure with CS coating was fabricated to remove metal ions. Reproduced with permission from the ref. [36], Springer Nature, copyright 2019

Oceanimonas that accomplish the goal of 96% ammonia removal by nitrification, thus highlighting the viability of programmable biodegradation platforms [38].

Nevertheless, the reuse and durability are still primary issues. CS monoliths treated with silane maintain 78% of their dye removal capacity after four reuse cycles but lose about 12% of their mass to structural erosion [39]. The water resistance and charge density (5.0 mmol/g) of nanocellulose-CS composites can be enhanced by dehydrothermal crosslinking with citric acid at 120 °C, thus increasing enzyme immobilization capacity. Polyurea crosslinked Alg aerogels have demonstrated an adsorption capacity of 2023 g/kg and full desorption of uranium by using Na_2CO_3, but they are not resistant to acid degradation [40].

Applications of these biofilms are also determined by material-specific properties. As an example, the adsorption capacity of oxidized *Luffa cylindrica* in combination with Alg has a maximum capacity of 404.6 mg/g of uranium in the inner-sphere complexation, which is higher than plain alginate (335.6 mg/g) but has a better performance when the pH is lower than 4 [41]. On the other hand, the polycaprolactone (PCL)/Alg composites show stable copper adsorption (90 mg/g) in a wide pH range [3–10], and the hydrophobic PCL layer of the composites provides structural stability [31]. Recent developments use machine-learning models that have been trained on more than 15,000 datasets to forecast the best biofilm formulations, with an error margin of less than 6% on new pollutants [34].

In spite of these technological advances, scaling up these hydrogels is a bottleneck. The existing printing rates (10–100 mL/h) and resolution (over 50 μm) are barriers to large-scale implementation, but new technologies like roll-to-roll printing and multi-nozzle systems are promising solutions to industrial process production [34]. Life-cycle assessments (LCA) show that up to 60% of the energy is saved as compared to the traditional separation methods, but the sourcing of Alg in the marine environment is a matter of concern in terms of sustainability. It is expected that CRISPR-based safety switches will be developed in the future to prevent genetic contamination, and novel polymerization methods will be developed to produce sub-100 μm features. With the onward convergence of synthetic biology and additive manufacturing, these engineered polysaccharide biofilms are set to become game changers in solving complex environmental pollution [38].

3 Fabrication Techniques for 3D-Printed Polysaccharide Biofilms

The production of polymer-based materials for biofilms is based on the use of naturally occurring polymers to mimic or enhance the extracellular matrix that supports the microbial communities. Developed fabrication protocols, including electrospinning, solvent casting, and 3D printing, give perfect control over structural framework and porosity. These synthetic matrices not only allow the formation of biofilms, but also allow controlled nutrient diffusion, removal of waste, and functionalization with antimicrobial or signaling compounds. Ultimately, these materials provide versatile platforms for biomedical, environmental, and industrial applications of biofilms.

3.1 Additive Manufacturing Technologies

3D printing is a relatively new technology that has quickly spread as a multipurpose and flexible approach to fabrication. It starts with the development of a digital model through the computer-aided design (CAD) software that is then translated into a stereolithography (STL) file. This STL file is then fed through slicing software, which allows the printer to read and print the build layer by layer. Once this has been done, temporary supports are eliminated, or separate parts are built to create more complicated forms or operational equipment [42–44]. Biopolymers-based 3D printing technologies are generally categorized into a number of different types depending on the control mechanism used: fused deposition modelling (FDM), direct ink writing (DIW), inkjet printing (IP), stereolithography (SLA), selective

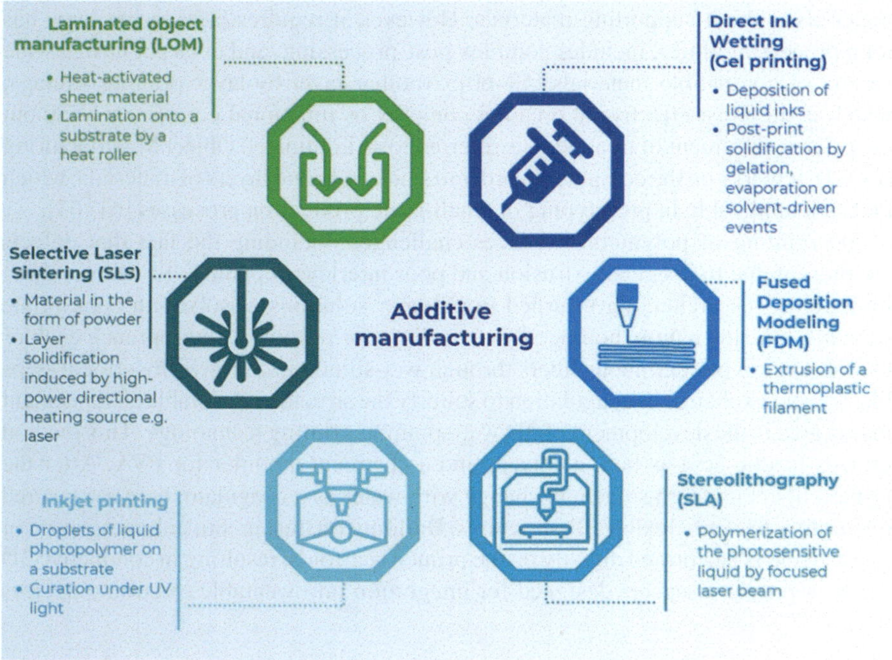

Fig. 3 Illustration mentioned the additive manufacturing techniques. Reproduced with permission from the ref. [45], Elsevier, copyright 2022

laser sintering or melting (SLS/SLM), and laminated object manufacturing (LOM) (Fig. 3).

FDM is one of the techniques that can be singled out due to its relatively low cost and ease of use, as it does not require the use of toxic solvents. However, the FDM-printed specimens usually face the problem of warping in the cooling process, and this may affect the mechanical integrity and lead to rough textures on the surface [46–49]. Conversely, DIW allows shaping of viscoelastic materials at room temperature, eliminating the use of heat, UV light, or lasers. One of the disadvantages of DIW is that the printed items tend to shrink during drying, and this process may give inaccurate size and geometry. Although the inkjet printing process is similar to DIW in terms of the equipment and materials used, it is different in that it does not use a continuous flow but piezoelectric processes, dispersing the ink droplets [50–52]. SLA involves the use of a laser to cure layers of a resin to create complex structures in a high-precision, even to the nanoscale. Although SLA has some accuracy, it has the disadvantage of prolonged processing time, increased costs, and environmental issues due to the use of harmless chemicals in some light-sensitive resins [53–56].

SLS is a 3D printing method, whereby the layers of finely spread powder material are selectively fused with the laser beam across the build area (Fig. 4) [57]. A new cross-sectional layer is added and fused onto the previous layer, creating the object step by step. SLS is beneficial because of its ability to produce highly structured materials without supporting materials. However, it requires laser technology, has long processing times, includes complex post-processing, and does not have a wide variety of compatible materials [58–60]. Another layer-by-layer printing strategy, which is the most effective at creating complex or functional components without the prior requirement of treatment, is referred to as Laminated Object Manufacturing (LOM). It works on the computer in order to stack and bond layers of materials, which makes it applicable in prototyping or small-scale production processes [61–63].

3D printing of polymers still faces challenges, including the fact that defects are present due to irregular extrusion and poor interlayer bonding. To address these limitations, researchers have turned to polymer solutions dissolved in appropriate solvents, offering a more homogeneous medium for printing. This approach enables the fabrication of continuous fibers through wet-spinning, followed by critical steps like solvent exchange or coagulation to solidify the structure. A notable advancement in this area is the development of 3D wet-spinning printing technology. This method utilizes formic acid as both a solvent and a chemical modifier for PVA. After the printed fibers undergo solvent exchange with water as a coagulant, they are layered into tightly packed, flexible 3D structures. Building on this, in-situ polymerization of polypyrrole is performed directly on the printed scaffolds, resulting in conductive 3D strain or pressure sensors designed for integration into wearable electronic devices [64].

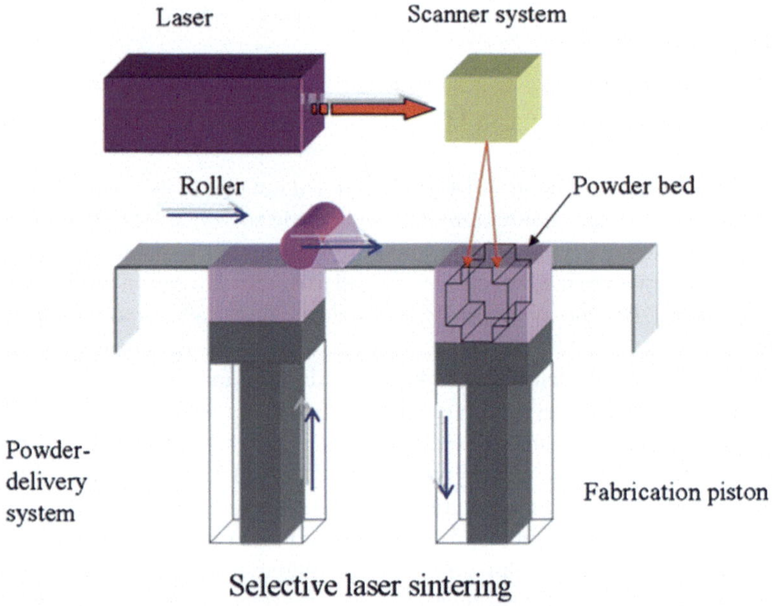

Fig. 4 Schematic representation of selective laser sintering (SLS) technique. Reproduced with permission from the ref. [57], Taylor & Francis, 2009

Polymer substrates commonly used in 3D printing are thermoplastic resins, photosensitive resins, rubber, and biomacromolecules. Recently, polymer solutions have gained traction as a promising alternative, particularly for innovative 3D wet-spinning printing techniques [65]. Over the past few years, polymer solutions have turned out to be a feasible alternative, especially when it comes to novel 3D wet-spinning printing methods [65, 66].

Cellulose is the most common and the most prevalent polysaccharide, which is hydrophilic, biodegradable, biocompatible, and crystalline [67]. The second most common polysaccharide on earth is chitin and is widely found on cell walls of various organisms, including arthropods, mollusks, annelids, protozoa, coelenterates, fungi, and algae [68]. Deacetylation of chitin gives CS as an alkaline polysaccharide [69]. Alg is an acidic polymer derived as a cell wall product of brown algae and is commonly used in a wide range of biomedical and industrial applications [70, 71].

3.2 Optimization of 3D Printing Parameters

The optimization of printing parameters of polysaccharide-based biofilms requires a careful balance between the rheological properties, mechanical load management, and the biological compatibility to ensure structural fidelity without losing the viability of microbes [24]. The viscosity of the bioink should be optimized depending on the deposition method: extrusion-based fabrication usually requires high viscosities between 6 and 30 $\times$ 10^7 mPa·s to guarantee extrusion and precise dimensions, whereas inkjet printing requires much lower viscosities of 3.5–12 mPa·s to prevent clogging of the nozzle [24]. The development of Alg-based biofilms requires Alg concentrations of 2–5% w/v. Even though high levels of algae enhance mechanical resilience, they also raise the shear stress, which may adversely affect microorganisms encapsulated within them during extrusion [72].

The diameter of the Nozzle is critical to the choice of printing resolution and cell survival. Smaller nozzles (100–500 μm) are more precise but can cause a loss in bacterial viability due to shear, to a maximum of 30%. To address this, temperature-controlled surroundings that are kept to a range of 4–25 °C are generally utilized in order to lower the thermal energy during deposition [24, 72, 73]. The crosslinking approaches are necessary to provide stability in structure and post-printing functionality. Alg-based systems are typically ionically crosslinked with $CaCl_2$, typically in the 50–200 mM range. The diffusion of calcium ions is regulated to slow down the gelation, leading to stronger interlayer binding in multilayer constructs. Improved approaches use dual-crosslinking, which combines faster ionic crosslinking with slower covalent crosslinking of methacrylate-functionalized polysaccharides, and affords compressive moduli ranging from 10 to 50 kPa, while maintaining cell viability over 90% in two weeks [24, 72, 73].

The speed of printing is also a necessary consideration and will be in line with the gelation kinetics. The extrusion-based systems usually take the best speeds ranging between 5 and 15 mm/s to avoid deformation and maintain effective deposition speed.

Another important parameter is the cellular density of the bioink; concentrations of 10^6–10^8 cells/mL are considered optimal as they allow enough metabolic activity without interfering with the characteristics of the flow or consistency. An increase in the levels more than 10^9 cells/mL can result in early biofilm formation, thus interfering with the rheological characteristics of the ink. Engineered strains of *E. coli* with the ability to form self-assembled curli fibers have been developed to stabilize the bioink under lower cell densities and allow post-printing reorganization of the biofilm matrix [72, 74]. In the generation of complex structures, the parameters like height of layers and infill geometry need to be tuned carefully. Patterns consisting of grids that vary in size, where the size of the pore is between 100–500 μm, are used to support mechanical strength and to ensure efficient transport of nutrients. These designs are able to reach compression strengths that are similar to the native cartilage (15–25 kPa) [24, 72].

Post-printing conditions have significant effects on the long-term functionality of printed biofilms. The natural production of extracellular polymeric substance (EPS) by incubation at 37 °C in a high-humidity condition (95%) occurs progressively to replace the synthetic polysaccharide scaffold over 72–96 h. In addition, EPS production can be augmented by dynamic cultivation systems with mechanical stimulation and nutrient delivery, by up to 40%, enhancing the viscoelastic characteristics of tissue engineering and other applications. A combination of these optimized parameters enhances the production of 3D-printed polysaccharide biofilms with less than 10% differences between computer-generated designs and long-term microbial survival over a period of over 21 days, which are generally applicable in bioremediation and biomedical engineering [24, 72–74].

4 Applications in Environmental Remediation

3D printed biofilms are important in environmental remediation due to their natural ability to adsorb, degrade, and immobilize pollutants. Their porous hydrophilic structure improves the removal of heavy metals, dyes, and organic pollutants from wastewater streams. NPs or carbon-based functionalization of these films also enhances adsorption efficiency and catalytic activity, making them effective to treat industrial effluents. Because of their biodegradability and eco-friendly nature, polysaccharide biofilm systems offer sustainable and cost-effective solutions for large-scale environmental cleanup. Table 1 summarizes the polysaccharide-based biofilms for environmental remediation applications.

4.1 Organic Pollutant and Heavy Metal Adsorption

Organic dyes, oils, heavy metals, and a wide range of toxic substances are presence in the wastewater just often and are a significant threat to the environment and

Table 1 3D printed polysaccharide-based biofilms for environmental remediation

Polysaccharide-based biofilms	Targeted pollutants	Removal performance	Methods and applications	Ref
Calcium alginate (3D-printed CA scaffold)	U (VI)	117.3 mg/g (max, pH 2.5), 90% removal in 30 min	3D printing produced macroporous scaffolds with high accessible sites for ion-exchange	[32]
3D GO / calcium-alginate hybrid (3D CA/GO)	U (VI), Pb^{2+}, cu	Enhanced adsorption vs pure alginate; high Pb uptake	GO added oxygenated groups + structural stability; hierarchical pores improved kinetics	[75–77]
Chitosan / ZIF-67 (3D printed composite)	Pb (II)	5.47 mg/g	MOF embedded in chitosan offered micro porosity and chelation, suitable for flow-through removal	[33, 78]
3D-printed PCL / sodium alginate composite	Cu (II)	90 mg/g	Heat-extruded PCL/ SA filaments improved mechanical strength and hydrophobicity	[79]
Chitosan diacrylate / Pluronic F-127 hydrogel	Pb (II), cu (II), cd (II), hg (II)	0.61 mg/g Pb, 0.58 mg/g cu removal	Printable geometries influenced adsorption capacity and surface exposure	[80]
3D-printed cellulose acetate mesh	Oil/water separation	> 95% separation efficiency, high flux	Superhydrophilic mesh with controlled pore size enabled gravity-driven separation	[81]
Alg-CS composites (with GO or nanocellulose)	Mixed heavy metals, dyes, microplastics	Up to 200 mg/g for U (VI)	Synergistic surface groups improved metal adsorption and dispersibility	[82]
3D-printed living polysaccharide biofilms	Microplastics, dyes, oils	90–92% capture/ removal efficiency	Living films using engineered microbes (laccase, curli fibers) degraded pollutants efficiently	[26, 83]

general health. Adsorption is considered as the best method of remediation that is used to eliminate dissolved aqueous contaminants. It is commonly used alongside other remediation methods. It has found wide usage due to its simple operation specifications, low cost, high levels of removal, and low chances of secondary contamination. Nanofibrils of CS of about 50 nm diameter are obtained through successive mechanical processes, such as homogenization and wet milling. The products of these processes are flexible, and can be used in a wide field of applications. The network of nanofibrils has been creatively developed into colloidal chelating agents and foam-based column adsorbents. The prepared CS-based nanofibrils exhibit significant properties, including adsorption, recyclability, and renewability. These nanofibrils can be used to remove aromatic compounds and heavy metals from water, even at low levels. Special attention is drawn to the fact that they are susceptible and selective to heavy-metal cations, such as Cu^{2+}, Pb^{2+}, Cr^{3+}, and Cd^{2+}, which helps in water remediation. Their ion-chelating capability is efficient, which shows their potential as effective agents in environmental remediation [84–87].

A model of a cylindrical filter membrane was created using 3D Max software and served as a schematic basis for the 3D printing of polysaccharide-based adsorbents and filtration components. After the polymer blends have been formulated, they are loaded into the 3D printer cartridge, and the printing process is carried out on a tailor-made CAD design to create the required structures. For example, Gayan et al. [88] developed a chitosan diacrylate-Pluronic F-127 hydrogel to adsorb heavy metals. This hydrogel was converted into different shapes, including sheets, cylinders, and solid blocks. Following a 30 min exposure to the metal-ion solutions, the concentration of the heavy metals was decreased to lower hazardous levels, and the adsorption capacities were approximately 0.61, 0.58, 0.57, and 0.55 mg/g of Pb^{2+}, Cu^{2+}, Cd^{2+}, and Hg^{2+}, respectively.

In another study, Liakos et al. [31] produced a 3D-printed film using a blend of sodium alginate and bio-based polycaprolactone thermoplastic, achieving a maximum adsorption capacity for copper ions of nearly 90 mg/g. A separate advancement involved formulating a printable nanocomposite hydrogel ink composed of SA, montmorillonite, and acrylic acid. This ink was subjected to electron beam irradiation, inducing in situ crosslinking during printing and eliminating the need for post-processing. The irradiation preserved the structural integrity, crystallinity, and thermal stability of the hydrogel, enabling efficient removal of Fe (III), Cr (III), Co (II), and Ni (II) ions from wastewater, with rapid adsorption kinetics [89].

Organic matter, including oil, can be separated from another solvent like water. Polysaccharide-based separation membranes have been developed for this purpose. Superwetting textiles, particularly those coated with polydopamine (PDA), have emerged as advanced functional materials for efficiently separating oil from oily wastewater and crude oil spills. These textiles are designed with specific surface roughness and chemical compositions to achieve superhydrophobic (SHB) or superhydrophilic/underwater superoleophobic (SHL/UWSOB) properties. PDA coatings are advantageous due to their low cost, versatility, biodegradability, and simple fabrication, enhancing adhesion and surface roughness while allowing for diverse post-functionalization [90]. Similarly, an innovative filter design integrated SA, graphene oxide, and nanosilver into a composite 3D-printed membrane. The structure was freeze-dried and ionically cross-linked with calcium chloride before being dialyzed. This solvent-free membrane performed excellent wastewater purification performance, achieving over 95% separation efficiency for oil-water mixtures and a 62.4% reduction in Biochemical Oxygen Demand (BOD), with adsorption equilibrium reached within two hours. These sustainable, eco-friendly membranes highlight the promise of 3D-printed polysaccharide composites for advanced sewage treatment with minimal environmental impact [89].

It is interesting to note that 3D CA constructs with mesh-like structures were also produced by direct ink writing (DIW), thus proving to be promising in terms of efficient oil-water separation (Fig. 5a). When pre-wetting is done, the CEL meshes act as selective barriers and block the entry of oil, unless the exerted pressure surpasses the intrusion threshold. When water is introduced, it gradually penetrates the mesh and is collected at the bottom of the container, and oil is held by the hydrophilic nature of CEL, which facilitates the preferential contact with water over oil (Fig. 5b). These properties enable the synthetic meshes to achieve separation efficiencies of more than 95% when the average pore diameter is smaller than 280 μm (Fig. 5c). With an impressive water flux reaching up to 160,000 $L.m^{-2}.h^{-1}$, these cellulose structures maintain separation efficiency around 99% across a broad spectrum of oil viscosities (ranging from 1 to 97 cP). In addition to their high performance, the meshes are chemically resilient in both acidic and alkaline environments and exhibit anti-oil fouling and self-cleaning behaviors, making them well-suited for practical oil-water separation tasks [91].

In another research, the authors found a new membrane fabrication process with super-hydrophilic and underwater super-oleophobic properties using DIW printing [92]. These membranes were produced using a specially developed viscose ink, which consisted of CA, PVA, and silica nanoparticles. The resulting structures have a structured porous network that provides gravity-based oil-water separation with a performance of approximately 99.0%. The membranes exhibit good durability, remaining mechanically stable after 30 min of ultrasonic agitation and 100 repeated bending cycles. Additionally, they exhibit high fouling resistance, and the separation effectiveness remains high even after 50 cycles of use. The characteristics highlight their potential for use at an industrial level in the treatment of wastewater and environmental remediation.

Altogether, the 3D-printed polysaccharide-based adsorbents with the incorporation of functional amino and hydroxyl groups have become sustainable and environmentally friendly products that could be used in various environmental applications. The major benefits of these materials include environmental friendliness, low cost, and high adsorption capacities. Because of their chemical structure, they are easily able to form coordination complexes with heavy-metal ions or react with organic pollutants, including oil, naphthalene, phenol, Green B, methyl orange, and methylene blue [52]. Moreover, these adsorbents can be completely biodegraded after undergoing wastewater treatment, which does not require organic solvents and avoids the formation of secondary pollutants, including waste products, toxic effluents, and emissions. They are appealing in various environmental remediation

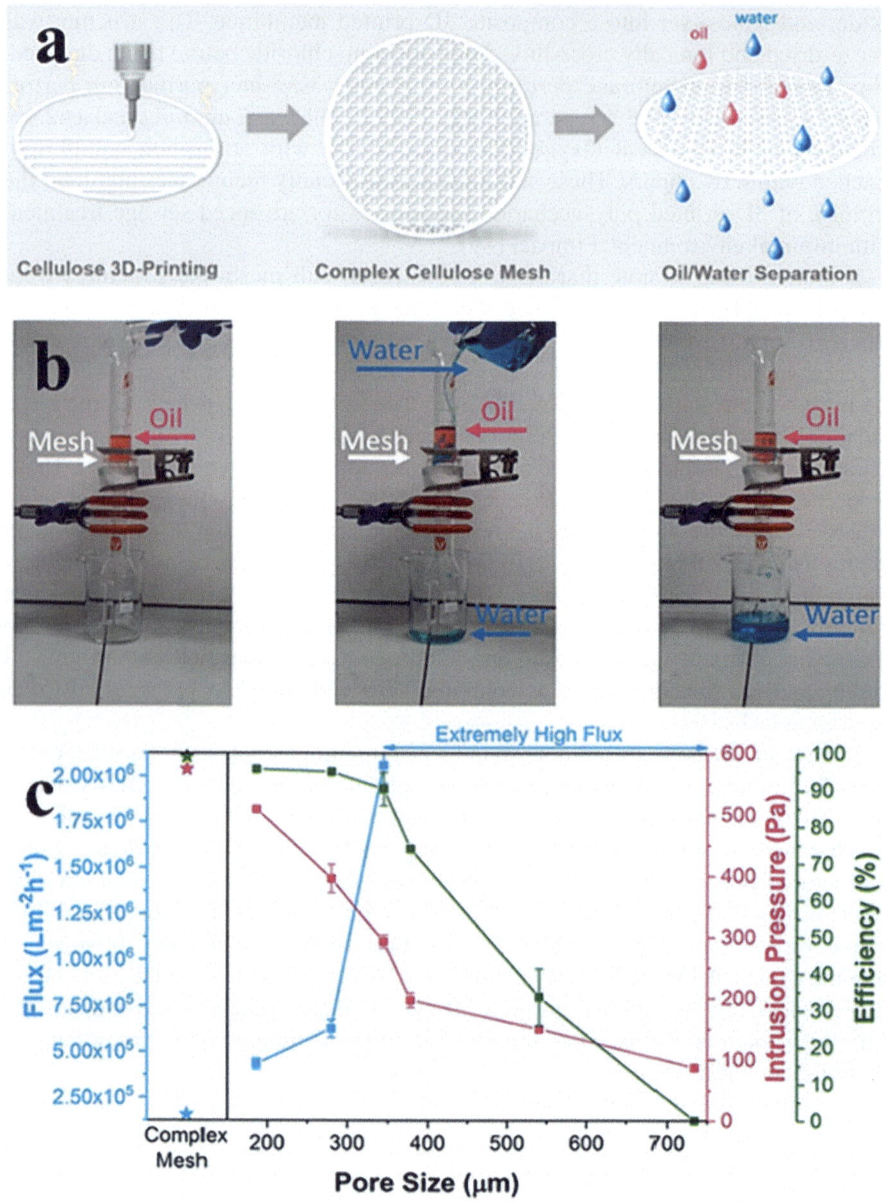

Fig. 5 3D-printed polysaccharide-based adsorbents for water purification. **a** Schematic illustration of the fabrication process for 3D-printed adsorbents and their operational mechanism for sewage treatment. **b** selective separation behavior of the cellulose mesh in oil/water mixtures. **c** relationship between average pore size and key performance metrics, water flux, intrusion pressure, and separation efficiency of the printed cellulose mesh structures. Reproduced with permission from the ref. [91], Elsevier, copyright 2024

applications due to their high surface area, tunable structural properties, mechanical stability, and reusability. They provide a solvent-free, post-treatment-free solution and a low-ecological impact solution [31, 88, 89].

4.2 Microplastic Capture

The 3D-printed polysaccharide biofilms are an innovative approach to reduce microplastic pollution that takes advantage of the adhesive capability of microbial extracellular polymeric substances (EPS) and the structural accuracy of additive manufacturing. These biofilms are generally prepared with the help of Alg—or CS-based bioink, which are ionically cross-linked on calcium-enriched surfaces to create hydrogels. These hydrogels serve as a scaffold of engineered bacteria, which are able to produce EPS with a binding capacity for microplastic particles. As an example, *Pseudomonas aeruginosa* biofilms that overexpress Psl polysaccharides show an increased adhesion of microplastics, trapping particles as small as 1 μm through hydrophobic and electrostatic forces [34, 93].

Nozzle diameters of 200–400 μm and bioink viscosities of 10–15 kPa·s are optimized 3D printing parameters that balance acceptable resolution of 50–100 μm and more than 80% bacterial viability during extrusion [20]. The microplastic capture mechanism involves physical capture of the microplastic by the polysaccharide matrix together with particular molecular interactions. The strains of engineered *E. coli* produce rigid meshes with structural integrity up to 0.5 Pa of fluid shear [24]. The inducible production of EPS can be achieved with the help of programmable genetic circuits that are induced by the presence of microplastics. One intriguing design also involves arabinose-stimulated phosphodiesterases to break down cyclic-di-GMP, which allows biofilm to be dissolved and captured microplastics to be released in a centralized way. This two-phase system has recorded an efficiency of 92% in removing microplastics in seawater with low amounts of residual contamination [94].

The biofilm has also been improved through material innovations. Compressive strength and adhesion of the hybrid bioink through hydrogen bonding are enhanced 300% and doubled, respectively, by the use of 1–5% w/v cellulose nanocrystals. Poly(N-isopropylacrylamide) domains in thermoresponsive variants allow temperature-controlled swelling (20–40 °C) of microplastics into aggregates that can be easily sedimented. In the field tests of the sea, these biofilms were active up to 21–28 days, and there were difficulties with the photodegradation of polysaccharides under UV-light conditions [24, 95].

Researchers developed biocomposite filters using pPLA reinforced with TEMPO-oxidized cellulose nanofibers (TCNF) or chitin nanofibers (ChNF) through a four-step process involving melt blending, phase separation, freeze-drying, and extrusion (Fig. 6) [100]. 3D printed in various shapes and pore structures, these filters were engineered to optimize contaminant contact and filtration efficiency. The nanofiber-reinforced versions demonstrated superior print quality, mechanical strength, and

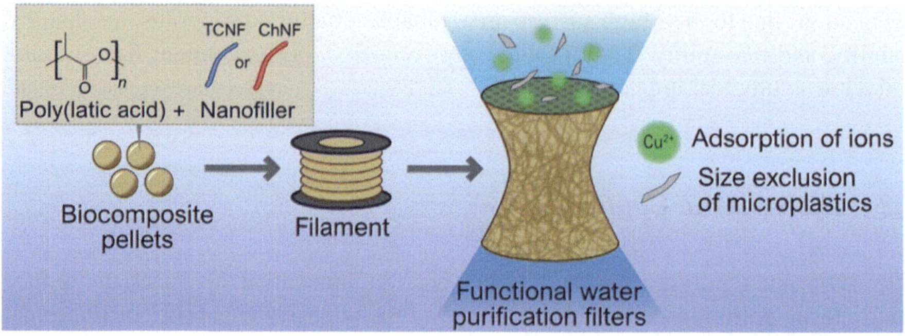

Fig. 6 Illustration showing 3D printed filters used for microplastics removal. Reproduced with permission from the ref. [96], under creative commons attribution (CC BY) license, 2023

water throughput compared to pure PLA filters. They also showed significantly higher copper ion adsorption capacities and improved microplastic removal performance. The study highlights the potential of combining sustainable materials with 3D printing for affordable, customizable water purification solutions in regions lacking clean water access.

Scalability remains a bottleneck, with current printing throughput limited to 10–100 mL/h. To address this, researchers are developing roll-to-roll systems featuring conveyor-based calcium baths and multi-nozzle arrays for parallelized production. Computational fluid dynamics guides the design of optimized geometries such as gyroid infill patterns, which increase contact probability with microplastics by 40%. Post-printing maturation with alternating nutrient pulses promotes native EPS replacement of the initial Alg matrix, enhancing durability [24].

Applications extend beyond microplastic capture. Biofilms embedded with laccase enzymes (0.5–2 U/mg) can bind and oxidatively degrade polyethylene, achieving 15% mass loss over 14 days. In wastewater treatment, 3D-printed biofilm modules installed in secondary clarifiers exhibit 68% higher microplastic retention than sand filters, though biofouling necessitates weekly backflushing. New research explores co-printed microbial consortia capable of microplastic trapping and pollutant degradation, forming multifunctional bioremediation platforms [20, 34].

Despite many innovations, there still remain a number of knowledge gaps. Interpretation of atomic force microscopy (AFM) shows that layers of EPS of 2–5 nm thickness develop on the surfaces of polyethylene and polypropylene within 6 h, suggesting that the side-chain permeates the crystalline structure of polymers. By contrast, polyurethane and nylon have 30–50% less adhesion, indicating that optimization of materials is necessary. Models based on machine-learning prediction applied to over 15,000 datasets are currently used to predict the ideal biofilm composition on specific plastic surfaces [24, 34].

Engineered strains contain auxotrophic or CRISPR-based containment systems, but there is a limitation in gene transfer into native microbes. Horizontal gene transfer to resident microorganisms is still a concern. According to life-cycle assessments, such biofilms might save up to 60% of energy used to remove microplastics through centrifugation. However, the environmental impact of harvesting large-scale alginates should be taken into consideration. This balance between performance, safety, and environmental impact will be very important as this technology moves into pilot-scale applications in wastewater treatment and marine remediation [24, 34, 95].

4.3 Mechanisms of Pollutant Removal

3D printed biofilms are used to clean the environment by physical adsorption and chemical reactivity. The porous structure based on the materials, like Alg or CS, offers them a high surface-area-to-volume ratio (between 200 and 500 m^2/g), which allows the physical capturing of contaminants through different mechanisms, including van der Waals forces, electrostatic forces, and hydrophobic interactions [34, 38]. Polar pollutants, specifically the heavy metals, react with hydroxyl and carboxyl groups through the hydrogen bonding reaction, but non-polar pollutants, like polyaromatic hydrocarbons, are caught by hydrophobic areas that are improved by additives, including cellulose nanocrystals.

At the chemical level, functional groups of the biofilm matrix aid in pollutant removal by covalent interaction. In a study, the amine group in CS forms a coordinate complex with metal ions like Pb^{2+} and Cd^{2+}, with removal efficiencies of 85–95% in water. The biofilms are further enhanced by genetic engineering [38, 97]. Phenolic pollutants are broken down by strains of the pseudomonad species, *P. putida*, which expresses laccase enzymes and breaks down the pollutant by radical-driven oxidation. Whereas, organophosphate compounds are broken down by strains of the bacillus, *B. subtilis* biofilms, which express TasA-HisTag nanofibers and catalyze the reaction by nucleophilic hydrolysis [23].

These systems can be modified to respond to environmental changes by incorporation of responsive materials, e.g., thermosensitive polymers. For example, biofilms modified with gold nanoparticles have the ability to reduce Cr (VI) to the less toxic trivalent form Cr (III) through redox reactions, approximately 40% more rapidly than free microbial cells. Through the integration of the physical adsorption capacity of structured polysaccharide scaffolds coupled with a controllable chemical reactivity, the engineered biofilms provide scalable and selective solutions to a wide range of pollutants, including microplastics and chemical industrial waste [98].

4.4 Physical Adsorption

3D printed polysaccharide biofilms are used to enhance the surface interactions and structural properties. Additive manufacturing enables fine control of porosity, and gyroid and lattice structures permit a surface-to-volume ratio that is an order of magnitude higher than conventional materials [34, 103]. Biofilms containing 2–5% w/v Alg, crosslinked with Ca^{2+} ions, have pore sizes ranging between 50 and 200 nm. These biofilms can trap the solid particles through size-based exclusion [38]. Van der Waals forces created between hydrophobic methyl and acetyl groups incorporated into the polysaccharide matrix and the polymer surfaces determine the adsorption of hydrophobic pollutants, such as polyethylene microplastics. Similarly, the electrostatic interactions play a key role in the elimination of heavy metals. The carboxylate (–COO^-) groups of Alg are negatively charged, and they form ionic bonds with the positively charged metal ions, including Pb^{2+} and Cu^{2+}, and adsorb the metal ions with adsorption capacities of up to 150–300 mg/g [38, 99].

These mechanisms of adsorption are further maintained by the 3D printing process, which controls the rheological properties of the bioink. The best particle contact is observed when viscosities are 10–15 kPa·s and nozzle diameters are 200–400 μm, with no harm to microbial viability of 80–90%. Multi-layered printed structures improve the performance of pollutant removal through the incorporation of a macroporous top layer (pore sizes: 500–1000 μm) to provide initial filtration and a finer bottom layer (pore sizes: 50–100 μm) to provide more in-depth adsorption, which in flow-through setups removes up to 92% of microplastics. Adding 1–5% w/v cellulose nanocrystals enhances nanoscale surface roughness, which forms fine crevices and increases the binding of nanoplastics smaller than 100 nm in diameter by 40% relative to smoother surfaces [38, 99].

4.5 *Chemical Interactions*

The innate chemical characteristics of polysaccharides and the incorporation of designed biological systems control the chemical remediation in these 3D-printed biofilms. Alg is composed of the units of α-L-guluronic acid containing reactive hydroxyl (–OH) and carboxyl (–COOH) groups that can establish a coordination bond with the metal ions through the ligand-exchange mechanisms. These interactions have been established by X-ray photoelectron spectroscopy (XPS), which has identified Pb-O binding energies at 138.9 eV in metal-laden biofilms [99]. Conversely, CS-based matrices employ protonated amine groups in acidic conditions to exchange ionically with Cr (VI) oxyanions to remove aqueous efficiencies of over 95% by a combination of electrostatic interactions and redox conversion [100].

Strains of the *P. putida*, transduced onto printed biofilms, express inducible laccase enzymes that oxidize phenolic pollutants (bisphenol A) through radical-mediated mechanisms, reaching a rate of 0.8 μM/min. Advanced dual-crosslinked networks, which combine ionic calcium bridges and covalently bonded methacrylate groups, have pH-sensitive swelling behavior. The hydrogel skeleton swells at alkaline pH (about 9), allowing the diffusion of molecules like paraoxon to accelerate and pesticide adsorption to increase up to 70% [38].

Embedded strains of *B. subtilis* use quorum sensing to regulate the EPS production and thus dynamically change the chemical microenvironment. The released proteins are capable of chelating the already adsorbed metals. FTIR spectra were used to show that the amide I band changed to 1650–1635 cm^{-1} because of Cd^{2+} displacement, which is a sign of the formation of metalloprotein complexes. To be applicable in photocatalytic degradation, biofilms loaded with 0.1–0.5% w/w TiO_2 NPs have been identified to have double functionality in adsorbing and breaking down the antibiotics (tetracycline) through Fenton-like reaction. These systems produce hydroxyl radicals at a rate of 2.3×10^{-6} M/s under the UV irradiation [38, 100].

The chemical functionalities are made possible by the synergy created by the combination of 3D printing and synthetic biology. For instance, concentric layers of Alg and CS structures are used to sequentially remove positively and negatively charged dyes by ion exchange and π-π stacking interactions. This combined, multimodal approach enables 3D-printed polysaccharide biofilms to integrate physical trapping with tailored chemical interactions, and results in a robust solution to complex environmental remediation problems.

5 Field Deployment and Case Studies

The current global landscape, with the fast growth of the population, increased industrialization, high waste output, reduction in natural resources, and the growing energy and consumption needs, requires the urgent implementation of the strategies of sustainable development. The circular bio-economy (CBE) is one of them, which is becoming more and more common. CBE is the meeting point of the two paradigms of the circular economy and bioeconomy, which is based on the creation of value-added products based on renewable biological resources using techno-economic biotechnologies, thus decreasing the reliance on non-renewable resources and improving the resilience of the ecosystem [101].

CBE structure is also gaining both regional and global momentum because it is focused on social, ethical, and environmental goals. The increasing sensitivity to the negative impact of chemical and synthetic compounds, specifically flocculants and adsorbents that are used in the treatment of water, has resulted in the introduction of stiffer regulatory measures in many countries. As a result, a growing interest in the replacement of these materials with sustainable and high-performance biomaterials.

Biopolymers have emerged as promising alternatives due to their:

(i) Biodegradability
(ii) Non-toxicity
(iii) Cost-effectiveness
(iv) Availability from renewable sources such as agricultural, industrial, and residential waste

Their production enables the valorization of waste streams and contributes to reducing the industrial carbon footprint by substituting petroleum-based polymers.

In addition, biopolymers possess a wide range of thermal and mechanical characteristics, which have the ability to be used in a variety of fields, such as biomedicine, food packaging, pharmaceuticals, and nutritional sciences. In order to be sustainable and efficient in manufacturing biopolymers, CBE needs to be integrated with zero liquid discharge (ZLD) and zero solid waste discharge. This guarantees that there is a minimum effect on the environment, coupled with the most material recovery and reuse [102].

However, the widespread adoption of CBE faces several challenges [103], including:

(a) Environmental and social impacts associated with replacing petroleum-based polymers (PBPs) with biopolymers.
(b) Many countries have economic dependency on established synthetic polymer industries.
(c) The rapid evolution of urban-centric business models often competes with recycling and reuse initiatives.

To mitigate these issues, it is important to have aligned policy responses, industrialization of novel processes, and adjustment in consumer behavior, as well as the implementation of scalable and cost-efficient technologies to produce biopolymers. Various biopolymers have been thoroughly studied as wastewater treatment indicators because of their remarkable selectivity regarding hazardous compounds and heavy metals. These materials are characterized by large adsorption capacities, which make them highly effective in eliminating various contaminants in contaminated water bodies. However, one major weakness of pure biopolymers is the fact that they have quite low shelf life, hence limiting their usefulness in practice and their usage in the long term. In order to overcome this shortcoming, chemical methods have been used to improve mechanical strength, degradation resistance and ability to remove pollutants, including cross-linking and grafting. The biodegradability is retained in these chemically modified biopolymers but the biopolymers are of better performance and durability thus making them more appropriate to be used over a long period. Despite these advancements, several challenges hinder the industrial-scale deployment of modified biopolymers:

(a) The use of toxic or hazardous chemicals during synthesis raises health and safety concerns.
(b) Operational complexity of the modification processes.
(c) High energy demands and the need for cost-intensive equipment.

To overcome these hurdles, it is required to undertake extensive research and development initiatives that would help to streamline the synthesis pathways, reduce environmental and health hazards, and create economic production systems. It is important to prove that modified biopolymers are technically and economically viable on a large scale to ensure their commercial application and the application of biopolymers in sustainable wastewater treatment technologies.

The cost of biological polymers is difficult to assess because of variations in their physicochemical properties, such as origin, molecular weight, purity, extent of deacetylation or esterification, and guluronic/mannuronic acid ratios, which have a substantial impact on the production costs. Although the price of raw materials is still one of the decisive factors, the use of energy and the sustainability criteria are becoming more significant, especially when the regulatory frameworks are changed and now focus on environmentally responsible procedures.

The material grade needed is strongly determined by the required properties of the application: biomedical, pharmaceutical, and food industries require high-purity biopolymers, which increases the cost of production, whereas industrial applications like wastewater treatment do not require high-grade biopolymers. However, variability of physicochemical characteristics poses a significant scaling-up challenge, which frequently constrains the application of laboratory-based innovations to commercial processes. In the example of CS, the market is growing at a very rapid rate, with the current growth rate of the compound annual growth rate (CAGR) approximated at 19.9% followed by an increase of 22% in the next five years [104], mostly through the use of the compound in high-value products in the cosmetics and biomedicine sectors [105]. This high demand is still a main issue in the cost of production.

In a case study conducted by Riofrio et al. [106], the environmental and economic performance of the CS extraction was evaluated in Ecuador, where shrimp waste was abundant. Their life cycle analysis (LCA), which was compared to Spain and India, showed that the cost of production in Ecuador was significantly lower at $8.39 USD/kg compared to that of Spain at about $14 USD/kg. With a processing capacity of 5000 tons/year, the net present value was calculated as $14 USD/kg. Nonetheless, these values are earlier than recent increases in energy prices throughout the world that might change profitability.

In the case of Alg, Saji et al. [107] conducted a comprehensive review relating extraction approaches to sustainability results. Despite the exploration of innovative methods (ultrasound, microwave, and enzymatic extraction), it has also not proven to be of any great economic benefit compared to conventional methods. The main ideas to alleviate sustainability and economic benefits are to grow algae, minimize the use of water and solvents, and value by-products. Alg markets have a CAGR of more than 3.3% [108], meaning that high-value applications dominate the market, and thus it is more difficult to source a wastewater treatment process cost-effectively. Therefore, algal by-products with low grade are becoming more and more the target of valorization [109]. The increase in Sargassum biomass which is commonly linked to environmental nuisance [110], can also provide economic alternatives of developing Alg-based products especially in climate-influenced coastal areas.

In the case of polyhydroxyalkanoates (PHAs), the price range is 8–12 USD/kg, depending on the quality of the product and its use (e.g., packaging or biomedical) [111]. Chavez et al. [111] explored alternative production techniques and feedstocks, including CO_2-derived carbon investigated the possibility of using other methods of production and feedstocks, such as carbon obtained through CO_2, finding costs ranging between $5–11 USD/kg. Their results highlight the need to optimize costs and sustainability of PHA production by ensuring that the feedstock is chosen and that the process is optimized.

6 Sustainability and Environmental Impact

Life Cycle Assessment (LCA) is one of the standardized tools for analyzing the overall environmental impact of a product during its entire life cycle, including the procurement of the raw material and manufacturing up to its final waste disposal [112]. LCA encourages sustainable development, as it takes into consideration such important aspects as the use of natural resources, human health, and the transparency of every stage of the production process. It also supplements circular economy (CE) measures, which are designed to keep resources in closed circles, thus reducing the impacts related to environmental damage and enhancing the material recovery and reuse [113]. In this context, biopolymers manufactured through additive manufacturing (AM) to be used in water treatment are a more environmentally friendly option. These materials make the resources efficient and minimize the chances of secondary pollution during use and disposal.

CS produced about 6–eight million tons of seafood globally every year, specifically from crab, shrimp, and lobster shells [114]. Hence, making CS an inexpensive and abundant natural material [115]. Its purification comprises of several chemical procedures such as deproteinization and deacetylation with alkalis and demineralization with acids [116].

The LCA model by Muñoz et al. [117] gives the following results: 1.4 kg of chitin, 5 kg of alkali, 250 L of water, 1.06 kWh of electricity, and 31 MJ of wood fuel are used to make 1 kg of CS. By-products of this material, nitrogen-based fertilizer, may be used back in agriculture, and the neutralized wastewater can be disposed of safely in the landfill [117].

Brown seaweed is used to produce SA through acid lixiviation, dewatering, alkaline treatment, and addition of sodium carbonate, and finally dried [118]. According to Langlois et al. [119], the utilization of electricity and chemicals is the major cause of environmental burden in the production of Alg. Peteiro [120] also expressed concerns about the depleting supply of brown seaweed because of the rising demand for Alg. In order to solve this, there has been a suggestion that seaweed can be grown in controlled environments in order to improve sustainability and decrease ecological impact.

CEL is the most natural polymer, which can be used to obtain the biodegradable and rather cheap CA. Old methods of CA production involve the use of sulfuric acid and other chemicals [121], although more green processes are emerging. An example is Araujo et al. [122], who obtained cellulose from corncob and produced CA by the treatment in water, which was followed by the use of the solvent-free acetylation procedure. Their LCA showed a much smaller environmental footprint than traditional methods, which supports the argument for sustainable alternatives. PLA is commonly based on renewable materials, like corn or wheat, that may also be obtained as food waste, therefore, making PLA a greener material compared to fossil-based plastics in the first place. PLA products at the end of life may be reused through various recycling options. They may be recycled in a mechanical manner, chemically, solvent-based, or composted.

Such processes are always more sustainable than the incineration of wastes [123]. Mechanical recycling is the least harmful to the environment, with chemical recycling and composting coming next [124]. It is worth noting that Zhao et al. [125] have shown that PLA in fused deposition modelling (FDM) can be recycled and reprinted at least two more times in 3D printing, which can improve its sustainability profile for use in water purification.

7 Challenges and Future Prospects

The high production cost and scalability of 3D-printed polysaccharide biofilms restrict their application in the environmental cleanup market. Bioinks, which are produced between 50 and 200 USD per liter, are typically low throughput in most extrusion-based bioprinting systems, with a throughput of 10 to 100 mL/h [126, 127]. Though consumer-grade (modified) 3D printers (priced under $500) have been found to work with great effectiveness in laboratory research, switching to industrial-sized production necessitates an improved system such as the thermal-electric aerosol (TEA) printer, which is capable of continuous roll-to-roll printing at a print length of up to 8600 mm per day [127]. Scaling up to see print volumes in cubic-meter scale, however, is associated with new engineering problems: shear stress with high-speed extrusion causes a 15–30% decrease in microbial survival rates, and ambient humidity control of post-processing at large scale is required to maintain print fidelity. There is also an obstacle of material cost in terms of deployment on a municipal level because the price of Alg is between 100 and 500 USD/kg, depending on its purity. To deal with this, the focus of research is shifting to sustainable alternatives like pectin purified in citrus waste and cellulose nanocrystals in wood pulp [128].

In addition to technical constraints, regulatory obstacles also present another challenge, particularly when biofilms are used that have genetically modified organisms (GMOs) tailored to exert effects such as pollutant degradation or mechanical reinforcement. The environmental release of GMOs under the EU Directive 2001/18/EC has to be subjected to full risk evaluation, and it must have a biosafety system, e.g., CRISPR-engineered kill switches in response to altered conditions of temperature or nutrient concentration. Likewise, a changing policy on the 3D-printed medical products in the U.S. indicates that the safety testing of the environmental usage might eventually be required to follow similar patterns, although the contexts are not the same [129]. There is still a significant discrepancy in standardized evaluation procedures, with field performance proving hard to compare with, since most laboratory assessments claim 60 90% of microplastic capture, varying testing conditions such as being tested under different pH levels (4–9), salinity levels (0–35 ppt), and UV irradiation. The problem of intellectual property also adds to the complexity of commercialization: the three global corporations hold 78% of the patents on Alg-modifications, which leaves open-source innovation, which has flourished in the case of low-cost bioprinter development, at risk of being stifled [34].

The emerging material innovations are assisting in solving some of these problems through better functionality and mechanical resilience. Hydrogen bonding between Alg's β-(1 → 4)-D-mannuronic acid and the crystalline CEL areas has led to hybrid bioinks to increase compressive strength threefold without altering printability [130]. The addition of conductive additives, 0.52% carbon nanotubes or polyaniline, has allowed the development of electroactive biofilms that can both adsorb and degrade simultaneously to remove up to 85% of diesel hydrocarbons in soil through microbial electrogenesis. Also, new materials like Alg grafted poly(N-isopropylacrylamide) provide temperature-responsive swelling (20–40 °C), which compresses microplastics into compact clumps, which can be easily collected [131]. These capabilities are being extended by the introduction of 4D printing, of dual cross-linked xanthan gum/Alg systems that polymerize under pH conditions and unfold into high-surface-area geometries (1.2–1.8 m^2/g) to bind heavy metal ions [34].

Looking ahead, the future of this field lies at the intersection of synthetic biology, AI-driven design, and advanced bioprinting technologies. Machine learning algorithms, trained on more than 15,000 experimental datasets, can now predict ideal nozzle diameters (150–350 μm) and layer thicknesses (100–200 μm) tailored to specific microbial consortia, cutting down trial-and-error by 70% [34]. Automated bioreactor platforms equipped with real-time Raman spectroscopy allow for closed-loop feedback systems that adjust nutrient inputs to increase extracellular polymeric substance (EPS) production by 40%. Roll-to-roll manufacturing at scale may reduce biofilm unit costs by up to 90% by 2030, though this depends on solving challenges such as bioink waste recovery. Current support bath recycling methods only reclaim 60–75% of unused materials. Regulatory frameworks are beginning to adapt, with initiatives like the EU's Horizon 2020 BioRoboost working toward standardized safety measures for deploying engineered biofilms in open settings [132].

Sustainability concerns are also prompting deeper analysis of polysaccharide sourcing. Although kelp-derived Alg is often considered carbon-neutral, increasing its harvest could disrupt coastal ecosystems through nutrient leaching and over-farming. Circular economy approaches are being explored, such as integrating biofilm fabrication into wastewater treatment facilities, where captured microplastics could be repurposed as structural fillers in construction materials (comprising 5–10% by weight), while recovered heavy metals could be reused in industry. Yet, public acceptance remains a potential barrier; surveys reveal that 62% of respondents are hesitant about the environmental use of GMO-based technologies, even when containment safeguards are in place [133]. Addressing this complex mix of technological, ecological, and social issues will require close cooperation among synthetic biologists, engineers, and regulators. Through such collaboration, 3D-printed polysaccharide biofilms could evolve from experimental innovations into pivotal tools for global sustainability.

8 Conclusion

3D-printed polysaccharide biofilms are one of the advancements in environmental remediation, where natural polymers, including Alg and CS, are combined with the accuracy of additive manufacturing to form functional and living materials. Such biofilms entrap pollutant-degrading microbes, achieving more than 80% microbial viability and robust mechanical stability under stress. Genetically engineered bacteria increase their resilience and pollutant binding capacity, enabling them to capture up to 92% of microplastics and 15% of polyethylene degradation during experiments. Although promising, it does not overcome the challenges of scalability, such as low printing rates and nozzle resolutions, which inhibit the diffusion of nutrients. It requires hybrid materials and biosafety, including CRISPR-based kill switches, to address environmental risks, such as UV degradation and possible gene transfer. Sourcing of sustainable bioink is also important to circumvent the ecological effects of the massive production of Alg. Future studies would focus on ideal biofilm design with machine learning and computational modeling, and advanced printing can enhance access to nutrients. The use of multi-functional biofilms, which comprise microbial consortia, could enable the concurrent capture and degradation of pollutants. Energetic culture systems would increase the lifespan of operations to out of bounds levels. Finally, these designed biofilms represent sustainable directions in water purification, bioremediation of soil, and carbon capture.

References

1. Achari A. Hydrology for engineers. Educohack Press; 2025.
2. De Luca DA, Cocca D, Egidio E, Lasagna M. Earthwater resources: a quantitative evaluation and future challenges. Water reuse and unconventional water resources: a multidisciplinary perspective. Springer; 2024. p. 13–38.
3. Tahir F, Ashfaq H, Khan AZ, Amin M, Akbar I, Malik HA, et al. Emerging trends in algae farming on non-arable lands for resource reclamation, recycling, and mitigation of climate change-driven food security challenges. Rev Environ Sci Biotechnol. 2024;23(3):869–96.
4. Bustos-Terrones YA, Quevedo-Castro A, Bandala ER, Kurniawan TA, Loaiza JG. Advancing water quality monitoring in lentic ecosystems: innovations for freshwater protection. Earth Syst Environ. 2025:1–37.
5. Prakash O, Tripathy P, Zade A, Sharma A, Juneja C, Hiwrale I, et al. Multifaceted dimensions of greywater recycling in advancing sustainable development goals: a comprehensive review. Environ Dev Sustain. 2024:1–30.
6. Patel M, Ahammed MM. Characteristics and removal of organic micropollutants in reverse osmosis concentrate from urban wastewater treatment plants: a review. Waste Manag Bull. 2024;2(4):249–60.
7. Zhang H, Duan L, Li S, Gao Q, Li M, Xing F, et al. Simultaneous wastewater treatment and resources recovery by forward osmosis coupled with microbial fuel cell: a review. Membranes. 2024;14(2):29.
8. Shen Y, Wei N, Fan K, Qi W, Feng J, Lai Z. Chemical-enhanced electrokinetic geosynthetics (EKG) electro-osmosis combined with vacuum preloading for consolidation and copper remediation in contaminated dredged sludge. Geotext Geomembr. 2025;53(6):1600–9.

9. Foorginezhad S, Zerafat MM, Ismail AF, Goh PS. Emerging membrane technologies for sustainable water treatment: a review on recent advances. Environ Sci: Adv. 2025;4(4):530–70.
10. Li J, Cao Y, Ding K, Ye J, Li F, Ma C, et al. Research progress of industrial wastewater treatment technology based on solar interfacial adsorption coupled evaporation process. Sci Total Environ. 2024;931:172887.
11. Si Z, Li Z, Zhuang X, Zhou D. Experimental investigation on the hybrid system of mechanical vapor recompression and hollow fiber vacuum membrane distillation applied for wastewater treatment. J Environ Manag. 2024;350:119633.
12. Arshad N, Asghar MS, Irshad MS, Alwadie N, Li J, Shah MY, et al. Spatial patterned solar evaporator for effective thermal management in enhanced solar-driven evaporation, and wastewater remediation. Chem Eng J. 2024;500:157068.
13. Vanaei HR, Khelladi S, Tcharkhtchi A. 3D printing as a multidisciplinary field. Industrial strategies and solutions for 3D printing. Appl Optim. 2024:1–24.
14. Adeleke AK, Montero DJP, Lottu OA, Ninduwezuor-Ehiobu N, Ani EC. 3D printing in aerospace and defense: a review of technological breakthroughs and applications. World J Adv Res Rev. 2024;21(2):1149–60.
15. Pethappachetty P, Kumar B, Govindaraj A, Prakash M, Kumari BV, Venkatesan KB, et al. 3D printing in personalized medicine: revolutionizing drug delivery and healthcare applications. Biomed Mater Devices. 2025:1–22.
16. Singh S, Kumar M, Kumar D, Kumar S, Chopra S, Bhatia A. Therapeutic precision: unveiling the potential of 3D printing in drug delivery, tissue engineering, and regenerative medicine. 3D Print Addit Manuf. 2024;
17. Al-Amri AM. Printed sensors for environmental monitoring: advancements, challenges, and future directions. Chem. 2025;13(8):285.
18. Idrees S, Abbas N, Asghar A, Khan MH, Hussain N. Development of Carboxymethyl cellulose and poly (acrylic acid) based composites for wastewater treatment: a review. Green energy and environmental. Technology. 2025;
19. Miao T, Wang J, Zeng Y, Liu G, Chen X. Polysaccharide-based controlled release systems for therapeutics delivery and tissue engineering: from bench to bedside. Adv Sci. 2018;5(4):1700513.
20. Aghajani M, Garshasbi HR, Naghib SM, Mozafari M. 3D printing of hydrogel polysaccharides for biomedical applications: a review. Biomedicine. 2025;13(3):731.
21. Kim YS, Shin YS. Surface functionalization of 3D-printed bio-inspired scaffolds for biomedical applications: a review. Biomimetics. 2024;9(11):703.
22. Shahbazi M, Jäger H, Chen J, Ettelaie R. Construction of 3D printed reduced-fat meat analogue by emulsion gels. Part II: printing performance, thermal, tribological, and dynamic sensory characterization of printed objects. Food Hydrocoll. 2021;121:107054.
23. Balasubramanian S, Aubin-Tam ME, Meyer AS. 3D printing for the fabrication of biofilm-based functional living materials. ACS Synth Biol. 2019;8(7):1564–7.
24. Lazarus E, Meyer AS, Ikuma K, Rivero IV. Three dimensional printed biofilms: fabrication, design and future biomedical and environmental applications. Microb Biotechnol. 2024;17(1):e14360.
25. Aghajani M, Garshasbi HR, Naghib SM, Mozafari MR. 3D printing of hydrogel polysaccharides for biomedical applications: a review. Biomedicines (Internet). 2025;13(3)
26. Hall DC Jr, Palmer P, Ji HF, Ehrlich GD, Król JE. Bacterial biofilm growth on 3D-printed materials. Front Microbiol. 2021;12:646303.
27. Naranda J, Bračič M, Vogrin M, Maver U. Recent advancements in 3D printing of polysaccharide hydrogels in cartilage tissue engineering. Materials (Internet). 2021;14(14)
28. Bhatia SK, Ramadurai KW. 3D printing and bio-based materials in global health. Switzerland: Springer International Publishing AG; 2017.
29. Fernandes EM, Pires RA, Mano JF, Reis RL. Bionanocomposites from lignocellulosic resources: properties, applications and future trends for their use in the biomedical field. Prog Polym Sci. 2013;38(10):1415–41.

30. Moreira MF, Kanaan AF, Piedade AP. Indirect additive manufacturing: a valid approach to modulate sorption/release profile of molecules from chitosan hydrogels. Polymers (Basel). 2022;14(13)
31. Liakos IL, Mondini A, Dottore ED, Filippeschi C, Pignatelli F, Mazzolai B. 3D printed composites from heat extruded polycaprolactone/sodium alginate filaments and their heavy metal adsorption properties. Mater Chem Front. 2020;4(8):2472–83.
32. Fuxiang S, Na W, Qiangqiang Z, Jie W, Bin L. 3D printing calcium alginate adsorbents for highly efficient recovery of U(VI) in acidic conditions. J Hazard Mater. 2022;440:129774.
33. Wu T, Karimi-Maleh H, Li Y, Zhang D, Zhang Z, Zhong N, et al. 3D printed porous chitosan/metal—organic framework composites as effective adsorbents to remove heavy metals from wastewater. Chem Eng J. 2024;493:152780.
34. Finny AS. 3D bioprinting in bioremediation: a comprehensive review of principles, applications, and future directions. PeerJ. 2024;12:e16897.
35. Song F, Niu Y, Wang N, Chen W, Xie W, Liu B. Fabrication of three-dimensional printed hybrid graphene oxide/calcium alginate network by 3D printing technology with outstanding U(VI) recycling capacity in acidic solution. Chem Eng J. 2023;457:141245.
36. Zhang D, Xiao J, Guo Q, Yang J. 3D-printed highly porous and reusable chitosan monoliths for cu(II) removal. J Mater Sci. 2019;54(8):6728–41.
37. Shojaeiarani J, Shirzadifar A, Bajwa DS. Robust and porous 3D-printed multifunctional hydrogels for efficient removal of cationic and anionic dyes from aqueous solution. Microporous Mesoporous Mater. 2021;327:111382.
38. Pu X, Wu Y, Liu J, Wu B. 3D bioprinting of microbial-based living materials for advanced energy and environmental applications. Chem Bio Engineering. 2024;1(7):568–92.
39. Husna MYN, Chong CH, Wong V-L, Cheah KH, Wan YK. 3D-printed PEGDA monolith with robust silane-grafted chitosan for enhanced textile wastewater treatment. J Environ Chem Eng. 2022;10(6):108581.
40. Georgiou E, Raptopoulos G, Papastergiou M, Paraskevopoulou P, Pashalidis I. Extremely efficient uranium removal from aqueous environments with Polyurea-cross-linked alginate aerogel beads. ACS Appl Polym Mater. 2022;4(2):920–8.
41. Ayiotis A, Georgiou E, Ioannou PS, Pashalidis I, Krasia-Christoforou T. 3D composite U(VI) adsorbents based on alginate hydrogels and oxidized biochar obtained from Luffa cylindrica. Materials (Internet). 2023;16(19)
42. Zalesskiy SS, Kitson PJ, Frei P, Bubliauskas A, Cronin L. 3D designed and printed chemical generators for on demand reagent synthesis. Nat Commun. 2019;10(1):5496.
43. Sochol RD, Sweet E, Glick CC, Wu S-Y, Yang C, Restaino M, et al. 3D printed microfluidics and microelectronics. Microelectron Eng. 2018;189:52–68.
44. Park BJ, Choi HJ, Moon SJ, Kim SJ, Bajracharya R, Min JY, et al. Pharmaceutical applications of 3D printing technology: current understanding and future perspectives. J Pharm Investig. 2019;49(6):575–85.
45. Fijoł N, Aguilar-Sánchez A, Mathew AP. 3D-printable biopolymer-based materials for water treatment: a review. Chem Eng J. 2022;430:132964.
46. Chen Z, Sun X, Shang Y, Xiong K, Xu Z, Guo R, et al. Dense ceramics with complex shape fabricated by 3D printing: a review. J Adv Ceram. 2021;10(2):195–218.
47. Shao H, Zhao D, Lin T, He J, Wu J. 3D gel-printing of zirconia ceramic parts. Ceram Int. 2017;43(16):13938–42.
48. Ngo TD, Kashani A, Imbalzano G, Nguyen KTQ, Hui D. Additive manufacturing (3D printing): a review of materials, methods, applications and challenges. Compos Part B. 2018;143:172–96.
49. Mott EJ, Busso M, Luo X, Dolder C, Wang MO, Fisher JP, et al. Digital micromirror device (DMD)-based 3D printing of poly(propylene fumarate) scaffolds. Mater Sci Eng C. 2016;61:301–11.
50. Chae C, Kim Y-B, Lee SS, Kim T-H, Jo Y, Kim H-S, et al. All-3D-printed solid-state microsupercapacitors. Energy Storage Mater. 2021;40:1–9.

51. Chu T, Park S, Fu K. 3D printing-enabled advanced electrode architecture design. Carbon Energy. 2021;3(3):424–39.
52. Kotikian A, Truby RL, Boley JW, White TJ, Lewis JA. 3D printing of liquid crystal elastomeric actuators with spatially programed Nematic order. Adv Mater. 2018;30(10):1706164.
53. Li H, Liu Y-S, Liu Y-S, Zeng Q-F, Liang J-J. Microstructure and mechanical properties of 3D printed ceramics with different vinyl acetate contents. Rare Metals. 2021;40(11):3241–50.
54. Ndinisa SS, Whitefield DJ, Sigalas I. Fabrication of complex shaped alumina parts by gelcasting on 3D printed moulds. Ceram Int. 2020;46(3):3177–82.
55. Negi S, Dhiman S, Sharma RK. Investigating the surface roughness of SLS fabricated glass-filled polyamide parts using response surface methodology. Arab J Sci Eng. 2014;39(12):9161–79.
56. Li K, Zhao Z. The effect of the surfactants on the formulation of UV-curable SLA alumina suspension. Ceram Int. 2017;43(6):4761–7.
57. Peltola SM, Melchels FPW, Grijpma DW, Kellomäki M. A review of rapid prototyping techniques for tissue engineering purposes. Ann Med. 2008;40(4):268–80.
58. Wu Z, Sun D, Shi C, Chen S, Tang S, Li Y, et al. Moisture-thermal stable, Superhydrophilic alumina-based ceramics fabricated by a selective laser sintering 3D printing strategy for solar steam generation. Adv Funct Mater. 2023;33(45):2304897.
59. Ghouse S, Babu S, Van Arkel RJ, Nai K, Hooper PA, Jeffers JRT. The influence of laser parameters and scanning strategies on the mechanical properties of a stochastic porous material. Mater Des. 2017;131:498–508.
60. Yan M, Tian X, Peng G, Cao Y, Li D. Hierarchically porous materials prepared by selective laser sintering. Mater Des. 2017;135:62–8.
61. Zhou G, Wang KP, Liu HW, Wang L, Xiao XF, Dou DD, et al. Three-dimensional polylactic acid@graphene oxide/chitosan sponge bionic filter: highly efficient adsorption of crystal violet dye. Int J Biol Macromol. 2018;113:792–803.
62. Zou R, Xia Y, Liu S, Hu P, Hou W, Hu Q, et al. Isotropic and anisotropic elasticity and yielding of 3D printed material. Compos Part B. 2016;99:506–13.
63. Li H, Tan C, Li L. Review of 3D printable hydrogels and constructs. Mater Des. 2018;159:20–38.
64. Qian J, Xiao R, Su F, Guo M, Liu D. 3D wet-spinning printing of wearable flexible electronic sensors of polypyrrole@polyvinyl formate. J Ind Eng Chem. 2022;111:490–8.
65. Jandyal A, Chaturvedi I, Wazir I, Raina A, Ul Haq MI. 3D printing—a review of processes, materials and applications in industry 4.0. Sustain Oper Comput. 2022;3:33–42.
66. Vithani K, Goyanes A, Jannin V, Basit AW, Gaisford S, Boyd BJ. An overview of 3D printing technologies for soft materials and potential opportunities for lipid-based drug delivery systems. Pharm Res. 2018;36(1):4.
67. Teow YH, Kam LM, Mohammad AW. Synthesis of cellulose hydrogel for copper (II) ions adsorption. J Environ Chem Eng. 2018;6(4):4588–97.
68. Hajji S, Ghorbel-Bellaaj O, Younes I, Jellouli K, Nasri M. Chitin extraction from crab shells by bacillus bacteria. Biological activities of fermented crab supernatants. Int J Biol Macromol. 2015;79:167–73.
69. Parkatzidis K, Chatzinikolaidou M, Kaliva M, Bakopoulou A, Farsari M, Vamvakaki M. Multiphoton 3D printing of biopolymer-based hydrogels. ACS Biomater Sci Eng. 2019;5(11):6161–70.
70. Zuo W, Liu X, Qiu J, Zhang D, Xiao Z, Xie J, et al. Engineering Na^{+}-layer spacings to stabilize Mn-based layered cathodes for sodium-ion batteries. Nat Commun. 2021;12(1):4903.
71. Bozkurt Y, Karayel E. 3D printing technology; methods, biomedical applications, future opportunities and trends. J Mater Res Technol. 2021;14:1430–50.
72. Teixeira MC, Lameirinhas NS, Carvalho JPF, Silvestre AJD, Vilela C, Freire CSR. A guide to polysaccharide-based hydrogel bioinks for 3D bioprinting applications. Int J Mol Sci. 2022;23(12)
73. Duraj-Thatte AM, Manjula-Basavanna A, Rutledge J, Xia J, Hassan S, Sourlis A, et al. Programmable microbial ink for 3D printing of living materials produced from genetically engineered protein nanofibers. Nat Commun. 2021;12(1):6600.

74. McCarthy RR, Ullah MW, Booth P, Pei E, Yang G. The use of bacterial polysaccharides in bioprinting. Biotechnol Adv. 2019;37(8):107448.
75. Wang N, Song F, Niu Y, Chen W, Liu B, Xie W. Three-dimensional-printed calcium alginate/graphene oxide porous adsorbent with super-high lead ion adsorption ability in aqueous solution. Sep Purif Technol. 2023;326:124757.
76. Wang N, Xu Y, Wei Y, Chen Q, Song F, Liu B. Proportion-optimized 3D-printed calcium alginate/graphene oxide adsorbent for efficient uranium separation from aqueous solutions. Sep Purif Technol. 2025:134199.
77. Pan L, Wang Z, Yang Q, Huang R. Efficient removal of lead, copper and cadmium ions from water by a porous calcium alginate/graphene oxide composite aerogel. Nano. 2018;8(11):957.
78. Omer AM, Abd El-Monaem EM, Abd El-Latif MM, El-Subruiti GM, Eltaweil AS. Facile fabrication of novel magnetic ZIF-67 MOF@ aminated chitosan composite beads for the adsorptive removal of Cr (VI) from aqueous solutions. Carbohydr Polym. 2021;265:118084.
79. Liakos IL, Mondini A, Del Dottore E, Filippeschi C, Pignatelli F, Mazzolai B. 3D printed composites from heat extruded polycaprolactone/sodium alginate filaments and their heavy metal adsorption properties. Mater Chem Frontiers. 2020;4(8):2472–83.
80. Pelegrino MT, de Araujo LB, Do Nascimento MH, Lombello CB, Brocchi M, Seabra AB. Biocompatible and antibacterial nitric oxide-releasing pluronic F-127/chitosan hydrogel for topical applications. Polymers. 2018;10(4):452.
81. Hernandez-Ruiz LD, Hassan M, Wang T, Mohanty AK, Misra M. Sustainable composites from microcrystalline cellulose and cellulose acetate: 3D printing and performance optimization. Composites Part C: Open Access. 2025;100606.
82. Naseem K, Tahir MH, Farooqi F, Manzoor S, Khan SU. Strategies adopted for the preparation of sodium alginate–based nanocomposites and their role as catalytic, antibacterial, and antifungal agents. Rev Chem Eng. 2023;39(8):1359–91.
83. Balasubramanian S, Yu K, Cardenas DV, Aubin-Tam M-E, Meyer AS. Emergent biological endurance depends on extracellular matrix composition of three-dimensionally printed Escherichia coli biofilms. ACS Synth Biol. 2021;10(11):2997–3008.
84. Ganglo C, Rui J, Zhu Q, Shan J, Wang Z, Su F, et al. Chromium (III) coordination capacity of chitosan. Int J Biol Macromol. 2020;148:785–92.
85. Liu D, Zhu Y, Li Z, Tian D, Chen L, Chen P. Chitin nanofibrils for rapid and efficient removal of metal ions from water system. Carbohydr Polym. 2013;98(1):483–9.
86. Liu D, Li Z, Zhu Y, Li Z, Kumar R. Recycled chitosan nanofibril as an effective cu(II), Pb(II) and cd(II) ionic chelating agent: adsorption and desorption performance. Carbohydr Polym. 2014;111:469–76.
87. Ma Z, Liu D, Zhu Y, Li Z, Li Z, Tian H, et al. Graphene oxide/chitin nanofibril composite foams as column adsorbents for aqueous pollutants. Carbohydr Polym. 2016;144:230–7.
88. Appuhamillage GA, Berry DR, Benjamin CE, Luzuriaga MA, Reagan JC, Gassensmith JJ, et al. A biopolymer-based 3D printable hydrogel for toxic metal adsorption from water. Polym Int. 2019;68(5):964–71.
89. Shahbazi M, Jäger H, Ahmadi SJ, Lacroix M. Electron beam crosslinking of alginate/nanoclay ink to improve functional properties of 3D printed hydrogel for removing heavy metal ions. Carbohydr Polym. 2020;240:116211.
90. Azad AK, Abu-Thabit NY, Elella MHA. Sustainable and durable superwetting textiles coated with polydopamine for oil/water separation. RSC Sustain. 2025;
91. Koh JJ, Lim GJH, Zhou X, Zhang X, Ding J, He C. 3D-printed anti-fouling cellulose mesh for highly efficient oil/water separation applications. ACS Appl Mater Interfaces. 2019;11(14):13787–95.
92. Jonhson W, Xu X, Zhang D, Chua WT, Tan YH, Liu X, et al. Fabrication of 3D-printed ceramic structures for portable solar desalination devices. ACS Appl Mater Interfaces. 2021;13(19):23220–9.
93. Romero M, Carabelli A, M RS, M IS. Fluid dynamics and cell-bound Psl polysaccharide allows microplastic capture, aggregation and subsequent sedimentation by Pseudomonas aeruginosa in water. Environ Microbiol. 2022;24(3):1560–72.

94. Liu SY, Leung MM-L, Fang JK-H, Chua SL. Engineering a microbial 'trap and release' mechanism for microplastics removal. Chem Eng J. 2021;404:127079.
95. Dhall A, Ramjee R, Oh MJ, Tao K, Hwang G. A 3D-printed customizable platform for multiplex dynamic biofilm studies. Adv Mater Technol. 2022;7(7)
96. Fijoł N, Aguilar-Sánchez A, Ruiz-Caldas M-X, Redlinger-Pohn J, Mautner A, Mathew AP. 3D printed polylactic acid (PLA) filters reinforced with polysaccharide nanofibers for metal ions capture and microplastics separation from water. Chem Eng J. 2023;457:141153.
97. Vo LH, Hong S, Stepler KE, Liyanaarachchi SM, Yang J, Nemes P, et al. Mapping protein-exopolysaccharide binding interaction in Staphylococcus epidermidis biofilms by live cell proximity labeling. bioRxiv. 2023.
98. Sun XL, Xiang H, Xiong HQ, Fang YC, Wang Y. Bioremediation of microplastics in freshwater environments: a systematic review of biofilm culture, degradation mechanisms, and analytical methods. Sci Total Environ. 2023;863:160953.
99. Melo A, Quintelas C, Ferreira EC, Mesquita DP. The role of extracellular polymeric substances in micropollutant removal. Front Chem Eng. 2022;4
100. Saini S, Tewari S, Dwivedi J, Sharma V. Biofilm-mediated wastewater treatment: a comprehensive review. Mater Adv. 2023;4(6):1415–43.
101. Corrado S, Sala S. Bio-economy contribution to circular economy. Des Sustain Technol, Prod Policies: Sci Innov. 2018:49–59.
102. Gurjar R, Behera M. Biopolymer: production from biomass. In: Verma P, editor. Biorefineries: a step towards renewable and clean energy. Singapore: Springer Singapore; 2020. p. 371–90.
103. Shah SA, Ahmad H, Alhazmi H, Anwar MK, Iqbal F. Utilization of self-consolidated green material for sustainable development: an environment friendly waste materials application for circular economy. Polym Int. 2021;13(17)
104. Chitosan Global Market Report 2025.
105. Sarkar MSI, Hasan MM, Hossain MS, Khan M, Islam AA, Paul SK, et al. Exploring fish in a new way: a review on non-food industrial applications of fish. Heliyon. 2023;9(12)
106. Riofrio A, Alcivar T, Baykara H. Environmental and economic viability of chitosan production in Guayas-Ecuador: a robust investment and life cycle analysis. ACS Omega. 2021;6(36):23038–51.
107. Saji S, Hebden A, Goswami P, Du C. A brief review on the development of alginate extraction process and its sustainability. Sustainability (Internet). 2022;14(9)
108. Ruales-Salcedo AV, Grisales-Díaz VH, Morales-Rodriguez R, Fontalvo J, Prado-Rubio OA. Production of high-added value compounds from biomass. In: Biofuels and biorefining. Elsevier; 2022. p. 381–445.
109. Ballesteros-Plata D, Zhang Y, Rodríguez-Castellón E, Vincent T, Guibal E. Brown algal residue for the recovery of metal ions—application to La(III), cd(II), and Ni(II) sorption. Adv Sustainable Syst. 2023;7(3):2200420.
110. Li H, Geng Y, Shi H, Wu C, Yu Z, Zhang H, et al. Biological mechanisms of invasive algae and meta-analysis of ecological impacts on local communities of marine organisms. Ecol Indic. 2023;146:109763.
111. Alvarez Chavez B, Raghavan V, Tartakovsky B. A comparative analysis of biopolymer production by microbial and bioelectrochemical technologies. RSC Adv. 2022;12(25):16105–18.
112. Gómez IDL, Escobar AS. The dilemma of plastic bags and their substitutes: a review on LCA studies. Sustain Prod Consum. 2022;30:107–16.
113. Peña L. ECVP Celebrates Its 25th Anniversary! VetY Pathol. 2021;58(1):6.
114. Yan N, Chen X. Sustainability: Don't waste seafood waste. Nature. 2015;524(7564):155–7.
115. Maddaloni M, Vassalini I, Alessandri I. Green Routes for the Development of Chitin/Chitosan Sustainable Hydrogels. Sustainable Chemistry (Internet). 2020;1(3):325–44.
116. Arbia W, Arbia L, Adour L, Amrane A. Chitin extraction from crustacean shells using biological methods–a review. Food Technol Biotechnol. 2013;51(1):12–25.
117. Muñoz I, Rodríguez C, Gillet D, M. Moerschbacher B. Life cycle assessment of chitosan production in India and Europe. Int J Life Cycle Assess. 2018;23(5):1151–60.

118. Hernández-Carmona G, McHugh DJ, López-Gutiérrez F. Pilot plant scale extraction of alginates from Macrocystis pyrifera. 2. Studies on extraction conditions andmethods of separating the alkaline-insoluble residue. J Appl Phycol. 1999;11(6):493–502.
119. Langlois J, Fréon P, Delgenès J-P, Steyer J-P, Helias A, editors. Life cycle assessment of alginate production. In: Proceedings of the 8th International Conference on LCA in the Agri-Food Sector, Saint-Malo, France; 2012.
120. Peteiro C. Alginate production from marine macroalgae, with emphasis on kelp farming. In: Rehm BHA, Moradali MF, editors. Alginates and their biomedical applications. Singapore: Springer Singapore; 2018. p. 27–66.
121. Fischer S, Thümmler K, Volkert B, Hettrich K, Schmidt I, Fischer K. Properties and applications of cellulose acetate. Macromol Symp. 2008;262(1):89–96.
122. Araújo D, Castro MCR, Figueiredo A, Vilarinho M, Machado A. Green synthesis of cellulose acetate from corncob: physicochemical properties and assessment of environmental impacts. J Clean Prod. 2020;260:120865.
123. Maga D, Hiebel M, Thonemann N. Life cycle assessment of recycling options for polylactic acid. Resour Conserv Recycl. 2019;149:86–96.
124. Cosate de Andrade MF, Souza PMS, Cavalett O, Morales AR. Life cycle assessment of poly(lactic acid) (PLA): comparison between chemical recycling, mechanical recycling and composting. J Polym Environ. 2016;24(4):372–84.
125. Zhao P, Rao C, Gu F, Sharmin N, Fu J. Close-looped recycling of polylactic acid used in 3D printing: an experimental investigation and life cycle assessment. J Clean Prod. 2018;197:1046–55.
126. Kahl M, Gertig M, Hoyer P, Friedrich O, Gilbert DF. Ultra-low-cost 3D bioprinting: modification and application of an off-the-shelf desktop 3D-printer for biofabrication. Front Bioeng Biotechnol. 2019;7
127. Ioannidis K, Danalatos RI, Champeris Tsaniras S, Kaplani K, Lokka G, Kanellou A, et al. A custom ultra-low-cost 3D bioprinter supports cell growth and differentiation. Front Bioeng Biotechnol. 2020;8
128. Putri KNA, Intasanta V, Hoven VP. Current significance and future perspective of 3D-printed bio-based polymers for applications in energy conversion and storage system. Heliyon. 2024;10(4):e25873.
129. Senderovich N, Shah S, Ow TJ, Rand S, Nosanchuk J, Wake N. Assessment of staphylococcus Aureus growth on biocompatible 3D printed materials. 3D Print Med. 2023;9(1):30.
130. Jessop ZM, Al-Sabah A, Gao N, Kyle S, Thomas B, Badiei N, et al. Printability of pulp derived crystal, fibril and blend nanocellulose-alginate bioinks for extrusion 3D bioprinting. Biofabrication. 2019;11(4):045006.
131. Liu M, Song X, Wen Y, Zhu J-L, Li J. Injectable Thermoresponsive hydrogel formed by alginate-g-poly(N-isopropylacrylamide) that releases doxorubicin-encapsulated micelles as a smart drug delivery system. ACS Appl Mater Interfaces. 2017;9(41):35673–82.
132. Market R-t-RP. Roll-to-Roll printing market for flexible electronics by printing technology (screen, inkjet, gravure, flexographic), application (displays, sensors, batteries, RFID, lighting), material, end-use industry, and geography—global forecast to 2025; 2025. Available from: https://www.marketsandmarkets.com/Market-Reports/roll-to-roll-printing-market-131 73263.html.
133. Ehlers U. Kill switches; 2025. Available from: https://2020.igem.org/Team:TU_Darmstadt/Project/Kill_Switch.

3D-Engineered Polysaccharide Sponge Formulations for Water and Wastewater Treatment

Aula A. Alwattar and Athir Haddad

Abstract The worldwide water emergency requires immediate solutions because pollution and water shortages have reached critical levels. The current water treatment methods fail to eliminate multiple contaminants including heavy metals and organic substances and nutrients and pathogens while producing additional waste products and requiring significant energy consumption. The chapter investigates 3D-engineered polysaccharide sponge formulations as an advanced water and wastewater treatment system for the future. The combination of natural abundance and biodegradability and multiple functional groups (–OH, $-NH_2$, –COOH) in biopolymers such as chitosan and cellulose and alginate and carrageenan enables sponge structure enhancement through freeze-drying and 3D printing and electrospinning methods. The 3D engineering process transforms traditional polysaccharide materials by creating porous structures that enhance mass transfer rates and adsorption capacity and mechanical strength. The performance of these sponges receives additional enhancement through the addition of functional additives which include nanoparticles (Fe_3O_4, TiO_2, AgNPs) and carbon nanomaterials that deliver catalytic and antimicrobial and magnetic capabilities for specific pollutant elimination and self-cleaning and recovery operations. The multifunctional sponges show excellent performance in heavy metal removal through adsorption and ion exchange processes and organic pollutant degradation through photocatalysis and nutrient capture through biofilm support and pathogen elimination through filtration and disinfection. The research examines how these materials can be integrated into current treatment systems through pilot-scale implementations and identifies future development paths for smart responsive sponges and affordable production methods. The

A. A. Alwattar · A. Haddad (✉)
Department of Chemistry, College of Science, University of Basrah, Basrah, Iraq
e-mail: athir.haddad@uobasrah.edu.iq

A. A. Alwattar
e-mail: Ula.Jumah@uobasrah.edu.iq

A. A. Alwattar
Department of Chemistry, College of Science, University of Manchester, Basrah, Iraq

M. H. Abu Elella et al. (eds.), *3D Engineered Polysaccharides*,
https://doi.org/10.1007/978-3-032-26154-0_7

3D-engineered polysaccharide sponge system offers a scalable eco-friendly solution which demonstrates high effectiveness for securing worldwide water resources despite facing regulatory hurdles and stability concerns.

Keywords 3D-engineered sponge polysaccharide · Water treatment · Wastewater remediation · Biopolymer scaffolds · Pollutant removal · Sustainable filtration

Abbreviations

AgNPs	Silver nanoparticles
BCI	Blood clotting index
BET	Brunauer–Emmett–Teller method (specific surface area by gas adsorption)
$CaCl_2$	Calcium chloride
Cd^{2+}	Cadmium ion
CNF-PEI	Cellulose nanofibers–polyethyleneimine
COD	Chemical oxygen demand
Cr(VI)	Hexavalent chromium
CuO NPs	Copper oxide nanoparticles
DLP	Digital light processing (3D printing)
DN	Dextran network
EGF	Epidermal growth factor
Fe_3O_4	Magnetite (iron oxide)
FTIR	Fourier-transform infrared spectroscopy
G'	Storage modulus
GBR	Guided bone regeneration
GTR	Guided tissue regeneration
HA	Hyaluronic acid
MIP	Mercury intrusion porosimetry
MOFs	Metal–organic frameworks
OHA	Oxidized hyaluronic acid
Pb^{2+}	Lead ion
PCL	Polycaprolactone
PEI	Polyethyleneimine
PLA	Polylactic acid
PVF	Polyvinyl formal
QCS	Quaternized chitosan
SA@CDs	Sodium alginate beads doped with carbon dots
SEM	Scanning electron microscopy
SiO_2	Silicon dioxide
STMP	Sodium trimetaphosphate
TEM	Transmission electron microscopy
TiO_2	Titanium dioxide

Ts-Car/Alg	Tosyl-carrageenan/alginate composite
UV	Ultraviolet
WOPW	Waste-office paper waste (wastepaper-derived aerogel)
XPS	X-ray photoelectron spectroscopy
ZnO NPs	Zinc oxide nanoparticles
ε-PL	Epsilon-polylysine

γ-irradiationGamma irradiation

1 Introduction

1.1 The Global Water Crisis: An Urgent Need for Advanced Treatment Technologies

Water scarcity is an urgent challenge worldwide, made worse by growing demand and increasing pollution that threaten both the quantity and quality of available water. The issue has worsened due, to growth and urban development coupled with the impact of climate change. This has resulted in rainfall cycles and extended periods of weather as well as the pollution of water sources by substances like heavy metals and microplastics alongside organic compounds and harmful pathogens [1]. Moreover, agricultural runoff and unprocessed industrial waste also play a part in the decline of water purity by introducing substances such, as pesticides, a pharmaceutical residues and chemicals that disrupt balance presenting long term dangers to both human wellbeing and natural habitats. In this situation traditional methods, like coagulation flocculation, sedimentation and filtration have long been vital in enhancing water safety. With today's pollution issues these old techniques are proving insufficient. For example, coagulation flocculation struggles with pollutants and results in significant sludge build up while chlorination may create harmful, by products. Fails to remove all chemical impurities or emerging micropollutants [2]. The adsorption capacity of activated carbon is. The process of reverse osmosis delivers excellent desalination results yet its high energy requirements and resulting brine waste need immediate environmental solutions. The adsorption capacity of activated carbon exists as a natural limitation while its regeneration process shows low efficiency. The desalination process of reverse osmosis achieves high success rates but requires large amounts of energy and produces concentrated brine waste that needs immediate sustainable environmental solutions [3]. These obstacles underscore the pressing demand for water treatment technologies that can effectively eliminate a range of impurities with improved efficiency and sustainability measures. These constraints highlight the need for water treatment technologies capable of removing a broader range of contaminants with greater efficiency and sustainability [4]. Promising alternatives

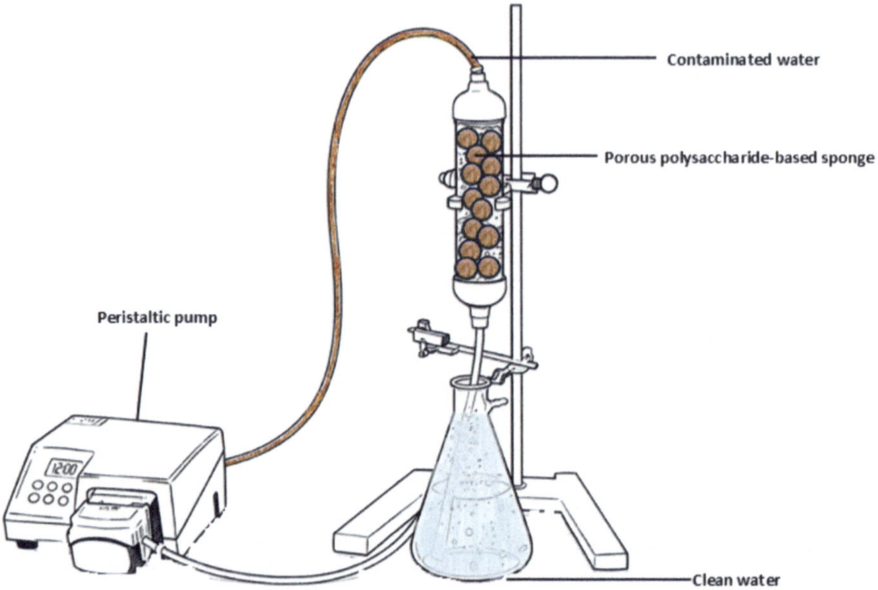

Fig. 1 Polysaccharide-based sponges for water treatment

like membrane filtration systems along, with oxidation processes including treatments and bio-based adsorbents are emerging as potential solutions. Materials made from polymers like polysaccharide-based sponges (Fig. 1) are getting noticed because of their porosity and surface features along, with being biodegradable and environmentally friendly products [5]. These materials can be designed to trap pollutants such as metals and organic dyes from polluted water sources providing a sustainable and budget friendly solution. With the increasing need, for water it is crucial to develop and apply creative and environmentally friendly water purification technologies at a large scale to ensure global water security and the well-being of the public. There are various water treatment applications of polysaccharides which include their use for pollutant elimination, purification and sustainable filtration processes, emphasizing their roles in pollutant removal, water purification, and sustainable filtration processes.

1.2 Polysaccharides: Versatile Biopolymers for Environmental Applications

Polysaccharides represent naturally occurring biopolymers, which consist of long monosaccharide chains that form through glycosidic bonds. Earth contains polysaccharides as one of its most common organic compounds because they originate from plants, algae, fungi, bacteria and crustaceans. Cellulose, starch, chitosan, alginate, pectin and xylan represent six common examples that show distinct structural and functional properties suitable for industrial and environmental applications. The combination of biocompatibility and biodegradability and low toxicity with natural abundance makes polysaccharides an attractive sustainable replacement for synthetic materials across medical applications, food technology, agriculture, and environmental remediation [6].

The natural abundance of polysaccharides enables continuous and economical supply which supports their extensive use in large-scale applications. The planet's largest biopolymer reserve consists of cellulose which occurs in plant cell walls and specific algae species while starch derives from corn and wheat and potato crops. The chitin shells of crustaceans like shrimp and crab serve as the source for chitosan which exhibits superior adsorption properties. The high ion-exchange properties together with gel-forming capabilities of alginate extracted from brown seaweed make it suitable for water treatment technologies [6].

The reactive functional groups (e.g., –OH, $-NH_2$, –COOH) in polysaccharides enable their use as effective adsorbents, flocculants, coatings and polymer matrices to remove pollutants from the environment. The porous structure of these materials can be altered through modification techniques or combined with other materials to boost their mechanical strength and selectivity while improving reusability capabilities. The environmentally friendly nature of these materials ensures that their disposal after use causes no significant damage to the environment thus fulfilling the guidelines of green chemistry and circular economy.

The advantages of polysaccharides make them suitable candidates for creating sustainable environmental solutions particularly in water purification and pollution control. These materials lead the development of future sustainable resources because of their flexible nature and regenerative properties that promote clean water accessibility and ecological sustainability [6].

1.3 3D-Engineered Scaffolds: Enhancing Material Performance

3D-engineered scaffold development has transformed material design within biomedical engineering and energy storage and environmental remediation applications. The fabrication process involving advanced techniques like freeze-drying, electrospinning, 3D printing and foaming enables users to control their architecture and porosity as well as their mechanical properties. 3D-engineered scaffolds with their unique interconnected porous network outperform traditional bulk materials by providing superior functional performance in applications needing efficient mass transfer and high surface area and structural stability [7].

Sponge-like structures represent a standout category of 3D-engineered materials because they combine superior porosity with extensive surface area and optimal permeability features. Their outstanding features make them suitable for applications that need fast liquid absorption and efficient contaminant transport. Sponge architectures maintain open-cell pores between micrometers and millimeters in size which enables free fluid movement and superior material-pollutant contact. The combination of lightweight construction and mechanical strength and reusability enables sponge architectures to perform well in extensive and extended operations [8].

Polysaccharides including chitosan, cellulose, alginate and starch when integrated with 3D-engineered scaffolds serve as a powerful approach to develop improved water treatment systems. The natural occurrence of polysaccharides combined with their degradability and diverse functional groups (–OH, $-NH_2$, –COOH) allows them to selectively capture heavy metals and organic dyes and pharmaceutical residues along with other pollutants. Native biopolymers show limited mechanical strength together with disorganized structure which hinders their effectiveness in dynamic water environments. The combination of 3D engineering methods allows researchers to create polysaccharide sponges with optimized pore structures that enhance both surface contact and structural stability which results in improved adsorption performance and longer material lifespan [9].

The union between natural polymer capabilities and advanced structural engineering enables polysaccharide-based 3D scaffolds to achieve exceptional water purification results. These materials represent an environmentally friendly and scalable and cost-effective answer to water pollution and scarcity issues which makes them leaders in future environmental remediation materials. Figure 2 shows a realistic 3D view of a bioprinting lab, where a high-precision printer creates a porous sponge from polysaccharide-based bio-ink, designed for advanced water filtration and adsorption.

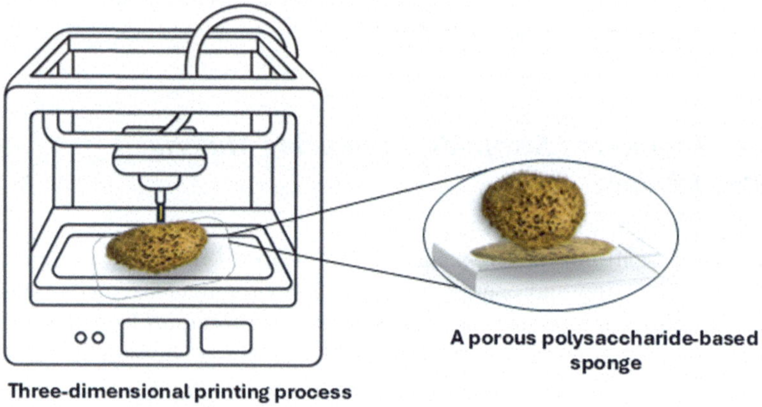

Fig. 2 3D rendering printing setup illustrating a porous polysaccharide-based sponge for filtration and adsorption applications

2 Water Treatment and Wastewater Applications of Polysaccharides

2.1 Polysaccharides' Methods for Removing Pollutants from Water

Polysaccharides demonstrate great potential for water and wastewater treatment applications because they are abundant naturally and biodegradable while showing minimal toxicity and various functional groups that enable contact with various pollutants. The water-treatment process utilizes biopolymers through four major methods: adsorption and flocculation and coagulation and membrane filtration and ion exchange [9].

2.1.1 Adsorption

Polysaccharides use adsorption as a common technique to trap contaminants through their surface or structural areas. The adsorption effectiveness depends on reactive functional groups of hydroxyl (–OH), carboxyl (–COOH), amino ($-NH_2$), and ether (C–O–C) present in chitosan, cellulose, alginate, and pectin. The functional groups of these molecules develop bond connections through ionic bonds coordination complexes and hydrogen bonds to pollutants that include heavy metals (Pb^{2+}, Cd^{2+}, Cr^{6+}) and organic dyes and pharmaceutical residues and pesticides [10].

The amino groups of chitosan become protonated during acidic conditions which creates electrostatic bonds with anionic pollutants. The porous structure of polysaccharide-based materials including sponges and hydrogels creates extensive surface area that allows multiple binding sites for better adsorption properties and faster adsorption rates [9, 10].

2.1.2 Flocculation and Coagulation

The natural properties of polysaccharides function as effective coagulants and flocculants that enable wastewater to eliminate suspended solids and colloidal particles and organic matter. These biopolymers create electrostatic interactions that eliminate charged particles and form larger flocs which either settle or become easy to filter out.

Polysaccharide-based coagulants outperform traditional synthetic chemicals such as alum and polyacrylamides because they are toxic-free and decompose naturally and produce minimal amounts of sludge. Starch derivatives together with modified cellulose and cationic chitosan serve as primary polysaccharide materials for treating textile effluents and food processing wastewater as well as municipal sewage [11].

2.1.3 Membrane Filtration (Polysaccharide-Based Membranes)

Membrane technology now serves as a fundamental method in advanced water purification because polysaccharides are used to develop environmentally friendly membrane solutions for microfiltration (MF) and ultrafiltration (UF) and nanofiltration (NF) and reverse osmosis (RO). Scientists produce these membranes by extracting them from cellulose while also creating hybrid materials from polysaccharides with other polymers to boost membrane strength and separation capabilities [12].

The operation of polysaccharide-based membranes depends on size exclusion together with electrostatic repulsion and specific adsorption that result from membrane surface charge and pore structure characteristics. The desalination and dye removal industry frequently uses cellulose acetate membranes, yet chitosan-coated membranes demonstrate antimicrobial properties while efficiently removing heavy metals and organic contaminants [13, 14].

2.1.4 Ion Exchange

Polysaccharides act as ion exchange materials because they possess ionizable functional groups that enable them to extract charged substances from solutions. Heavy metal extraction becomes more efficient through the ion exchange process because alginate and chitosan contain –COOH and $–NH_2$ groups that trade protons with Cu^{2+}, Zn^{2+} and Ni^{2+}.

The process of ion exchange happens through physical adsorption or complexation based on solution pH levels and ionic concentrations together with metal ion characteristics. Industrial wastewater treatment becomes possible through the improved ion exchange capacity and selectivity provided by modified polysaccharides including cross-linked and graft copolymers.

The platform of polysaccharides provides multiple pollutant removal capabilities through adsorption and flocculation/coagulation and membrane filtration and ion exchange mechanisms. Their chemical diversity enables customization of applications which target wide-ranging contaminants in water systems thus establishing them as environmentally friendly substitutes for synthetic materials. The development of efficient and sustainable water treatment systems based on natural biopolymers requires knowledge of their fundamental operational mechanisms.

2.2 Common Polysaccharides Used in Water Treatment

Natural water and wastewater treatment processes use polysaccharides because they occur in large amounts are non-toxic and degrade easily and contain reactive sites that bond with pollutants. The most widely used polysaccharides include chitosan together with cellulose and alginate and carrageenan and pectin and starch and xylan.

The particular chemical makeup and properties of each substance enable them to serve particular environmental functions.

2.2.1 Chitosan

The process of deacetylation of chitin results in chitosan since chitin exists in crustacean exoskeletons of shrimp and crab species.

The chemical structure consists of a linear sequence of β-(1 → 4)-linked D-glucosamine and N-acetyl-D-glucosamine molecules.

The combination of amino (–NH_2) and hydroxyl (–OH) functional groups in this material results in strong binding capacity for anionic pollutants. The substance maintains positive charge at acidic pH because the protonated –NH_2 groups make it superior for capturing heavy metals as well as dyes and negatively charged pathogens. Its antimicrobial properties allow it to work effectively in preventing biofouling on membranes. Applications: Heavy metal removal, dye adsorption, antimicrobial membranes, coagulant/flocculant [15].

2.2.2 Cellulose

The most abundant natural polymer; extracted from plant cell walls, algae, and bacteria.

The chemical structure of cellulose consists of β-(1 → 4)-linked D-glucose units arranged in a linear chain.

Multiple hydroxyl (–OH) groups present in this substance make it suitable for diverse chemical modifications. The derivative forms of this material possess high mechanical strength and thermal stability such as microcrystalline cellulose and cellulose nanocrystals. The hydrophilic nature allows the substance to absorb water and swell significantly. Applications: Adsorbent materials, membrane fabrication, filtration media [16].

2.2.3 Alginate

The extraction of brown seaweed *Laminaria* and *Macrocystis* leads to the production of alginate.

The chemical structure contains β-D-mannuronic acid (M) and α-L-guluronic acid (G) that form a linear chain through 1 → 4 glycosidic bonds.

The carboxylate (–COO^-) groups in this material enable strong metal ion binding through its interaction with divalent cations (e.g., Ca^{2+}). The substance creates hydrogels quickly through mild conditions which makes it suitable for encapsulation applications and controlled release mechanisms. Biocompatible and non-toxic. Applications: Heavy metal removal, drug delivery, flocculation, hydrogel beads for pollutant capture [17].

2.2.4 Carrageenan

The sulfated polysaccharide originates from red seaweed which includes *Chondrus crispus* and *Eucheuma cottonii* species.

The structure consists of linear sulfated galactans with alternating α-1,3- and β-1,6-glycosidic linkages.

The sulfate ester groups in this substance create an anionic character which makes it effective for removing cationic species. The gelling and thickening properties of this material enable the creation of composite materials. Applications: Coagulant aid, heavy metal adsorbent, emulsifier in oil–water separation [18].

2.2.5 Other Polysaccharides

Heavy metal adsorption properties make pectin suitable because it originates from plant cell walls and contains abundant carboxylic acid groups. Starch exists in cereal grains and tubers yet scientists modify it to boost its functional capabilities. The hemicellulosic polysaccharide xylan exists in hardwoods and grasses and serves as a material for composite products and biosorbent applications [9]. Table 1 provides a concise summary of commonly used polysaccharides in water treatment, highlighting their sources, applications, and key advantages.

2.3 Modifications to Enhance Performance

Polysaccharides need chemical or physical modification to address their main limitations which include poor solubility and weak mechanical strength and limited adsorption capacity.

Polyacrylamide and polyaniline polymers are grafted onto polysaccharide backbones to boost their adsorption abilities and selectivity according to [9].

Glutaraldehyde and epichlorohydrin together with genipin and other natural crosslinkers serve as agents to enhance mechanical stability and resistance to degradation [19].

The process of functionalization introduces specific functional groups such as amine, carboxyl, thiol to enhance reactivity with target contaminants [20].

The combination of inorganic materials such as clay and graphene oxide with metal oxides and synthetic polymers results in improved performance along with enhanced durability [19].

Nanocellulose/Nanostructuring: The creation of nano-scale derivatives including cellulose nanocrystals and nanofibrils boosts both surface area and reactivity [21].

Table 1 Common polysaccharides used in water treatment

Polysaccharide	Source	Chemical structure	Key functional groups/ properties	Applications
Chitosan	Derived from chitin, found in the exoskeletons of crustaceans (e.g., shrimp, crab)	Linear copolymer of β-(1 → 4)-linked D-glucosamine and N-acetyl-D-glucosamine	Contains $-NH_2$ and –OH groups; positively charged at acidic pH due to protonation; exhibits antimicrobial activity	Heavy metal removal, dye adsorption, antimicrobial membranes, coagulant/ flocculant
Cellulose	The most abundant natural polymer; extracted from plant cell walls, algae, and bacteria	Linear chain of β-(1 → 4)-linked D-glucose units	Rich in –OH groups; can be chemically modified; offers high mechanical strength and thermal stability in derivative forms (e.g., microcrystalline cellulose, cellulose nanocrystals); hydrophilic with high swelling capacity	Adsorbent materials, membrane fabrication, filtration media
Alginate	Extracted from brown seaweed such as *Laminaria* and *Macrocystis*	Linear copolymer of β-D-mannuronic acid (M) and α-L-guluronic acid (G) linked via 1 → 4 glycosidic bonds	Contains $-COO^-$ groups that bind divalent cations (e.g., Ca^{2+}); forms hydrogels under mild conditions; biocompatible and non-toxic	Heavy metal removal, drug delivery, flocculation, hydrogel beads for pollutant capture

(continued)

Table 1 (continued)

Polysaccharide	Source	Chemical structure	Key functional groups/ properties	Applications
Carrageenan	Sulfated polysaccharide extracted from red seaweed (*Chondrus crispus*, *Eucheuma cottonii*)	Linear sulfated galactans with alternating α-1,3- and β-1,6-glycosidic linkages	Contains sulfate ester groups, giving it an anionic character, effective for removing cationic species; possesses gelling and thickening properties	Coagulant aid, heavy metal adsorbent, emulsifier in oil–water separation
Other polysaccharides	Pectin: found in plant cell walls Starch: present in cereal grains and tubers Xylan: hemi cellulosic polysaccharide found in hardwoods and grasses	Pectin: rich in carboxylic acid groups Starch: composed of glucose units with variable branching Xylan: β-(1 → 4)-linked xylose units	Pectin: good for heavy metal adsorption Starch: often chemically modified to improve functionality Xylan: used in composite materials and biosorbents	Pectin: metal ion removal Starch: modified adsorbents Xylan: composite development and biosorption applications

2.4 Limitations of Traditional Polysaccharide-Based Materials

The practical use of traditional polysaccharide-based materials for water treatment faces multiple practical barriers because of their benefits such as biodegradability and non-toxicity and abundant functional groups. The main practical limitations which restrict the widespread use and efficiency of these materials consist of mechanical weaknesses together with limited surface area and challenges in separation and regeneration.

2.4.1 Mechanical Weakness

The main disadvantage of natural polysaccharides arises from their inherent mechanical fragility which becomes most evident when they exist in aqueous solutions or when transformed into hydrogels foams or membranes. Chitosan films have brittle properties which lead to cracking behaviour [22]. The soft nature of Alginate gels makes them collapse under minimal pressure [23]. The structures made from cellulose maintain dry strength but become weak when exposed to moisture [23].

Their inability to maintain mechanical stability prevents them from functioning in dynamic filtration systems and fixed-bed reactors as well as applications needing repeated handling and reuse. These materials need either supporting matrices or chemical crosslinking as a result which complicates the manufacturing process and may decrease their biocompatibility and pollutant-binding abilities [24].

2.4.2 Limited Surface Area

The specific surface area of traditional polysaccharide forms such as powders, granules and simple films remains low which restricts the number of active pollutant binding sites available. The structural limitation creates a direct impact on water treatment performance since it slows down adsorption kinetics through reduced surface exposure which results in slower contaminant uptake rates. The restricted surface area of materials creates a limit on the quantity of pollutants that can be adsorbed by each unit weight of material which reduces their overall adsorption capacity. These materials show weak performance when treating dilute contaminant solutions since the low pollutant concentration makes effective removal even more difficult. The improvement of surface area and porosity in polysaccharide-based materials has become a central goal to enhance their effectiveness in environmental remediation processes [9].

The natural polysaccharides that do exist with porous structures do not have sufficient micro- or mesoporosity for achieving high adsorption and ion exchange efficiency. Researchers are now investigating three methods to solve this problem: nanostructuring and porogen-assisted fabrication and composite formation with high-surface-area materials (such as activated carbon or graphene oxide) [25].

2.4.3 Difficulty in Separation and Regeneration

The practical implementation of water treatment methods faces a major challenge in recovering and reusing polysaccharide-based materials after they finish removing pollutants. The main problem arises from separation difficulties which affect powder and gel bead forms because settling or filtration operations become inefficient and time-consuming especially during large-scale applications. The regeneration of these materials faces substantial challenges because pollutant desorption requires methods that preserve both the structural integrity and active binding site capacity [25].

Polysaccharides show a tendency to experience structural damage throughout multiple adsorption-desorption cycles especially when they encounter harsh pH conditions. Material replacement needs to happen frequently because binding sites cannot be regenerated and biofouling occurs which drives up operational expenses while reducing sustainability. Research now focuses on developing magnetic composites and immobilized forms and 3D-printed scaffolds to solve these problems by enabling simple magnetic recovery while improving material reuse and sustaining functional performance for more sustainable water treatment systems [9].

3 3D-Engineering of Polysaccharide Sponge Formulations (Design and Fabrication)

3.1 Design Principles for 3D-Engineered Sponge Scaffolds

The development of 3D-engineered polysaccharide sponges requires collaboration between materials science and chemical engineering and environmental technology professionals to build water treatment materials with customized features. The fabrication methods that include freeze-drying, foaming, electrospinning and 3D printing allow precise control of sponge microstructure and mechanical behaviour and functional performance [26].

3.1.1 Control of Porosity, Pore Size, and Interconnectivity

Among the most important structural features affecting sponge-like material efficiency in water treatment are porosity, pore size and interconnectivity since they affect adsorption capacity, fluid permeability and mass transfer efficiency.

The volume fraction of void space in the scaffold known as porosity functions as a key determinant for contaminant storage capacity in sponge materials. Sponge materials with porosity exceeding 70% demonstrate better accessibility for pollutants and greater contaminant uptake. The pore size affects which contaminants a sponge can effectively capture because micropores (less than 2 nm) work best for ion exchange and molecular-level adsorption and mesopores (2–50 nm) facilitate the movement of larger organic pollutants and macropores (greater than 50 nm) enhance fluid flow and mechanical stability. Interconnectivity stands equally important for sponge matrices because it creates unbroken pathways that enable efficient water and contaminant transport across the sponge structure [27, 28].

The manufacturing parameters for precise structural control involve freeze-drying freezing rates which create aligned large pores and the use of porogens including salt particles or ice crystals with their respective sizes and the gas foaming parameters that include pressure and temperature and surfactant addition and the printing parameters for additive manufacturing that control nozzle diameter and layer thickness for customized pore architectures [29].

3.1.2 Tailoring Mechanical Strength and Stability

The natural softness of polysaccharides combined with their weak wet-weather durability makes them unsuitable for durable water treatment applications. Different methods have been developed to improve both mechanical stability and long-term performance and reusability of these materials.

A widely applied method for enhancing the polymer network involves using chemical agents such as glutaraldehyde or genipin or natural crosslinkers to create stronger bonds which decrease water solubility and boost structural resistance. The formation of composites by mixing polysaccharide matrices with nanocellulose, carbon nanotubes, graphene oxide or clay produces substantial compressive strength along with elastic properties and mechanical robustness improvements. The design of hybrid structures with different mechanical properties in various layers enables the creation of flexible-rigid combinations which suit operational needs [22].

Modern scaffold engineering techniques have produced anisotropic or hierarchically porous architectural designs which were inspired by biological systems to achieve better mechanical stability with preserved high porosity and functionality. The combination of these strategies enables polysaccharide sponges to endure dynamic aqueous environments and multiple uses for sustainable water treatment applications [30].

3.1.3 Incorporation of Functional Additives

Different functional additives added to the matrix of polysaccharide sponges extend their practical use in water treatment applications. The addition of metal oxides including Fe_3O_4, TiO_2, and ZnO provides both photocatalytic degradation functionality and magnetic separation ability for efficient sponge recovery and pollutant removal [31].

The addition of conductive polymers including polypyrrole and polyaniline enhances electron transfer properties which enables redox-based contaminant removal. A common long-term application challenge known as biofouling can be addressed through surface embedding of antimicrobial agents that prevent microbial growth on the sponge surface. Magnetic nanoparticles enable sponge retrieval from aqueous solutions through external magnets which enhances their effectiveness in batch treatment procedures [32].

The modification of nanomaterials like carbon nanotubes and mesoporous silica enhances surface area while creating specific binding regions for heavy metal and organic pollutant removal. The improved performance of polysaccharide sponges from advanced modifications enables their development into multifunctional materials which can perform adsorption along with catalysis and filtration and self-cleaning through mechanisms such as photocatalysis and magnetic recovery thus establishing them as versatile tools for sustainable water purification technologies [33].

3.2 *Fabrication Techniques for Polysaccharide Sponge Scaffolds*

Advanced fabrication techniques allow researchers to create polysaccharide sponge scaffolds used in water treatment through precise manipulation of structural and functional properties. The production methods focus on developing parameters such as porosity together with pore size distribution along with interconnectivity and mechanical strength and surface functionality which affect sponge performance in water pollution removal. The fabrication techniques freeze-drying (lyophilization) and emulsion templating and 3D printing (bioprinting) and solvent casting with particulate leaching and electrospinning serve different applications based on their specific features.

3.2.1 Freeze-Drying (Lyophilization)

Freeze-drying or lyophilization stands as a widely used method for building highly porous polysaccharide sponge structures. A solution or hydrogel of polysaccharides undergoes freezing before vacuum-based ice crystal sublimation creates a porous scaffold structure. The final structure of the scaffold depends heavily on both freezing speed and the direction of freezing process. The freezing speed determines sponge pore size and arrangement because slow freezing produces aligned large pores but fast freezing creates smaller pores with random distribution. The technique produces sponge structures with more than 80% porosity along with excellent pore interconnectivity thus making it suitable for applications requiring fast pollutant capture and efficient fluid movement. Materials including chitosan, alginate, cellulose and starch become easy to process into sponge structures with customized pore arrangements through this method. These sponges lack sufficient mechanical strength when exposed to water because they lack additional crosslinking or reinforcement [33].

3.2.2 Emulsion Templating

The creation of porous structures happens through emulsion templating which uses emulsion systems of water-in-oil or oil-in-water as templates to generate controlled pore sizes. The emulsion achieves stability and solidification after which the selected phase can be extracted to produce a porous matrix. The parameters which determine pore size and uniformity in this process include surfactant concentration together with polymer content and phase ratios. The technique shows excellent effectiveness in making micro- to nanoscale pores that improve surface area and adsorption capacity. Emulsion templating provides good control of pore shape but its optimization needs precise formulation and processing steps and may require organic solvents or chemical crosslinkers that complicate large-scale manufacturing [34, 35].

3.2.3 3D Printing (Bioprinting)

The development of 3D printing technology or bioprinting enables scientists to create accurate printable polysaccharide inks for layer-by-layer deposition. The additive manufacturing method allows researchers to construct intricate geometries with adjustable pore networks which match the requirements of filtration and adsorption applications. The final structure properties including mechanical strength and functionality depend heavily on printing speed, nozzle diameter and layer thickness and ink rheology. Adding reinforcing agents and functional nanoparticles to the printing ink enables the production of stronger sponges with improved selectivity. The advanced design capabilities of 3D printing make it ideal for specific applications yet its specialized equipment requirements and optimized formulations restrict its use mainly to niche or high-value markets [36, 37].

3.2.4 Solvent Casting/Particulate Leaching

The combination of solvent casting with particulate leaching produces sponges which have adjustable porosity and pore size through a straightforward method. A polysaccharide solution undergoes casting into a mold containing leachable particles which become dissolvable salt crystals or sugar granules that create a porous structure after removal. The dimensions of pores along with total porosity in the scaffold depend on the amount and size of the porogen. The method allows for easy scalability and provides a simple manufacturing process that makes it suitable for industrial use. High pore interconnectivity remains difficult to achieve without proper arrangement of the porogen or additional post-processing modifications to the structure [38].

3.2.5 Electrospinning

The electrospinning technique produces micro- to nanoscale fibers through the creation of ultrafine fibres. Polysaccharide solutions processed by this method generate nonwoven mats which achieve beneficial adsorption and filtration through their elevated surface area-to-volume ratio. Traditional electrospun mats show reduced fluid permeability because their dense fibre arrangement yet their bulk porosity remains limited until they become three-dimensional structures or receive additional porous supports. Electrospinning serves primarily to enhance base sponge materials through hybrid membrane and surface coating applications rather than producing standalone sponge structures [39, 40].

3.3 Characterization of 3D-Engineered Sponge Scaffolds

The performance and suitability of 3D-engineered polysaccharide sponge scaffolds for water treatment applications are heavily dependent on their structural, mechanical, and chemical properties. Therefore, comprehensive characterization techniques are essential to evaluate these features and ensure that the materials meet the required functional criteria. A variety of analytical tools are employed to assess morphology, porosity, mechanical behaviour, chemical composition, and water absorption capacity, each providing critical insights into the material's design and effectiveness.

3.3.1 Morphological Analysis (SEM, TEM)

Morphological analysis is typically carried out using scanning electron microscopy (SEM) or transmission electron microscopy (TEM), which allow for detailed visualization of the sponge's internal architecture at micro- and nano-scale resolutions. SEM is particularly useful for examining surface topography, pore structure, and overall scaffold organisation. These imaging techniques help confirm whether the fabrication process has yielded the intended pore size distribution, interconnectivity, and homogeneity, all of which are crucial for efficient pollutant adsorption and transport [41].

3.3.2 Porosity and Pore Size Distribution

To quantify the porosity and pore size distribution, advanced techniques such as mercury intrusion porosimetry (MIP) and gas adsorption analysis (e.g., BET method) are commonly used. Mercury porosimetry involves forcing mercury into the pores under pressure, allowing for the determination of pore volume and size across a wide range, especially in the meso- to macropore range. In contrast, BET gas adsorption, typically using nitrogen, provides data on specific surface area and microporosity, which are important for understanding the material's adsorption capacity and interaction with contaminants at the molecular level [42].

3.3.3 Mechanical Testing (Compression, Tensile)

Mechanical testing is another vital aspect of characterization, especially when considering the practical handling and durability of sponge scaffolds in real-world applications. Techniques such as compression testing and tensile strength analysis provide quantitative measurements of the scaffold's ability to withstand physical stress without deformation or failure. This is particularly important for sponges that may be subjected to repeated compression during use or regeneration cycles. Enhanced mechanical properties can be achieved through crosslinking, composite formation, or structural optimisation, and mechanical testing helps validate the success of such modifications [43].

3.3.4 Chemical Characterization (FTIR, XPS)

Chemical characterization plays a key role in confirming the functional groups, surface chemistry, and molecular interactions within the sponge matrix. Fourier-transform infrared spectroscopy (FTIR) is widely used to identify the presence of specific functional groups—such as hydroxyl (–OH), carboxyl (–COOH), and amino ($-NH_2$)—which are responsible for binding pollutants like heavy metals, dyes, and organic compounds. Additionally, X-ray photoelectron spectroscopy (XPS) offers more detailed information about the elemental composition and oxidation states of atoms on the surface, which is valuable for understanding adsorption mechanisms and verifying the successful incorporation of functional additives or modifications [44].

3.3.5 Water Absorption Capacity

Lastly, the water absorption capacity of a sponge scaffold is a fundamental parameter that influences its efficiency in water treatment processes. This property is typically assessed by immersing the dry sponge in water and measuring the weight gain over time. High water uptake indicates good hydrophilicity and swelling ability, which enhances pollutant accessibility and diffusion into the sponge matrix. It also reflects the material's compatibility with aqueous environments and its potential for applications such as oil–water separation, dye removal, or controlled release systems [27].

Together, these characterization techniques offer a comprehensive profile of the 3D-engineered polysaccharide sponge scaffolds, enabling researchers to correlate structural and chemical features with functional performance. This data-driven approach supports the iterative design and optimisation of sponge materials tailored for specific water treatment needs, ultimately contributing to the development of efficient, sustainable, and scalable environmental remediation solutions.

4 Applications in Water and Wastewater Treatment (Performance Evaluation)

Figure 3 illustrates the key applications of polysaccharide sponges in water treatment, showcasing their role in removing microorganisms, heavy metals, organic pollutants, and excess nutrients to support cleaner, safer water through sustainable, bio-based filtration.

4.1 Removal of Heavy Metals

The presence of heavy metals in water poses a major threat to environmental health because these metals exhibit toxic properties alongside their ability to build up in living organisms and resist natural decomposition processes including lead (Pb^{2+}), cadmium (Cd^{2+}), chromium (Cr^{6+}/Cr^{3+}), mercury (Hg^{2+}), and arsenic (As^{3+}/

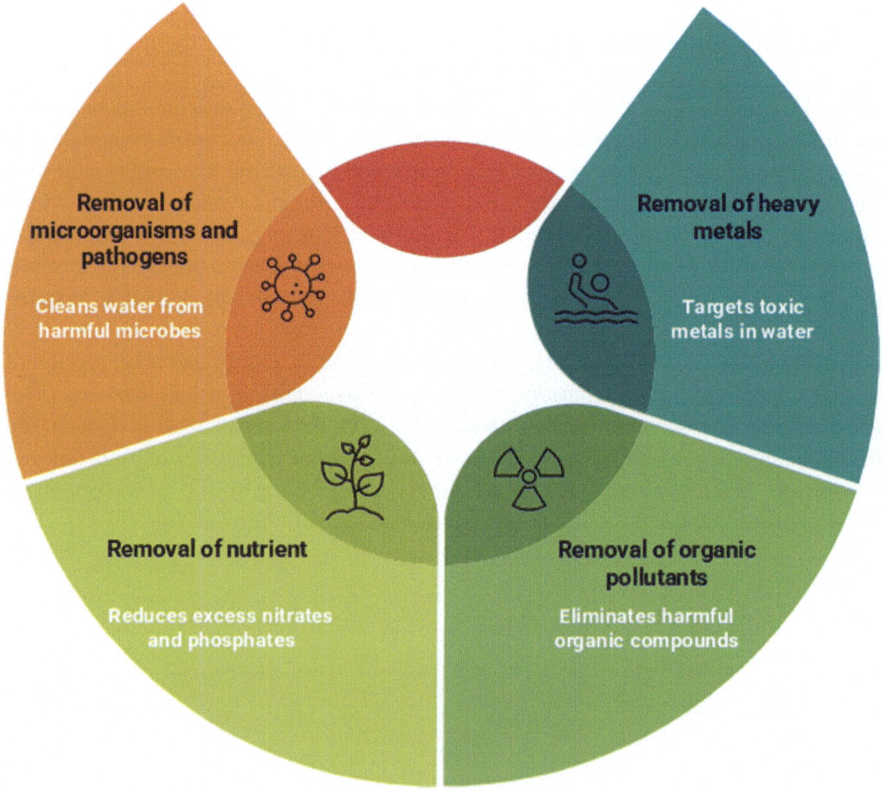

Fig. 3 Visual highlights the main ways polysaccharide sponges are used in water treatment

As^{5+}). The porous structure and numerous reactive functional groups of polysaccharide sponge scaffolds enable effective heavy metal removal from water through surface area expansion and adsorption alongside ion exchange and complexation and electrostatic binding.

The research shows that polysaccharide-based sponges effectively capture heavy metals from water thus establishing their potential as efficient water treatment materials. Chitosan-based sponges have been found to demonstrate outstanding lead ion (Pb^{2+}) adsorption capabilities that make them a highly promising natural polymer adsorbent for heavy metal removal [45]. The removal efficiency of cadmium (Cd^{2+}) by alginate beads reaches high levels within a brief 60-min contact time which proves their effectiveness in fast heavy metal elimination [46].

The combination of electrostatic attraction and redox reactions enables strong removal of hexavalent chromium (Cr^{6+}) by modified starch and pectin sponges. The performance of these sponges depends on multiple operational parameters which require adjustment for peak effectiveness. Most polysaccharides show better metal binding when the solution pH ranges from 4 to 6 because protonated functional groups become available for ion exchange and complexation reactions [47, 48]. The adsorption time needed to reach equilibrium varies between tens of minutes and several hours based on the specific system conditions.

The time period during which metal and adsorbent interact directly affects the adsorption process until the system reaches its equilibrium point. The adsorption capacity of the system grows when initial metal concentrations rise until the system reaches saturation. The adsorption kinetics improves through better diffusion rates at moderate temperature increases although high heat causes damage to the sponge structure [49, 50].

Selective modifications become essential to manage the presence of competitive ions such as Ca^{2+} and Mg^{2+} which reduce active binding site availability in complex water matrices. The research shows that maximizing heavy metal removal efficiency from contaminated water depends heavily on both material design and process optimization for polysaccharide sponges [9].

A vital requirement for sponge implementation in practice includes methods to restore and reuse the material. The process of effective regeneration maintains both operational cost-effectiveness and operational sustainability throughout extended periods. Different methods of regeneration include dilute acid solutions (HCl, HNO_3) and chelating agents (EDTA) and basic solutions based on the metal-polymer interaction characteristics. Research indicates that engineered polysaccharide sponges can be reused multiple times (5–10 cycles) while maintaining their effectiveness especially when treated with graphene oxide or magnetic nanoparticles for structural support [27].

4.2 Removal of Organic Pollutants

Organic pollutants constitute a major class of water and wastewater contaminants which include synthetic dyes and pharmaceutical residues and pesticides and endocrine-disrupting compounds and volatile organic chemicals. These substances create major environmental and health threats because they are toxic and long-lasting and tend to build up in biological systems. The removal of pollutants from water is becoming increasingly feasible through the use of polysaccharide-based sponge scaffolds which work by adsorption and photocatalytic degradation when functional additives are incorporated.

The functional groups present in chitosan, cellulose, alginate and pectin-derived sponges enable these materials to capture organic pollutants effectively because of their hydroxyl (–OH), carboxyl (–COOH) and amino ($-NH_2$) groups that establish interactions with dyes and organics through hydrogen bonding, electrostatic attraction, π–π interactions and van der Waals forces [9, 27].

The strong adsorption properties of chitosan-based sponges enable them to efficiently remove anionic dyes such as Congo Red under optimal conditions [51]. The adsorption capabilities of alginate beads make them suitable for tetracycline antibiotic removal [52] where they demonstrated high uptake efficiency for their use in pharmaceutical contaminant cleanup. Chitosan was functionalized with zinc ferrites ($Cs–ZnFe_2O_4$) to create a composite material. The formed composite material acted as an adsorbent for the removal of 2,2-dichlorovinyl dimethyl phosphate (DDVP) an insecticide from water-based solutions through batch adsorption tests [53].

The adsorption characteristics of these sponges can be evaluated using adsorption isotherms and kinetic models to determine the binding capacity and mechanism of adsorption.

Langmuir isotherm uses monolayer adsorption to determine maximum adsorption capacity and indicates favourable uptake when the correlation coefficient (R^2) is close to 1. The Freundlich isotherm describes heterogeneous surface adsorption and is commonly applied when multiple interaction mechanisms are involved. The adsorption rate can be determined using pseudo-first-order and pseudo-second-order kinetic models to distinguish between physical diffusion and chemical bonding. In most scenarios the pseudo-second-order model matches well indicating chemisorption dominates the adsorption process. These models enable researchers to optimize sponge design and operating conditions; such as contact time, initial pollutant concentration, and pH for maximal removal efficiency [49, 54, 55].

Some polysaccharide sponges receive engineered photocatalytic capabilities which allow them to degrade organic pollutants into less harmful byproducts by using light irradiation. The addition of semiconductor nanoparticles such as TiO_2, ZnO, graphene oxide and metal–organic frameworks (MOFs) into sponge matrices enables photocatalytic activity.

Nanocomposite hydrogels were synthesized via γ-radiation-induced copolymerization of chitosan, acrylic acid, and TiO_2 nanoparticles ($CS-PAAc/TiO_2$). The hydrogel showed effective removal of methylene blue dye from wastewater through

TiO_2 that improved physical adsorption interactions. The removal efficiency of dyes showed maximum increase at pH 10. Optimal adsorbent dosages were 0.20 g/L for CS-PAAc/TiO_2 and 0.15 g/L for CS-PAAc. The addition of TiO_2 nanoparticles improved dye removal performance which shows great potential for water treatment applications [56].

A hydrophilic ZnO nanoparticles@calcium alginate composite was produced through the incorporation of ZnO nanoparticles into calcium alginate. The ZnO NPs act as antibacterials by triggering oxidative stress and ROS production that results in bacterial apoptosis alongside calcium alginate using adsorption to remove organic impurities. The water purifier utilizing this composite material decreased aerobic bacteria to 9 cfu mL^{-1} and lowered turbidity to 0.51 NTU thus satisfying WHO requirements for safe drinking water consumption [57].

These photocatalytic sponges feature self-cleaning properties along with regeneration capabilities that do not require chemical elution making them suitable for continuous water treatment systems. The reuse of sponges after organic pollutant removal depends on the nature of the interaction. For adsorptive systems, desorption can be carried out using solvents, pH adjustment, or surfactants. Photocatalytic systems regenerate themselves in situ under light exposure thus eliminating the requirement for chemical regeneration in many scenarios [58].

The transition of these materials to industrial applications faces obstacles regarding structural maintenance and nanoparticle leaching prevention as well as cost-effectiveness concerns. Ongoing research into advanced modifications and composite formulations continues to enhance the efficiency and selectivity and durability of polysaccharide-based sponges for water organic pollutant removal.

The use of polysaccharide sponge scaffolds presents a flexible and environmentally responsible method for extracting numerous organic pollutants from water systems. These materials solve the increasing organic pollution problem in water systems through their adsorption mechanisms and kinetic and isotherm modelling and photocatalytic enhancement capabilities.

4.3 Removal of Nutrients (Nitrates and Phosphates)

Excessive nutrients in water environments primarily nitrates (NO_3^-) and phosphates (PO_4^{3-}) have caused widespread environmental problems which include eutrophication and its resulting harmful algal blooms, oxygen depletion and ecosystem disruption. Nutrients in water come mainly from agricultural runoff and sewage discharge and industrial effluents. Nutrient pollution management stands as a fundamental requirement for modern water and wastewater treatment operations [59].

4.3.1 Biological and Chemical Removal Methods

The removal of nutrients from water uses biological and chemical methods which operate at different efficiency levels.

Specific microbial communities including denitrifying bacteria and phosphate-accumulating organisms (PAOs) within natural and engineered systems break down nitrates and phosphates to produce biomass or gaseous byproducts (e.g., N_2). The method operates through constructed wetlands, biofilters and activated sludge processes because it provides energy efficiency together with low chemical consumption [60].

The chemical method of phosphate precipitation happens through metal salts (alum, ferric chloride) addition while ion exchange resins and adsorbent materials remove both nitrates and phosphates. These methods produce sludge as a byproduct while requiring additional costs for operations and potentially creating new contaminants within the system [61].

4.3.2 The Use of Modified Sponges to Support Bacterial Growth

Biological nutrient removal receives significant enhancement through the utilization of polysaccharide-based sponge scaffolds which function as biofilm carriers that support the development of nutrient-removing microorganisms. These materials offer optimal conditions for bacterial attachment because they have high porosity and large surface area together with hydrophilic properties.

The surface charge and nutrient accessibility along with microbial compatibility of sponges are enhanced through chemical modification and composite engineering [62]. For example:

Denitrifying bacteria adhere and proliferate better on chitosan-based sponges after they receive nitrogen-doped carbon dot or bioactive surface modifications which results in improved nitrate removal efficiency [63]. The combination of alginate-cellulose composite sponges with iron oxide nanoparticles provides both structural support for biofilms and electrostatic-based phosphate adsorption capabilities [64]. The manufacturing process of these sponges includes creating graded pore structures that produce simultaneous aerobic-anoxic zones which enable nitrification and denitrification processes inside the matrix.

These materials possess engineering features that preserve their structure under fluctuating flow conditions while resisting both biofouling and biodegradation throughout extended use periods. The biocompatibility of polysaccharides creates an environment which is harmless to microorganisms leading to continuous biological operations [65].

Biofilm-coated polysaccharide sponges have demonstrated nitrate removal rates higher than 80% and phosphate removal rates above 70% during weeks of continuous-flow operation. These materials excel in treating nutrient-rich wastewaters from agricultural and municipal sources because they combine physical adsorption with ion exchange and biological transformation mechanisms.

4.4 Removal of Microorganisms and Pathogens

Pathogenic microorganisms including bacteria (*Escherichia coli*, *Salmonella*, *Vibrio cholerae*) and viruses (norovirus, adenovirus) present in water threaten public health. The entry of contaminants into water systems occurs through sewage release as well as agricultural drainage and inappropriate waste management thus requiring effective microbial elimination procedures. Water treatment systems can effectively remove microbial contaminants through polysaccharide-based sponge scaffolds because of their high surface area and porous structure along with adjustable antimicrobial features [66].

4.4.1 Antimicrobial Properties of Polysaccharide Sponges

Chitosan represents a well-documented antimicrobial polysaccharide which disrupts bacterial cell membranes through electrostatic bonds. The positively charged amino groups ($-NH_3^+$) found in chitosan form electrostatic bonds with bacterial cell surfaces which leads to membrane disruption and cell death. The antimicrobial properties of chitosan-based sponges make them highly effective at stopping both Gram-positive and Gram-negative bacterial growth [67].

The antimicrobial properties of cellulose, alginate and pectin are usually absent but these polymers become antibacterial when chemically treated or engineered with bioactive substances such as quaternary ammonium compounds, essential oils or antimicrobial peptides [68]. The modified structure improves microbial capture efficiency while stopping microbial growth on sponge surfaces during extended periods of use.

4.4.2 Filtration of Bacteria and Viruses

Polysaccharide sponges function as physical barriers to remove microbial contaminants from water. The sponge matrices achieve efficient microorganism capture through their high porosity structure and adjustable pore size selection which filters based on size differences.

Bacterial cells ranging from 0.2 to 5 μm can be efficiently captured by sponge matrices when the pores are smaller than this range. Viruses need nanoporous structures or surface functionalization to achieve high retention efficiencies because they measure between 20 and 300 nm in size.

The continuous-flow filtration tests using electrospun cellulose-chitosan composite mats showed a > 99% *E. coli* removal efficiency. The pathogen-binding properties of freeze-dried alginate-chitosan sponges are enhanced by their cationic surface charge which improves electrostatic attraction with negatively charged microbes [65].

4.4.3 Disinfection Using Immobilized Nanoparticles

Polysaccharide sponges gain enhanced disinfection capabilities through antimicrobial nanoparticle functionalization beyond their physical filtration properties. Silver nanoparticles (AgNPs) exhibit wide-ranging antimicrobial action by producing ions that disrupt both microbial enzymes and DNA. Under light exposure ZnO NPs generate reactive oxygen species which induce oxidative stress in their environment. The antimicrobial properties of porous structures are effectively targeted by copper oxide nanoparticles (CuO NPs). Graphene-based materials use their mechanical properties to damage cell membranes while enhancing electron transfer which leads to improved microbial inactivation. The modifications applied to sponges create active pathogen elimination capabilities which enhances water safety and expands their use in advanced water treatment systems [57, 69].

Nanoparticles can be incorporated onto sponge surfaces or integrated into the matrix structure by applying methods such as sol-gel coating, layer-by-layer assembly or in situ synthesis. The technique of immobilization protects treated water from nanoparticle leakage thus ensuring both effectiveness and environmental protection [70]. Research shows that chitosan-AgNP composite sponges eliminate *E. coli* completely when in contact for 60 min [69]. Photocatalytic disinfection occurs in TiO_2-functionalized alginate sponges under UV light exposure which provides a sustainable microbial control method that does not produce chemical residuals [71].

4.5 Integration into Existing Water Treatment Systems

Water treatment applications of polysaccharide sponge scaffolds at scale depend on their ability to integrate into current systems along with scalability and performance comparison to traditional methods. During the last decade, pilot-scale research together with real-world field trials have received growing attention for assessing both practical usability, operational efficiency and sustainability of these biopolymer-based systems.

4.5.1 Pilot-Scale Studies and Real-World Applications

Research groups and industries have conducted pilot-scale evaluations to assess how polysaccharide sponges perform in dynamic flow systems with real-world fluctuating contaminant concentrations and complex wastewater matrices. The evaluation of essential operational parameters including adsorption capacity under continuous or semi-continuous flow requires these investigations to determine pressure drop, hydraulic resistance in filtration units, regeneration efficiency and long-term reusability. Scientists currently evaluate how conventional water treatment methods including coagulation, sedimentation and disinfection interact with these sponges [72].

Chitosan-cellulose composite sponges proved successful in municipal wastewater treatment when added to sand filtration systems to remove pharmaceutical residues and heavy metals without needing extensive infrastructure modifications [73]. A pilot plant in the textile sector used alginate-based sponges to treat post-dyeing effluent waste by removing more than 90% of color while reducing chemical oxygen demand (COD) [74]. Starch-chitosan hybrid sponges deployed in constructed wetlands efficiently captured nitrates and phosphates which reduced nutrient loading in downstream aquatic ecosystems. Real-world implementations of polysaccharide sponges in different settings demonstrate their expanding utility for sustainable water treatment systems [75].

Portable point-of-use water filters that use polysaccharide sponges have been deployed in rural areas and disaster zones to provide decentralized water treatment solutions for communities without centralized facilities [76]. The systems combine adsorption filtration and antimicrobial features into one device which delivers affordable safe drinking water while requiring minimal power consumption. Table 2 shows some types of polysaccharide, sponge fabrication method, application and performance.

5 Advanced Modifications and Future Directions (Innovation and Outlook)

5.1 Incorporation of Nanomaterials (e.g., Nanoparticles, Carbon Nanotubes)

Polysaccharide sponge scaffolds acquire next-generation water treatment capabilities through nanomaterial integration of nanoparticles, carbon nanotubes, graphene derivatives and metal–organic frameworks (MOFs).

The implementation of these nanomaterials with sponges enables scientists to increase the adsorption performance, selectivity and functional capabilities of biopolymer matrices.

The main purpose behind nanomaterial addition is to boost pollutant-binding capabilities of polysaccharide sponges. Nanomaterials show three main advantages in contaminant binding: they have large surface areas combined with many active sites and demonstrate strong attraction towards heavy metals, organic dyes and pharmaceutical compounds [31]. The adsorption of arsenic (As^{3+}/As^{5+}) and chromium (Cr^{6+}) by chitosan sponges increases substantially when iron oxide nanoparticles (Fe_3O_4) are embedded because they create strong electrostatic interactions and facilitate redox reactions [97].

Cellulose sponge activated carbon nanoparticles boost hydrophobic organic pollutant removal from pesticides and endocrine disruptors by enhancing hydrophobic surface contacts [98]. The phosphate ion-selective behaviour of Zirconium-based MOFs attached to alginate sponges makes them highly suitable for nutrient removal

Table 2 Polysaccharide-based sponges: fabrication, application, and performance

Polysaccharide	Sponge/fabrication method	Application	Performance	References
Alginate + organobentonite (fixed-bed column)	Alginate hydrogel beads embedding organobentonite; column adsorption	Cr(VI) removal from tannery effluent (continuous mode)	Fixed-bed removal to show results which included kinetic and column parameters through breakthrough curves	[77]
Carrageenan/alginate composite (Ts-Car/Alg)	Crosslinked tosyl-carrageenan blended with alginate into beads	Pb^{2+} removal from water	$q_{max} \approx 74$ mg/g (Langmuir); optimum at pH ~ 5.3, contact time 120 min	[78]
Sodium alginate (reviewed high-capacity SAH w/ carbon dots)	Alginate hydrogel beads (with carbon dots), various studies summarized	Heavy-metal removal (Pb^{2+})	The reported Pb^{2+} adsorption capacity reaches approximately 305 mg/g	[79]
Sodium alginate + lotus carbon dots (SA@CDs)	Ionotropic gelation → alginate hydrogel beads doped with carbon dots	Heavy-metal ion removal (Cd^{2+}, Pb^{2+}) from water	The Langmuir q_{max} values for Cd^{2+} and Pb^{2+} reached 179.35 mg/g and 291.23 mg/g respectively. The material maintained approximately 85% of its removal capacity throughout five consecutive cycles	[80]
Starch (superhydrophobic cryogel)	Starch-based cryogel with hydrophobic modification	Oil/water separation; oil uptake	The material shows oil absorption values between 4.79 and 13.05 g/g when tested with lab oils while achieving separation efficiency rates of up to 90.58% in the process	[81]
Chitosan (with CNTs and silica)	Porous chitosan sponge loaded with carbon nanotubes and SiO_2; superhydrophobic	Selective oil uptake and oil/water separation	The material demonstrates oil absorption capabilities of 18.24 g/g while showing selectivity in separation processes	[82]

(continued)

Table 2 (continued)

Polysaccharide	Sponge/fabrication method	Application	Performance	References
Cellulose (wood-based sponge)	Delignified/modified wood sponge with hydrophobic coating	Oil–water separation; crude oil/organic solvent uptake	The material demonstrates 12.72 g/g oil absorption capacity and achieves a separation efficiency of approximately 92% while being recyclable	[83]
Cellulose (wastepaper-derived aerogel)	Wastepaper pulp → freeze-drying aerogel (WOPW)	Mineral oil uptake/ oil–water separation	Absorption 6.1 g/g; reaches max in ~ 5 min	[84]
Cellulose (nanocellulose aerogel, CNF-PEI)	Freeze-dried cellulose nanofibers, polyethyleneimine (PEI) modification; hydrophobic aerogel	Oil/organic solvent absorption; oil/water separation	Sorption capacity 85.6–148.7 g/g (various oils/solvents)	[85]
Okara (soybean residue)	3D printing + freeze-drying	Wound hemostasis and antimicrobial	The antibacterial efficiency of the material reached 95.1% against *E. coli* while the BCI hemostatic value measured 24.5% lower than commercial PVF at 53.1%. The wound healing rate reached 87.8% by day 9	[86]
Okara cellulose	3D printing	Antioxidant and hemostatic sponge	The multifunctional sponge with wet elastic recovery features antibacterial and hemostatic and antioxidant properties for enhanced usage	[87]

(continued)

Table 2 (continued)

Polysaccharide	Sponge/fabrication method	Application	Performance	References
Cellulose	3D bioprinting	Solar desalination	A double-layered cellulosic evaporator made from waste cotton material functions as an efficient interfacial solar steam generator to produce clean water	[88]
Cellulose	3D printing	Oil/water separation	DLP-printed hydrophobic meshes made from fluorinated acrylates achieve adjustable wettability to separate complex emulsions effectively	[89]
Cellulose	3D printing	Sustainable filtration media	The material called reprogrammable sustainable cellulose hydroplastic allows users to create filters through on-demand fabrication while enabling control over pore size and functional properties	[90]
Cellulose	3D printing	Oil/water separation	A new method for creating hierarchical ceramic-cellulose structures through ultrafast emulsion treatment which enables simultaneous oil–water filtration with enhanced mechanical stability	[91]

(continued)

Table 2 (continued)

Polysaccharide	Sponge/fabrication method	Application	Performance	References
Acemannan (β-(1 → 4) glucomannan, O-acetylated) + other aloe vera polysaccharides (glucomannan, galactoglucomannan, pectins, etc.)	Freeze-drying of aloe vera gel (with/without gellan gum coating); composite scaffolds with chitosan, collagen, silk fibroin; PLA scaffolds coated with aloe vera	Wound healing dressings, tissue engineering applications for skin, cartilage, bone development, drug delivery systems with antimicrobial protection and sustained release capabilities	The material has a porosity of 72–77% with pores measuring 80 μm and expands by 270% while degrading 55% in 21 days and reaches compressive strength of 6.7 MPa when dry and maintains protein delivery for three weeks while being safe for fibroblasts and chondrocytes	[92]
Hyaluronic acid (HA)	Hydrogels or sponge formats; sometimes chemically modified	Soft and hard tissue regeneration in dental pulp/periodontal tissues	The material supports stem cell proliferation and migration and differentiation but lacks mechanical strength and degrades quickly unless it undergoes modification	[93]
Synthetic: PCL (polycaprolactone)	Porous scaffolds, membranes, combinations with other polymers/ceramics	Regeneration of periodontal ligament, cementum, alveolar bone; hard tissue repair	The material exhibits tensile strength between 10 and 50 MPa based on molecular weight, fabrication method while sustaining structural integrity and enabling cell attachment, differentiation processes and showing extended degradation periods	[93]

(continued)

Table 2 (continued)

Polysaccharide	Sponge/fabrication method	Application	Performance	References
Pullulan + dextran (primary chemically cross-linked network) ± alginate (second, ionically cross-linked network to form DN)	PuDAlg membranes were produced via STMP cross-linking and $CaCl_2$ ionization, freeze-dried to form a biphasic structure (smooth non-porous and porous surfaces), sterilized using UV and γ-irradiation	The GTR/GBR membranes function as barriers to stop soft tissue penetration while enabling bone tissue development; their effectiveness was proven through in vivo studies conducted on mice under the skin	The double-crosslinked PuDAlg-NaCl shows improved mechanical stability compared to the single network system because its hydrated storage modulus G' reaches 250 Pa whereas the pullulan–dextran network alone (PuD-NaCl) reaches only 70 Pa	[94]

(continued)

Table 2 (continued)

Polysaccharide	Sponge/fabrication method	Application	Performance	References
Oxidized hyaluronic acid (OHA) + ε-polylysine (ε-PL) + quaternized chitosan (QCS)	The material forms a gel network through Schiff-base bonds (OHA–ε-PL/QCS) which enables it to achieve self-healing properties, shear-thinning behaviour and injectability for wound coverage	The material supports hemostasis in liver and femoral injuries of rats and promotes angiogenesis, accelerates infected skin wound healing while reducing inflammation, enhancing collagen formation and enabling EGF encapsulation	The material undergoes rapid gelation within 50 s which makes it suitable for direct application in medical procedures. The material exhibits shear-thinning properties, self-healing ability, hemostatic properties for treating rat liver and femoral injuries. The addition of EGF resulted in complete wound closure of infected areas within 10 days. The material shows biocompatibility through cell growth promotion and angiogenesis induction and minimal inflammatory response and secure EGF storage	[95]
Protein/synthetic composites (e.g., HA/gelatin, chitosan/PLA, alginate/PEG)	3D bioprinting, freeze-drying, electrospun fibrous composites, nano-reinforced hydrogels; functionalized with peptides/GFs or nanoparticles	The research focuses on developing load-bearing cartilage/osteochondral constructs and translational scaffolds that improve both mechanical stability and biological functionality	The combination of optimal microstructure with functionalization results in materials that show both mechanical properties, biological activity (strong compressive strength, controlled degradation, enhanced chondrogenesis and tissue integration). The clinical application of these materials requires the development of scalable and consistent manufacturing methods	[96]

applications [99]. The modifications enable sponge developers to create targeted pollutant capture systems that match specific contaminants while preserving the natural advantages of polysaccharides including their biodegradable, non-toxic nature and renewable source materials.

5.2 Introducing Catalytic or Antimicrobial Properties

Sponges that receive nanomaterial integration can acquire both catalytic properties and antimicrobial capabilities which enhances their functionality for advanced water purification applications.

The photodegradation of organic pollutants occurs through the use of TiO_2, ZnO, or g-C_3N_4 nanoparticles which operate under UV or visible light irradiation [100]. Poly saccharide composites provide dual functions of contaminant elimination and destructive breakdown of pollutants into harmless substances which serves as a green alternative for traditional adsorbent regeneration techniques.

Silver nanoparticles (AgNPs) are commonly employed to give polysaccharide sponges wide-ranging antimicrobial capabilities. AgNPs embedded in chitosan and cellulose matrices produce silver ions that penetrate microbial cell walls and block DNA synthesis to render contaminated water free from bacteria and viruses [69].

The integration of carbon nanotubes or graphene oxide into sponge materials leads to enhanced electron transfer capabilities that support applications in electrochemical sensing and advanced oxidation processes (AOPs) for pollutant destruction [101].

The creation of magnetic nanocomposite sponges through Fe_3O_4 nanoparticles enables simple water treatment by magnet extraction which benefits batch or semi-continuous treatment operations [102].

The combination of nanomaterials with polysaccharide sponge scaffolds represents an important innovation for creating advanced water treatment materials. The combination of biopolymer sustainability with nanotechnology superiority enables hybrid systems to deliver superior adsorption capabilities and selectivity while performing photocatalytic degradation and microbial control. Research advancements in synthesis techniques alongside optimized nanoparticle loading and environmental safety protocols will transform these advanced materials into fundamental components for future water purification systems which will deliver efficient sustainable integrated solutions for global water challenges.

5.3 Smart and Responsive Polysaccharide Sponge Scaffolds

The development of smart or stimuli-responsive polysaccharide sponge scaffolds represents a cutting-edge advancement in water treatment technology. The materials possess an engineered ability to react dynamically to environmental triggers like pH, temperature, light and ionic strength to perform demand-based functions and select

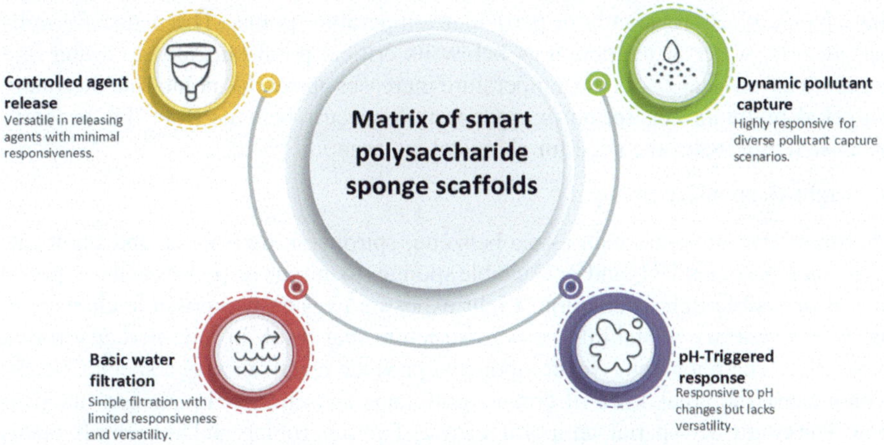

Fig. 4 An illustration of smart polysaccharide sponge scaffolds, highlighting their advanced functions such as controlled agent release, dynamic pollutant capture, basic filtration, and pH-triggered responses for adaptable water purification

pollutant capture and controlled agent release. The responsive nature of sponges increases both operational efficiency and versatility allowing them to work effectively in various complex water purification applications (Fig. 4).

5.3.1 Materials that Respond to pH Changes as well as Temperature Changes and Light Exposure

The modification or blending of polysaccharide sponges with responsive polymers allows their structure or function to respond to environmental conditions through introduction of stimuli-sensitive behaviour:

1. pH-Responsive Sponges

Polysaccharides contain naturally occurring ionizable functional groups such as –COOH in alginate and $-NH_2$ in chitosan that show pH responsiveness [103]. The protonation of chitosan-based sponges occurs at low pH which leads to positive charge development and increased attraction to anionic pollutants such as dyes and heavy metal anions [104]. The porosity and swelling of alginate sponges increase in alkaline solutions which enables better agent release and pollutant adsorption capabilities [17]. The inclusion of poly(acrylic acid) or poly(N-isopropylacrylamide) in sponge matrices enhances their pH sensitivity for self-regulating pollutant uptake or release based on acidity levels [105].

2. Temperature-Responsive Sponges

The grafting of thermos-responsive polymers such as poly(N-isopropylacrylamide) (PNIPAM) onto polysaccharide networks enables researchers to develop sponges that change phase reversibly at particular temperature points. The material swells and absorbs water at temperatures below its critical point but releases water and contaminant substances as the temperature increases above this point. The thermally triggered pollutant desorption or regeneration process benefits from this property because it eliminates the need for chemical intervention [106].

3. Light-Responsive Sponges

Photosensitive moieties such as azobenzene, spiropyran derivatives and photocatalytic nanoparticles TiO_2 and ZnO enable sponges to initiate responses when exposed to UV or visible light irradiation. Light exposure triggers two possible changes in the sponge surface wettability or pore structure that result in modified adsorption properties. The photocatalytic components produce reactive oxygen species (ROS) which cause the breakdown of organic pollutants and microbial pathogen inactivation. The controlled spatial, temporal release of antimicrobials and nutrients happens through light activation of loaded agents [107, 108].

5.3.2 Controlled Release of Functional Agents

The main benefit of smart sponges lies in their ability to act as delivery systems which release bioactive compounds and antimicrobial agents or nutrients in response to particular triggers. This characteristic proves especially important for uses that include:

1. The release of antimicrobial agents from responsive sponges becomes triggered when they detect microbial activity or environmental changes to perform water disinfection tasks [69].
2. The designed sponge system operates as an adsorbent for phosphate and nitrate during one set of conditions before it releases these nutrients at a controlled rate for agricultural fertilization purposes [109].
3. The controlled release of biocides and disinfectants occurs from point-of-use filters through smart sponge reservoirs in decentralized water purification systems [110].
4. The intelligent features enhance both operational control and resource efficiency while decreasing chemical waste, energy usage and manual labour needs which aligns with green and sustainable water treatment goals.

5.4 Development of Sustainable and Cost-Effective Fabrication Methods

Eco-friendly water treatment technology demands require the development of sustainable fabrication methods for polysaccharide sponge scaffolds that also reduce costs. Renewable biomass sources enable the production of materials which provide promising synthetic adsorbent alternatives. Their widespread adoption depends

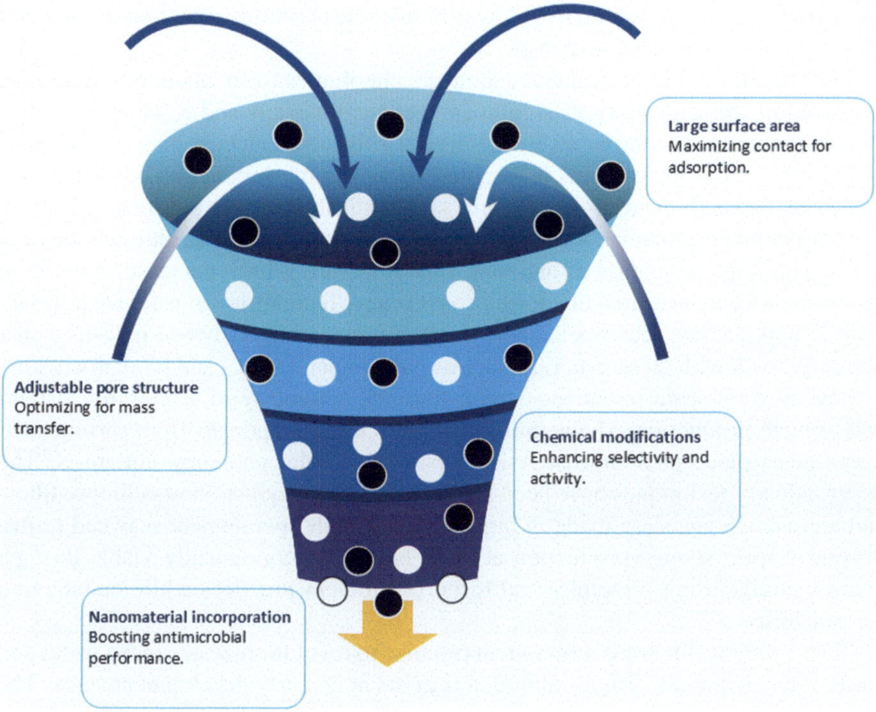

Fig. 5 Enhancing sponge performance for water treatment

on achieving large-scale production through green chemistry principles alongside low-energy processes and waste reduction measures (Fig. 5).

The main benefit of polysaccharide sponges comes from their natural occurrence in renewable abundant resources including chitosan from crustacean shells (seafood waste) as well as cellulose from wood, cotton or agricultural residues, alginate from brown seaweed, starch from corn, cassava or potatoes and pectin from citrus peels and apple pomace. The raw materials used in these materials are biodegradable while also being obtained from industrial byproducts which supports circular economy principles.

Researchers now dedicate more attention to developing environmentally friendly fabrication techniques for creating polysaccharide sponges. The use of green solvents like ionic liquids and deep eutectic solvents replaces toxic organic solvents and biodegradable crosslinkers genipin and citric acid substitute formaldehyde-based agents. The combination of freeze-drying with electrospinning and 3D printing techniques eliminates chemical usage while enhancing process security. Various waste valorisation methods exist for sponge management which enable composting and incineration without hazardous emissions as well as agricultural and soil remediation applications. The recycling of process water and functional additive recovery with optimized biomass extraction from low-cost materials minimizes both environmental effects and manufacturing expenses.

The transition of polysaccharide sponge technologies from laboratory success to industrial manufacturing faces significant technical barriers and economic hurdles. The development of continuous freeze-drying, roll-to-roll electrospinning and automated solvent casting systems is essential for achieving production consistency at industrial scales. Industrial use requires standardization of three critical elements: pore structure, mechanical strength and adsorption performance throughout each batch. The implementation of automated manufacturing lines decreases workforce expenses and enhances manufacturing consistency. Technology transfer for polysaccharide applications receives support through partnerships between industries that currently work with these materials such as pulp, paper, textiles and food processing.

Several pilot-scale initiatives demonstrate the feasibility of scaling up sustainable sponge production. The manufacturing of chitosan sponges from shrimp shell waste takes place in Southeast Asia as part of circular economy initiatives. The paper industry techniques have been adapted for mass production of cellulose filters and alginate sponges are made in biorefineries which operate near seaweed farms. Polysaccharide sponge production at scale becomes economically viable through green manufacturing principles and resource-efficient practices while maintaining sustainability.

3D printing technology shows great promise to revolutionize advanced water and wastewater treatment systems although it exists at an early development stage. The technology shows its ability to create polysaccharide-based sponges and meshes through precise engineering of pore structures which leads to enhanced adsorption, filtration and separation performance according to research findings. The technology enables users to create complex multifunctional structures through on-demand fabrication which traditional methods cannot achieve. The production of printed biopolymer materials faces ongoing challenges regarding scalability, cost and material durability but 3D printing technology enables the development of decentralized point-of-use systems and smart materials that respond to environmental stimuli. The combination of sustainable biomaterials with this technology establishes it as a fundamental element for developing precise environmental remediation systems of the future [86, 111, 112].

5.5 Challenges and Opportunities

To achieve sustainable water treatment through polysaccharide-based sponge scaffolds various technical regulatory and economic challenges need to be addressed for practical implementation and widespread adoption. The challenges represent opportunities for innovation together with interdisciplinary collaboration and sustainable development [27].

5.5.1 Long-Term Stability Requires Proper Solutions

Real-world water treatment system applications of polysaccharide sponges face major obstacles because they must maintain both stability and structural integrity during continuous flow operations and harsh environmental situations. The beneficial environmental aspect of biodegradation leads to decreased performance levels with time. Natural polysaccharides experience structural weakness when wet which causes their materials to disintegrate or break into pieces. The leaching of functional additives including nanoparticles, crosslinkers poses safety and efficiency risks to the system [25].

Scientists work to develop three main solutions: using non-toxic agents for crosslinking with genipin and citric acid while reinforcing structures with nanocellulose or graphene oxide and applying surface coatings for chemical resistance improvement. The created stable systems with biodegradable features should perform well during multiple cycles of use and regeneration to become suitable for practical large-scale implementation [113].

5.5.2 Regulatory Hurdles

Water treatment systems need to undergo strict regulatory reviews for their integration of novel biomaterials because they must ensure human health protection and ecological safety while fulfilling effluent discharge standards. The lack of standardized bio-based material certification protocols together with unresolved nanoparticle release uncertainties from metal, carbon-based nanomaterials and strict drinking water standards create challenges for validation of contaminant removal and byproduct safety.

The commercialization process requires researchers to team up with industry stakeholders and regulatory agencies for establishing evaluation frameworks. Life cycle assessments combined with toxicity studies will prove environmental sustainability while customized certification systems will advance circular economy-aligned green technologies. The successful management of regulatory obstacles stands essential for large-scale implementation of polysaccharide-based sponges in real-world water treatment applications.

5.5.3 Future Research Directions

The field of polysaccharide sponge engineering presents substantial opportunities for innovation beyond its present-day obstacles. The current research direction focuses on building intelligent sponge systems that change their functions according to environmental factors including pH, temperature and light exposure. The development of single-platform systems that combine adsorption capabilities with catalytic breakdown, disinfection functions and nutrient harvesting capabilities shows increasing interest. Machine learning integration with predictive modelling helps researchers discover new materials and enhance sponge functionality across various operational conditions.

The development of affordable mobile water filtration systems designed for unconnected rural communities and disaster zones represents an important research direction. These innovations support sustainable zero-waste fabrication through circular production models that use agricultural and industrial waste. Social acceptance receives increasing attention along with community engagement and policy alignment to achieve equitable technology access and maximize real-world water treatment effectiveness.

6 Conclusion

6.1 Recapitulation of the Advantages of 3D-Engineered Polysaccharide Sponge Formulations

3D-engineered polysaccharide sponge formulations show great promise for solving the increasing problems of water and wastewater treatment. These materials unite the natural properties of polysaccharides including biodegradability, non-toxicity, and abundant functional groups with modern structural engineering techniques which allow precise control of porosity, pore size, interconnectivity, and mechanical stability.

The sponges achieve versatility in pollutant treatment by using fabrication techniques such as freeze-drying, electrospinning, 3D printing, solvent casting and emulsion templating to handle various contaminants like heavy metals, organic dyes, pharmaceuticals, nutrients and pathogens. The combination of large surface area with adjustable pore structure in these materials boosts adsorption capacity, mass transfer and microbial interaction but additional chemical modifications and nanomaterial incorporation lead to enhanced selectivity, catalytic activity and antimicrobial performance.

These systems work with smart and responsive systems that include pH-, temperature- or light-responsiveness for controlled pollutant capture and release which enables new possibilities for water purification and regenerative filtration technologies.

6.2 Emphasizing Their Potential for Sustainable and Efficient Water Treatment

Polysaccharide-based sponge scaffolds function as sustainable replacement for traditional synthetic adsorbents and membranes because they use renewable materials that originate from agricultural, marine and industrial waste streams. The approach follows global circular economy principles by reducing waste while optimizing resource use.

The production of these materials on large scales through environmentally friendly methods together with their low energy requirements and minimal ecological footprint makes them suitable for both centralized and decentralized water treatment systems. These sponge systems serve as efficient contaminant removal solutions for industrial effluent treatment and municipal wastewater processing and point-of-use drinking water applications because they combine economic efficiency with environmental sustainability.

The platform's ability to combine adsorption with filtration and disinfection and nutrient recovery functions enables it to handle diverse pollution profiles effectively in various environmental settings.

6.3 Final Thoughts on the Future of This Technology

The increasing global need for clean water because of population growth and industrial development and climate change requires innovative sustainable solutions that can be scaled up. The future development of 3D-engineered polysaccharide sponge technology requires both performance enhancements and solutions for practical challenges that include extended stability issues and regulatory challenges and industrial production scalability.

These materials will achieve better selectivity and reusability and integration into existing infrastructure through ongoing research on hybrid composites and machine learning-driven design and smart water systems. Scientific, industrial and policy sector collaboration will speed up their implementation while ensuring fair access to clean water worldwide.

The material class of polysaccharide sponge scaffolds combines sustainable principles with efficient performance and innovative capabilities. These materials will become essential for future water treatment technologies when properly advanced and strategically deployed which will create a cleaner and healthier world with improved water security.

References

1. Han X, Boota MW, Soomro SEH, Ali S, Soomro SGH, Soomro NEH, et al. Water strategies and management: current paths to sustainable water use. Appl Water Sci. 2024;14(7). https://doi.org/10.1007/s13201-024-02214-2.
2. Belete B, Desye B, Ambelu A, Yenew C. Micropollutant removal efficiency of advanced wastewater treatment plants: a systematic review. Environ Health Insights. 2023;17. https://doi.org/10.1177/11786302231195158.
3. Farahbod F. A novel approach to desalination: producing potable water from seawater. Appl Water Sci. 2025;15(7). https://doi.org/10.1007/s13201-025-02540-z.

4. Ali ME, Hoque ME, Hossain SKS, Biswas MC. Nanoadsorbents for wastewater treatment: next generation biotechnological solution. Int J Environ Sci Technol. 2020;17(9):4095–132. https://doi.org/10.1007/s13762-020-02755-4.
5. Fatullayeva S, Tagiyev D, Zeynalov N, Mammadova S, Aliyeva E. Recent advances of chitosan-based polymers in biomedical applications and environmental protection. J Polym Res. 2022;29(7). https://doi.org/10.1007/s10965-022-03121-3.
6. Oladoja NA, Unuabonah EI, Amuda OS, Kolawole OM. Polysaccharides as a green and sustainable resources for water and wastewater treatment. 2017. https://doi.org/10.1007/978-3-319-56599-6.
7. Loh QL, Choong C. Three-dimensional scaffolds for tissue engineering applications: role of porosity and pore size. Tissue Eng Part B Rev. 2013;19(6):485–502. https://doi.org/10.1089/ten.teb.2012.0437.
8. Cui Y, Wang Y, Shao Z, Mao A, Gao W, Bai H. Smart sponge for fast liquid absorption and thermal responsive self-squeezing. Adv Mater. 2020;32(14). https://doi.org/10.1002/adma.201908249.
9. Na Y, Lee J, Lee SH, Kumar P, Kim JH, Patel R. Removal of heavy metals by polysaccharide: a review. Polym-Plast Technol Mater. 2020;59(16):1770–90. https://doi.org/10.1080/25740881.2020.1768545.
10. Zhang R, Tian Y. Characteristics of natural biopolymers and their derivative as sorbents for chromium adsorption: a review. J Leather Sci Eng. 2020;2(1). https://doi.org/10.1186/s42825-020-00038-9.
11. Singh RP, Nayak BR, Biswal DR, Tripathy T, Banik K. Biobased polymeric flocculants for industrial effluent treatment. Mater Res Innov. 2003;7(5):331–40. https://doi.org/10.1007/s10019-003-0273-z.
12. Dongre RS, Sadasivuni KK, Deshmukh K, Mehta A, Basu S, Meshram JS, et al. Natural polymer based composite membranes for water purification: a review. Polym-Plast Technol Mater. 2019;58(12):1295–310. https://doi.org/10.1080/25740881.2018.1563116.
13. Islam MD, Uddin FJ, Rashid TU, Shahruzzaman M. Cellulose acetate-based membrane for wastewater treatment—a state-of-the-art review. Mater Adv. 2023;4(18):4054–102. https://doi.org/10.1039/d3ma00255a.
14. Musarurwa H, Tavengwa NT. Application of polysaccharide-based metal organic framework membranes in separation science. Carbohydr Polym. 2021;275:118743. https://doi.org/10.1016/j.carbpol.2021.118743.
15. Guibal E, Van Vooren M, Dempsey BA, Roussy J. A review of the use of chitosan for the removal of particulate and dissolved contaminants. Sep Sci Technol. 2006;41(11):2487–514. https://doi.org/10.1080/01496390600742807.
16. Klemm D, Heublein B, Fink H, Bohn A. Cellulose: fascinating biopolymer and sustainable raw material. Angew Chem Int Ed. 2005;44(22):3358–93. https://doi.org/10.1002/anie.200460587.
17. Wang B, Wan Y, Zheng Y, Lee X, Liu T, Yu Z, et al. Alginate-based composites for environmental applications: a critical review. Crit Rev Environ Sci Technol. 2018;49(4):318–56. https://doi.org/10.1080/10643389.2018.1547621.
18. Ali KA, Hassan ME, Elnashar MMM. Development of functionalized carrageenan, chitosan and alginate as polymeric chelating ligands for water softening. Int J Environ Sci Technol. 2017;14(9):2009–14. https://doi.org/10.1007/s13762-017-1298-y.
19. Simsek M, Asiyanbi-Hammed TT, Rasaq N, Hammed AM. Progress in bioactive polysaccharide-derivatives: a review. Food Rev Int. 2021;39(3):1612–27. https://doi.org/10.1080/87559129.2021.1935998.
20. Hu Y, Li Y, Xu FJ. Versatile functionalization of polysaccharides via polymer grafts: from design to biomedical applications. Acc Chem Res. 2017;50(2):281–92. https://doi.org/10.1021/acs.accounts.6b00477.
21. Ansari F, Salajková M, Zhou Q, Berglund LA. Strong surface treatment effects on reinforcement efficiency in biocomposites based on cellulose nanocrystals in poly(vinyl acetate) matrix. Biomacromolecules. 2015;16(12):3916–24. https://doi.org/10.1021/acs.biomac.5b01245.

22. Wahba MI. Enhancement of the mechanical properties of chitosan. J Biomater Sci Polym Ed. 2019;31(3):350–75. https://doi.org/10.1080/09205063.2019.1692641.
23. Bao Z, Xian C, Yuan Q, Liu G, Wu J. Natural polymer-based hydrogels with enhanced mechanical performances: preparation, structure, and property. Adv Healthc Mater. 2019;8(17). https://doi.org/10.1002/adhm.201900670.
24. Abd TA, Fattah MY, Aswad MF. Strengthening of soft soil using caboxymethyl cellelouse biopolymer. IOP Conf Ser Earth Environ Sci. 2022;961(1):012030. https://doi.org/10.1088/1755-1315/961/1/012030.
25. Al-Hazmi HE, Łuczak J, Habibzadeh S, Hasanin MS, Mohammadi A, Esmaeili A, et al. Polysaccharide nanocomposites in wastewater treatment: a review. Chemosphere. 2023;347:140578. https://doi.org/10.1016/j.chemosphere.2023.140578.
26. Tsurkan D, Wysokowski M, Petrenko I, Voronkina A, Khrunyk Y, Fursov A, et al. Modern scaffolding strategies based on naturally pre-fabricated 3D biomaterials of poriferan origin. Appl Phys A. 2020;126(5). https://doi.org/10.1007/s00339-020-03564-9.
27. Liu C, Yu J, You J, Wang Z, Zhang M, Shi L, et al. Cellulose/chitosan composite sponge for efficient protein adsorption. Ind Eng Chem Res. 2021;60(25):9159–66. https://doi.org/10.1021/acs.iecr.1c01133.
28. Grosse J, Dietrich B, Garrido GI, Habisreuther P, Zarzalis N, Martin H, et al. Morphological characterization of ceramic sponges for applications in chemical engineering. Ind Eng Chem Res. 2009;48(23):10395–401. https://doi.org/10.1021/ie900651c.
29. Joukhdar H, Seifert A, Jüngst T, Groll J, Lord MS, Rnjak-Kovacina J. ICE templating soft matter: fundamental principles and fabrication approaches to tailor pore structure and morphology and their biomedical applications. Adv Mater. 2021;33(34). https://doi.org/10.1002/adma.202100091.
30. Alsaid Y, Wu S, Wu D, Du Y, Shi L, Khodambashi R, et al. Tunable sponge-like hierarchically porous hydrogels with simultaneously enhanced diffusivity and mechanical properties. Adv Mater. 2021;33(20). https://doi.org/10.1002/adma.202008235.
31. Liao X, Zhang Q. Mesoporous polymer nanosponges immobilized with functional polyols for rapid removal of boric acid and organic micropollutants. ACS Appl Polym Mater. 2019;1(8):2089–98. https://doi.org/10.1021/acsapm.9b00399.
32. Zhang Y, Xue Q, Li F, Dai J. Removal of heavy metal ions from wastewater by capacitive deionization using polypyrrole/chitosan composite electrode. Adsorp Sci Technol. 2019;37(3–4):205–16. https://doi.org/10.1177/0263617418822225.
33. Liu Y, An M, Qiu HX, Wang L. The properties of chitosan-gelatin scaffolds by once or twice vacuum freeze-drying methods. Polym-Plast Technol Eng. 2013;52(11):1154–9. https://doi.org/10.1080/03602559.2013.798818.
34. Zhang T, Sanguramath RA, Israel S, Silverstein MS. Emulsion templating: porous polymers and beyond. Macromolecules. 2019;52(15):5445–79. https://doi.org/10.1021/acs.macromol.8b02576.
35. Kimmins SD, Cameron NR. Functional porous polymers by emulsion templating: recent advances. Adv Funct Mater. 2010;21(2):211–25. https://doi.org/10.1002/adfm.201001330.
36. Shokrani H, Shokrani A, Seidi F, Mashayekhi M, Kar S, Nedeljkovic D, et al. Polysaccharide-based biomaterials in a journey from 3D to 4D printing. Bioeng Transl Med. 2023;8(4). https://doi.org/10.1002/btm2.10503.
37. Gao Y, Lalevée J, Simon-Masseron A. An overview on 3D printing of structured porous materials and their applications. Adv Mater Technol. 2023;8(17). https://doi.org/10.1002/admt.202300377.
38. Refifi J, Oudadesse H, Merdrignac-Conanec O, Feki HE, Lefeuvre B. Salt leaching using powder (SLUP) process for glass/chitosan scaffold elaboration for biomaterial applications. J Aust Ceram Soc. 2020;56(3):1167–78. https://doi.org/10.1007/s41779-020-00460-6.
39. Li Y, Wang J, Qian D, Chen L, Mo X, Wang L, et al. Electrospun fibrous sponge via short fiber for mimicking 3D ECM. J Nanobiotechnol. 2021;19(1). https://doi.org/10.1186/s12951-021-00878-5.

40. Gong X, Kalantari M, Aslanzadeh S, Boluk Y. Interfacial interactions and electrospinning of cellulose nanocrystals dispersions in polymer solutions: a review. J Dispers Sci Technol. 2020;43(7):945–77. https://doi.org/10.1080/01932691.2020.1847137.
41. Liu M, Jin J, Zhong X, Liu L, Tang C, Cai L. Polysaccharide hydrogels for skin wound healing. Heliyon. 2024;10(15):e35014. https://doi.org/10.1016/j.heliyon.2024.e35014.
42. Schlumberger C, Thommes M. Characterization of hierarchically ordered porous materials by physisorption and mercury porosimetry—a tutorial review. Adv Mater Interfaces. 2021;8(4). https://doi.org/10.1002/admi.202002181.
43. Karimi A, Navidbakhsh M. Measurement of the nonlinear mechanical properties of a poly(vinyl alcohol) sponge under longitudinal and circumferential loading. J Appl Polym Sci. 2013;131(10). https://doi.org/10.1002/app.40257.
44. Shi X, Lan Y, Peng S, Wang Y, Ma J. Green fabrication of a multifunctional sponge as an absorbent for highly efficient and ultrafast oil–water separation. ACS Omega. 2020;5(24):14232–41. https://doi.org/10.1021/acsomega.9b03736.
45. Hao D, Liang Y. Adsorption of Cu^{2+}, Cd^{2+} and Pb^{2+} in wastewater by modified chitosan hydrogel. Environ Technol. 2020;43(6):876–84. https://doi.org/10.1080/09593330.2020.1807612.
46. Kwiatkowska-Marks S, Wójcik M. Removal of cadmium(II) from aqueous solutions by calcium alginate beads. Sep Sci Technol. 2014;49(14):2204–11. https://doi.org/10.1080/01496395.2014.912223.
47. González-López ME, Laureano-Anzaldo CM, Pérez-Fonseca AA, Arellano M, Robledo-Ortíz JR. Chemically modified polysaccharides for hexavalent chromium adsorption. Sep Purif Rev. 2020;50(4):333–62. https://doi.org/10.1080/15422119.2020.1783311.
48. Sidlauskiene D, Klimaviciute R, Zemaitaitis A. Adsorption of hexavalent chromium on cationic starches with different degree of crosslinking. J Appl Polym Sci. 2007;107(6):3469–75. https://doi.org/10.1002/app.27461.
49. Miao MS, Wang YN, Kong Q, Shu L. Adsorption kinetics and optimum conditions for Cr(VI) removal by activated carbon prepared from luffa sponge. Desalin Water Treat. 2015;57(17):7763–72. https://doi.org/10.1080/19443994.2015.1015453.
50. Karthikeyan G, Anbalagan K, Andal NM. Adsorption dynamics and equilibrium studies of Zn(II) onto chitosan. J Chem Sci. 2004;116(2):119–27. https://doi.org/10.1007/bf02708205.
51. Raval NP, Shah PU, Ladha DG, Wadhwani PM, Shah NK. Comparative study of chitin and chitosan beads for the adsorption of hazardous anionic azo dye Congo Red from wastewater. Desalin Water Treat. 2015;57(20):9247–62. https://doi.org/10.1080/19443994.2015.1027959.
52. Luo H, Liu Y, Lu H, Fang Q, Rong H. Efficient adsorption of tetracycline from aqueous solutions by modified alginate beads after the removal of Cu(II) ions. ACS Omega. 2021;6(9):6240–51. https://doi.org/10.1021/acsomega.0c05807.
53. Adeogun AI, Osideko OA, Idowu MA, Akinloye OA, Ofudje EA. Synthesis, characterization and investigation of chitosan-functionalized $ZnFe_2O_4$ for the removal of dichlorvos from aqueous solution. J Dispers Sci Technol. 2021;44(1):51–60. https://doi.org/10.1080/01932691.2021.1930036.
54. Zhang Q, Xie M, Guo X, Zeng L, Luo J. Fabrication and adsorption behavior for Congo Red of chitosan and alginate sponge. Integr Ferroelectr. 2014;151(1):61–75. https://doi.org/10.1080/10584587.2014.899458.
55. Kong Q, Wang YN, Shu L, Miao MS. Isotherm, kinetic, and thermodynamic equations for cefalexin removal from liquids using activated carbon synthesized from loofah sponge. Desalin Water Treat. 2015;57(17):7933–42. https://doi.org/10.1080/19443994.2015.1052991.
56. Mahmoud GA, Sayed A, Thabit M, Safwat G. Chitosan biopolymer based nanocomposite hydrogels for removal of methylene blue dye. SN Appl Sci. 2020;2(5). https://doi.org/10.1007/s42452-020-2753-9.
57. Zhao WB, Du MR, Liu KK, Zhou R, Ma RN, Jiao Z, et al. Hydrophilic ZNO nanoparticles@calcium alginate composite for water purification. ACS Appl Mater Interfaces. 2020;12(11):13305–15. https://doi.org/10.1021/acsami.9b23458.

58. Zhu L, Ji J, Liu J, Mine S, Matsuoka M, Zhang J, et al. Designing 3D-MOS_2 sponge as excellent cocatalysts in advanced oxidation processes for pollutant control. Angew Chem Int Ed. 2020;59(33):13968–76. https://doi.org/10.1002/anie.202006059.
59. Nur T, Johir MAH, Loganathan P, Vigneswaran S, Kandasamy J. Effectiveness of purolite A500PS and A520E ion exchange resins on the removal of nitrate and phosphate from synthetic water. Desalin Water Treat. 2012;47(1–3):50–8. https://doi.org/10.1080/19443994.2012.696424.
60. Bhambri A, Karn SK. Biotechnique for nitrogen and phosphorus removal: a possible insight. Chem Ecol. 2020;36(8):785–809. https://doi.org/10.1080/02757540.2020.1777991.
61. Sengupta S, Nawaz T, Beaudry J. Nitrogen and phosphorus recovery from wastewater. Curr Pollut Rep. 2015;1(3):155–66. https://doi.org/10.1007/s40726-015-0013-1.
62. Fan X, Su Z, Zhang W, Huang H, He C, Wu Z, et al. An advanced chitosan based sponges dressing system with antioxidative, immunoregulation, angiogenesis and neurogenesis for promoting diabetic wound healing. Mater Today Bio. 2024;29:101361. https://doi.org/10.1016/j.mtbio.2024.101361.
63. Tabrizi NS, Yavari M. Fixed bed study of nitrate removal from water by protonated cross-linked chitosan supported by biomass-derived carbon particles. J Environ Sci Health A. 2020;55(7):777–87. https://doi.org/10.1080/10934529.2020.1741998.
64. Minh VX, Hanh LTM, Lan PT, Van Bien T, Dung NT. Synthesis of Fe_3O_4/alginate composite for dye removal from aqueous solution. Vietnam J Chem. 2020;58(2):185–90. https://doi.org/10.1002/vjch.201900151.
65. Ruan H, Aulova A, Ghai V, Pandit S, Lovmar M, Mijakovic I, et al. Polysaccharide-based antibacterial coating technologies. Acta Biomater. 2023;168:42–77. https://doi.org/10.1016/j.actbio.2023.07.023.
66. Cabral JPS. Water microbiology. Bacterial pathogens and water. Int J Environ Res Public Health. 2010;7(10):3657–703. https://doi.org/10.3390/ijerph7103657.
67. Raafat D, Von Bargen K, Haas A, Sahl HG. Insights into the mode of action of chitosan as an antibacterial compound. Appl Environ Microbiol. 2008;74(12):3764–73. https://doi.org/10.1128/aem.00453-08.
68. Tavakolian M, Jafari SM, Van De Ven TGM. A review on surface-functionalized cellulosic nanostructures as biocompatible antibacterial materials. Nano-Micro Lett. 2020;12(1). https://doi.org/10.1007/s40820-020-0408-4.
69. Wu Z, Zhou W, Deng W, Xu C, Cai Y, Wang X. Antibacterial and hemostatic thiol-modified chitosan-immobilized AGNPS composite sponges. ACS Appl Mater Interfaces. 2020;12(18):20307–20. https://doi.org/10.1021/acsami.0c05430.
70. Giese EC, Silva DDV, Costa AFM, Almeida SGC, Dussán KJ. Immobilized microbial nanoparticles for biosorption. Crit Rev Biotechnol. 2020;40(5):653–66. https://doi.org/10.1080/07388551.2020.1751583.
71. Wei S, Zhang X, Zhao K, Fu Y, Li Z, Lin B, et al. Preparation, characterization, and photocatalytic degradation properties of polyacrylamide/calcium alginate/TiO_2 composite film. Polym Compos. 2014;37(4):1292–301. https://doi.org/10.1002/pc.23295.
72. Kirishima Y, Choeisai P, Khotwieng W, Hatamoto M, Watari T, Choeisai K, et al. Efficiency of high rate treatment of low-strength municipality sewage by a pilot-scale combination system of a sedimentation tank and a down-flow hanging sponge reactor. Environ Technol. 2021;43(16):2457–66. https://doi.org/10.1080/09593330.2021.1882584.
73. Biswas S, Fatema J, Debnath T, Rashid TU. Chitosan–clay composites for wastewater treatment: a state-of-the-art review. ACS EST Water. 2021;1(5):1055–85. https://doi.org/10.1021/acsestwater.0c00207.
74. Sharmila A, Mardera MS, Soundarya HP, Shreyas M. Treatment of textile wastewater using sodium alginate beads. J Phys Conf Ser. 2021;1979(1):012005. https://doi.org/10.1088/1742-6596/1979/1/012005.
75. Turunen J, Karppinen A, Ihme R. Effectiveness of biopolymer coagulants in agricultural wastewater treatment at two contrasting levels of pollution. SN Appl Sci. 2019;1(3). https://doi.org/10.1007/s42452-019-0225-x.

76. Correa S, Grosskopf AK, Hernandez HL, Chan D, Yu AC, Stapleton LM, Appel EA. Translational applications of hydrogels. Chem Rev. 2021;121(18):11385–457. https://doi.org/10.1021/acs.chemrev.0c01177.
77. Muñoz-Martinez MX, Macías-Quiroga IF, Sanabria-González NR. Adsorption of Cr(VI) using organoclay/alginate hydrogel beads and their application to tannery effluent. Gels. 2024;10(12):779. https://pmc.ncbi.nlm.nih.gov/articles/PMC11675422/
78. Awed M, Mohamed RR, Kamal KH, Sabaa MW, Ali KA. Tosyl-carrageenan/alginate composite adsorbent for removal of Pb^{2+} ions from aqueous solutions. BMC Chem. 2024;18(1). https://doi.org/10.1186/s13065-023-01103-0.
79. Felicia WXL, Rovina K, Supri S, Matanjun P, Amin SFM, Rahman MNA. Next-generation sodium alginate hydrogels for heavy metal ion removal: properties, dynamic adsorption–desorption mechanisms, and sustainable application potential. Polym Bull. 2025. https://doi.org/10.1007/s00289-025-05956-4.
80. Younis MY, Li W, Khan J, Wei Y. Lotus carbon dots doped sodium alginate hydrogel beads for effective adsorptive removal of Cd^{2+} and potential application in Pb^{2+} removal from wastewater. Int J Biol Macromol. 2025:146638. https://pubmed.ncbi.nlm.nih.gov/40780334/
81. Mao L, Chen H, Lian Y, Cao Y, Wang X, Chen Z, Zhang Y. Degradable starch-based superhydrophobic cyrogel for efficient oil/water separation. Compos Commun. 2025:102525. https://doi.org/10.1016/j.coco.2025.102525.
82. Xu C, He H, Wang Y, Huang Y, Zhang TC, Yuan S. Superhydrophobic sponge-like chitosan/CNTs/silica composite for selective oil absorption and efficient separation of water-in-oil emulsion. Carbohydr Polym. 2025;353:123256. https://doi.org/10.1016/j.carbpol.2025.123256.
83. Sun Y, Zhang Y, Tian Z, Liu H, Chen S, Huang X, Cai Z, Fang B. Research progress of biomass-based porous oil-absorbing materials. J Renew Mater. 2024;12(11):1837–56. https://doi.org/10.32604/jrm.2024.055933
84. Romero-Montero A, Rosas-Melendez SA, Valencia-Bermúdez JL, Nuñez-Tapia I, Piña-Barba MC, Melgoza-Ramírez LJ, Leyva-Gómez G, Del Prado-Audelo ML. Oil/water separation by super-hydrophobic wastepaper cellulose-candelilla wax cryogel: a circular material-based alternative. Front Mater. 2023;10. https://doi.org/10.3389/fmats.2023.1308094.
85. Wu Z, Shu S, Hu X, Luo Z, Gong H, Fang D, Kou Y, Zhang X, Li X, Deng Z. Hydrophobic cellulose nanofiber based aerogel for high-efficiency oil–water separation. Int J Biol Macromol. 2025;305:141173. https://doi.org/10.1016/j.ijbiomac.2025.141173.
86. Fang J, Wu Z, Fu H, Xu C, Li X, Yu H, Wang X. 3D printed porous okara-based sponge with hemostatic and antimicrobial effects for wound healing. ACS Sustain Chem Eng. 2024;12(30):11231–43. https://doi.org/10.1021/acssuschemeng.4c02407.
87. Li Z, Zhao J, Fang J, Li X, Wang X. Okara cellulose-based multifunctional sponge with wet elastic deformation recovery for antibacterial, hemostatic, and antioxidant properties. Int J Biol Macromol. 2025;321:146273. https://doi.org/10.1016/j.ijbiomac.2025.146273.
88. Zhou Z, Luo L, Feng X, Mao Z, Rong L, Wang B. Double-layered cellulosic interfacial evaporator via upcycling of waste cotton fabrics for efficient solar desalination. Sep Purif Technol. 2024;349:127817. https://doi.org/10.1016/j.seppur.2024.127817.
89. Lee WH, Haddleton D. Digital light processing (DLP) 3D printing fabrication of hydrophobic meshes incorporating fluorinated and silicone-based acrylates combined with surface engineering: comparison of their oil–water separation efficiency. ACS Omega. 2024;9(50):49463–9. https://doi.org/10.1021/acsomega.4c07193.
90. Koh JJ, Koh XQ, Chee JY, Chakraborty S, Tee SY, Zhang D, Lai SC, Yeo JCC, Soh JWJ, Li P, Tan SC, Thitsartarn W, He C. Reprogrammable, sustainable, and 3D-printable cellulose hydroplastic. Adv Sci. 2024;11(29). https://doi.org/10.1002/advs.202402390.
91. Jonhson W, Xu X, Bian K, Xun Y, Tan YH, Chen Z, Zhang D, Ding J. 3D-printed hierarchical ceramic architectures for ultrafast emulsion treatment and simultaneous oil–water filtration. ACS Mater Lett. 2022;4(4):740–50. https://doi.org/10.1021/acsmaterialslett.2c00147.
92. Matei CE, Visan AI, Cristescu R. Aloe vera polysaccharides as therapeutic agents: benefits versus side effects in biomedical applications. Polysaccharides. 2025;6(2):36. https://doi.org/10.3390/polysaccharides6020036.

93. Lesko L, Jungova P, Culenova M, Thurzo A, Danisovic L. Polymer-based scaffolds as an implantable material in regenerative dentistry: a review. J Funct Biomater. 2025;16(3):80. https://doi.org/10.3390/jfb16030080.
94. Maillard M, Dujardin C, Aprile P, Aid R, Letourneur D, Simon-Yarza T. Double-network polysaccharide hydrogel for guided tissue repair. Biomater Sci. 2025. https://doi.org/10.1039/d5bm00117j.
95. Zheng H, Zhang X, Wang W, Xin H, Yu B, Wang L, Xie H, Zhang Y, Song Y, Ma Q. Oxidized hyaluronic acid/ε-polylysine/quaternized chitosan hydrogel with shear thinning injectability and self-healing properties for full-time and multipurpose wound healing. Int J Biol Macromol. 2025;293:139347. https://doi.org/10.1016/j.ijbiomac.2024.139347.
96. Zhang B, Zhang M, Jiang C, Yan W, Pan Y, Meng F. Engineered polysaccharide scaffolds for cartilage regeneration: mechanisms, functionalization, and clinical prospects. Colloids Surf B Biointerfaces. 2025:115134. https://doi.org/10.1016/j.colsurfb.2025.115134.
97. Ociński D. Optimization of hybrid polymer preparation by ex situ embedding of waste Fe/Mn oxides into chitosan matrix as an effective As(III) and As(V) sorbent. Environ Sci Pollut Res. 2019;26(25):26026–38. https://doi.org/10.1007/s11356-019-05856-x.
98. Kong X, Zhou S, Strømme M, Xu C. All-cellulose-based freestanding porous carbon nanocomposites and their versatile applications. Compos Part B Eng. 2021;232:109602. https://doi.org/10.1016/j.compositesb.2021.109602.
99. Mazloomi S, Yousefi M, Nourmoradi H, Shams M. Evaluation of phosphate removal from aqueous solution using metal organic framework; isotherm, kinetic and thermodynamic study. J Environ Health Sci Eng. 2019;17(1):209–18. https://doi.org/10.1007/s40201-019-00341-6.
100. Ramesh N, Lai CW, Johan MRB, Mousavi SM, Badruddin IA, Kumar A, et al. Progress in photocatalytic degradation of industrial organic dye by utilising the silver doped titanium dioxide nanocomposite. Heliyon. 2024;10(24):e40998.
101. Shao P, Yu S, Duan X, Yang L, Shi H, Ding L, et al. Potential difference driving electron transfer via defective carbon nanotubes toward selective oxidation of organic micropollutants. Environ Sci Technol. 2020;54(13):8464–72. https://doi.org/10.1021/acs.est.0c02645.
102. Makarchuk OV, Dontsova TA, Astrelin IM. Magnetic nanocomposites as efficient sorption materials for removing dyes from aqueous solutions. Nanoscale Res Lett. 2016;11(1). https://doi.org/10.1186/s11671-016-1364-2.
103. Gemeinhart RA, Chen J, Park H, Park K. pH-sensitivity of fast responsive superporous hydrogels. J Biomater Sci Polym Ed. 2000;11(12):1371–80. https://doi.org/10.1163/156856200744390.
104. Lee M, Hong K, Shin-Ya Y, Kajiuchi T. Adsorption of hexavalent chromium by chitosan-based polymeric surfactants. J Appl Polym Sci. 2005;96(1):44–50. https://doi.org/10.1002/app.21356.
105. Papaphilippou P, Christodoulou M, Marinica OM, Taculescu A, Vekas L, Chrissafis K, et al. Multiresponsive polymer conetworks capable of responding to changes in pH, temperature, and magnetic field: synthesis, characterization, and evaluation of their ability for controlled uptake and release of solutes. ACS Appl Mater Interfaces. 2012;4(4):2139–47. https://doi.org/10.1021/am300144w.
106. Jiang S, Helfricht N, Papastavrou G, Greiner A, Agarwal S. Low-density self-assembled poly(N-isopropyl acrylamide) sponges with ultrahigh and extremely fast water uptake and release. Macromol Rapid Commun. 2018;39(8). https://doi.org/10.1002/marc.201700838.
107. Zhai C, Azhar U, Yue W, Dou Y, Zhang L, Yang X, et al. Preparation and insights of smart foams with phototunable foamability based on azobenzene-containing surfactants. Langmuir. 2020;36(50):15423–9. https://doi.org/10.1021/acs.langmuir.0c03028.
108. Zhou YN, Li JJ, Zhang Q, Luo ZH. Light-responsive smart surface with controllable wettability and excellent stability. Langmuir. 2014;30(41):12236–42. https://doi.org/10.1021/la501907w.
109. Zhang L, Tang S, Guan Y. Excellent adsorption–desorption of ammonium by a poly(acrylic acid)-grafted chitosan and biochar composite for sustainable agricultural development. ACS Sustain Chem Eng. 2020;8(44):16451–62. https://doi.org/10.1021/acssuschemeng.0c05070.

110. Pokhrel LR, Dean RL, Jacobs ZL, Burrows WB. Nanotechnology-enabled point-of-use (POU) filters for drinking water disinfection. In: Springer eBooks. 2019. p. 187–216. https://doi.org/10.1007/978-3-662-59600-5_7.
111. Koh JJ, Lim GJH, Zhou X, Zhang X, Ding J, He C. 3D-printed anti-fouling cellulose mesh for highly efficient oil/water separation applications. ACS Appl Mater Interfaces. 2019;11(14):13787–95. https://doi.org/10.1021/acsami.9b01753.
112. Firmanda A, Fahma F, Syamsu K, Suprihatin S, Purnawati R, Mahardika M, Suryanegara L, Saito Y, Wood K, Sinaga R. Cellulose and its composite for sustainable oils/water (O/W) separation: from cellulose sponge to 3D printed nanocellulose. J Environ Chem Eng. 2023;11(5):110359. https://doi.org/10.1016/j.jece.2023.110359.
113. Tu H, Zhu M, Duan B, Zhang L. Recent progress in high-strength and robust regenerated cellulose materials. Adv Mater. 2020;33(28). https://doi.org/10.1002/adma.202000682.

Superwetting 3D Porous Polysaccharides for Oil/Water Separation

Nedal Y. Abu-Thabit, Mahmoud H. Abu Elella, Riham R. Mohamed, and Zubaida Rukhsana Usha

Abstract The increasing occurrence of oil spills and oily wastewater discharges poses severe environmental and economic challenges, highlighting the urgent need for efficient and sustainable separation technologies. In recent years, three-dimensional (3D) porous polysaccharide-based materials have emerged as eco-friendly and versatile candidates for oil/water separation. Derived from natural polymers such as cellulose, chitosan, alginate, and starch, these materials combine biodegradability, biocompatibility, and functional tunability with unique structural advantages, including high porosity, large surface area, and adjustable wettability. This chapter provides a comprehensive overview of superwetting 3D polysaccharide materials, sponges, aerogels, hydrogels, and 3D-printed scaffolds, engineered for efficient separation of stratified and emulsified oil/water mixtures. The discussion highlights fabrication methods such as freeze-drying, supercritical drying, chemical modification, and additive manufacturing, emphasizing how these strategies enhance wettability, mechanical stability, recyclability, and anti-fouling properties. Examples demonstrate separation efficiencies exceeding 98%, high oil absorption capacities, and strong reusability across diverse environmental conditions. Notably, the chapter underscores the sustainability benefits of polysaccharide-based systems, including the valorization of biomass and waste feedstocks into high-value functional materials. Finally, current limitations, such as scale-up challenges, long-term durability, and the trade-off between mechanical strength and porosity, are critically assessed. Future research opportunities lie in multifunctional composites, bio-inspired designs, and advanced 3D printing techniques, paving the way toward greener, cost-effective, and durable separation platforms for environmental remediation.

N. Y. Abu-Thabit (✉)
Chemical Engineering Department, Jubail Industrial College, Jubail Industrial City, Saudi Arabia
e-mail: abuthabit_nidal@yahoo.com

M. H. Abu Elella · R. R. Mohamed
Chemistry Department, Faculty of Science, Cairo University, Giza, Egypt

Z. R. Usha
Institute of Solid-State Physics, Hefei Institutes of Physical Science, Chinese Academy of Sciences, Hefei, China

M. H. Abu Elella et al. (eds.), *3D Engineered Polysaccharides*,
https://doi.org/10.1007/978-3-032-26154-0_8

Keywords Superhydrophobic · Superhydrophilic · Biopolymer · 3D printing · Aerogel · Hydrogel · Sponge

Abbreviations

3D	Three-Dimensional
AM	Additive Manufacturing
BC	Bacterial Cellulose
CA	Cellulose Acetate
CMC	Carboxymethyl Cellulose
CNT	Carbon Nanotube
DIW	Direct Ink Writing
EC	Ethyl Cellulose
FDM	Fused Deposition Modeling
FSP	Filament Spacing
GO	Graphene Oxide
HDI	Hexamethylene Diisocyanate
HDTMS	Hexadecyltrimethoxysilane
MDI	1,4-Phenyl Diisocyanate
OTMS	Octadecyltrimethoxysilane
PDMS	Polydimethylsiloxane
PEI	Polyethyleneimine
PLA	Polylactic Acid
POSS	Polyhedral Oligomeric Siloxane
PVA	Poly(vinyl alcohol)
SA	Sodium Alginate
SHB	Superhydrophobic
SHL	Superhydrophilic
SHB/SOL	Superhydrophobic/Superoleophilic
SHL/UWSOB	Superhydrophilic/Underwater Superoleophobic
SLA	Stereolithography
SLM	Sheet Lamination
UWSOB	Underwater Superoleophobic
UV-DIW	Ultraviolet-Assisted Direct Ink Writing
WCA	Water Contact Angle

1 Introduction

The increasing industrial and economic growth has led to frequent oily wastewater discharges and offshore oil spills, endangering both human life and the environment [1–5]. A significant example is the 2010 Deepwater Horizon oil spill, which caused severe ecological and community damage [6–8]. Traditional oily wastewater treatment methods, physical, chemical, and biochemical, include techniques like filtration, centrifugation, incineration, and anaerobic treatment [9–11]. While effective, they often have low efficiency, high costs, and the potential for secondary pollution. An alternative and promising technology for the treatment of oily wastewater is the use of advanced smart superwetting materials [12–16].

Superwetting materials are created by introducing surface roughness and altering the surface energy [17, 18]. Materials with high surface energy chemical composition (e.g., hydrophilic polymers, oxides, hydrogels) produce superhydrophilic/underwater superoleophobic (SHL/UWSOB) surfaces, which are regarded as (water-removing/oil-blocking) materials [19–21]. In contrast, materials designed with low surface energy chemical composition (e.g., hydrophobic polymers, polydimethylsiloxane) enable the fabrication of superhydrophobic/superoleophilic (SHB/SOL) surfaces, which are regarded as (oil-removing/water-blocking) materials [22–27]. In general, oil/water separation has been achieved by filtration or adsorption mechanisms using various materials, such as sponges, foams, metal meshes, textiles, and filter paper [28, 29]. Filtration methods provide rapid separation and consume fewer materials, whereas adsorption methods are slow and demand large amounts of superwetting materials, leading to higher cost [13, 15]. Membranes designed with SHB/SOL wettability suffer from two main shortcomings [28]. First, the coated materials may be gradually leached out over time. Second, the membrane can become fouled by oils, particularly those with high viscosity. In contrast, SHL/UWSOB membranes prevent oil contamination and resist biofouling, extending their lifespan. The underwater superoleophobicity of these membranes depends on their water hydration ability, with stronger hydration leading to lower oil adhesion.

While superwetting 3D porous scaffolds derived from non-biodegradable synthetic polymers have been proposed as innovative materials for oil/water separation [30], increasing attention has been directed toward porous materials derived from natural polysaccharides as hydrogels, aerogels, sponges, and 3D-printed scaffolds, owing to their biocompatibility, biodegradability, and ease of functional modification [31–39]. Hydrogels, created by physically or chemically crosslinking polysaccharides such as alginate or chitosan, absorb significant amounts of water, making them suitable for creating porous superwetting structures with SHL/UWSOB for separating emulsified oil/water mixtures [40–44]. Aerogels are produced by removing liquid from hydrogels via supercritical or freeze-drying, yielding ultra-lightweight, highly porous solids (e.g., from cellulose or pectin) that are suited for oil/water separation via adsorption [45–48]. Sponges emerge from freeze-drying polysaccharide gels, forming macroporous structures used as absorbents for oil/water separation [49–52]. Researchers are now utilizing 3D printing to fabricate intricate structures from polysaccharides like chitosan and cellulose, creating highly tailored porous materials explicitly designed for efficient oil/water separation [53–56]. These printed constructs, often forming lattices or grids sensiwith precisely controlled geometry and surface chemistry, leverage the inherent biocompatibility and adsorption properties of polysaccharides to effectively separate oil from water while offering potential for reusability and environmental sustainability [57, 58]. The properties of these materials, whether derived from neutral polysaccharides (cellulose, starch) or charged ones (anionic alginate, cationic chitosan), can be precisely tailored through their functional groups, enabling specific interactions and applications. This chapter highlights recent advances in fabricating superwetting 3D porous polysaccharide-based scaffolds for oil/water separation, including sponges/foams (Sect. 2), aerogels (Sect. 3), hydrogels (Sect. 4), and 3D-printed porous structures (Sect. 5). The current limitations and future perspectives are highlighted at the end of the chapter (Sect. 6) to reflect future research directions in this area.

2 3D Polysaccharide Sponges

Chitosan sponges, derived from natural chitin, have been modified to become water-repellent but oil-absorbent (SHB/SOL), making them ideal for separating oil from water in spills or industrial wastewater. Their high porosity, biodegradability, and tunable surface properties have enabled the efficient separation of stratified and emulsified oil/water mixtures (Table 1). Liang et al. reported the fabrication of an elastic, superhydrophobic, and biodegradable chitosan (CS) sponge for oil/water separation [59] (Fig. 1a). The sponge was fabricated through a multi-step process involving cross-linking with lemon juice and heptanal, *in-situ* loading of Fe_3O_4/polydopamine (PDA) particles, and coating with polylactic acid (PLA). The modified chitosan sponge exhibited high elasticity, superhydrophobicity, biodegradability, and magnetic recyclability. The as-prepared sponge achieved an absorption capacity of oils/organic solvents up to 50 times its weight and efficient emulsion separation with a flux of 875.3 L m^{-2} h^{-1}. Xu et al. reported the fabrication of a SHB/SOL sponge-like chitosan-based composite by incorporating carbon nanotubes (CNTs) and depositing SiO_2 nanoparticles (SH-SiO_2@3CNTs/CS) [60] (Fig. 1b). This was followed by functionalization with a low-surface-energy octadecyltrimethoxysilane (OTMS) organosilane to achieve superhydrophobic performance (WCA = 152°). The prepared SHB composite featured a 3D porous structure, enhanced mechanical strength (1.15 MPa at 3 wt.% CNTs), and superhydrophobicity. The fabricated sponge demonstrated high oil absorption capacity (8.71–18.24 g/g), efficient separation of oil–water mixtures, and water-in-oil emulsions. The fabricated sponge exhibited outstanding chemical durability in the pH range 1–13, high thermal stability, mechanical abrasion durability, and good recyclability.

Cellulose sponges have been fabricated via dissolution-regeneration of cellulose powder. By using this strategy and without chemical modifications, Wang et al. fabricated a bilayer sponge by a simple dissolution, regeneration, and freeze-drying

Table 1 3D-polysacharide sponges and aerogels with different wettability for oil/water separation

Type of 3D structure	Polysaccharide type/source	Materials for controlling surface energy/roughness	Wettability	Flux (L m^{-2} h^{-1})	Maximum oil absorption capacity for organic solvents/oils (g/g)	References
Sponge	Chitosan	Heptanal	SHB/SOL	875.3	50	[59]
	Chitosan	CNTs, SiO_2 NPs, OTMS	SHB/SOL	–	18	[60]
	Cellulose	–	SHL/UWSOB	91	–	[61]
	Luffa	POSS	SHB/SOL	–	18	[84]
	Luffa	Fluorinated SiO_2 NPs	SHB/SOL	–	3–4	[85]
	Discarded jute bags	TEOS/HDTMS	SHB/SOL	–	33.55	[68]
	Shola pith	None	SHB/SOL	–	40–60 166 (silicone oil)	[69]
Aerogel	Cellulose (seaweed solid waste)	PEI + MDI for hydrophobic modification	SHB	–	11–30	[92]
	Cellulose (hardwood pulp)	Carbonization at 400 °C	SHB	117,188	65–133	[93]
	Chitosan/starch/cellulose nanofiber	Zein for hydrophobic modification	SHB	5595	–	[104]
	Chitosan/collagen	Graphene oxide (GO)	Hydrophobic	–	227.53 ± 9.35	[105]
	Sodium alginate (SA)	Directional freezing + CVD coating	SHB	667	–	[141]
	Sodium alginate (SA)	GO and SiO_2	SHB	–	18–44	[142]
	Cellulose	GO/ n-dodecyltriethoxysilane (DDTS)	SHB	–	80–197	[143]
	Cellulose	GO, silica	SHB	–	39–68	[144]

(continued)

Table 1 (continued)

Type of 3D structure	Polysaccharide type/source	Materials for controlling surface energy/roughness	Wettability	Flux (L m^{-2} h^{-1})	Maximum oil absorption capacity for organic solvents/oils (g/g)	References
	Chitosan/cellulose	Trimethylchlorosilane (TMCS)	SHB	–	14–28	[102]
	Chitosan	Methyl trimethoxysilane (MTMS)	SHB	–	31–63	[145]
	SA/ N,N′-methylenebisacrylamide (MBA)	Cold plasma treatment	Hydrophobic	–	10.20–13.25	[146]
	SA/CMC	Cold plasma treatment	Hydrophobic	–	10.98–13.98	[146]
	SA/cellulose	PDMS/MTCS	Hydrophobic	–	17–34	[147]
	SA/Ca	$ZrOCl_2 \cdot 8H_2O$ (ion-induced hydrophobicity)	Hydrophobic	–	11.2–25.9	[148]
	Cellulose/chitosan	Citric acid crosslinking + MTMS	SHL/UWSOB	–	–	[149]

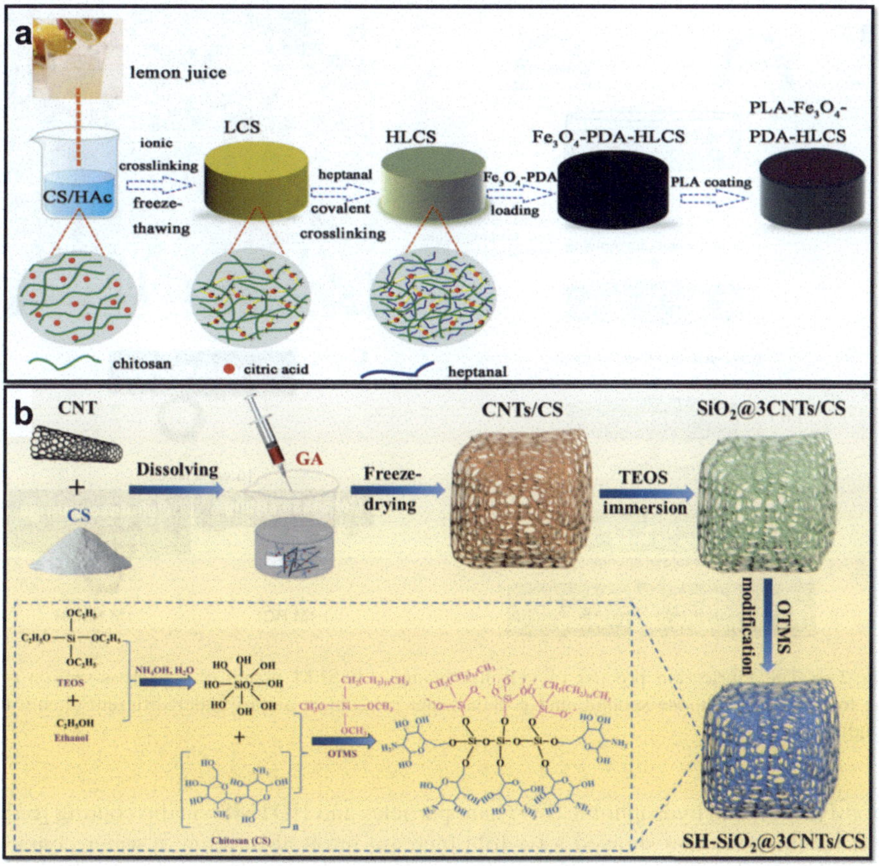

Fig. 1 Schematic illustration for the fabrication of **a** elastic SHB chitosan sponges [59], and **b** SHB composite chitosan/CNTs/SiO_2 sponge [60]

procedure [61] (Fig. 2a). Cellulose powder was dissolved in 67% $ZnCl_2$ solution at 80 °C, sodium sulfate ($Na_2SO_4 \cdot 10H_2O$) was used as a pore-forming agent, and 2.4% fibers were added as reinforcing fiber matrix. The resulting cellulose sponge exhibited an asymmetric porous structure (Fig. 2b) and SHL/UWSOB wettability, which enabled the separation of surfactant-stabilized toluene-in-water emulsion with > 99% separation efficiency and a flux of 91 L m^{-2} h^{-1}. The fabricated cellulose sponge exhibited high chemical durability under acidic, alkaline, or saline conditions (Fig. 2c). Chen et al. fabricated a rough-surface cellulose sponge, grafting it with isocyanate and F-PEG to attain SHL/UWSOB properties [62]. The fabricated sponge separated surfactant-stabilized toluene-in-water emulsions at 99.83% efficiency and a flux of 297 L m^{-2} h^{-1}.

Lu et al. developed a durable superhydrophobic sponge by reinforcing a cross-linked ethyl cellulose (EC) sponge with silanized carbon nanotubes (S-CNTs) and

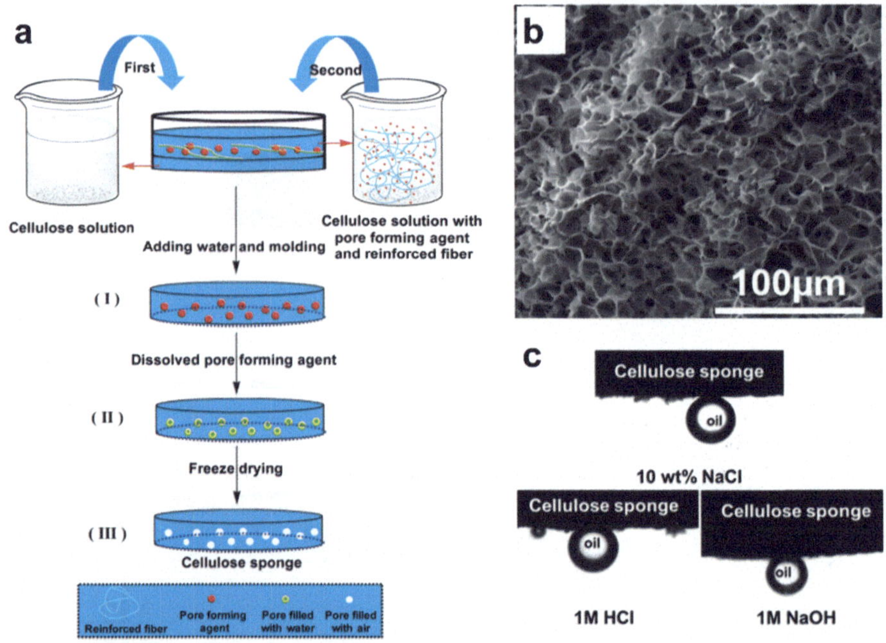

Fig. 2 **a** The fabrication process of cellulose sponge, **b** SEM image of the cross-section of the freeze-dried cellulose sponge, and **c** underwater oil contact angles under different chemical conditions [61]

modifying it with hydrophobic SiO_2 nanoparticles and HDTMS via dip coating [63]. The resulting sponge showed a 64-fold increase in oil absorption, a water contact angle of 158.2°, rapid oil uptake (~ 16 s), 90% separation efficiency, and reusability up to 50 times. It also exhibited low density (< 20 mg/cm^3), high porosity (> 98%), strong mechanical resistance (up to 28.6 kPa), and stability in harsh chemical environments. Recent research highlights increasing efforts to upcycle lignocellulosic waste into useful materials for oil/water separation [64–67]. Ahuja et al. converted discarded jute bags into cellulose nanofibers to create a superhydrophobic cellulose sponge (SCS) using Tetraethylorthosilicate/Hexadecyltrimethoxy Silane (TEOS/HDTMS) as a low surface energy material [68]. The SCS achieved high separation efficiencies of 98.5% for diesel and 97.2% for motor oil, with absorption capacities of 33.55 g/g and 31.37 g/g, respectively.

Mondal has reported the exploration of naturally occurring porous, hydrophobic/oleophilic material (WCA = 135°), called Shola, for oil sorption and emulsion separation applications [69] (Fig. 3a). Shola is a soft, lightweight, milky-white shrub from the *Aeschynomene* genus, native to the wetlands of Bengal and eastern India. The Shola pith contains ≈ 70 wt.% cellulose, higher than typical lignocellulosic materials, along with hemicellulose, lignin, pectin, wax, and protein in smaller amounts. The pith can absorb 40–60 g/g of various oils, with up to 166 g/g for silicone oil, and effectively separates oil from both water surfaces and oil-in-water emulsions. The material remains highly effective after seven absorption cycles, and ≈ 95% of the absorbed oil can be recovered using solvent extraction. Loofahs and luffa sponges, while often used interchangeably, refer to the same natural sponge made from the dried fruit of the luffa gourd. The terms are often used to describe the dried, fibrous part of the luffa gourd used for cleaning and exfoliation. Luffa sponge is a class of natural 3D polysaccharide sponges that are composed of cellulose (55–90%), hemicelluloses (8–22%), lignin (10–23%), extractives (~ 3.2%), and ash (0.4%) [70]. The luffa sponge fibers exhibit complex chemical composition and crosslinking between lignin and hemicellulose [71]. Functionalized and composite luffa sponges have been used as smart superwetting 3D polysaccharide membranes for oil/water separation via adsorption and/or filtration [72–83]. Natural luffa sponge is intrinsically hydrophilic due to the presence of hydroxyl groups of cellulose components. Wang et al. reported a simple one-step coating procedure using low-surface-energy polyhedral oligomeric siloxane (POSS) [84]. The modified luffa sponge exhibited a SHB surface wettability (WCA = 155°), an oil absorption capacity of 8–12 times its weight, and good recyclability. Lin et al. reported the fabrication of SHB luffa sponge, which was modified with fluorinated silica nanoparticles (F-SiO_2) and hexamethylene diisocyanate (HDI), which was used as a coupling agent [85] (Fig. 3b). The superhydrophobic performance is attributed to the simultaneous increase in surface roughness and decrease in surface energy of the coated luffa sponge. However, the absorption capacity for various oils was in the range 300–400% by weight, which is equivalent to 3–4 g of oil per one gram of the dry luffa sponge.

3 3D Polysaccharide Aerogels

Superwetting aerogel-based materials, with their ultra-lightweight, highly porous, and surface-tunable properties, provide efficient and environmentally friendly solutions for oil–water separation, excelling in oil spill cleanup and wastewater treatment [86] (Table 1). Aerogels are produced using three primary drying methods, each with distinct advantages and disadvantages, as listed in Table 2. Aerogels derived from natural and renewable polysaccharide resources such as cellulose, sodium alginate, and chitosan excel in oil–water separation due to their high porosity, low density, and excellent adsorption capacity [87–91]. These materials are also non-toxic, biodegradable, and biocompatible, aligning perfectly with the principles of sustainable chemistry. Dai et al. developed a SHB cellulose aerogel from algae waste, crosslinked with polyethyleneimine (PEI) and modified with 1,4-phenyl diisocyanate (MDI), demonstrating excellent elasticity (82% performance retention at − 40 °C) and an oil adsorption capacity in the range of 11–30 g/g [92]. Ma et al. created a carbonized cellulose aerogel from hardwood pulp, achieving SHB performance, high oil adsorption capacity (65–133 g/g), and remarkable separation flux ≈ 120,000 L m^{-2} h^{-1} for continuous oil–water separation [93].

As illustrated in Fig. 4a, Chen et al. reported the fabrication of a hydrophobic (WCA ≈ 146°) bacterial cellulose (BC)-based aerogel reinforced with sodium

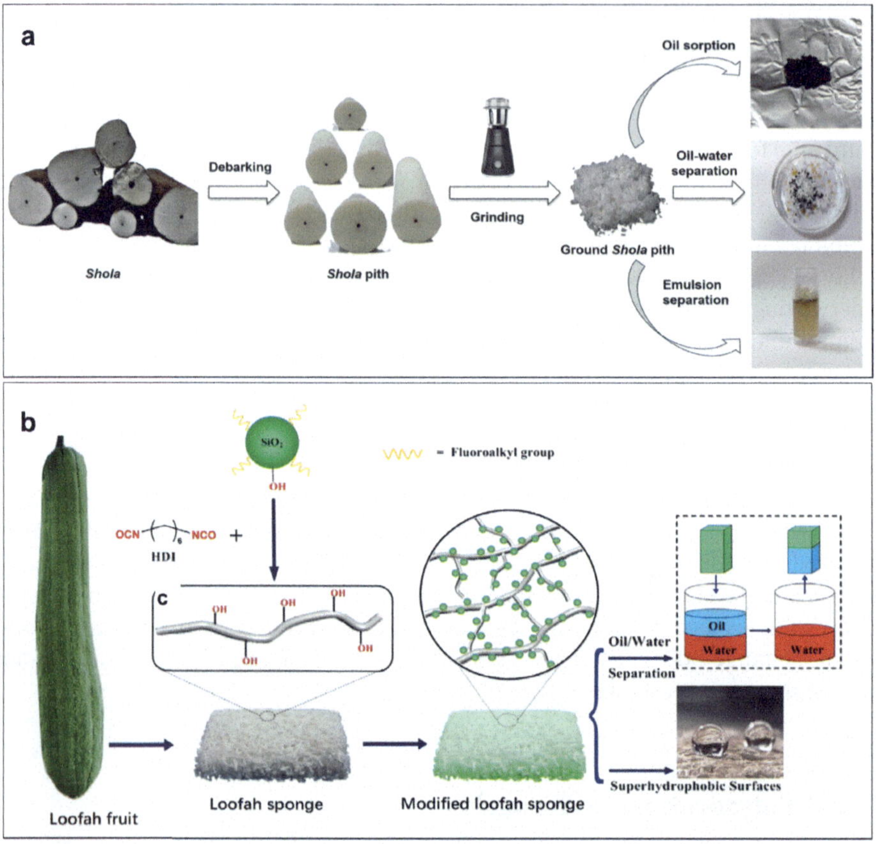

Fig. 3 **a** Schematic illustration of the preparation and application of the hydrophobic Shola pith [69], and **b** schematic illustration of the method for fabricating 3D SHB material using a loofah sponge as a substrate and the performance evaluation of the modified loofah sponge [85]

carboxymethyl cellulose (CMC) and polydimethylsiloxane (PDMS) by a mixing/freeze-drying strategy [94]. The as-prepared aerogel demonstrated high porosity (96%), low density (0.097 g/cm^3), and excellent mechanical strength (0.70 MPa at 80% strain). It showed a remarkable dye adsorption capacity of ≈ 0.50 g/g for methylene blue, achieving ≈ 97% removal within 12 h. The prepared aerogel exhibited a high oil sorption capacity of 186 g/g for petroleum ether, with rapid separation in less than 5 s, indicating strong potential as a sustainable material for industrial wastewater remediation. Shang et al. reported the fabrication of hydrophobic (WCA ≈ 136°) CNF/TA/ICO aerogel, made from cellulose nanofibers (CNF), tannic acid (TA), and castor oil-derived siloxane (ICO), via a co-precipitation and freeze-drying green strategy [95] (Fig. 4b). The aerogel showed high oil and solvent adsorption capacity (53.2–113.8 g/g) with over 90% efficiency after 10 cycles, and achieved

Table 2 Various methods are used for the fabrication of aerogels

	Supercritical drying	Ambient pressure drying	Freeze-drying
Process	Solvent replaced with liquid CO_2; heated above the critical point to avoid pore collapse	Solvent exchange + chemical modification (e.g., silanization); slow evaporation	Solvent frozen → sublimated under vacuum
Key applications	High-performance insulation (NASA spacecraft, cryogenic pipelines), Cherenkov radiation detectors, acoustic dampening, transparent thermal barriers (e.g., silica aerogels)	Oil–water separation (hydrophobic sponges for spills), flexible composites (e.g., wearable sensors), and low-cost thermal insulation (building materials)	Biomedical (drug delivery scaffolds, wound dressings), food packaging (antimicrobial aerogels), lightweight biodegradable materials (e.g., cellulose aerogels for eco-friendly absorbents), and oil absorption
Porosity/ density	Ultra-high (95–99.8%), ~ 0.001 g/cm^3	85–95%, ~ 0.05–0.1 g/cm^3	80–95%, 0.01–0.2 g/cm^3 (ice crystals may enlarge pores)
Mechanical strength	Brittle (unless reinforced)	Improved (due to chemical crosslinking)	Moderate (depends on biopolymer)
Advantages	Best pore preservation, highest thermal resistance (~ 0.02 W/ m K)	Cost-effective, scalable for industrial use	Ideal for heat-sensitive biomaterials
Limitations	Expensive equipment, energy-intensive	Lower porosity, potential shrinkage	Less uniform pores, slower production

water-in-oil emulsion separation efficiency above 94.4% and a flux in the range of $\approx$ 123–474 L m^{-2} h^{-1}.

Chitosan is a natural, positively charged polysaccharide from chitin found in crustaceans, fungi, and plants [96–99]. Its rich hydroxyl and amino groups enable strong bonding, adsorption, and easy chemical modification, enhancing uses like oil–water separation, making it biodegradable, biocompatible, and ideal for environmental remediation [100]. As an attempt to achieve a green fabrication process, Takeshita et al. developed an aldehyde-free, transparent chitosan aerogel using urea's thermal decomposition to induce gelation, followed by ethanol exchange and supercritical drying [101]. Low urea concentrations ($\leq$ 25 g/L) yielded transparent mesoporous aerogels with > 400 m^2/g surface area and 3.5 cm^3/g pore volume, while higher concentrations ($\geq$ 30 g/L) produced opaque macroporous versions, demonstrating urea-dependent control over microstructure formation through chitosan coagulation timing, offering a non-toxic alternative to formaldehyde-crosslinked aerogels with comparable optical and structural properties.

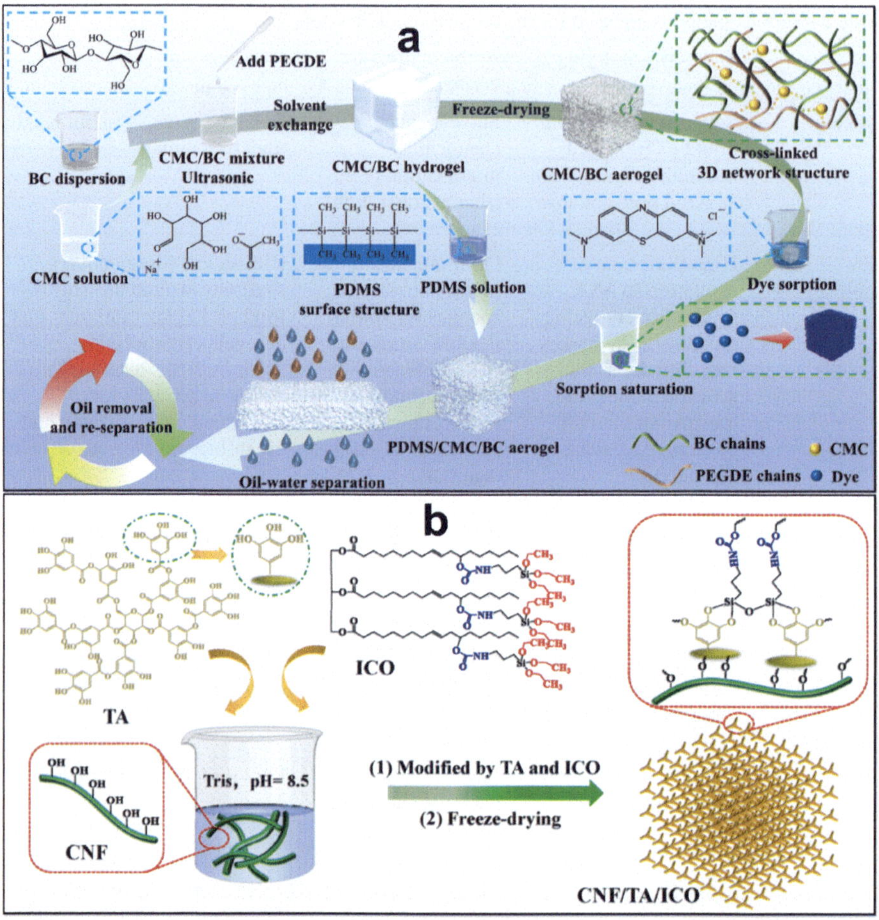

Fig. 4 **a** Schematic illustration for the fabrication process of the PDMS/CMC/BC aerogel [94], and **b** schematic illustration for the fabrication of the hydrophobic CNF/TA/ICO aerogel [95]

Researchers fabricated an innovative cellulose aerogel with exceptional hydrophobicity (WCA ≈ 153°) and oil absorption capacity of 13.77–28.20 g/g through oxidation and plasma modification [102]. The material maintains performance over 50 compression cycles, follows pseudo-second-order kinetics, and offers a sustainable, economical option for oil spill cleanup and industrial separation processes. Frequent oil spills have created urgent demand for sustainable solutions, leading to the development of an innovative all-biomass aerogel. Tuo et al. reported the fabrication of a series of hydrophobic CW&BW@DCGA aerogels through multiple dynamic covalent bonding interactions, utilizing carboxymethyl chitosan (CMCS) as the base material, modified gelatin (Gel-ADH) as the reinforcing component, and dialdehyde β-cyclodextrin (Da-β-CD) as a non-toxic crosslinking agent [103]. This eco-friendly aerogel demonstrates exceptional properties, WCA ≈ 149°, 31.7–62.5 g/g oil absorption capacity, and remarkable elasticity (60% compressive strain at optimal 7:3 CMCS: Gel-ADH ratio), offering a green, high-performance solution for marine oil spill remediation.

Fu et al. reported the fabrication of a composite aerogel (ZOMA) using chitosan, starch, and carbonylated cellulose nanofibers, with zein as a low-surface-energy agent [104]. The modified composite aerogel exhibited a low density of 0.11 g cm^{-3}, a high separation flux of 5595 L m^{-2} h^{-1}, and an impressive oil–water separation efficiency of 99.8%, while maintaining an adsorption capacity of over 90% after 10 reuse cycles. Further improvements were achieved by Ramalingam et al., who enhanced surface wettability and antibacterial activity of chitosan/collagen by incorporating graphene oxide (GO), resulting in the production of GO-collagen-chitosan sponge (GCCS) via a cryogenic freeze-drying strategy [105]. The as-prepared aerogel sponges exhibited a maximum oil adsorption of 227.53 ± 9.35 g/g, dye adsorption of 204.17 ± 7.49 g/g, making it a versatile method for treating various environmental pollutants.

4 3D Polysaccharide Hydrogels

Hydrogels are hydrophilic, 3D network polymeric materials capable of absorbing and retaining large amounts of water. These materials have been explored as promising advanced SHL/UWSOB materials for oil/water separation, water purification, and photocatalysis [106]. Xue et al. were the first to use a superhydrophilic polyacrylamide-coated surface for separating oil/water mixtures. Since then, extensive research has focused on developing hydrogel-based coatings for this purpose. Hydrogels also maintain their superhydrophilicity in highly saline environments, making them suitable for marine applications, such as combating oil fouling and cleaning crude oil spills. The superwetting performance of hydrogels depends on their 3D structure and water swelling properties, with lower crosslinking causing dissolution and higher crosslinking reducing water retention. Cellulose, the most abundant renewable natural polymer, is a linear polysaccharide composed of β-(1 → 4) linked D-glucose units [107]. Due to its water insolubility, water-soluble and hydrophilic cellulose derivatives (e.g., cellulose acetate, carboxymethyl cellulose) have been investigated for constructing superwetting 3D hydrogels for oil/water separation. Alternatively, hydrogels have been fabricated through cellulose dissolution and regeneration [108]. The study produced robust cellulose hydrogels via a physical dissolution/regeneration procedure using LiOH/urea and $LiBF_4$. The hydrogels exhibit a notable compressive strength of 14.5 MPa, surpassing previously reported cellulose-based hydrogels. Coated onto stainless steel meshes, they afforded underwater superoleophobic surfaces for efficient oil/water separation, with > 98.9% efficiency, along with high flux (up to 3806 L m^{-2} h^{-1}). The coatings exhibited excellent reusability and saltwater resistance, demonstrating potential for sustainable water treatment.

Lubricants are specialized oils that reduce friction and wear in mechanical systems; they are usually paraffinic petroleum-based oils with additives. Their lubrication efficiency degrades once water enters from leaks, condensation, or exposure to humidity. Thus, lubricant oils with water content greater than 0.2% (2000 ppm) lead to a loss of oil performance due to reduced viscosity, decreased electrical resistance, and accelerated oxidation. Estevam et al. reported the fabrication of crosslinked cellulose hydrogel for removing water from turbine engine lubricants [109]. The hydrogel was tested in batch and continuous-flow systems using a custom fixed-bed apparatus (Fig. 5). The prepared hydrogel effectively reduced water content from 412 to 197 mg/kg, meeting industry standards. Offering sustainability, cost-effectiveness, and industrial adaptability, hydrogel treatment shows promise, with future research needed to enhance efficiency and scalability.

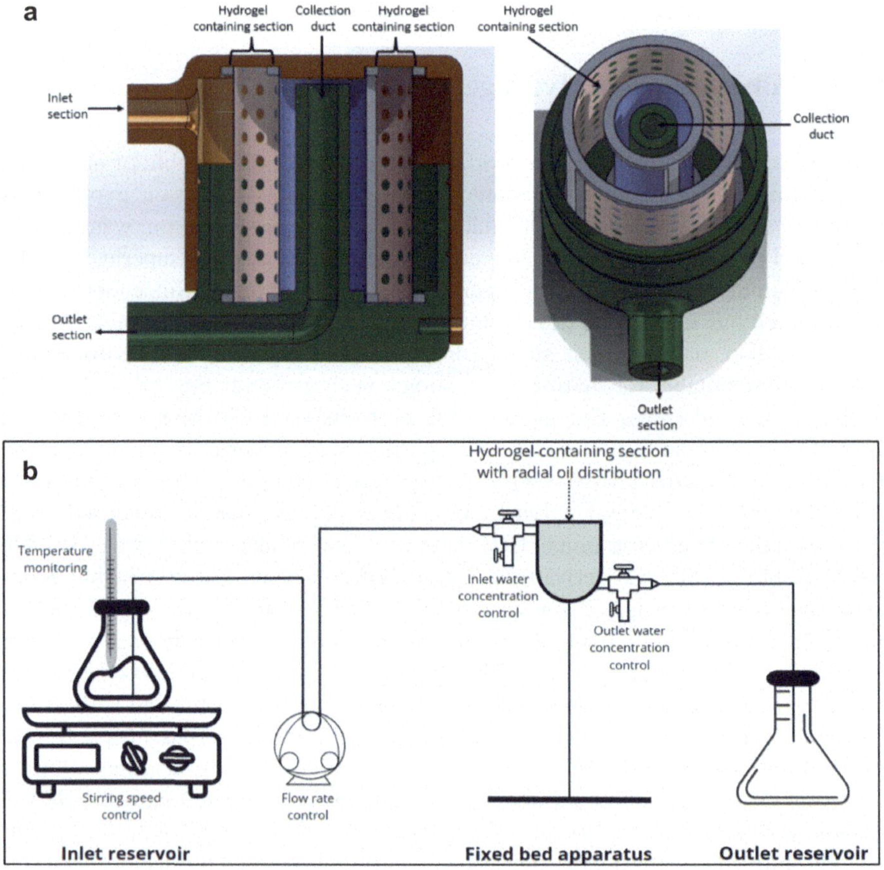

Fig. 5 Front and top view of continuous flow apparatus used in oil treatment (**a**) and schematic representation of the continuous flow operation system (**b**) [109]

5 3D-Printed Polysaccharides

Additive manufacturing (AM), or 3D printing, has recently emerged as a promising technology due to advances that offer high-resolution printing, improved energy efficiency, faster production speeds, and compatibility with a wide range of materials and applications [110–112]. These advancements have led researchers to explore AM for water remediation, wastewater treatment, and sorption applications, where it offers precise control over porosity and structure, enabling high water flux without external energy while maintaining mechanical strength for filter regeneration and replacement [40, 113–122]. As shown in Fig. 6, the additive manufacturing (AM) process begins with designing a 3D CAD model, which is converted into an STL file and sliced into 2D layers to generate printer instructions [123]. The object is then built layer by layer, followed by removal, cleaning, and post-processing steps such as support removal or curing, ultimately resulting in a finished part ready for use [124]. Currently, various 3D printing technologies are classified according to ISO/ASTM 52921:2013, including material extrusion, jetting, binder jetting, powder bed fusion, direct energy deposition, vat photopolymerization, and sheet lamination. Depending on the technique and ink formulation, objects are built layer by layer using methods like extrusion, chemical binding, or laser sintering, with detailed classifications and materials summarized in Fig. 7. The main AM technologies used for 3D printing of polysaccharide hydrogels includes Direct Ink Writing (DIW), Fused Deposition Modelling (FDM), Inkjet Printing (IJP), and Stereolithography (SLA) [125].

DIW or robocasting is an extrusion-based 3D printing method that uses air pressure to deposit viscoelastic inks at mild temperatures, making it suitable for biomaterials such as polysaccharides [128, 129]. Successful DIW inks exhibit shear-thinning

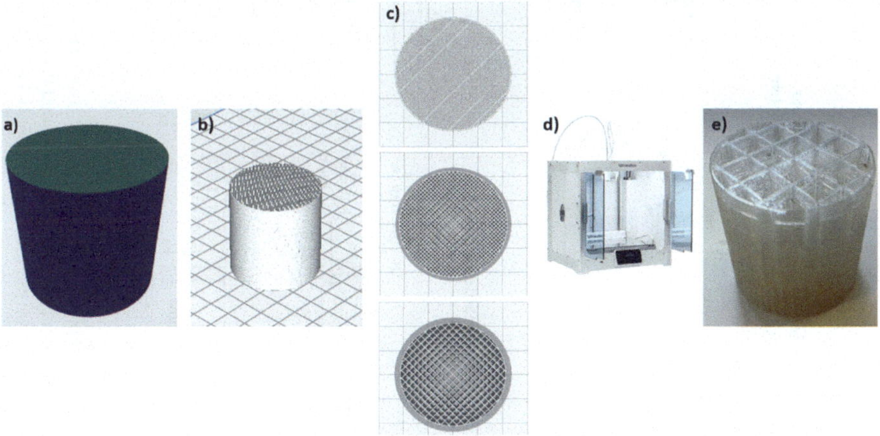

Fig. 6 Overview of the additive manufacturing procedure for preparation of a water purification filter: **a** CAD model, **b** adjusted.STL file, **c** model slicing, **d** 3D printing, **e** application. Reproduced with permission from Ref. [126]

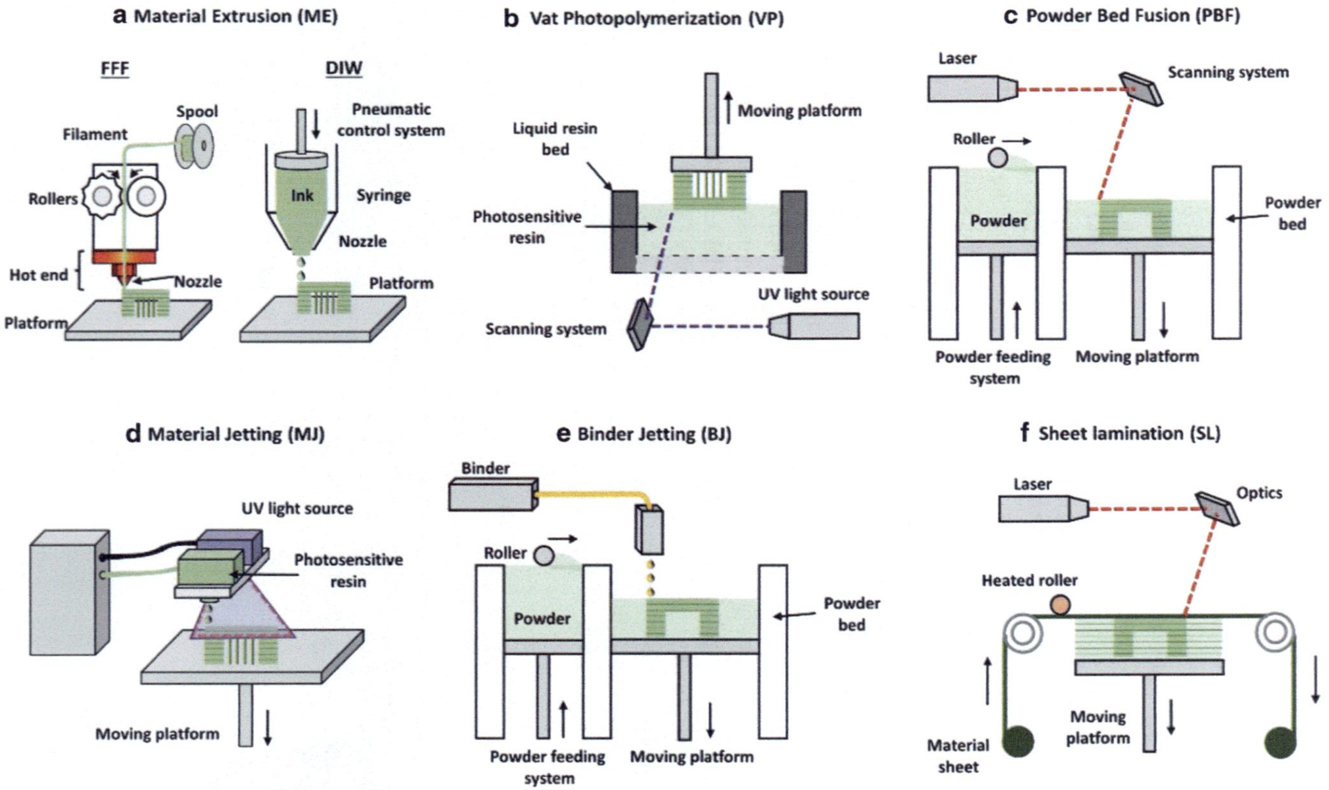

Fig. 7 Schematics of 3D printing techniques. **a** In material extrusion (ME), filament fused fabrication (FFF) is a 3D printing method in which the solid filament is melted, extruded through the nozzle, and deposited layer by layer. Direct ink writing (DIW) employs viscoelastic ink, and the ink is extruded through a small nozzle under controlled flow rates. **b** Vat photopolymerization (VP) uses UV light to cure the photopolymer resin in the liquid bath by forming crosslinked networks. **c** Powder bed fusion (PBF) utilizes thermal energy to melt powdered material and prints out a 3D object. **d** Material jetting (MJ) prints out 3D products by selectively depositing the droplets of build materials. **e** Binder jetting (BJ) drops liquid binder onto powder materials, which are the main components in final objects. **f** Sheet lamination (SL) uses thin sheets of material and bonds together layer by layer to fabricate a 3D object. Reprinted with permission from Ref. [127]

behavior, elastic recovery, and a controlled yield stress, enabling smooth extrusion and shape stability [130]. DIW enables delicate structures and supports various materials, finding use in biomedical, energy, sensing, and food fields. Polysaccharides like starch, cellulose derivatives, chitosan, and alginate offer ideal rheological properties for these applications [131–134]. Li et al. reported the fabrication of 3D-printed aerogel using TEMPO-oxidized cellulose nanofibers via a direct ink writing (DIW) strategy [135, 136]. The fabricated aerogel exhibited high porosity (98%) and a density of 26 mg/cm^3, indicating potential applications in diverse areas, including oil/water separation. Li et al. reported the fabrication of a SHL/UWSOB 3D printed composite membrane using cellulose acetate (CA), poly(vinyl alcohol) (PVA), and silica nanoparticles (CA/PVA/Si) [137]. A homogeneous CA/PVA/SiO_2 ink was formulated by dissolving the components in DMSO at 90 °C for 6 h, followed by centrifugation (10,000 rpm, 30 min) to eliminate air bubbles. The ink comprised 28 wt.% polymers (CA/PVA in varying ratios) and 0–16 wt.% % SiO_2 nanoparticles. Using a custom 3D, the ink was extruded through a 200 μm nozzle at 9.0 bar and 5.0 mm/s. The printer executed CAD-designed paths (G-code) to regulate filament spacing (Fsp), thereby allowing for adjustable pore sizes. After printing, the membranes were cured in deionized water for 60 min and stored in a wet state. The fabricated membrane exhibited a water contact angle of 18.14° in air and an underwater oil contact angle of 159.14°. The membrane achieved 99% efficiency in gravity-driven oil/water separation and maintained strong mechanical stability, anti-fouling ability, and high performance after repeated use. The printed membrane outperforms CA/PVA/Si composite steel mesh in mechanical stability and oil/water separation efficiency, even after sonication or bending. It demonstrates high efficiency, stability, and reusability for various oil/water separation applications. Feng et al. reported the fabrication of a bio-based 3D printed membrane with an ordered porous structure using UV-assisted DIW from methacrylated chitosan (CSMA), methacrylated gelatin (GELMA), and nanographene oxide (nGO) [138]. The optimized membrane, M−CSMA/GELMA/nGO-0.7, exhibited a water contact angle of 0° in air, an underwater oil contact angle of 161.5°, and maintained over 99.5% separation efficiency with a high-water flux above 38,300 L m^{-2} h^{-1} after 20 cycles in 1 M NaCl, 1 M HCl, or 1 M NaOH. Hence, UV-DIW 3D printing bio-inks have been demonstrated as a straightforward method for fabricating fully integral membranes with enhanced mechanical durability and superwetting performance, leading to extended service life, high water flux, and superior anti-fouling properties. Wang et al. reported the use of DIW 3D printing for the fabrication of SA/GO/Ag multifunctional aerogel (SGA) to separate oil, silt, and bacteria from sewage through a simple gravity-driven filtration process [139]. The SA/GO composite aerogel intercepts solids and separates oil/water mixtures, while silver nanoparticles kill harmful bacteria like *Escherichia coli* and *Staphylococcus aureus*. This energy-efficient system works without external power, making it suitable for large-scale water treatment.

Koh et al. reported the fabrication of a 3D-printed SHL cellulosic filtration mesh membrane using DIW for oil/water separation [140]. The 3D printed mesh was fabricated from cellulose acetate/ethyl acetate solution, followed by cellulose regeneration via alkaline hydrolysis, and finally drying at room temperature (Fig. 8a).

The regenerated cellulose mesh achieved over 95% separation efficiency with pore sizes under 280 μm, and around 99% performance with a high-water flux of about 160,000 L m^{-2} h^{-1}. The cellulose meshes separated oils of different viscosities, from viscous PDMS (~ 97 cP) to low-viscosity cyclohexane (~ 1 cP), without chemical treatments. The cellulose mesh exhibited superior anti-oil-fouling property, which is attributed to its superhydrophilic feature both in air and underoil as compared to the cellulose acetate (CA) mesh, plasma-treated stainless-steel mesh (SSM), and PVA-coated SSM (Fig. 8b). Although neat PVA is hydrophilic, contains hydroxyl groups, and is water-soluble, PVA-coated meshes are not hydrophilic in air due to their exposed hydrophobic methylene backbone groups when dried. In oily environments, PVA surfaces become more hydrophobic due to the oil fouling. Plasma-treated SSM is hydrophilic in air but loses this property in oil due to surface energy changes caused by oil fouling. The 3D-printed CA mesh initially exhibits hydrophobicity due to the presence of acetyl groups. After hydrolysis, the CA mesh becomes superhydrophilic, with a water contact angle below the detection limit. This is attributed to the rigid structure of cellulose chains, reinforced by intramolecular hydrogen bonding and β-1,4′ glycosidic linkages. Cellulose's rigidity prevents conformational changes during processing, keeping hydrophilic groups exposed. Unlike plasma-treated SSM, cellulose mesh stays superhydrophilic under oil. This can be explained as follows: water forms strong hydrogen bonds with cellulose's hydroxyl groups, creating a hydration shell. This reduces *van der Waals* interactions between oil and cellulose, allowing water to penetrate the loose cellulose regions, weakening oil/cellulose interactions and promoting water spreading at the three-phase contact line (TPL). As shown in Fig. 8c, the 3D-printed regenerated cellulose mesh has lower crystallinity than natural cellulose, enhancing water penetration and anti-oil-fouling performance. When swollen, its mass increases by about 100 wt.%. The mesh must also possess adequate mechanical strength. The 3D-printed cellulose meshes exhibit exceptional flexibility when swollen in water (Fig. 8d).

6 Conclusions and Future Perspectives

Mass production with 3D printing has challenges. Limitations in resolution, due to nozzle size, printer type, material, and technique, can cause defects like interlayer gaps, porosity, and weak bonding. The layer-by-layer method may lead to microstructural inconsistencies affecting durability. CAD inaccuracies, slow speeds, and high costs also make it less competitive than traditional methods. However, combining 3D printing with techniques like ALD, VPD, track-etching, and electrospinning could help. The current 3D printing methods face limitations in achieving sub-micron resolution for efficient separation, especially of emulsions, and must compromise between printability, superwetting function, and durability. Scaling up is difficult due to slow layer-by-layer production, while required surface properties often necessitate costly and complex post-processing. Additionally, designing robust, fouling-resistant porous structures for real-world use remains a significant challenge.

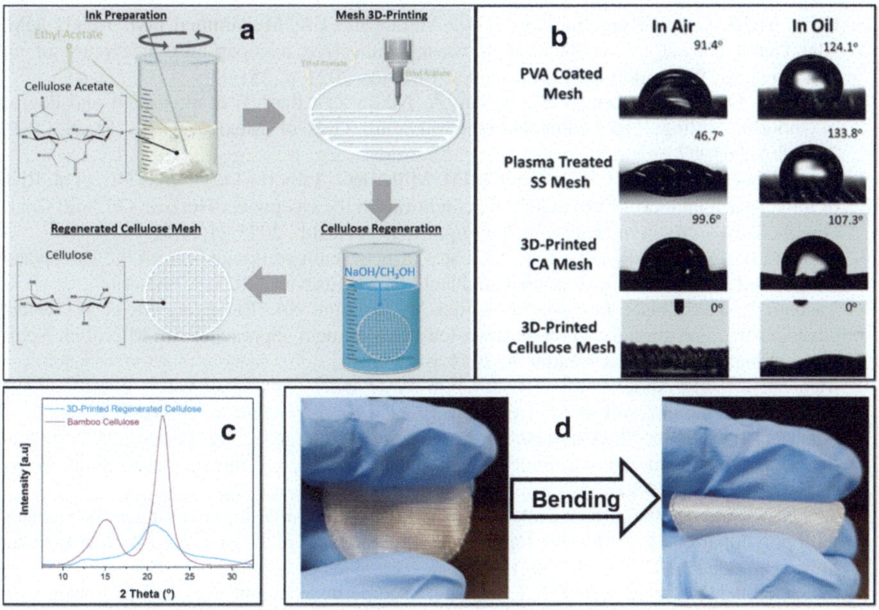

Fig. 8 **a** Fabrication procedure of the 3D-printed cellulose mesh, **b** experimental images of water contact angle on various meshes in air and oil environment, **c** XRD diffractogram of 3D-printed regenerated cellulose and neat bamboo cellulose, and **d** bending test to demonstrate the flexibility of the swelled 3D-printed cellulose mesh. Reprinted with permission from Ref. [140]

Progress in this field depends on high-resolution, multi-material printing for complex nano/micro-features, improved control of wettability, and new functional polymer composites. AI-driven design will accelerate biomimetic membrane development, while priorities include sustainable polymers, in-situ functionalization, and scalable, cost-effective production methods like modular or parallel printing.

References

1. Jayalath GNT, Ratnayake AS. Overview of coastal and marine pollution: sources, impacts, and challenges. In: Coastal and marine pollution: source to sink, mitigation and management. 2025. p. 1–25.
2. Samsuria NNC, Ismail WZW, Nazli MNWM, Aziz NAA, Ghazali AK. Problems, effects, and methods of monitoring and sensing oil pollution in water: a review. Water (Basel). 2025;17(9):1252.
3. Gunasekara MI, Mahawaththa I, Madhubhashini D, Amarasena K. Pollution from land-based sources: industrial and urban runoff. In: Coastal and marine pollution: source to sink, mitigation and management. 2025. p. 27–44.
4. Elella MHA, Castro-Riquelme CL, López-Maldonado EA, Abady MM, Mohammed DM, Abu-Thabit NY, et al. Synthesis of NPs by wastewater and bio-sludge. In: Waste-derived nanoparticles. Elsevier; 2024. p. 125–55.

5. Elella MHA, Castro-Riquelme CL, López-Maldonado EA, Mohammed DM, Abady MM, Abu-Thabit NY, et al. Application of waste-synthesized nanoparticles in capture of air pollutants. In: Waste-derived nanoparticles. Elsevier; 2024. p. 251–80.
6. Michel J, Owens EH, Zengel S, Graham A, Nixon Z, Allard T, et al. Extent and degree of shoreline oiling: Deepwater Horizon oil spill, Gulf of Mexico, USA. PLoS ONE. 2013;8(6):e65087.
7. Denic-Roberts H, Engel LS, Buchanich JM, Miller RG, Talbott EO, Thomas DL, et al. Risk of longer-term endocrine and metabolic conditions in the Deepwater Horizon Oil Spill Coast Guard cohort study–five years of follow-up. Environ Health. 2025;24(1):12.
8. Radović JR, Adhikari PL, Overton EB. Fates of petroleum compounds from the Deepwater Horizon oil spill. In: Oil spill science and technology. Elsevier; 2025. p. 661–99.
9. Yazdani R, Harirchi S, Lak M, Mirshafiei M, Nojoumi SA, Ramezani M, et al. Recent advances in conventional and modern wastewater treatment approaches. In: Microbial nexus for sustainable wastewater treatment. 2024. p. 83–115.
10. Mansour MSM, Abdel-Shafy HI, Ibrahim AM. Petroleum wastewater: environmental protection, treatment, and safe reuse: an overview. J Environ Manage. 2024;351:119827.
11. Abuhasel K, Kchaou M, Alquraish M, Munusamy Y, Jeng YT. Oily wastewater treatment: overview of conventional and modern methods, challenges, and future opportunities. Water (Basel). 2021;13(7):980.
12. Azad AK, Abu-Thabit N, Abu Eella M. Sustainable and durable superwetting textiles coated with polydopamine for oil/water separation. RSC Sustain. 2025. [Internet]. Available from: https://doi.org/10.1039/D5SU00041F.
13. Abu-Thabit NY, Uwaezuoke OJ, Elella MHA. Superhydrophobic nanohybrid sponges for separation of oil/water mixtures. Chemosphere. 2022;294:133644.
14. Abu Elella MH, Abu-Thabit NY, Uwaezuoke OJ, Azad AK. Superwetting cotton textiles for separation of oil/water mixtures, vol. 30, cellulose. Springer Science and Business Media B.V.; 2023. p. 7427–62.
15. Abu-Thabit NY, Elella MHA, Azad AK, Ratemi E, Hakeem AS. Superwetting metal mesh membranes for oil/water separation: a comprehensive review. Sep Purif Technol. 2025;132016.
16. Elella MHA, Hassan SS, Abdallah HM, Mostafa MS, Abu-Thabit NY. Metal chalcogenides and their nanocomposites in water purification systems. In: Metal-chalcogenide nanocomposites. 2024. p. 59–82.
17. Abu-Thabit N, Azad AK, Mezghani K, Hakeem A, Qasem D, Akhtar S, et al. Facile and green fabrication of superhydrophobic polyacrylonitrile nonwoven fabric with iron hydroxide nanoparticles for efficient oil\/water separation. ACS Appl Polym Mater. 2022;4:8450–60.
18. Abu-Thabit NY, Azad AK, Mezghani K, Akhtar S, Hakeem AS, Drmosh QA, et al. Rapid, sustainable, and versatile strategy towards fabricating superhydrophobic cotton textile membranes for separation of emulsified and stratified oil/water mixtures. Sep Purif Technol. 2024;330:125352.
19. Xu Y, Wu Q, Yang D, Qiu B, Mo X, You B, et al. Transparent superhydrophilic cellulose-based coating with anti-fogging, anti-biofouling and oil/water separation properties. Sustain Mater Technol. 2025;43:e01278.
20. Naseem M, Sultan M, ul Islam M, Kareem A, Ali N, Khan I, et al. Superhydrophilic wettability porous materials from construction tool/water separation applications. Fuel. 2025;386:134195.
21. Liao J, Fu J, Sun Q, Yang A, Zhang J, Peng J, et al. Bioinspired superwetting materials on emerging organic contaminants removal and oil/water separation: a review. J Mol Liq. 2025;128105.
22. Luo D, Wang L, Liu Y, Mei C, Yang T. Processing hydrophilic poplar wood into durably superhydrophobic materials with high stability and self-cleaning properties. ACS Sustain Chem Eng. 2025;13(15):5737–47.
23. Zhu MX, Luo LY, Li MT, Xu XK, Ren JR. Robust high-surface-insulating and superhydrophobic materials by constructing nanoparticles decorated porous structures. Compos Sci Technol. 2025;261:111033.

24. Xue Y, Verdross P, Liang W, Woodward RT, Bismarck A. Breaking the ice: applications of photothermal superhydrophobic materials for efficient deicing strategies. Adv Colloid Interface Sci. 2025;103489.
25. Zhao J, Lv J, Otgonbayar Z, Srikrajang S, Zhou C, Chen Y, et al. Recent advance in superhydrophobic materials for biomedical applications: comprehensive review. J Drug Deliv Sci Technol. 2025;107018.
26. Chu H, Xu Q, Liu Z, Xu N, Zhang H. Eco-friendly photothermal superhydrophobic coatings: recently advances in icephobic mechanisms, efficient photothermal conversion, and sustainable anti/de-icing technologies. Adv Colloid Interface Sci. 2025;103565.
27. Wang W, Wang Z, Li Z, Chen C, Wei Y, Xu X, et al. Bioinspired bionic superhydrophobic coating based on rosin acid and TiO_2 nanoparticles for oil–water separation and anti-icing. ACS Appl Nano Mater. 2025;8(1):881–93.
28. Jiang X, Yang F, Guo Z. Superwetting surfaces for filtration separation of high-viscosity raw petroleum/water mixtures. J Mater Chem A Mater. 2022;10(27):14273–92.
29. Yu J, Cao C, Pan Y. Advances of adsorption and filtration techniques in separating highly viscous crude oil/water mixtures. Adv Mater Interfaces. 2021;8(16):2100061.
30. Guan Y, Cheng F, Pan Z. Superwetting polymeric three dimensional (3D) porous materials for oil/water separation: a review. Polymers (Basel). 2019;11(5):806.
31. Yagoub H, Zhu L, Shibraen MHMA, Altam AA, Babiker DMD, Liang S, et al. Complex aerogels generated from nano-polysaccharides and its derivatives for oil–water separation. Polymers (Basel). 2019;11(10):1593.
32. Fu Y, Guo Z. Natural polysaccharide-based aerogels and their applications in oil–water separations: a review. J Mater Chem A Mater. 2022;10(15):8129–58.
33. Fouladi M, Kavousi Heidari M, Tavakoli O. Development of porous biodegradable sorbents for oil/water separation: a critical review. J Porous Mater. 2023;30(3):1037–53.
34. Pandey M, Pradhan N, Elella MHA, Makhado E, Samy M, Botlhoko OJ, et al. Frontiers in polymer/carbon dots nanocomposites: recent advances in synthesis and diverse applications. Surf Interfaces. 2025;106101.
35. Pandey M, Nazar R, Elella MHA, Praharaj S, Makhado E, Rout D, et al. A comprehensive review of recent developments in biomedical materials based on graphene-modified bionanocomposites. Bionanoscience. 2025;15(1):125.
36. Elella MHA, Abdallah HM, Ali EA, Makhado E, Abd El-Ghany NA. Recent developments in conductive polysaccharide adsorbent formulations for environmental remediation: a review. Int J Biol Macromol. 2025;140915.
37. Elella MHA, Kamel AM, López-Maldonado EA, Uzondu SW, Abdallah HM. A review of recent progress in alginate-based nanocomposite materials for tissue engineering applications. Int J Biol Macromol. 2025;139840.
38. Abu-Thabit NY. Nanomaterials for flexible transparent conductive films and optoelectronic devices. In: Advances in smart coatings and thin films for future industrial and biomedical engineering applications. Elsevier; 2020. p. 619–43.
39. Abu-Thabit N. Optical colorimetric sensing label for monitoring food freshness. Eng Proc. 2023;48(1):16.
40. Zhang Z, Lu Y, Gao S, Wu S. Sustainable and efficient wastewater treatment using cellulose-based hydrogels: a review of heavy metal, dye, and micropollutant removal applications. Separations. 2025;12(3):72.
41. Yu X, Ji J, Wu QY, Gu L. Direct-coating of cellulose hydrogel on PVDF membranes with superhydrophilic and antifouling properties for high-efficiency oil/water emulsion separation. Int J Biol Macromol. 2024;256:128579.
42. Zhang Z, Lu Y, Zhao Y, Cui L, Xu C, Wu S. Current developments in chitosan-based hydrogels for water and wastewater treatment: a comprehensive review. ChemistrySelect. 2025;10(6):e202404061.
43. Guo Z, He Y, Xu C, Huang J, Liu W. Hydrogel: a versatile application for efficient and stable oil-water separation. J Mater Chem A Mater. 2025.

44. Visan AI, Negut I. Environmental and wastewater treatment applications of stimulus-responsive hydrogels. Gels. 2025;11(1):72.
45. Croitoru AM, Niculescu AG, Bîrcă AC, Mihaiescu DE, Rădulescu M, Grumezescu AM. Nanostructured aerogels for water decontamination: advances, challenges, and future perspectives. Nanomaterials. 2025;15(12):901.
46. Zhang Y, Wang Y. Construction of three-dimensional aerogels from electrospun cellulose fibers as highly efficient and reusable oil absorbents. Sep Purif Technol. 2025;353:128604.
47. Segneanu AE, Herea DD, Buema G, Bradu IA, Cepan M, Grozescu I. Advanced aerogels for water remediation: unraveling their potential in fats, oils, and grease sorption—a comprehensive review. Gels. 2025;11(4):268.
48. Zhang X, Dong K, Liu Z, Xiong K, Ren L. Preparation and characterization of reusable sodium alginate/enteromorpha cellulose micro-nanofibers composite aerogels for efficient oil absorption. Colloids Surf A Physicochem Eng Asp. 2025;137015.
49. Jiang J, Wan S, Wen C, Tang L, Xu N. Frontiers in innovative materials and technologies for oil-water separation. Polymers (Basel). 2025;17(12):1635.
50. Pandey A, Arputharaj A, Vigneswaran N, Raja ASM, Kumar TS, Patil S, et al. Advancement in biodegradable foam for packaging, filtration, thermal insulation, and medical application: a review. Waste Biomass Valorization. 2025;1–21.
51. Nayl AA, Abd-Elhamid AI, Mosnáčková K, Arafa WAA, Alanazi AH, Ahmed IM, et al. Advances in the modification and applications of cellulosic-based materials for wastewater remediation: a review. Appl Water Sci. 2025;15(7):177.
52. Jin R, Gong J, Xiang B, Li J. Biomass materials with special wettability: a sustainable solution for efficient oil-water separation. J Mater Chem A Mater. 2024.
53. Firmanda A, Syamsu K, Sari YW, Cabral J, Pletzer D, Mahadik B, et al. 3D printed cellulose based product applications. Mater Chem Front. 2022;6(3):254–79.
54. Ghosal P, Gupta B, Ambekar RS, Rahman MM, Ajayan PM, Aich N, et al. 3D printed materials in water treatment applications. Adv Sustain Syst. 2022;6(3):2100282.
55. Huang J, Li L. 3D printing of cellulose-based biomaterials: a review. BME Horiz. 2025;3(1):138.
56. Kacem E, Suk HD, Ponnamma D, Hassan MK, Hawari A, Alshammari BA, et al. Exploring 3D printing and electrospinning technologies for advanced porous membrane fabrication: a review. Adv Mater Technol. 2025;70032.
57. Tong Y, Jiang C, Ji C, Liu W, Wang Y. Innovative applications of nanocellulose in 3D printing: a review. Small. 2025;21(4):2407956.
58. Amjad M, Nguir M, Ma X, Wen D. Superhydrophobic 3D-printed microstructures: applications, challenges, and prospects. Prog Addit Manuf. 2025;1–28.
59. Liang F, Hou T, Li S, Liao L, Li P, Li C. Elastic, super-hydrophobic and biodegradable chitosan sponges fabricated for oil/water separation. J Environ Chem Eng. 2021;9(5):106027.
60. Xu C, He H, Wang Y, Huang Y, Zhang TC, Yuan S. Superhydrophobic sponge-like chitosan/CNTs/silica composite for selective oil absorption and efficient separation of water-in-oil emulsion. Carbohydr Polym. 2025;353:123256.
61. Wang G, He Y, Wang H, Zhang L, Yu Q, Peng S, et al. A cellulose sponge with robust superhydrophilicity and under-water superoleophobicity for highly effective oil/water separation. Green Chem. 2015;17(5):3093–9.
62. Chen J, Zhang Y, Chen C, Xu M, Wang G, Zeng Z, et al. Cellulose sponge with superhydrophilicity and high oleophobicity both in air and under water for efficient oil–water emulsion separation. Macromol Mater Eng. 2017;302(9):1700086.
63. Lu Y, Yuan W. Superhydrophobic/superoleophilic and reinforced ethyl cellulose sponges for oil/water separation: synergistic strategies of cross-linking, carbon nanotube composite, and nanosilica modification. ACS Appl Mater Interfaces. 2017;9(34):29167–76.
64. Li X, Jin X, Wu Y, Zhang D, Sun F, Ma H, et al. A comprehensive review of lignocellulosic biomass derived materials for water/oil separation. Sci Total Environ. 2023;876:162549.
65. Zhang W, Liu Y, Tao F, An Y, Zhong Y, Liu Z, et al. An overview of biomass-based oil/water separation materials. Sep Purif Technol. 2023;316:123767.

66. Qasim M, Alnaser AS. Transforming cellulose waste into advanced superhydrophobic membranes for effective oil/water separation. ACS Omega. 2025.
67. Kang L, Wang B, Zeng J, Cheng Z, Li J, Xu J, et al. Degradable dual superlyophobic lignocellulosic fibers for high-efficiency oil/water separation. Green Chem. 2020;22(2):504–12.
68. Ahuja D, Dhiman S, Rattan G, Monga S, Singhal S, Kaushik A. Superhydrophobic modification of cellulose sponge fabricated from discarded jute bags for oil water separation. J Environ Chem Eng. 2021;9(2).
69. Mondal K, Advani K, Ghosh S, Shanmugnathan K, Kulsi G, Sivaram S, et al. Shola: a 3D porous hydrophobic–oleophilic lignocellulosic material for efficient oil/water separation. J Mater Chem A Mater. 2025;13(3):1983–95.
70. Chen Y, Su N, Zhang K, Zhu S, Zhu Z, Qin W, et al. Effect of fiber surface treatment on structure, moisture absorption and mechanical properties of luffa sponge fiber bundles. Ind Crops Prod. 2018;123:341–52.
71. Chen Y, Su N, Zhang K, Zhu S, Zhao L, Fang F, et al. In-depth analysis of the structure and properties of two varieties of natural luffa sponge fibers. Materials. 2017;10(5):479.
72. Liu Z, Pan Y, Shi K, Wang W, Peng C, Li W, et al. Preparation of hydrophilic luffa sponges and their water absorption performance. Carbohydr Polym. 2016;147:178–87.
73. Alvarado-Gómez E, Tapia JI, Encinas A. A sustainable hydrophobic luffa sponge for efficient removal of oils from water. Sustain Mater Technol. 2021;28:e00273.
74. Zhang F, Wang C, Zhou J, Wu J, Gu H, Lin W. Hydrophobic sponge derived from natural loofah for efficient oil/water separation. Sep Purif Technol. 2024;330:125519.
75. Dan H, Wang Y, Liu C, Yin W, Yue Q, Gao B, et al. Graphene-wrapped loofah sponge with tunable surface wettability for solar-driven oil and freshwater recovery. Sep Purif Technol. 2025;133365.
76. Souare M, Dong C, Xing T, Zhang J, Hu X. Efficient separation of oil/water by a biodegradable and superhydrophobic composite based on loofah and rice straw. Membranes (Basel). 2024;14(11):243.
77. Lei Y, Yang W, Yuan Y, Wei G, Guo L. Functionalization of loofah sponge as a natural cellulose-based material for water environment remediation and sustainability assessment. Carbohydr Polym. 2025;123514.
78. Yang W, Li W, Lei Y, He P, Wei G, Guo L. Functionalization of cellulose-based sponges: design, modification, environmental applications, and sustainability analysis. Carbohydr Polym. 2024;122772.
79. Xing T, Dong C, Hu X, Zheng Q, Zhang J, Zhao Y, et al. Cellulose membranes via a top-down approach from loofah for oil/water separation. Biomass Convers Biorefin. 2024;14(15):17167–75.
80. Chen Z, Weng P, Song Y, Zheng L, Tan Y, Yin X. Loofah-inspired sodium alginate/carboxymethyl cellulose sodium-based porous frame for all-weather super-viscous crude oil adsorption and wastewater treatment in harsh environment. Carbohydr Polym. 2024;323:121450.
81. Liu Z, Gao B, Zhao P, Fu H, Kamali AR. Magnetic luffa/graphene/$CuFe_2O_4$ sponge for efficient oil/water separation. Sep Purif Technol. 2024;337:126347.
82. Gundu S, Sahi AK, Kumari P, Vishwakarma NK, Mahto SK. Assessment of various forms of cellulose-based *Luffa cylindrica* (mat, flakes and powder) reinforced polydimethylsiloxane composites for oil sorption and organic solvents absorption. Int J Biol Macromol. 2023;240:124416.
83. Alhijazi M, Safaei B, Zeeshan Q, Asmael M, Eyvazian A, Qin Z. Recent developments in luffa natural fiber composites. Sustainability. 2020;12(18):7683.
84. Wang Z, Ma H, Chu B, Hsiao BS. Super-hydrophobic modification of porous natural polymer "luffa sponge" for oil absorption. Polymer (Guildf). 2017;126:470–6.
85. Lin Y, Yi P, Yu M, Li G. Fabrication and performance of a novel 3D superhydrophobic material based on a loofah sponge from plant. Mater Lett. 2018;230:219–23.
86. Li J, Liu Q, He J, Zhao Y, Mu L, Liu X, et al. A review of superwetting aerogel-based oil-water separation materials. Mater Today Sustain. 2024;100741.

87. Zhai Y, Yuan X, Weber C, Varley RJ, Henderson LC. Review of plant cellulose-based aerogel materials for oil/water mixture separation. J Environ Chem Eng. 2024;113716.
88. Li A, Huber T, Barker D, Nazmi AR, Zadeh HN. An overview of cellulose aerogels and foams for oil sorption: preparation, modification, and potential of 3D printing. Carbohydr Polym. 2024;343:122432.
89. Stark FW, Thue PS, Missio AL, Machado FM, Delucis R de A, Andreazza R. Cellulose-based aerogels for environmentally sustainable applications: a review of the production, modification, and sorption of environmental contaminants. Polymers (Basel). 2025;17(2):236.
90. Muhammad S, Albadn YM, Yahya EB, Nasr S, Khalil HPSA, Ahmad MI, et al. Trends in enhancing the efficiency of biomass-based aerogels for oil spill clean-up. Giant. 2024;100249.
91. Ashori A, Chiani E, Shokrollahzadeh S, Madadi M, Sun F, Zhang X. Cellulose-based aerogels for sustainable dye removal: advances and prospects. J Polym Environ. 2024;32(12):6149–81.
92. Dai Q, Li D, Sun Y, Wang H, Lu Y, Yang D. Low temperature-resistant superhydrophobic and elastic cellulose aerogels derived from seaweed solid waste as efficient oil traps for oil/water separation. Chemosphere. 2023;336:139179.
93. Ma X, Zhou S, Li J, Xie F, Yang H, Wang C, et al. Natural microfibrils/regenerated cellulose-based carbon aerogel for highly efficient oil/water separation. J Hazard Mater. 2023;454:131397.
94. Chen J, Chen J, Zheng L, Zheng H. Cellulose-based aerogels for efficient dye sorption and oil-water separation in textile wastewater treatment. Int J Biol Macromol. 2025;143612.
95. Shang Q, Cheng J, Hu L, Bo C, Yang X, Hu Y, et al. Bio-inspired castor oil modified cellulose aerogels for oil recovery and emulsion separation. Colloids Surf A Physicochem Eng Asp. 2022;636:128043.
96. Schäfer HL, Gandras L, Schneider L, Witthohn M, Troidl K, Muffler K, et al. Analysis, properties, and applications of insect-derived chitosan: a sustainable path to functional polysaccharide materials. Gels. 2025;11(4):291.
97. Scarcelli E, Catalano A, Iacopetta D, Ceramella J, Sinicropi MS, Aiello F. Chitin, chitosan and its derivatives: antimicrobials and/or mitigators of water. Macromol. 2025;5(2):15.
98. Pandey R, Mathur G. Current trends in chitosan functionalization methods and their applications. Starch-Stärke. 2025;77(1):2300248.
99. Iber BT, Kasan NA, Torsabo D, Omuwa JW. A review of various sources of chitin and chitosan in nature. J Renew Mater. 2022;10(4):1097.
100. Fu C, Yao L, Guo Z. Biomass chitosan-based complexes with superwettability for oil-water separation. Mater Today Chem. 2024;40:102265.
101. Takeshita S, Zhao S, Malfait WJ. Transparent, aldehyde-free chitosan aerogel. Carbohydr Polym. 2021;251:117089.
102. Li Z, Shao L, Hu W, Zheng T, Lu L, Cao Y, et al. Excellent reusable chitosan/cellulose aerogel as an oil and organic solvent absorbent. Carbohydr Polym. 2018;191:183–90.
103. Tuo Z, Pan Y, Cai P. Facile and green fabrication of biodegradable aerogel from chitosan derivatives/modified gelatin as absorbent for oil removal. Int J Biol Macromol. 2025;298:139949.
104. Fu Y, Ai S, Guo Z, Liu W. Nature-inspired polysaccharide-based aerogel for oil–water separation. J Bionic Eng. 2023;20(5):1956–66.
105. Ramalingam B, Das SK. Biomimetic strategy for fabrication of bifunctional graphene oxide-biomaterial aerogel as highly porous antifouling material for oil/water separation. Chem Eng J. 2023;475:145906.
106. Kumar N, Gusain R, Pandey S, Ray SS. Hydrogel nanocomposite adsorbents and photocatalysts for sustainable water purification. Adv Mater Interfaces. 2023;10(2):2201375.
107. Ngwuluka NC, NY Abu-Thabit, Uwaezuoke OJ, Erebor JO, Ilomuanya MO, Mohamed RR, et al. Natural polymers in micro- and nanoencapsulation for therapeutic and diagnostic applications: part I: lipids and fabrication techniques. In: Nano- and microencapsulation—techniques and applications. IntechOpen; 2021.
108. Kassem I, Kassab Z, Khouloud M, Sehaqui H, Bouhfid R, Jacquemin J, et al. Phosphoric acid-mediated green preparation of regenerated cellulose spheres and their use for all-cellulose cross-linked superabsorbent hydrogels. Int J Biol Macromol. 2020;162:136–49.

109. Estevam BR, Perez ID, Gonçalves HL, Zago GP, Tsuruta KM, Neto RMF, et al. Dehydration of turbine engine lubricant oil using cellulose hydrogel. Colloids Surf A Physicochem Eng Asp. 2025;704:135424.
110. Li Z, Ruan H, Li M, Yu C, Jia Q, Wang J, et al. Additive manufacturing: prospects and diverse applications. Metals (Basel). 2025;15(2):158.
111. Hossain MJ, Tabatabaei BT, Kiki M, Choi JW. Additive manufacturing of sensors: a comprehensive review. Int J Precis Eng Manuf-Green Technol. 2025;12(1):277–300.
112. Hamza A, Bousnina K, Dridi I, Ben YN. Revolutionizing automotive design: the impact of additive manufacturing. Vehicles. 2025;7(1):24.
113. Yong GTX, Koh WB, Yusoff NHBM, Norman A, Chong CH, Cheah KH, et al. 3D-printed SiOC ceramics grafted with chitosan-graphene oxide composite for enhanced dye adsorption. Sep Purif Technol. 2025;362:131669.
114. Aslam M, Shahzad T, Hussain S, Waseem M, Muzammil S, Afzal M, et al. Implementation of 3D-printed materials for water purification applications. In: Sustainable 3D printing for innovative biopolymer production and applications. 2025;191–213.
115. Zheng J, Robertson M, Patil NA, Park JH, Qiang Z. Additive manufacturing and functionalization of hollow polypropylene sorbents for water remediation. Macromol Rapid Commun. 2025;2400879.
116. Li N, Wang Y, Chang X, Gao W, Kong L, Yan B, et al. Revolutionizing catalytic water treatment: a critical review on the role of 3D printed catalysts. Sep Purif Technol. 2025;132194.
117. da Silva Fortes AG, de Abreu IR, de Sousa Nascimento Júnior R, Sampaio AAS, Leitão LV, Reis ALT, et al. Biodegradable polymeric membranes via additive manufacturing for methylene blue adsorption. J Polym Environ. 2025;33(4):2029–57.
118. Ali S, Aljarrah MT, Al-Otoom A, Abdelaziz N. Development and testing of robust 3D printed ZnO/clay photocatalysts for sustainable wastewater treatment. ACS Omega. 2025;10(16):16156–68.
119. Wilson CB, Parolari AJ, Hiestand A, Mayer BK, Murray A. Characterizing phosphorus removal and recovery performance of a 3D printed iron-embedded polylactic acid composite. J Environ Eng. 2025;151(2):04024076.
120. Estévez R, Quijada-Maldonado E, Romero J, Abejón R. Additive manufacturing and chemical engineering: looking for synergies from a bibliometric study. Appl Sci. 2025;15(6):2962.
121. Abu-Zurayk R, Walvekar R, Choo HL, Waleed H, Luo C, Khalaf A, et al. 3D printed polymeric membranes for oil/water separation: a comprehensive review. Prog Addit Manuf. 2025;1–26.
122. Kumar P, Singh J. 3D printed materials in environmental remediation. In: Nanomaterials in environmental remediation. CRC Press; 2025. p. 283–307.
123. Hernandez-Tenorio F, Múnera-Gutiérrez E, Miranda AM, Sáez AA, Marín-Palacio LD, Giraldo-Estrada C. 3D printing of polysaccharide-based formulations: opportunities for innovation. Bioprinting. 2024;e00383.
124. de Pastre MA, Quinsat Y, Lartigue C. Effects of additive manufacturing processes on part defects and properties: a classification review. Int J Interact Des Manuf (IJIDeM). 2022;16(4):1471–96.
125. Liu J, Sun L, Xu W, Wang Q, Yu S, Sun J. Current advances and future perspectives of 3D printing natural-derived biopolymers. Carbohydr Polym. 2019;207:297–316.
126. Fijoł N, Aguilar-Sánchez A, Mathew AP. 3D-printable biopolymer-based materials for water treatment: a review. Chem Eng J. 2022;430:132964.
127. Park S, Shou W, Makatura L, Matusik W, Fu KK. 3D printing of polymer composites: materials, processes, and applications. Matter. 2022;5(1):43–76.
128. Abas M, Jan Z, Rahman K. Direct ink writing. In: Innovative development in micromanufacturing processes. CRC Press; 2023. p. 188–209.
129. Li ZL, Zhou S, Saiz E, Malik R. Ink formulation in direct ink writing of ceramics: a meta-analysis. J Eur Ceram Soc. 2024.
130. Lewis JA. Direct ink writing of 3D functional materials. Adv Funct Mater. 2006;16(17):2193–204.

131. Patadiya J, Kandasubramanian B. Biopolymer-chitin products by direct ink writing (DIW): a review. Hybrid Adv. 2024;5:100115.
132. Ho M, Ramirez AB, Akbarnia N, Croiset E, Prince E, Fuller GG, et al. Direct ink writing of conductive hydrogels. Adv Funct Mater. 2025;2415507.
133. Ajdary R, Reyes G, Kuula J, Raussi-Lehto E, Mikkola TS, Kankuri E, et al. Direct ink writing of biocompatible nanocellulose and chitosan hydrogels for implant mesh matrices. ACS Polym Au. 2021;2(2):97–107.
134. Ramirez Caballero SS, Saiz E, Montembault A, Tadier S, Maire E, David L, et al. 3-D printing of chitosan-calcium phosphate inks: rheology, interactions and characterization. J Mater Sci Mater Med. 2019;30:1–15.
135. Li VCF, Mulyadi A, Dunn CK, Deng Y, Qi HJ. Direct ink write 3D printed cellulose nanofiber aerogel structures with highly deformable, shape recoverable, and functionalizable properties. ACS Sustain Chem Eng. 2018;6(2):2011–22.
136. Li VCF, Dunn CK, Zhang Z, Deng Y, Qi HJ. Direct ink write (DIW) 3D printed cellulose nanocrystal aerogel structures. Sci Rep. 2017;7(1):8018.
137. Li X, Shan H, Zhang W, Li B. 3D printed robust superhydrophilic and underwater superoleophobic composite membrane for high efficient oil/water separation. Sep Purif Technol. 2020;237:116324.
138. Feng Z, Xu Y, Ding W, Li Q, Zhao X, Wei X, et al. Nano graphene oxide creates a fully biobased 3D-printed membrane with high-flux and anti-fouling oil/water separation performance. Chem Eng J. 2024;485:149603.
139. Wang Z, Liao X, Wang X, Bai Y, Huang H, Shen K, et al. Converting complex sewage containing oil, silt, and bacteria into clean water by a 3D printed multiscale and multifunctional filter. ACS Appl Bio Mater. 2021;4(12):8509–21.
140. Koh JJ, Lim GJH, Zhou X, Zhang X, Ding J, He C. 3D-printed anti-fouling cellulose mesh for highly efficient oil/water separation applications. ACS Appl Mater Interfaces. 2019;11(14):13787–95.
141. Tian N, Wu S, Han G, Zhang Y, Li Q, Dong T. Biomass-derived oriented neurovascular network-like superhydrophobic aerogel as robust and recyclable oil droplets captor for versatile oil/water separation. J Hazard Mater. 2022;424:127393.
142. Yang Y, Chen X, Li Y, Yin Z, Bao M. Construction of a superhydrophobic sodium alginate aerogel for efficient oil absorption and emulsion separation. Langmuir. 2021;37(2):882–93.
143. Mi HY, Jing X, Politowicz AL, Chen E, Huang HX, Turng LS. Highly compressible ultra-light anisotropic cellulose/graphene aerogel fabricated by bidirectional freeze drying for selective oil absorption. Carbon N Y. 2018;132:199–209.
144. Mi HY, Jing X, Huang HX, Peng XF, Turng LS. Superhydrophobic graphene/cellulose/silica aerogel with hierarchical structure as superabsorbers for high efficiency selective oil absorption and recovery. Ind Eng Chem Res. 2018;57(5):1745–55.
145. Yi L, Yang J, Fang X, Xia Y, Zhao L, Wu H, et al. Facile fabrication of wood-inspired aerogel from chitosan for efficient removal of oil from Water. J Hazard Mater. 2020;385:121507.
146. Cheng Y, Lu L, Zhang W, Shi J, Cao Y. Reinforced low density alginate-based aerogels: preparation, hydrophobic modification and characterization. Carbohydr Polym. 2012;88(3):1093–9.
147. Yang J, Xia Y, Xu P, Chen B. Super-elastic and highly hydrophobic/superoleophilic sodium alginate/cellulose aerogel for oil/water separation. Cellulose. 2018;25:3533–44.
148. Wang Y, Feng Y, Yao J. Construction of hydrophobic alginate-based foams induced by zirconium ions for oil and organic solvent cleanup. J Colloid Interface Sci. 2019;533:182–9.
149. Varamesh A, Abraham BD, Wang H, Berton P, Zhao H, Gourlay K, et al. Multifunctional fully biobased aerogels for water remediation: applications for dye and heavy metal adsorption and oil/water separation. J Hazard Mater. 2023;457:131824.

3D Engineered Polysaccharides for Advanced Engineering Applications

Electrically Conductive Polysaccharide-Based 3D Hydrogels: Fabrication and Applications

Edwin Makhado, Wilson Marema Seleka, and Mpitloane Joseph Hato

Abstract A novel multifunctional material that integrates the electroactivity of conductive elements with the structural versatility and biocompatibility of natural polymers is electrically conductive polysaccharide-based 3D hydrogels. Polysaccharide matrices may achieve conductivity by the incorporation of conductive polymers, carbon nanotubes, and metal nanoparticles; this chapter explores the materials design and fabrication methods that facilitate this process. The methodologies influencing network design and functional performance are being, 3D bioprinting, and advanced crosslinking procedures. This chapter reviews significant structure-property correlations, including electrical, mechanical, and swelling for specific applications. The subjects addressed include the challenges of scalability, repeatability, and functional integration, along with their contemporary applications in bioelectronics, soft robotics, and tissue interfaces. This chapter aims to inform material scientists about the synthesis-structure-function paradigm driving advancements in conductive hydrogel systems.

Abbreviations

Acrylamide	AAm
Aluminium oxide	Al_2O_3
Carbon black	CB
Carbon nanotubes	CNTs
Carbon quantum dots	CQDs
Cerium dioxide	CeO_2
Copper	Cu
Copper oxide	CuO
Copper phthalocyanine-3,4′,4″,4‴-tetrasulfonic acid tetrasodium salt	CuPcTs
Cyclic voltammetry	CV

E. Makhado (✉) · W. M. Seleka · M. J. Hato
Nanotechnology Research Lab, Department of Chemistry, School of Physical and Mineral Sciences, University of Limpopo (Turfloop), Polokwane, South Africa
e-mail: edwin.makhado@ul.ac.za

M. H. Abu Elella et al. (eds.), *3D Engineered Polysaccharides*,
https://doi.org/10.1007/978-3-032-26154-0_9

Differential pulse voltammetry	DPV
Electrically conductive hydrogels	ECHs
Electrochemical impedance spectroscopy	EIS
Extracellular matrix	ECM
Gold	Au
Graphene oxide	GO
Iron oxides	Fe_2O_3
Low critical solution temperature	LCST
Magnetite iron oxide	Fe_3O_4
Manganese dioxide	MnO_2
Metal nanoparticles	MNPs
Metal oxide nanoparticles	MONPs
Molybdenum disulfide	MoS_2
N,N′-methylene bis-acrylamide	MBA
Palladium	Pd
Platinum	Pt
Poly(3,4-ethylenedioxythiophene)	PEDOT
Polyaniline	PANi
Polypyrrole	Ppy
Potassium persulfate	KPS
Reactive oxygen species	ROS
Silver	Ag
Single-walled carbon nanotube	SWCNT
Tetraaniline	TA
Three-dimensional	3D
Titanium dioxide	TiO_2
Upper critical solution temperature	UCST
Zero-dimensional	0D
Zinc oxide	ZnO

1 Introduction

A hydrogel is a network of hydrophilic polymers that can hold a great deal of water or biological fluids—sometimes hundreds of times their dry weight—while keeping their structure intact. It is a three-dimensional (3D) network [1]. Hydrogels exhibit mechanical softness and significant water content; electrically conductive hydrogels (ECHs) represent a unique hybrid system that integrates these properties with those of conductive materials [2]. This dual functioning is essential for signal transmission, stimulus response, and dynamic interaction with biological systems. Polysaccharides serve as a robust and versatile foundation for hydrogel manufacturing, being biopolymers that exhibit diverse structural properties and are abundant in nature [3]. Numerous hydrogels, particularly those derived from natural polymers (e.g., alginate, chitosan, cellulose, and xanthan gum), synthetic polymers (such as polyacrylamide and polyacrylic acid), and hybrid compositions (integrating natural and synthetic polymers), exhibit biocompatibility and, in some instances, biodegradability [4–8]. These materials serve as ideal bioactive scaffolds that mimic the extracellular matrix (ECM) within 3D hydrogel networks, owing to their chemical activity, biocompatibility, and malleability. When imbued with electrical conductivity, these hydrogels can link inactive biomaterials to interactive electronic interfaces [9].

This chapter focusses on the design, synthesis, structure-property interactions, and applications of electrically conductive polysaccharide-based 3D hydrogels. The focus is on fabrication methods that integrate conductive fillers into polysaccharide matrices via in situ polymerisation, 3D printing, and self-assembly. Conductive fillers comprise carbon nanostructures, metal nanoparticles, and intrinsically conductive polymers such as Ppy and polyaniline. The metrics employed to evaluate these methods include the impact on network design, the reliability of mechanical components, and the behaviour of charge transfer. This chapter aims to equip researchers with a comprehensive understanding of how molecular structure, processing conditions, and composite formulation influence the performance of conductive hydrogel systems, emphasising fundamental materials science concepts. Next, we will examine emerging applications of this technology, including implanted electronics and soft actuators, alongside existing challenges that require resolution, such as enhancing long-term stability, achieving uniform conductivity, and facilitating scalability. This study elucidates the critical material choices driving innovation in next-generation conductive soft materials.

2 Polysaccharides as Hydrogel Matrices

Long-chain carbohydrates called natural polysaccharides are made up of repeated monosaccharide units that are joined by glycosidic linkages. Many natural polysaccharides exist, and they are sourced from plants, animals, fungi, or microbes. Some examples include cellulose, chitosan, starch, alginate, carrageenan, agarose, and hyaluronic acid. These polysaccharides have different structures, such as linear or branching, and different functions. They may be changed to have different qualities, such as viscosity, gelation, water retention, and biodegradability [10]. They are perfect for use in food, drugs, cosmetics, and biomedical industries since they are biocompatible, non-toxic, and can create hydrogels. Natural polysaccharides make up the structures that allow for the formation of 3D networks in hydrogel systems. These networks can contain a lot of water and respond to stimuli [11]. Their chemical or physical properties can be changed by additional features like pH sensitivity, antibacterial activity, or electrical conductivity. Such functionality makes them more useful for medication delivery, wound healing, tissue engineering, and cleaning up the environment [12].

Polysaccharides have various molecular architectures, with linear, branched, or networked structures that determine their hydrogel-forming abilities—linear chains like cellulose form rigid networks through strong interchain hydrogen bonds, whereas branched polysaccharides like dextran promote softer, more flexible gels. Their many functional groups (hydroxyl, carboxyl, amino, and sulphate) allow for hydrogen bonding, ionic interactions, and crosslinking, which are all required for stable hydrogel formation [10]. Additionally, the chain conformation, which ranges from helices in carrageenan to random coils in alginate, has a direct impact on gelation behaviour, porosity, and mechanical quality. Finally, the proportion of crystalline and amorphous areas in these polymers determines the swelling capacity, mechanical strength, and degradation rates of the resulting hydrogels, allowing for fine customisation for particular applications [13].

Natural polysaccharides exhibit biocompatibility and biodegradability, rendering them suitable for applications in tissue engineering, wound healing, and drug delivery while minimising the risk of toxicity or chronic inflammation [1, 10]. Their hydrophilic properties facilitate significant water absorption and retention, which is crucial for the formation of swollen three-dimensional hydrogel networks that replicate biological tissues [1]. Numerous polysaccharides demonstrate significant gelation capabilities, characterised by swift and reversible gel formation, as exemplified by ionically crosslinked alginate and thermally gelled agarose, which provide ease of processing and versatility [14]. The mechanical properties of these hydrogels can be precisely adjusted by modifying molecular weight, crosslinking density, or polymer blends to attain the required stiffness and elasticity for particular applications. Their responsiveness to stimuli facilitates reactions to variations in pH, temperature, ionic strength, or enzymatic activity, thereby enabling controlled drug release and smart sensing capabilities [15]. Natural polysaccharides exhibit inherent bioactivity, exemplified by the antimicrobial properties of chitosan. They can also be chemically modified to enhance characteristics such as solubility, targeted delivery, or electrical conductivity, thereby expanding their potential applications [16].

Natural polysaccharides are recognised for their superior biocompatibility, enabling safe interactions with live tissues without provoking notable immunological or inflammatory reactions [10, 16]. This renders them exceptionally appropriate for biological applications such as wound dressings, medication delivery systems, and tissue engineering scaffolds. The biodegradability of polysaccharide-based hydrogels guarantees their slow decomposition into non-toxic, naturally occurring byproducts that the body may metabolise or excrete, obviating the need for surgical removal and mitigating their long-term consequences. The disintegration rate may be adjusted by altering the polysaccharide's molecular weight, crosslinking density, or chemical changes, enabling precise regulation of the hydrogel's functionality lifetime in applications like controlled drug release or temporary tissue scaffolds. The synergy of biocompatibility and biodegradability renders polysaccharide-based hydrogels appealing and sustainable materials for many medicinal and environmental applications [17].

3 Fabrication Techniques for 3D Conductive Hydrogels

3.1 Formation of a Hydrogel Through the Utilisation of a Conductive Infill Suspension

To make hydrogels that are electrically conductive, it is possible to use a conductive filler solution. Through a one-step synthesis procedure, the biopolymer, initiator, cross-linker, and cross-linking agents are combined, and the polymerisation process is initiated by heat. Gelling a hydrogel monomer inside a dispersion of conductive fillers, such as conducting polymers, metallic nanoparticles, and carbon-based materials, is a key method for manufacturing an electrochemical hologram (ECH). To improve the hydrogel's mechanical and thermal durability, as well as to impart electrical conductivity to the hydrogel matrix, graphene, graphene oxide, reduced graphene oxide, and carbon nanotubes have been included in the backbone of the hydrogel (Fig. 1a).

3.2 Polymerisation of the Polymer within the Hydrogel Monomer Backbone

The formation of a non-conductive hydrogel is the first step in the process of in-situ radical polymerisation, which is an efficient approach for the production of double-network epoxy composite hydrogels (ECH). This procedure is followed by the insertion of conductive polymers such as polyaniline (PANi). This approach enables the overlaying of π-bonds in conductive polymers with non-conductive fillers, hence increasing the mobility of electrons. A very elastic PANi-PAAm-GOCs hydrogel with outstanding mechanical strength and skin compatibility was created by Jin et al. This hydrogel was synthesised using a freeze-thaw technique and subsequent polymerisation utilising acrylamide and ammonium persulfate [18]. The incorporation of graphene oxide (GO) into the hydrogel resulted in the enhancement of its mechanical and electrical characteristics. This process was accomplished by expanding the number of active sites for π stacking and therefore increasing the electron density [18].

Li et al. [19] investigated the production of PANi, cellulose, and polyacrylamide hydrogels, revealing the potential of these materials for use in high-performance supercapacitors and motion sensors, among other applications [19]. By combining cellulose with ECH and activating the pre-gel solution with heat, they were able to obtain excellent dispersion. This step resulted in the formation of a cellulose/polyacrylamide hydrogel, which was subsequently polymerised with aniline. In addition, Hu et al. [20] reported on composite hydrogels that were created from magnetite iron oxide (Fe_2O_3), molybdenum disulphide (MoS_2), and polyaniline (PANi) [20].

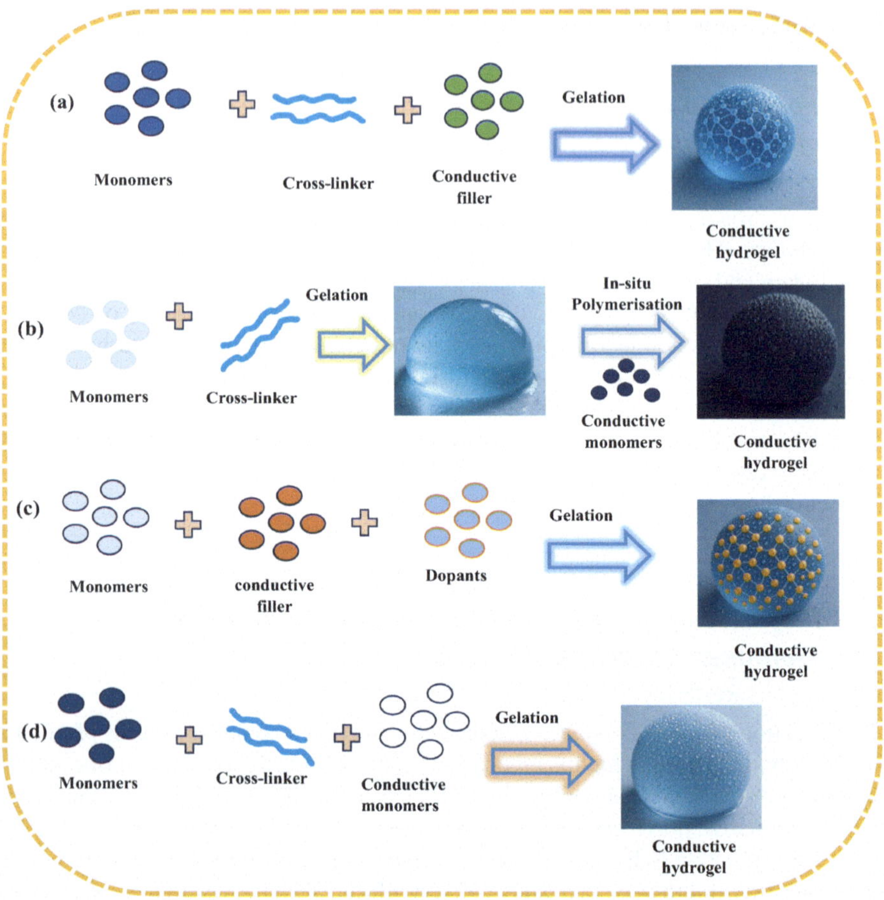

Fig. 1 Diverse methodologies for the development of conductive hydrogels **a** formation of a hydrogel through the utilisation of a conductive infill suspension, **b** polymerisation of the polymer within the hydrogel monomer backbone; crosslinking of electrically conductive polymers by dopant molecules; and molecular self-assembly of graphene hydrogel through supramolecular interaction

These hydrogels exhibited remarkable conductivity, magnetism, toughness, and flexibility. To create the final composite by radical polymerisation, the synthesis was carried out using a hydrothermal technique, which was then followed by immersion in an aniline solution [20].

3.3 Dopant Molecules Crosslink Electrically Conductive Polymers

The conductivity of the ECH can be improved by incorporating a dopant, such as phytic acid, into the conductive polymer integrated within the non-conductive hydrogel matrix. A complex three-dimensional crosslinked microstructure is produced when phytic acid protonates amino groups, which in turn promotes the bonding of numerous PANi chains. Angstroms (10^{-10}) to nanometres (10^{-9}) and beyond to microns (10^{-6}) are the pore widths shown by this structure. Graphene, graphite, graphene oxide, metal nanoparticles, and conductive monomers like PANi and polypyrrole (Ppy) are integrated into a non-conductive framework to create a conducting composite, as shown in Fig. 1c according to the procedures outlined in the previous section. Using a supramolecular technique that integrates a dopant, Wang and colleagues [21] outlined an easy way to synthesise a conductive hydrogel composed of PPy-CuPcTs. Copper phthalocyanine-3,4′,4″,4-tetrasulfonic acid tetrasodium salt (CuPcTs) was used to crosslink the conductive monomer PPy. By acting as both a crosslinker and a dopant molecule, CuPcTs facilitate the self-assembly of a nanostructured hydrogel from PPy monomer. The CuPcT functional groups and the lengthy chain of PPy were able to synthesise the hydrogel via electrostatic interactions and hydrogen bonding. A self-sorting mechanism is used to align the polypyrrole (PPy) chain with the hydrogel, resulting in the formation of a cohesive one-dimensional matrix. In contrast to the traditional methods of conducting monomer and nanofibre synthesis, this method presents a new way to create hierarchically interconnected nanofibre monolithic structures in situ.

3.4 The Molecular Self-Assembly of Graphene Hydrogel Through Supramolecular Interaction

The development of ECH using graphene-based nanocomposites is performed by the hydrothermal method, which involves exposing graphene-based composites to elevated temperatures (Fig. 1d). Xu and his colleagues synthesised a self-assembled graphene hydrogel via a one-step method [22]. Following the synthetic methodology, just the graphene solution was introduced into a Teflon-lined autoclave, where the reaction mixture was maintained at 180 °C for 12 h. The graphene hydrogel was synthesised using hydrothermal reduction and supramolecular bonding mechanisms, including electrostatic interactions and π-π stacking. The characteristics of graphene hydrogel depend on the precursor, indicating that hydrogel formation is rare at lower concentrations. At elevated concentrations, the hydrogel may effectively develop owing to enough contacts within the two-dimensional network. The SGH demonstrates exceptional specific capacitance, in addition to its electrical conductivity, mechanical strength, and thermal stability. Jiang and associates created a hybrid hydrogel electrode made of chitosan and graphene oxide, including a porous network intended to improve the ultrahigh energy density of a flexible supercapacitor [23]. The hybrid hydrogel was synthesised in a single step via a microwave hydrothermal technique. The M-Cs and H-GO were integrated into the Teflon-lined autoclave using the hydrothermal technique to produce a hybrid hydrogel composed of chitosan and graphene oxide. The resultant compound was labelled MCHG; the solid substance was cast using liquid nitrogen (N_2) and then freeze-dried for additional characterisation, culminating in MCHG-D. The results indicated that chemical pretreatments produce binary hydrogels with conductive porous networks and heteroatomic functional groups. During the acylation of chitosan, amides and carboxyl groups are added to improve its solubility in water, promoting an efficient reaction. Moreover, the oxidation etching process of graphene oxide in the defect region using H_2O_2 promotes the development of in-plane nanopore networks, which provide a short ion diffusion channel and a substantial active surface area. The hydrogel's surface is abundant in nitrogen and oxygen functional groups, resulting in a fast enhancement in capacitance. The hierarchically porous network and in-plane nanopores enhance wettability, charge transfer, and charge storage capacity.

Zhang and associates created a composite hydrogel composed of nanocellulose and N_2O co-doped graphene, intended for symmetric supercapacitors, demonstrating exceptional gravimetric and volumetric performance [24]. The extraction technique produced a nanocellulose (NC) solution generated from maize stalks. The solution obtained via NC was further subjected to rotary evaporation, followed by the incorporation of GO via a one-step hydrothermal reaction in a Teflon-lined autoclave. Consequently, the final product, NGCHs, was obtained, and the graphene hydrogel was synthesised under the same circumstances as the final composite by the addition of ammonia (NH_3). The preparation demonstrated that the substantial concentration of GO, attributed to the π–π stacking interaction, was the principal cause leading to the reduced density of NGCHs. Ammonia functioned as a nitrogen dopant, a reducing agent, and a modifier of the material's structure. NC functioned as a spacer, successfully inhibiting the excessive aggregation of graphene by the electrostatic contact generated between its oxygen-rich surface and the GO sheet.

4 Fundamentals of Electrically Conductive Hydrogels

Electrically conductive hydrogels (ECHs) represent a distinctive category of soft materials, combining the high water content and three-dimensional polymer structure of traditional hydrogels with the ability to conduct electricity. This combination allows for the conduction of electricity while preserving the flexible, hydrated, and biocompatible characteristics of hydrogels [25]. Electrochemical capacitors can be created by integrating conductive materials—like conductive polymers (PANi, Ppy, Poly(3,4-ethylenedioxythiophene) (PEDOT)), carbon-based nanostructures (graphene, carbon nanotubes (CNTs)), or metallic nanoparticles—into hydrophilic polymer matrices [25].

Integrated conductive phases in ECHs facilitate the transmission of electrical impulses, hence enhancing applications in sensors, actuators, bioelectronics, and energy devices. ECHs demonstrate a broad conductivity spectrum (10^{-6}–10^{2} S/m) influenced by the kind and arrangement of conductive fillers [26]. Their elevated water content (70–95%) provides soft, tissue-like mechanical characteristics and facilitates ion or molecule movement [27]. These hydrogels possess elasticity, flexibility, and self-healing properties, making them appropriate for wearable and implantable devices. Biocompatible polymers such as cellulose, chitosan, and gelatin provide secure biomedical applications. Electrochemical cells transmit electricity via ionic or electronic channels, or a combination of both, hence improving performance. They may react to stimuli characteristics by modifications in polymer composition, crosslinking, filler content, and ambient conditions [4, 27].

Electrical conductivity in hydrogels is facilitated by two primary mechanisms that stem ionic conductivity, which involves charge transport through the movement of ions in the hydrogel's aqueous phase, and electronic conductivity, which is enabled by embedded conductive materials such as conductive polymers, carbon nanomaterials, and metal nanoparticles that allow electron flow [28]. Certain hydrogels have mixed conductivity, integrating ionic and electronic channels, therefore synergistically improving their electrochemical performance. The prevailing mechanism is contingent upon the nature of the conductive fillers, their dispersion within the polymer matrix, and the hydration status of the hydrogel. These methods ascertain the hydrogel's appropriateness for applications like biosensors, energy storage devices, and bioelectronics [25].

The 3D network structure of hydrogels is what makes them special and allows them to be used in so many different ways. A 3D structure contains interconnected porous channels that allow water to flow in rapidly, enable ions and biomolecules to diffuse swiftly, and facilitate quick swelling or deswelling [4, 10]. This structure lets hydrogels act like the extracellular matrix of real tissues, which is important for tissue engineering and regenerative medicine because it helps cells stick together, grow, and move [29]. The 3D network also helps the hydrogel stay strong, flexible, and adjustable in stiffness, which are all important for soft robotics, medication delivery systems, sensors, and actuators. You may carefully tune the porosity and mechanical behaviour of the 3D structure to satisfy specific functional needs by changing the crosslinking density and polymer composition [29].

5 Strategies for Imparting Electrical Conductivity

5.1 Electrical Percolation

Electrical percolation refers to the process by which an insulating matrix (such as a polymer) attains electrical conductivity upon the incorporation of conductive fillers (including carbon black (CB), CNTs, graphene, and metal particles) beyond a certain critical concentration known as the percolation threshold [30, 31]. Below this threshold, conductive particles are segregated: electrons cannot transfer between them; hence, the composite remains an insulator. Above it, sufficient particles are in contact or sufficiently proximate to facilitate the passage of electrons across an interconnected network, therefore creating a "percolating" channel. This results in a rapid, pronounced escalation of electrical conductivity. Percolation theory, derived from statistical physics, elucidates the emergence of connection in stochastic systems. In three-dimensional systems, the threshold is contingent upon the distribution, morphology, and dimensions of the filler. The electrical conductivity σ at the threshold behaves like a power law [32]:

$$\sigma \propto (\phi - \phi_c)^{\mathrm{t}}$$

where, ϕ is volume fraction of conductive filler, ϕ_c is percolation threshold, t is critical exponent (varies with system dimensionality; typically ranges from 1.3 to 2.0 in 3D composites).

There are several things that may affect the percolation threshold, which is the point at which a system changes from being an insulator to a conductor because a linked network forms. These parameters fluctuate based on the particular system under consideration such as(i) aspect ratio of fillers: long, slender fillers (nanotubes and fibres) establish networks at far lower loadings than spherical particles, thereby substantially decreasing the percolation threshold, (ii) dispersion and distribution of fillers: evenly distributed fillers effectively create networks, (iii) agglomeration increases the percolation threshold, hence diminishing conductivity, (iv) interfacial interactions: optimal compatibility between the matrix and filler enhances dispersion and reduces the percolation threshold, and dimensionality and orientation: two-dimensional or aligned networks (e.g., aligned carbon nanotubes) may percolate at lower thresholds compared to random 3D dispersions. The majority of conductive hydrogel studies failed to provide a percolation threshold [33]. The individuals providing commentary on percolation threshold concentration are organised in the following section.

5.2 *Electrical Percolation Studies in Hydrogels*

Electrical percolation studies in hydrogels focus on how the electrical conductivity of hydrogel composites changes with the concentration of conductive fillers. These studies are crucial for designing hydrogels with tunable electrical properties for use in flexible electronics, biosensors, actuators, and tissue engineering [25]. It is well established that hydrogels are 3D polymeric networks capable of absorbing large amounts of water [1]. On their own, they are usually poor conductors. However, incorporating conductive nanofillers turns them into conductive hydrogels, which can respond to electrical stimuli or transmit signals. For example, the electrical percolation of the hydrogel occurs at a low permeability threshold, and the hydrogel electrode exhibits a flexibility of 288 kPa alongside a high conductivity of 135.75 S/m [34].

At the critical filler concentration, nanocomposite hydrogels form a continuous conductive network, greatly improving electrical conductivity. The threshold relies on nanofiller, aspect ratio, hydrogel matrix dispersion, and polymeric network type. High-aspect-ratio CNTs may form conductive channels at modest filler loadings. Agarose–double-walled CNT hydrogels at 1% w/w had no percolation threshold, suggesting that the nanotubes were poorly connected or disseminated to form a network [35]. Macdonald et al. reported no percolation threshold in collagen–single-walled CNT (SWNT) composites at 0.008% w/w, suggesting that the protein matrix's structural and interfacial properties may hinder network growth [36]. These findings demonstrate that conductive hydrogels are challenging to make and that filler–matrix interactions, dispersion techniques, and hydrogel design must be optimised for percolation at actual filler loadings [36].

Research on solid nanocomposite polymers indicates a consistent electrical percolation threshold. This phenomenon can be elucidated by four variables. (1) There is a limited body of research on nanocomposite hydrogels. (2) Carbon-based materials exhibit low percolation thresholds, are hydrophobic, and tend to aggregate, resulting in poor dispersions in aqueous environments. (3) Polymer chains may encase nano-objects, thereby restricting direct contact between the conductive phase. (4) Hydrogels represent more intricate systems, comprising at least three components. This impedes research on percolation thresholds. Increasing the content of nanofillers in various nanocomposite hydrogels results in a modest increase in conductivity, rather than a significant enhancement by multiple orders of magnitude [37]. The following part discusses the common conductive fillers found in hydrogel matrices (Table 1).

Table 1 Common hydrogel conductive fillers

Filler type	Representative materials	Typical electrical conductivity range	Typical percolation threshold	Mechanism of conductivity
Conductive polymers	PPy, PANi, PEDOT	10^{-2} to 10^2 S/cm (depending on doping and morphology)	15–30 wt% (when used as dispersed phase)	Intrinsic conductivity within polymer chains
Carbon-based	MWCNTs, graphene	10^2 to 10^4 S/cm (MWCNTs), 10^2 to 10^5 S/cm (graphene)	0.1–1.5 vol%	Electron transfer and quantum tunnelling
Metallic nanomaterials	Ag, Au, Cu NPs	10^5 to 10^6 S/cm (bulk)	0.05–0.5 vol%	Metallic conductivity
MXenes	$Ti_3C_2T_x$ Nb_2C, V_2C	10^3 to 10^4 S/cm (few-layer flakes)	1–5 wt%	Enhanced surface conductivity

6 Conductive Polymers

The incorporation of conductive polymers into polysaccharide-based hydrogels has created new opportunities for the advancement of electroactive biomaterials applicable in tissue engineering, biosensors, and drug delivery. Conductive polymers, including Ppy, PANi, PEDOT, and others, possess inherent electrical conductivity that facilitates signal transmission and electrochemical reactivity inside typically insulating polysaccharide matrices [2, 38]. When integrated with natural polysaccharides, these conductive polymers yield composite hydrogels that combine the biocompatibility, hydrophilicity, and biodegradability of polysaccharides with the electrical properties of synthetic polymers [1, 10]. The inclusion may be accomplished by in situ polymerisation, blending, or layer-by-layer assembly, yielding 3D networks that facilitate electrical conduction while preserving elevated water content and mechanical flexibility. These hybrid hydrogels provide a synergistic foundation for the development of next-generation soft electronics [28].

6.1 Polypyrrole

Polypyrrole (PPy) is a prominent intrinsically conductive polymer valued for its high electrical conductivity, environmental stability, and biocompatibility. Despite its useful properties, PPy is inherently brittle and lacks mechanical flexibility, limiting its broader application. To address these limitations, composite materials integrating PPy with natural polysaccharides have been developed [39]. These PPy–polysaccharide composites combine the electrical functionality of PPy with the biodegradability, hydrophilicity, and mechanical reinforcement provided by the polysaccharide matrix. The resulting materials exhibit enhanced processability and biological compatibility, making them suitable for diverse applications including biosensors, drug delivery systems, tissue engineering scaffolds, flexible electronics, and energy storage devices [40]. Fabrication is typically achieved via in situ polymerization or blending techniques, allowing for structural versatility (e.g., films, coatings, hydrogels). While promising, the development of these composites requires careful control of conductivity, dispersion, and long-term stability, particularly in biological environments [41]. PPy–polysaccharide composites present a promising class of green, functional materials with a wide range of uses from bioelectronics to energy storage and biomedicine. Their combination unites electrical functionality with natural, renewable, and biodegradable support matrices, offering a pathway toward sustainable advanced materials [42].

6.2 Polyaniline

Polyaniline (PANi) has become one of the most adaptable conducting polymers in recent decades, owing to its distinctive amalgamation of electrical conductivity, environmental stability, and relative simplicity of production. As a result, aniline has attracted significant scientific attention, especially in polymer chemistry, where its incorporation with polysaccharides has broadened its practical uses [43]. The integration of PANi with naturally sourced polysaccharides boosts the mechanical and structural attributes of the composite while concurrently augmenting its electrical performance and environmental resilience. Researchers are jointly evaluating this composite as a potential energy storage system, sensor, and corrosion-resistant covering [44]. This section provides a comprehensive review of the chemical and physical properties of PANi, the synthesis methods of PANi-derived polysaccharides, the electrical conductivity characteristics of these materials, their potential uses, and an assessment of their environmental durability. The discussion seeks to include international researchers in polymer chemistry, concentrating on viral details and critical evaluations derived only from relevant literature studies. PANi is characterised by several oxidative states that significantly affect its electrical and chemical characteristics. The three primary oxidation states of aniline are fully reduced leucoemeraldine, partially reduced emeraldine, and completely oxidised pernigraniline [45]. The half-reduced emeraldine form of PANi demonstrates conductivity, with its conductive and electroactive properties being affected by the benzenoid and quinoid units.

6.3 Poly(3,4-ethylenedioxythiophene)

Poly(3,4-ethylenedioxythiophene) (PEDOT) is an extensively researched conductive polymer, known for its significant electrical conductivity, remarkable environmental stability, and optical transparency when doped. The properties of PEDOT render it particularly suitable for use in electronics, sensors, and environmental remediation [46, 47]. Nevertheless, pristine PEDOT exhibits limitations, including poor water solubility, brittleness, and restricted mechanical robustness, which constrain its application in aqueous and flexible systems. Researchers have investigated the integration of PEDOT with natural polysaccharides, including cellulose, chitosan, alginate, and starch, leading to hybrid composites that exhibit improved physicochemical and functional properties [48]. The PEDOT-based polysaccharide materials integrate the high conductivity of PEDOT with the biocompatibility, biodegradability, and structural tunability of polysaccharides, resulting in multifunctional platforms for environmental remediation [49].

Polysaccharides fulfil multiple essential functions: they serve as dispersing matrices to enhance the processability of PEDOT, offer mechanical reinforcement, and introduce functional groups (e.g., hydroxyl, amino, and carboxyl) that promote pollutant binding. Hybrid materials have shown significant efficacy in the removal of various environmental contaminants, such as heavy metals, organic dyes, and pharmaceutical residues [48]. PEDOT-based hydrogels and aerogels demonstrate significant potential owing to their high porosity, extensive surface area, and electroactive network structure. The identified characteristics facilitate swift pollutant adsorption and improve regeneration cycles. The combination of PEDOT's conductive properties with the adsorption capabilities of the polysaccharide matrix improves electron transfer and catalytic activity in electrochemical applications [50].

7 Carbon-Based Nanomaterials

The integration of carbon-based nanoparticles into polysaccharide-based hydrogels has facilitated the development of multifunctional materials [51]. The combination of the inherent biocompatibility and biodegradability of natural polysaccharides with the mechanical strength, electrical conductivity, and chemical tunability of carbon-based nanomaterials yields hybrid hydrogels with exceptional properties, rendering them suitable for biomedical, environmental, and electronic applications [52].

Carbon-based materials—such as graphene, CNTs, CB, and CQDs—have attracted significant attention as functional nanofillers in polymeric systems owing to their exceptional electrical conductivity, mechanical strength, thermal stability, and extensive surface area. The integration of these carbonaceous nanomaterials into polysaccharide matrices or hydrogel networks markedly improves the physicochemical properties of the composite, resulting in enhanced performance in applications including sensors, energy storage, flexible electronics, and environmental remediation [53, 54]. Graphene and GO: Monolayer or few-layer carbon sheets featuring sp^2-hybridized networks. Graphene provides exceptional conductivity, whereas GO incorporates oxygen-containing functional groups (–OH, –COOH, −epoxy) that enhance dispersion and adhesion within hydrophilic polysaccharide matrices [55]. On the other hand, CNTs are cylindrical nanostructures formed by the rolling of graphene sheets, which can be either single-walled or multi-walled. CNTs demonstrate exceptional tensile strength and electron mobility, making them extensively used in the production of mechanically reinforced, conductive polymer composites [56]. CB: an amorphous, cost-effective conductive particle commonly used in extensive composites. CB improves conductivity and mechanical rigidity, particularly in bulk hydrogels. CQDs: Zero-dimensional (0D) fluorescent nanoparticles that integrate conductivity with adjustable optical characteristics. CQDs are frequently integrated for concurrent sensing and imaging applications in conductive biogels [57].

Polysaccharides have reactive functional groups that facilitate strong physical or chemical interactions with carbon nanofillers. These interactions ensure efficient dispersion and interfacial adhesion, essential for uniform conductivity and mechanical enhancement. The incorporation of carbon fillers into polysaccharide matrices improves the composite's electrical and thermal conductivity, elasticity, swelling capacity, and responsiveness to external stimuli (e.g., pH, strain, temperature), which is crucial for sensor and actuator design [58]. Hydrogels are three-dimensional, hydrophilic polymer networks that can retain substantial quantities of water while preserving structural integrity. Pristine polysaccharide hydrogels generally have inadequate electrical conductivity, restricting their utility in electrochemical and bioelectronic applications. The integration of carbon-based fillers into these hydrogels addresses this issue by creating conductive channels, which allow for the establishment of percolating networks that enhance electron transport [59]. For instance, GO–chitosan hydrogels demonstrate improved conductivity, antibacterial efficacy, and pollutant adsorption capability [60]. CNT-cellulose hydrogels provide enhanced mechanical strength and electrochemical performance, suitable for supercapacitors and biosensors [61]. CQD-alginate hydrogels integrate fluorescence and electrical responsiveness for dual-mode sensing applications [62]. The ultimate composite characteristics are significantly influenced by dispersion quality, interfacial adhesion, filler content, and network architecture.

8 Metal/Metal Oxide Nanoparticles

8.1 Metallic Nanoparticles

Metallic nanoparticles have become a crucial category of nanomaterials owing to their distinctive physicochemical characteristics, such as a high surface area-to-volume ratio, superior electrical conductivity, adjustable optical properties, and catalytic efficacy. Widely researched metal nanoparticles (MNPs), including gold (Au), silver (Ag), copper (Cu), platinum (Pt), and palladium (Pd), have been widely used in several domains such as biosensing, catalysis, drug delivery, and environmental remediation [10]. Their nanoscale size promotes improved surface interactions with chemical substances, making them suitable for pollution detection, adsorption, and degradation purposes.

Notwithstanding their advantageous characteristics, the stability and dispersion of metallic nanoparticles in aqueous settings provide a problem owing to their propensity to agglomerate. To resolve this issue, researchers have concentrated on integrating magnetic nanoparticles into polymeric matrices, especially conductive polymers, which offer a stable and conductive milieu to sustain nanoparticle dispersion and accessibility. The collaboration of magnetic nanoparticles and conductive polymers produces hybrid materials with improved electrochemical, mechanical, and adsorption properties. Su and colleagues [63] synthesised a HAPAM/CNCs@CNTs nanocomposite hydrogel, inspired by the structure of human tissues, using partially crystalline fiber-like cellulose nanocrystals and carbon nanotubes (CNCs@CNTs). This research chooses carbon nanotubes as the conductive element because to their fibrous structure and exceptional conductivity. Concurrently, highly crystalline amphiphilic CNCs are used to distribute CNTs, producing a homogeneous CNCs@CNTs aqueous nanohybrid. This structure comprises elongated, linear carbon nanotubes (CNTs) completely surrounded by rod-shaped cellulose nanocrystals (CNCs), significantly improving dispersion stability and forming a partially crystalline conductive network with a nanoscale architecture, thereby greatly enhancing the hydrogel's conductivity [63].

Polysaccharides, enhance the functioning of these composites by providing a biocompatible, biodegradable, and hydrophilic support matrix. Their plethora of hydroxyl, amino, and carboxyl groups enhance robust binding interactions with metallic nanoparticles and target contaminants, while augmenting the processability and structural integrity of the hybrid material. These multifunctional composites are especially beneficial for environmental remediation, where the concurrent identification and removal of pollutants is often necessary. PEDOT is an extensively researched conductive polymer, known for its superior electrical conductivity, remarkable environmental resilience, and optical transparency in its doped form. The aforementioned qualities render PEDOT very appealing for use in electronics, sensors, and environmental cleanup [64]. Nevertheless, virgin PEDOT has drawbacks like inadequate water solubility, brittleness, and restricted mechanical resilience, which constrain its use in watery and flexible systems [64].

To address these issues, researchers have investigated the amalgamation of PEDOT with natural polysaccharides—namely cellulose, chitosan, alginate, and starch—yielding hybrid composites with improved physicochemical and functional attributes [48]. These PEDOT-based polysaccharide materials integrate the exceptional conductivity of PEDOT with the biocompatibility, biodegradability, and structural adaptability of polysaccharides, resulting in multifunctional platforms for environmental remediation.

8.2 Metal Oxide Nanoparticles

Metal oxide nanoparticles (MONPs) have garnered considerable interest in various scientific fields owing to their remarkable physicochemical characteristics, including elevated surface area, adjustable bandgap, redox activity, and photocatalytic capabilities. Prevalent metal oxide nanoparticles (MONPs) encompass zinc oxide (ZnO), titanium dioxide (TiO_2), iron oxides (Fe_3O_4, Fe_2O_3), copper oxide (CuO), manganese dioxide (MnO_2), cerium dioxide (CeO_2), and aluminium oxide (Al_2O_3), all of which demonstrate multifunctional properties suitable for applications in sensing, catalysis, environmental remediation, and biomedical technologies [65]. The nanoscale size of MONPs facilitates increased surface contacts with contaminants, enhanced adsorption kinetics, and superior dispersion within composite matrices. Standalone MONPs frequently encounter difficulties like aggregation, instability in aquatic environments, and challenges in post-use recovery. Recent advancements aim to mitigate these limitations by integrating MONPs into conductive polymer and polysaccharide frameworks, yielding hybrid materials with complementary structural and functional benefits [66]. MONPs demonstrate distinctive semiconducting and photocatalytic properties. ZnO and TiO_2 nanoparticles are recognised for their effective UV absorption and production of reactive oxygen species (ROS), which facilitate the breakdown of organic contaminants [67]. Iron oxide nanoparticles possess superior magnetic characteristics for their efficient separation and recovery from treated water, whereas MnO_2 and CuO nanoparticles enhance oxidative and redox-based removal mechanisms for organic and inorganic pollutants [68].

The inherent features of MONPs render them exceptional candidates for incorporation into intelligent composite materials for environmental and electrochemical applications. Conductive polymers have been extensively used as matrices to stabilise MONPs. These polymers improve electrical conductivity, flexibility, and stability of the composite, whereas MONPs augment surface area, catalytic activity, and pollutant adsorption [69]. The resultant hybrid materials exhibit enhanced efficacy for electrochemical sensing, capacitive deionisation, and photocatalytic pollutant degradation. TiO_2–PANI composites have enhanced photocatalytic activity under visible light owing to the reduced bandgap and effective charge separation between the polymer and metal oxide [70]. Fe_3O_4–PPy nanocomposites are employed for the removal of magnetic pollutants and the electrochemical detection of heavy metals, owing to their redox properties and elevated conductivity [71]. Chitosan is commonly used for its intrinsic antibacterial properties and its ability to chelate metal ions. When integrated with MONPs and conductive polymers, it functions as a stabiliser, crosslinker, and functional scaffold. In an alkaline solution, a sensor based on chitosan–Fe_3O_4–PANI hydrogel nanocomposites showed a decreased sensitivity of 16.64 μA M with a detection limit of 3.58 μM and response and recovery times of 0.6 s [52].

9 MXenes

MXenes represent a novel category of two-dimensional (2D) nanomaterials, garnering considerable interest for their distinctive properties, including electrical conductivity, mechanical flexibility, and chemical tunability [72]. The synthesis of these materials generally involves the selective etching of the A element from MAX phases, which are layered ternary carbides or nitrides characterised by the general formula $M_{n+1}AX_n$, where n ranges from 1 to 4. In this structure, M denotes an early transition metal (e.g., Ti, V, Nb), A signifies a group 13 or 14 element (e.g., Al, Si), and X represents carbon, nitrogen, or both [73]. The etching process eliminates the A-layer, resulting in ultrathin 2D layers, typically comprising 3–9 atomic layers in thickness, with lateral dimensions varying from several microns to tens of microns [72, 74]. The delaminated sheets exhibit a significant surface area and adjustable surface chemistry, typically ending with functional groups like –OH, –O, or –F, resulting from the etching process. The surface terminations of MXenes exhibit inherent hydrophilicity, which enhances their dispersion in aqueous media and compatibility with hydrophilic polymer networks [74].

MXenes demonstrate a wide spectrum of electrical conductivities, influenced by their composition and surface functionalisation. Certain compositions, including $Ti_3C_2T_n$, demonstrate metallic characteristics with conductivities reaching 2.4×10^6 S·m^{-1}, exceeding those of numerous other 2D materials [75]. The exceptional conductivity, along with their significant aspect ratio and hydrophilicity, positions MXenes as suitable candidates for integration into electrically conductive nanocomposite hydrogels, especially in the contexts of soft electronics and biomedical sensing platforms [74]. Recent studies demonstrate the successful integration of MXenes into hydrogel matrices for the development of flexible and stretchable strain sensors capable of monitoring subtle human motions or physiological signals [74]. The MXene–hydrogel composites exhibit enhanced properties due to the combined mechanical elasticity of the hydrogel and the rapid signal transduction facilitated by the MXene phase. One notable limitation of MXenes is their metastable nature in aqueous environments. Exposure to water and oxygen results in gradual oxidation and degradation, particularly in monolayer or few-layer forms, thereby compromising their long-term electrical performance and biocompatibility [72]. Consequently, current research aims to enhance environmental stability through methods such as encapsulation, surface modification, and the incorporation of antioxidant additives in hydrogel systems. Li et al. [76] engineered a flexible, self-adhesive, and conductive hydrogel by integrating SF-coated MXene into a polyacrylamide (PAM) matrix. The PAM/(MXene-SF) hydrogel has a pliable structure defined by MBAA-covalently crosslinked PAM chains. The existence of many non-covalent reversible interactions between the SF functional groups on MXene nanosheets and PAM chains enhanced the interfacial bonding between MXene and the hydrogel network, leading to the PAM/(MXene-SF) hydrogel demonstrating exceptional stretchability and strength. Thermal compensation tests used the PAM/(MXene-SF) hydrogel on two copper electrodes, which were secured using VHB tape. The PAM/(MXene-SF) hydrogel has significant electrical conductivity (0.25 S/m) and strain sensitivity throughout a wide spectrum of stresses, making it appropriate for the detection of both substantial and little human motions. This research illustrates the prospective uses of conductive MXene hydrogel-based epidermal sensors in wearable electronics, diagnostics, and health monitoring [76].

10 Technological Applications of Conductive Polysaccharides

10.1 Tissue Engineering

Tissue engineering is an innovative interdisciplinary field that uses living cells, chemical and physical bioelements, and biodegradable materials to create tissue-like structures. The artificial scaffold can be employed in vivo in tissue engineering to repair injuries or replace dysfunctional tissues, organs, and congenital defects. Conductive hydrogels can mimic the electrical characteristics of tissue, attracting significant attention. In this context, various conductive materials, such as metals, conductive polymers, and organonanoparticles, have been utilised to create electroactive hydrogels. The interest in developing tissue engineering scaffolds using chitosan stems from its remarkable physicochemical and biological properties. The characteristics encompass in vitro and in vivo cytocompatibility, tunable mechanical and biodegradation properties through modification, mild processing conditions, the development of interconnected porous structures, and antibacterial activity linked to its cationic nature. The presence of functional side groups, such as amine and hydroxyl, promotes attachment to other molecules, thereby enhancing interaction with adhesive proteins and receptors [77].

Huang et al. [78] investigated the effects of electrical stimulation on Schwann cells derived from the sciatic nerves of neonatal Sprague–Dawley rats, utilising an electrically conductive scaffold composed of 2.5 wt% PPy and 97.5 wt% CS. PPy nanoparticles were synthesised through a microemulsion polymerisation technique, utilising dodecylbenzene sulfonic acid and sodium salt as the emulsifier and $FeCl_3$ as the oxidising agent. The PPy/CS scaffold was synthesised by processing a solution in a 1% (v/v) acetic acid aqueous medium. The scaffold produced exhibits an electrical conductivity of 10^{-3} S cm^{-1} [78]. Chitosan has been identified as a viable scaffold material for bone regeneration based on its in vitro properties. Chitosan scaffolds exhibit promise as viable options for cardiac tissue engineering and regeneration through the precise modulation of their mechanical properties [79].

10.2 Cardiac and Muscle Tissue Repair

Globally, cardiovascular diseases, particularly myocardial infarction (MI), are the primary causes of morbidity and mortality. Cell delivery has been investigated for its potential to regenerate cardiac tissue. The inadequate post-transplantation survival rate of administered cells complicates cell-based treatment. The majority of cells perish upon injection into the infarcted myocardium due to the adverse environment characterised by elevated inflammatory cytokines, oxidative stress, and insufficient nutrients [80]. Biocompatible cell encapsulation has potential for enhancing cell viability and therapeutic efficacy. Encapsulation safeguards cardiomyocytes from detrimental factors and facilitates a gradual, controlled release at the site of injury [81]. Therapeutic cells are enclosed inside hydrogels, scaffolds, and cardiac patches. The high water content, tissue-like elasticity, and extracellular matrix mimicking of hydrogels make them attractive [82].

Neat (cell-free) hydrogels improve endogenous heart cell adhesion, proliferation, and tissue remodelling during myocardial infarction, as well as facilitate cell transport. These hydrogels may facilitate cellular infiltration and tissue regeneration as transient matrices. Designing a hydrogel for cardiac tissue engineering necessitates consideration of the physiological milieu and the mechanical and electrical requirements of heart tissue. The myocardium is subjected to mechanical stress from contraction and relaxation cycles, as well as electrical communication that regulates synchronised heartbeats. Consequently, CTE hydrogels must possess electrical conductivity, mechanical strength, biocompatibility, and degradability. They must facilitate functional integration with the host cardiac tissue and maintain structural integrity under cyclic stress [83].

Biomimetic, multifunctional hydrogels that emulate the normal cardiac environment are essential for regenerative therapies in heart disease. Lv et al. [84] recently reviewed the design and application of electroconductive hydrogels (ECHs) for the treatment of MI, a primary contributor to heart failure [84]. After myocardial infarction, the electrical conductivity of the heart is compromised, leading to impaired cardiac function. Researchers are developing electrically conductive hydrogels that integrate biocompatibility and flexibility to restore electrical signalling and facilitate tissue regeneration. Recent studies indicate that ECHs containing therapeutic agents yield better results in the treatment of MI than non-drug-loaded ECHs, as illustrated in Fig. 2. The authors systematically elucidate that ECHs provide a promising platform for cardiac tissue engineering, particularly in myocardial infarction repair, by restoring electrical function and supporting cardiac regeneration. Challenges persist regarding long-term safety, standardisation, and clinical translation [84].

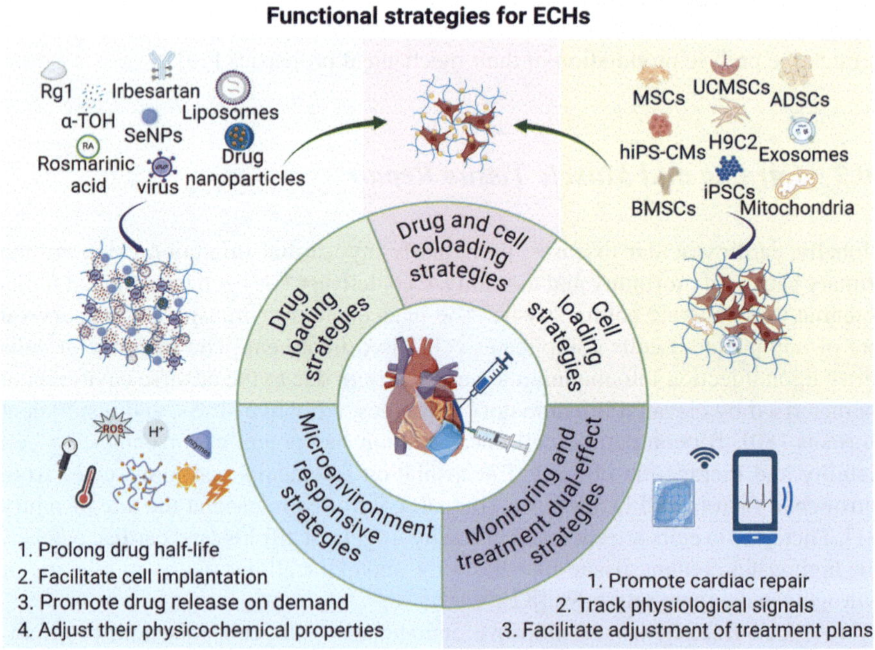

Fig. 2 A schematic representation of the functional strategies and implantation techniques for engineered cardiac tissues in MI repair [84]

Recent MI treatments use antioxidant compounds in electroconductive hydrogels (ECHs) to mend the heart. Integrating conductive and antioxidant aniline oligomer tetraaniline (TA) with selenium nanoparticles (SeNPs) and α-tocopherol (Vitamin E) may minimise oxidative stress. SeNPs and alginate-hyaluronic acid (HA)/TA hydrogels with DPCA@PDA and MMP-SP offer therapeutic benefits. TA-modified ECHs reduced infarct size by 33.8% with DPCA@PDA and 17.1% with SeNPs compared to unloaded controls [85]. TA, polydopamine (PDA), and MMP-SP suppress ROS, inflammation, and fibroblast proliferation, improving efficacy. The MMP-SP components influence angiogenesis by controlling hydrogel disintegration, DPCA release, and HIF-1α stabilisation [85]. A PDA/α-tocopherol-loaded PEDOT:PSS hydrogel improved left ventricular ejection fraction (EF) and reduced infarct size in rabbit models of MI compared to baseline. PEDOT:PSS, PDA, and α-tocopherol have anti-oxidative and anti-apoptotic capabilities, suggesting multifunctional ECHs have therapeutic promise in cardiac tissue regeneration [86].

This section addresses the intricacies of cardiac disease care, highlighting the need for a balanced strategy that takes into account clinical results, patient-centred considerations, and unique patient circumstances. The text addresses several illnesses, beginning with ischaemic heart disease, where evidence favours early medicinal treatment over invasive interventions for stable people and reserving revascularisation for those with ongoing symptoms or certain anatomical factors. In acute coronary syndromes, prompt reperfusion therapies are essential, especially for ST-elevation myocardial infarction. A thorough pharmacological regimen is necessary for heart failure with low ejection fraction, with devices acting as adjuncts. The therapy of atrial fibrillation emphasises stroke risk and symptomatology, advocating early catheter ablation in symptomatic individuals. This chapter emphasises the pivotal function of transcatheter aortic valve replacement in aortic stenosis and the need for mechanical circulatory support in severe instances. Ultimately, it promotes evidence-based treatments, personalised treatment plans, and the assessment of patient-centered outcomes in clinical decision-making.

10.3 Wound Healing and Skin Regeneration

Wound healing is the biological process of repairing damaged skin and tissue after an injury. The process consists of many stages: inflammation, tissue creation, and tissue remodelling, all of which seek to restore the skin's protective barrier and functioning. Conductive hydrogels used in wound healing are synthesised by creating a soft, biocompatible polymer matrix, usually integrating ingredients such as chitosan, gelatin, or hyaluronic acid, together with conductive components such Ppy, PANi, GO, or Ag NPs. The crosslinking of these hydrogels guarantees structural integrity while promoting electrical conductivity, thereby improving cell motility, proliferation, and tissue regeneration. Amino ($-NH_2$), hydroxyl (–OH), and carboxyl (–COOH) functional groups are crucial for their cellular connections, hydration augmentation, antibacterial support, and activation of healing-related cellular

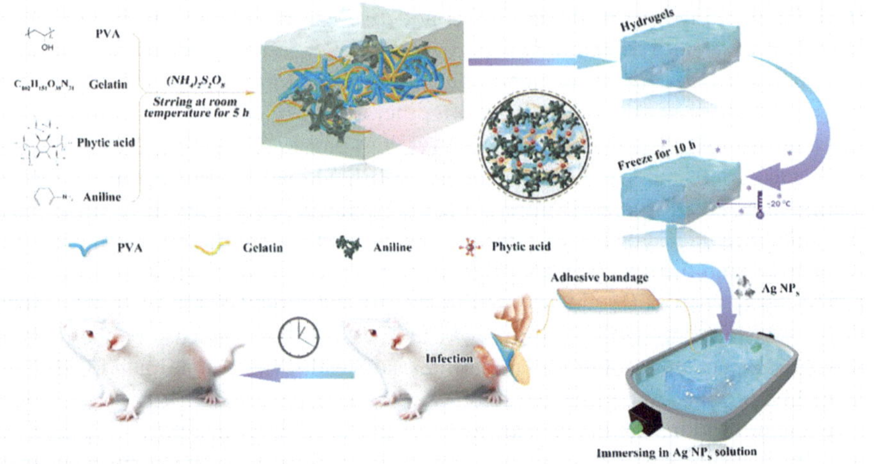

Fig. 3 Preparation of Ag NPs/CPH medical hydrogel and its use in a novel animal model for treating infected wounds [87]

signalling pathways. Xiao et al. [87] formulated a silver nanoparticles/collagen peptide hydrogel for wound healing (Fig. 3) [87]. This composite hydrogel integrates silver nanoparticles with a conductive polymeric hydrogel to enhance therapeutic dressings. Silver nanoparticles are potent antibacterial agents that inhibit microorganisms responsible for wound infections. The biocompatibility, mechanical strength, and electrical conductivity of the CPH matrix facilitate cell growth and migration. Tissue healing and epithelialisation need a moist wound environment, which the gel provides. It suppresses pro-inflammatory cytokines and modulates the immune response to lessen inflammation. These properties make Ag NPs/CPH hydrogel a promising wound treatment alternative since they stimulate tissue growth and repair.

Typical skin conductivity ranges from 10^{-7} to 2.6×10^{-3} S cm^{-1}. Research shows that after a skin defect, the positive charge in the wound interacts with the negative charge surrounding the skin to create an endogenous electric field that helps macrophages, neutrophils, and keratinocytes migrate during wound closure. However, winding defects typically make endogenous electric field transfer inefficient. Supplemental EF stimulation facilitates migration and proliferation of epithelial cells and fibroblasts by activating ion channels and transduction signals, promoting angiogenesis and immune regulation [88].

Cytotoxicity testing protocols are essential for evaluating the biocompatibility of conductive polysaccharide-based hydrogels, which are crucial for their clinical application. Testing adheres to international standards like ISO 10993 and ASTM F748–16 [88]. Initial assessments involve in vitro cytotoxicity assays, including MTT or Alamar Blue assays to measure cell metabolic activity, Live/Dead staining for cell viability, and LDH leakage assays for membrane integrity [89]. While polysaccharide hydrogels generally exhibit biocompatibility, the addition of conductive fillers can lead to varied cytotoxic responses, necessitating careful dose-response profiling [90]. For hydrogels intended for vascular applications, hemocompatibility tests, including hemolysis and platelet adhesion assays, are critical, alongside studies on macrophage activation and cytokine release to evaluate inflammatory responses. In vivo studies follow, assessing tissue integration and biodegradation through small-animal implantation, with histological analyses to quantify responses. Preclinical development must ensure physicochemical stability and compliance with toxicology guidelines before clinical trials, which are categorized into phases assessing safety, proof-of-concept, and comparative efficacy. Regulatory classification influences the evaluation stringency, with post-market surveillance focusing on long-term performance and compatibility.

10.4 Biosensors and Wearable Devices

The multifunctional polysaccharide-based hydrogel in wearable electronics provides excellent sensitivity, tissue adhesion, conductivity, and on-demand removability, hence improving healthcare monitoring devices in next-generation wearable technology. The wearable sensor, using a robust conductive polysaccharide-based hydrogel, has exceptional flexibility, self-repair capabilities, fatigue resistance, low detection threshold, high conductivity, and sensitivity for monitoring animals and human motion both in real-time and remotely [91]. These materials have considerable promise in electrochemical biosensors, with tuneable properties achieved by precise molecular engineering and microstructural configuration. Wearable biosensors hold significant appeal due to their ability to offer a non-invasive or minimally invasive and convenient method for monitoring physiological biomarkers present in bodily fluids. Human biofluids, including saliva, tears, urine, and sweat. Their hydrated matrix lets target molecules move quickly, and their conductive network lets electrons move quickly to the detecting electrode. Conductive polysaccharide-based hydrogels are facilitating sustainable, skin-compatible, and highly efficient biosensing and wearable devices, connecting soft biointerfaces with dependable electrical performance. Wearable sweat biosensors have developed as a groundbreaking device that facilitates the collection and analysis of sweat constituents on the body. These gadgets, often affixed to the body, monitor, communicate, and engage with individuals, providing a diverse array of applications from health monitoring to immersive experiences (Fig. 4).

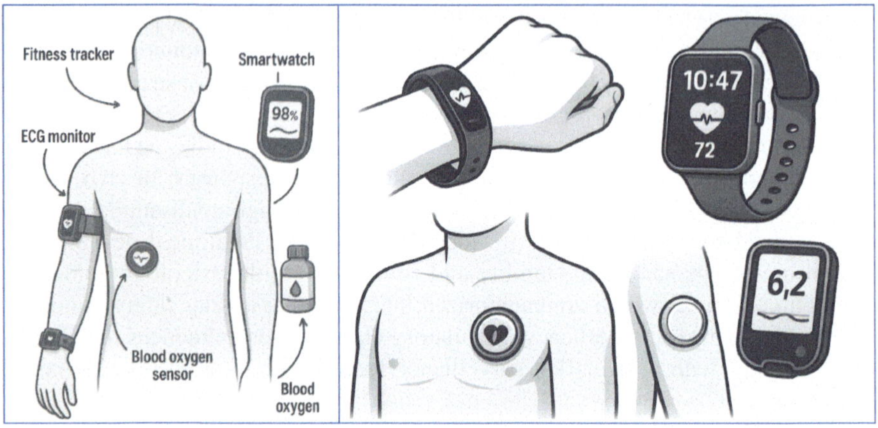

Fig. 4 Various wearable electronics devices used in healthcare

10.5 Electrochemical Sensors

Electrochemical sensors can be categorised into various types, including amperometric, potentiometric, and impedimetric, each tailored to meet specific analytical needs [92]. Amperometric sensors function by detecting variations in current, while potentiometric sensors operate on the principle of measuring changes in voltage within the system. Impedimetric sensors, on the other hand, are concerned with assessing alterations in electrical resistance and capacitance [93]. Common methods for detecting encompass cyclic voltammetry (CV), amperometry, differential pulse voltammetry (DPV), and electrochemical impedance spectroscopy (EIS) [93]. Wang et al. [95] formulated a sodium alginate/PAM composite hydrogel incorporating glucose oxidase for glucose detection in honey and a mineraliser, utilising EIS across a broad temperature range. The 3D composite hydrogel was designed to enhance the chemical stability of the sensor [94].

In recent years, gas sensors have found applications in ecological detection, food safety prevention, medical diagnostics, and industrial monitoring. Flexible gas sensors are increasingly favoured by researchers owing to their thinness, biocompatibility, portability, durability, transparency, and safety, making them suitable for wearable electronics. Flexible gas sensors using diverse sensing materials can identify hazardous gases such as ammonia (NH_3), nitrogen dioxide (NO_2), hydrogen sulphide (H_2S), nitric oxide (NO) and hydrogen (H_2) at ambient temperature. The potential decline in the detection capabilities of flexible gas sensors under certain conditions is a notable worry nowadays. Research demonstrates that conductive polysaccharide-based hydrogel may react to toxic gases at ambient temperature due to their chemical properties and exposed active regions. For example, Seleka et al. [52] developed a conductive hydrogel composed of carrageenan (Car), polyvinyl alcohol (PVA), and polypyrrole (PPy) that could be used as an electrochemical hydrogen gas sensor at ambient temperature. Adding PPy, the most conductive monomer in

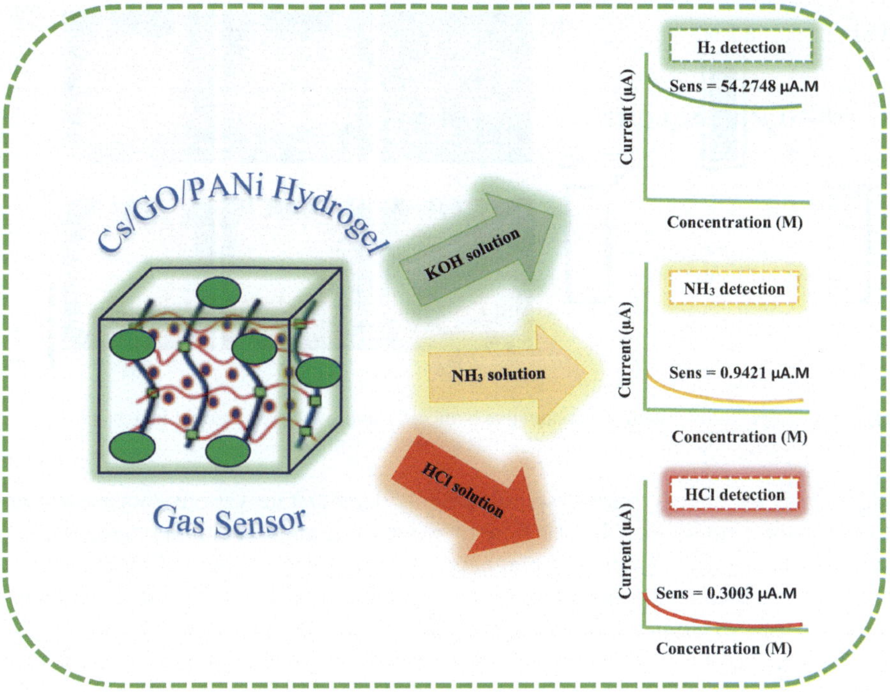

Fig. 5 Illustration of the Cs/GO/PANi hydrogel sensor material and its response sensitivity towards various hydrogen sources [96]

its class, greatly improved electrical conductivity and created a stable network for transferring electrons. This improvement made the sensor very sensitive, with a detection limit of 3.435 μM and a sensitivity of 58.468 $\mu A \cdot M^{-1}$. Car, PVA, and PPy worked together in a way that made the response and recovery times extremely rapid (1.1 s and 1.6 s, respectively), which is better than what you would get with regular hydrogel-based sensors [95]. In a similar research, Seleka et al. created a Cs/GO/PANi composite hydrogel for electrochemical hydrogen detection at ambient temperature using KOH solution. Polyaniline increased the number of active adsorption sites in the Cs/GO hydrogel, overcoming its site constraints. Exposure to HCl and NH_3 solutions caused partial blockage of active sites owing to impurities. Figure 5 shows that the composite had a hydrogen response sensitivity of 54.7248 $\mu A \cdot M^{-1}$, much greater than HCl (0.3003 $\mu A \cdot M^{-1}$) and NH_3 (0.9421 $\mu A \cdot M^{-1}$) [96].

10.6 Stimuli-Responsive Hydrogel

Stimuli-responsive hydrogels are revolutionising wearable electronics by offering adaptation to various environmental stimuli, hence improving flexibility, durability,

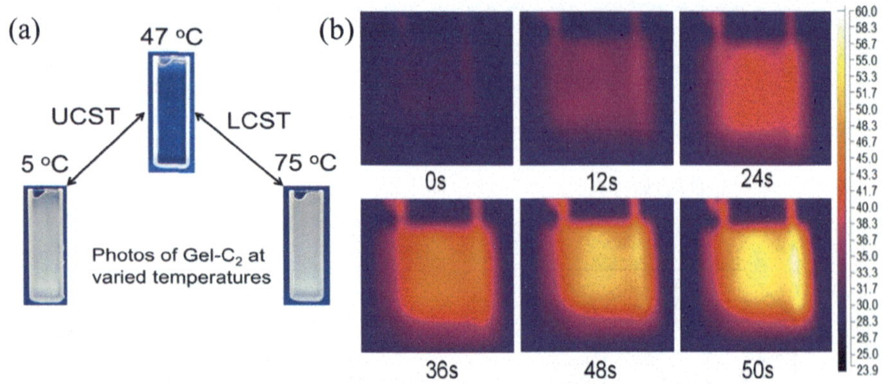

Fig. 6 **a** Temperature-responsive conductive hydrogels, and **b** infrared thermal images [98]

and overall usefulness. Responsive hydrogels demonstrate various morphological and structural alterations in reaction to external stimuli, including temperature, humidity, near-infrared light, pH, magnetic fields, and electrical signals. The changes encompass alterations in transparency, colour, and the extent of bending. Conductive hydrogels exhibiting responsive properties have significant potential for applications in actuators, smart displays, and controlled drug release [45]. A temperature-responsive conductive hydrogel by grafting methyl cellulose with acrylamide using ceric ion initiation [97]. This hydrogel exhibited both LCST (low critical solution temperature) and UCST (upper critical solution temperature) behavior, enabling reversible transitions between transparent (47 °C) and opaque states (5 and 75 °C) (Fig. 6a). The opacity changes were due to hydrogen bonding interactions at low and high temperatures. Additionally, they created a device for rapid, uniform electric heating of the hydrogel. Infrared imaging and transmittance tests showed that opacity switching occurred within 25 s at 20 V, demonstrating potential for smart display applications controlled by voltage Fig. 6b [98].

10.7 Wastewater Treatment

Environmental pollution has become a pressing worldwide concern, necessitating immediate action owing to its extensive effects on human health and ecological equilibrium. Water contamination has emerged as a particularly worrying kind of pollution in recent years. This decline is mostly due to fast industrialisation and increasing human densities, resulting in the ongoing release of toxic compounds into aquatic habitats. The quality of drinking water has dramatically declined. Prevalent pollutants—including heavy metal ions, synthetic colours, pharmaceutical residues, and phenolic compounds—are often non-biodegradable and persist in the environment. These compounds cannot be efficiently eliminated using standard water treatment techniques, and their buildup in biological organisms presents significant health hazards, including chronic and potentially fatal illnesses. Researchers globally are exerting significant effort to provide new, precise, and efficient analytical methods for the detection, measurement, and elimination of harmful pollutants from water systems. The progress of these technologies is crucial for protecting public health and maintaining the sustainability of natural resources. Numerous studies on conducting polysaccharides for water remediation have recently been published, demonstrating the intrinsic properties of conducting polymers and polysaccharides. Researchers' attention has lately been directed to the use of conductive polysaccharides in water treatment. Elella et al. [99] have emphasised the value of integrating recent developments in conductive polymers, including Ppy, PANi, and polythiophene, with natural polysaccharides for the objective of environmental remediation [99]. Their review explored the development and application of conductive polymer-modified polysaccharide-based materials, such as hydrogels, aerogels, nanofibers, and nanocomposites. Hybrid materials demonstrate considerable efficacy in the removal of various toxic pollutants, such as heavy metals, synthetic dyes, and pharmaceutical residues from contaminated environments. The study emphasised advancements in synthesis strategies, structural optimisation, and performance evaluation over the last decade. Key parameters, such as maximum adsorption capacities, adsorption mechanisms, kinetic models, and adsorption-desorption behaviours, are critically examined, as they are essential for understanding and improving the efficiency of these materials in practical environmental applications. Their study provided a significant foundation for directing future research and technological advancements in sustainable pollution management [99]. Liang and colleagues investigated the application of an MXene/PANi/sodium alginate composite gel for the removal of Cu(II) and Hg(II) from aqueous solutions across a wide pH range, attributed to the proton transfer effect characteristic of PANi [100]. The adsorption mechanism of the prepared adsorbent included physical adsorption, electrostatic attraction, ion exchange, complexation, and redox reaction. Upon integration into the adsorption site, Cu(II) and Hg(II) experienced reduction and complexation, aided by electron-donating groups including C=O, COH, C=N, $-NH_2$, and –NH–. The adsorption mechanisms outlined above are illustrated in Fig. 7.

11 Scalability and Long-Term Stability

The scalability and enduring stability of electrically conductive polysaccharide-based 3D hydrogels are essential for their advancement from laboratory settings to industrial and therapeutic applications. Critical considerations affecting these elements include polymer breakdown rates, shelf-life characteristics, and processing viability during upscaling. The deterioration of these hydrogels is determined by the hydrolytic or enzymatic vulnerability of the polysaccharide matrix and the stability of conductive elements such as polyaniline and graphene fillers. Although conductive fillers

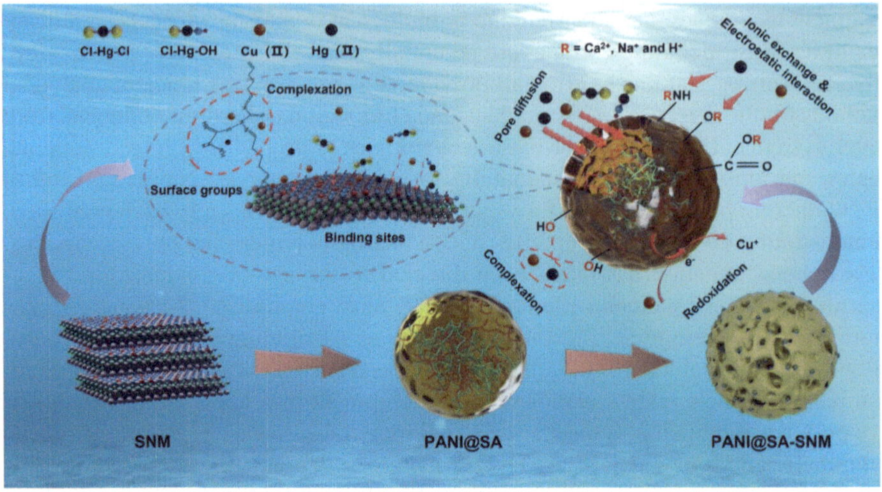

Fig. 7 Adsorption mechanism of the MXene/PANi/sodium alginate composite gel for the removal of Cu(II) and Hg(II) [100]

might improve stability by preventing degradation, extended exposure to physiological circumstances may destabilise these polymers, thereby impacting conductivity. Kinetic modelling suggests that degradation adheres to pseudo-first-order kinetics, with higher filler loads resulting in uneven degradation zones. Conductive hydrogels may maintain 60–80% of their original conductivity in saltwater over many weeks, surpassing unmodified gels.

The stability of shelf life is influenced by susceptibility to moisture and oxygen, with encapsulation and antioxidants aiding in its extension. Cryostabilisation techniques may maintain polysaccharide matrices; however, excessive freeze-thaw cycles may diminish conductivity. Optimised hydrogels often possess a shelf life of 6–12 months when stored under refrigeration. Increasing production encounters obstacles like material variability, control of crosslinking, and the need for uniform processing techniques such as 3D printing. Future research must concentrate on multi-scale stabilisation solutions, predictive deterioration modelling, and standardised ageing assessments to guarantee economic viability and regulatory compliance.

12 Conclusion

Electrically conductive polysaccharide-based 3D hydrogels signify notable progress in smart materials, integrating the eco-friendliness and biocompatibility of natural polymers with the electrical responsiveness of conductive elements. This chapter examines the integration of conductive elements, including conducting polymers, metallic and metal oxide nanoparticles, and carbon-based nanomaterials, into polysaccharide hydrogels, resulting in the development of multifunctional platforms exhibiting exceptional physicochemical, electrochemical, and mechanical properties. Hybrid hydrogels provide a distinctive blend of structural flexibility, ionic and electronic conductivity, adjustable porosity, and responsiveness to stimuli, rendering them applicable in various fields such as biosensing, tissue engineering, drug delivery, energy storage, and environmental remediation. Recent advancements in fabrication techniques, including in situ polymerisation, freeze-drying, and 3D printing, have enhanced control over the architecture and functionality of these materials, thereby broadening their range of applications. Despite these advancements, challenges persist regarding scalability, long-term stability, mechanical durability, and cost-effectiveness, particularly during the transition from laboratory research to practical applications. Furthermore, achieving a balance between electrical conductivity, biodegradability, and biocompatibility necessitates meticulous material design and processing. Future research should concentrate on designing fully bio-based conductive systems, developing self-healing and recyclable hydrogel matrices, and integrating these materials into flexible electronic and sensing devices. The integration of material science, nanotechnology, and bioengineering will play a crucial role in developing the next generation of eco-friendly, high-performance conductive hydrogels, addressing the increasing needs in healthcare, environmental monitoring, and sustainable electronics.

Future prospective

The future of electrically conductive polysaccharide-based 3D hydrogels depends on progress in materials chemistry, polymer physics, and bioengineering to improve electronic and ionic transport. Essential domains for the forthcoming investigation include the systematic design of molecular structures and crosslinking techniques to achieve adjustable conductivity and mechanical durability, all while preserving biocompatibility. Understanding how charge moves through materials using sophisticated characterisation methods is essential for material optimisation. The use of standardised performance assessment parameters will enable cross-study comparisons and expedite the deployment of these materials in industrial and biomedical contexts. Next-generation hydrogels must demonstrate long-term stability and have multifunctional characteristics for use in implantable devices and flexible energy storage. Collaboration among chemists, materials scientists, and biomedical engineers, in conjunction with collaborations with doctors and industry, is crucial for connecting laboratory advances with scalable production. Ultimately, an emphasis on sustainable and bioinspired development, using renewable polysaccharide sources and eco-friendly manufacturing techniques, will integrate forthcoming research with global sustainability objectives. By tackling these problems, the discipline may progress towards practical applications that enhance both healthcare and technology.

Acknowledgments The authors extend their appreciation to the Department of Chemistry at the University of Limpopo. The authors are appreciative of the financial assistance provided by the National Research Foundation of South Africa through the Thuthuka program (Grant No. TTK240408212927).

References

1. Makhado E, Pandey S, Nomngongo PN, Ramontja J. New horizons in wastewaters management: emerging monitoring and remediation, hydrogel nanocomposites: innovations in nanotechnology for water treatment. Hauppauge, NY: Nova Science Publishers; 2019. p. 87–113.
2. Ghosh S, Kumar N, Chattopadhyay S. Electrically conductive "SMART" hydrogels for on-demand drug delivery. Asian J Pharm Sci. 2025;20(1):101007.
3. Li Y, Ding X, Hu H, Xu FJ. Stimulus-responsive polysaccharide-based hydrogels: from design to biomedical applications. Prec Med Eng. 2024;1(1):100001.
4. Bashir S, Hina M, Iqbal J, Rajpar AH, Mujtaba MA, Alghamdi NA, Wageh S, Ramesh K, Ramesh S. Fundamental concepts of hydrogels: synthesis, properties, and their applications. Polymers. 2020;12(11):2702.
5. Karchoubi F, Ghotli RA, Pahlevani H, Salehi MB. New insights into nanocomposite hydrogels; a review on recent advances in characteristics and applications. Adv Industr Eng Polym Res. 2024;7(1):54–78.
6. Jiang M, Fu Q, Li T, Hou R, Cheng K, Lv S, Yang X, Xue P. Natural polysaccharide and protein-based hydrogels: a novel class of materials for sustainable agricultural development. Int J Biol Macromol. 2025;314:144435.
7. Sharma G, Verma A, García-Peñas A, Kumar A, Dhiman P, Wang T, Amirian J. Polysaccharide-based biopolymeric magnetic hydrogels for remediation of antibiotics from aqueous solution. Int J Biol Macromol. 2024;283:137555.
8. Tang L, Xie S, Wang D, Wei Y, Ji X, Wang Y, Zhao N, Mou Z, Li B, Sun WR, Wang PY. Astragalus polysaccharide/carboxymethyl chitosan/sodium alginate based electroconductive hydrogels for diabetic wound healing and muscle function assessment. Carbohydr Polym. 2025;350:123058.
9. Zhou B, Jiang X, Zhou X, Tan W, Luo H, Lei S, Yang Y. GelMA-based bioactive hydrogel scaffolds with multiple bone defect repair functions: therapeutic strategies and recent advances. Biomater Res. 2023;27(1):86.
10. Pandey S, Makhado E, Kim S, Kang M. Recent developments of polysaccharide based superabsorbent nanocomposite for organic dye contamination removal from wastewater—a review. Environ Res. 2023;15(217):114909.
11. Junaid PM, Dar AH, Dash KK, Rohilla S, Islam RU, Shams R, Pandey VK, Srivastava S, Panesar PS, Zaidi S. Polysaccharide-based hydrogels for microencapsulation of bioactive compounds: a review. J Agric Food Res. 2024;15:101038.
12. Moradifar F, Sepahdoost N, Tavakoli P, Mirzapoor A. Multi-functional dressings for recovery and screenable treatment of wounds: a review. Heliyon 2025;11(1).
13. Malektaj H, Drozdov AD, deClaville CJ. Mechanical properties of alginate hydrogels cross-linked with multivalent cations. Polymers. 2023;15(14):3012.
14. Duceac IA, Stanciu MC, Nechifor M, Tanasă F, Teacă CA. Insights on some polysaccharide gel type materials and their structural peculiarities. Gels. 2022;8(12):771.
15. Rumon MM, Rahman MS, Akib AA, Sohag MS, Rakib MR, Khan MA, Yesmin F, Shakil MS, Rahman Khan MM. Progress in hydrogel toughening: addressing structural and crosslinking challenges for biomedical applications. Discov Mater. 2025;5(1):5.
16. Román-Doval R, Torres-Arellanes SP, Tenorio-Barajas AY, Gómez-Sánchez A, Valencia-Lazcano AA. Chitosan: properties and its application in agriculture in context of molecular weight. Polymers. 2023;15(13):2867.

17. Zhu T, Mao J, Cheng Y, Liu H, Lv L, Ge M, Li S, Huang J, Chen Z, Li H, Yang L. Recent progress of polysaccharide-based hydrogel interfaces for wound healing and tissue engineering. Adv Mater Interfaces. 2019;6(17):1900761.
18. Jin X, Jiang H, Li G, Fu B, Bao X, Wang Z, Hu Q. Stretchable, conductive PAni-PAAm-GOCS hydrogels with excellent mechanical strength, strain sensitivity and skin affinity. Chem Eng J. 2020;394:124901.
19. Li Y, Gong Q, Liu X, Xia Z, Yang Y, Chen C, Qian C. Wide temperature-tolerant polyaniline/cellulose/polyacrylamide hydrogels for high-performance supercapacitors and motion sensors. Carbohydr Polym. 2021;267:118207.
20. Hu H, Zhong X, Yang S, Fu H. Tough and stretchable Fe_3O_4/MoS_2/PAni composite hydrogels with conductive and magnetic properties. Compos Part B. 2020;182:107623.
21. Wang Y, Shi Y, Pan L, Ding Y, Zhao Y, Li Y, Shi Y, Yu G. Dopant-enabled supramolecular approach for controlled synthesis of nanostructured conductive polymer hydrogels. Nano Lett. 2015;15(11):7736–41.
22. Xu M, Wang A, Xiang Y, Niu J. Biomass-based porous carbon/graphene self-assembled composite aerogels for high-rate performance supercapacitor. J Clean Prod. 2021 Sep;15(315):128110.
23. Jiang C, Gao M, Zhang S, Huang L, Yu S, Song Z, Wu Q. Chitosan/graphene oxide hybrid hydrogel electrode with porous network boosting ultrahigh energy density flexible supercapacitor. Int J Biol Macromol. 2023;225:1437–48.
24. Zhang Y, Wei L, Liu X, Ma W, Lou C, Wang J, Fan S. Nanocellulose/N, O co-doped graphene composite hydrogels for high gravimetric and volumetric performance symmetric supercapacitors. Int J Hydrog Energy. 2022;47(79):33827–38.
25. Wang J, Yang B, Jiang Z, Liu Y, Zhou L, Liu Z, Tang L. Recent advances of conductive hydrogels for flexible electronics. Electron Mater. 2024;5(3):101–31.
26. Zhu Y, Chen B, Liu Y, Tan T, Gao B, Lu L, Zhu P, Mao Y. Recent advances in conductive hydrogels for electronic skin and healthcare monitoring. Biosensors 2025.
27. Shin Y, Lee HS, Jeong H, Kim DH. Recent advances in conductive hydrogels for soft biointegrated electronics: materials, properties, and device applications. Wear Electron. 2024.
28. Tang H, Li Y, Liao S, Liu H, Qiao Y, Zhou J. Multifunctional conductive hydrogel interface for bioelectronic recording and stimulation. Adv Healthc Mater. 2024;13(22):2400562.
29. El-Sherbiny IM, Yacoub MH. Hydrogel scaffolds for tissue engineering: progress and challenges. Global Cardiol Sci Pract. 2013;2013(3):38.
30. Antunes RA, De Oliveira MC, Ett G, Ett V. Carbon materials in composite bipolar plates for polymer electrolyte membrane fuel cells: a review of the main challenges to improve electrical performance. J Power Sources. 2011;196(6):2945–61.
31. Zare Y, Rhee KY. A power model to predict the electrical conductivity of CNT reinforced nanocomposites by considering interphase, networks and tunneling condition. Compos Part B. 2018;155:11–8.
32. Stauffer D, Aharony A. Introduction to percolation theory. Taylor & Francis; 2018.
33. Li C, Thostenson ET, Chou TW. Sensors and actuators based on carbon nanotubes and their composites: a review. Compos Sci Technol. 2008;68(6):1227–49.
34. Guan M, Han Z, Liu N, Zhou Z, Qu X, Zhang T, Chen S, Wang H. Electrical percolation network based on nano-cellulose template for flexible hydrogel bioelectrode. Carbohydr Polym. 2025;362:123693.
35. Guillet JF, Valdez-Nava Z, Golzio M, Flahaut E. Electrical properties of double-wall carbon nanotubes nanocomposite hydrogels. Carbon. 2019;146:542–8.
36. MacDonald RA, Voge CM, Kariolis M, Stegemann JP. Carbon nanotubes increase the electrical conductivity of fibroblast-seeded collagen hydrogels. Acta Biomater. 2008;4(6):1583–92.
37. Kougkolos G, Golzio M, Laudebat L, Valdez-Nava Z, Flahaut E. Hydrogels with electrically conductive nanomaterials for biomedical applications. J Mater Chem B. 2023;11(10):2036–62.

38. Li H, Cao J, Wan R, Feig VR, Tringides CM, Xu J, Yuk H, Lu B. PEDOTs-based conductive hydrogels: design, fabrications, and applications. Adv Mater. 2025;37(7):2415151.
39. Janata J, Josowicz M. Conducting polymers in electronic chemical sensors. Nat Mater. 2003;2(1):19–24.
40. Vafa ZJ, Zare EN, Eslam MR, Makvandi P. Development of ecofriendly, biodegradable electrically conductive double-layer bio-hydrogel nanocomposite for sustainable medical device applications. Adv Compos Hybrid Mater. 2025;8(2):174.
41. Cheng T, Xiang Y, He X, Pang J, Zhu W, Luo L, Cao Y, Pei R. Nanostructured conductive polymers: synthesis and application in biomedicine. J Mater Chem B. 2025.
42. Chowdhury MS, Oliullah MS, Islam RT, Hurayra MA, Al Mahmud MZ, Hasan N, Rahman MK, Islam MS, Khan J, Hossain N. Biomaterials for energy storage: synthesis, properties, and performance. Green Technol Sustainability. 2025;3(2):100152.
43. Goswami S, Nandy S, Fortunato E, Martins R. Polyaniline and its composites engineering: a class of multifunctional smart energy materials. J Solid State Chem. 2023 Jan;1(317):123679.
44. Masemola CM, Moloto N, Tetana Z, Linganiso LZ, Motaung TE, Linganiso-Dziike EC. Advances in polyaniline-based composites for room-temperature Chemiresistor gas sensors. PRO. 2025;13(2):401.
45. Seleka WM, Ramohlola KE, Modibane KD, Makhado E. Quaternary conducting Cs/GO/PANi hydrogel composites: a smart material for room temperature hydrogen sensing. Diam Relat Mater. 2024;146:111156.
46. Groenendaal L, Jonas F, Freitag D, Pielartzik H, Reynolds JR. Poly (3, 4-ethylenedioxythiophene) and its derivatives: past, present, and future. Adv Mater. 2000;12(7):481–94.
47. Kirchmeyer S, Reuter K. Scientific importance, properties and growing applications of poly (3, 4-ethylenedioxythiophene). J Mater Chem. 2005;15(21):2077–88.
48. Liu L, Wang Y, Li X, Li P, Jin Y, Yu H, Yan M, Hu Y. Recent developments in polysaccharide-based conductive hydrogels: a review and perspective. Carbohydr Polym. 2025;22:124453.
49. Shi H, Dai Z, Sheng X, Xia D, Shao P, Yang L, Luo X. Conducting polymer hydrogels as a sustainable platform for advanced energy, biomedical and environmental applications. Sci Total Environ. 2021;10(786):147430.
50. Bhat MA, Rather RA, Shalla AH. PEDOT and PEDOT: PSS conducting polymeric hydrogels: a report on their emerging applications. Synth Met. 2021 Mar;1(273):116709.
51. Radia ND, Aljeboree AM, Mhammed AA. Enhanced removal of crystal violet from aqueous solution using carrageenan hydrogel nanocomposite/MWCNTs. Inorg Chem Commun. 2024;167:112803.
52. Seleka WM, Makhado E, Kganyakgo LK, Mofokeng LE, Makwakwa D, Botlhoko OJ. Development of a rapid responsive conductive electrochemical sensor for sensitive hydrogen detection: chitosan-based GO/Fe_3O_4/PANi hydrogel nanocomposite. Int J Hydrog Energy. 2025.
53. Geim AK, Novoselov KS. The rise of graphene. Nat Mater. 2007;6(3):183–91.
54. Teffu DM, Makhafola MD, Ndipingwi MM, Makhado E, Hato MJ, Iwuoha EI, Modibane KD, Makgopa K. Interrogation of electrochemical performance of reduced graphene oxide/metal-organic framework hybrid for asymmetric Supercabattery application. Electroanalysis. 2020;32(12):2827–37.
55. Makhado E, Pandey S, Ramontja J. Microwave assisted synthesis of xanthan gum-cl-poly (acrylic acid) based-reduced graphene oxide hydrogel composite for adsorption of methylene blue and methyl violet from aqueous solution. Int J Biol Macromol. 2018;119:255–69.
56. De Volder MF, Tawfick SH, Baughman RH, Hart AJ. Carbon nanotubes: present and future commercial applications. Science. 2013;339(6119):535–9.
57. Guo J, Tsou CH, Yu Y, Wu CS, Zhang X, Chen Z, Yang T, Ge F, Liu P, Guzman MR. Conductivity and mechanical properties of carbon black-reinforced poly (lactic acid)(PLA/CB) composites. Iran Polym J. 2021;30(12):1251–62.
58. Khan B, Abdullah S, Khan S. Current progress in conductive hydrogels and their applications in wearable bioelectronics and therapeutics. Micromachines. 2023;14(5):1005.

59. Zhan TY, Wang S, Guo ZY, Hu YF. Preparation and application of a stretchable, conductive and temperature-sensitive dual-network nanocomposite hydrogel. J Macromol Sci A. 2022;59(1):72–82.
60. Verma A, Sharma G, Wang T, Kumar A, Dhiman P, Verma Y, Bhaskaralingam A, García-Penas A. Graphene oxide/chitosan hydrogels for removal of antibiotics. Environ Technol. 2025;19:1–31.
61. Sun Z, Qi H, Chen M, Guo S, Huang Z, Maganti S, Murugadoss V, Huang M, Guo Z. Progress in cellulose/carbon nanotube composite flexible electrodes for supercapacitors. Eng Sci. 2021;18(23):59–74.
62. Zhang J, Jin J, Wan J, Jiang S, Wu Y, Wang W, Gong X, Wang H. Quantum dots-based hydrogels for sensing applications. Chem Eng J. 2021;408:127351.
63. Su G, Cao J, Zhang X, Zhang Y, Yin S, Jia L, Guo Q, Zhang X, Zhang J, Zhou T. Human-tissue-inspired anti-fatigue-fracture hydrogel for a sensitive wide-range human–machine interface. J Mater Chem A. 2020;8(4):2074–82.
64. Dong R, Yang M, Zuo Y, Liang L, Xing H, Duan X, Chen S. Conducting polymers-based gas sensors: principles, materials, and applications. Sensors. 2025;25(9):2724.
65. Afzal S, Ullah S, Shahid M, Najam T, Shaaban IA, Shah SS, Nazir MA. Metal oxide nanoparticles: synthesis and applications in energy, biomedical, and environment sector. J Chin Chem Soc. 2025;72(2):96–123.
66. Liu Y, Fu S, Jin K, Cheng Y, Li Y, Zhao Y, Liu R, Tian Y. Advances in polysaccharide-based conductive hydrogel for flexible electronics. Carbohydr Polym. 2025;15(348):122836.
67. Sirelkhatim A, Mahmud S, Seeni A, Kaus NH, Ann LC, Bakhori SK, Hasan H, Mohamad D. Review on zinc oxide nanoparticles: antibacterial activity and toxicity mechanism. Nano-Micro Lett. 2015;7(3):219–42.
68. Aragaw TA, Bogale FM, Aragaw BA. Iron-based nanoparticles in wastewater treatment: a review on synthesis methods, applications, and removal mechanisms. J Saudi Chem Soc. 2021;25(8):101280.
69. Musa AA, Bello A, Adams SM, Onwualu AP, Anye VC, Bello KA, Obianyo II. Nano-enhanced polymer composite materials: a review of current advancements and challenges. Polymers. 2025;17(7):893.
70. Wang F, Min S, Han Y, Feng L. Visible-light-induced photocatalytic degradation of methylene blue with polyaniline-sensitized TiO_2 composite photocatalysts. Superlattice Microst. 2010;48(2):170–80.
71. Alzahrani FM, Alsaiari NS, Katubi KM, Amari A, Ben Rebah F, Tahoon MA. Synthesis of polymer-based magnetic nanocomposite for multi-pollutants removal from water. Polymers. 2021;13(11):1742.
72. Anasori B, Lukatskaya MR, Gogotsi Y. 2D metal carbides and nitrides (MXenes) for energy storage. In: MXenes. Jenny Stanford Publishing; 24 Aug 2023. pp. 677–722.
73. Naguib M, Kurtoglu M, Presser V, Lu J, Niu J, Heon M, Hultman L, Gogotsi Y, Barsoum MW. Two-dimensional nanocrystals produced by exfoliation of Ti_3AlC_2. In: MXenes. Jenny Stanford Publishing; 24 Aug 2023. pp. 15–29.
74. Rasool K, Pandey RP, Rasheed PA, Berdiyorov GR, Mahmoud KA. MXenes for environmental and water treatment applications. In: 2D metal carbides and nitrides (MXenes) structure, properties and applications. Cham: Springer International Publishing. 31 Oct 2019. pp. 417–444.
75. Shahzad F, Alhabeb M, Hatter CB, Anasori B, Man Hong S, Koo CM, Gogotsi Y. Electromagnetic interference shielding with 2D transition metal carbides (MXenes). Science. 2016;353(6304):1137–40.
76. Li N, Wang X, Liu Y, Li Y, Li J, Qin Z, Jiao T. Ultrastretchable, self-adhesive and conductive MXene nanocomposite hydrogel for body-surface temperature distinguishing and electrophysiological signal monitoring. Chem Eng J. 2024;483:149303.
77. Vandghanooni S, Eskandani M. Electrically conductive biomaterials based on natural polysaccharides: challenges and applications in tissue engineering. Int J Biol Macromol. 2019;141:636–62.

78. Huang J, Hu X, Lu L, Ye Z, Zhang Q, Luo Z. Electrical regulation of Schwann cells using conductive polypyrrole/chitosan polymers. J Biomed Mater Res Part A Off J Soc Biomater Japanese Soc Biomater Austr Soc Biomater Korean Soc Biomater. 2010;93(1):164–74.
79. Behrooznia Z, Nourmohammadi J, Mohammadi Z, Shabani F, Mashhadi R. Biological evaluation of 3D-printed chitosan-based scaffolds for tissue engineering. Carbohydr Res. 2025;551:109416.
80. Kamal NH, Heikal LA, Abdallah OY. The future of cardiac repair: a review on cell-free nanotherapies for regenerative myocardial infarction. Drug Deliv Transl Res. 2025;20:1–9.
81. Moreira Souza AC, Grabe-Guimarães A, Cruz JD, Santos-Miranda A, Farah C, Teixeira Oliveira L, Lucas A, Aimond F, Sicard P, Mosqueira VC, Richard S. Mechanisms of artemether toxicity on single cardiomyocytes and protective effect of nanoencapsulation. Br J Pharmacol. 2020;177(19):4448–63.
82. Zoeller K, To D, Bernkop-Schnuerch A. Biomedical applications of functional hydrogels: innovative developments, relevant clinical trials and advanced products. Biomaterials. 2025;312:122718.
83. Li P, Hu J, Wang J, Zhang J, Wang L, Zhang C. The role of hydrogel in cardiac repair and regeneration for myocardial infarction: recent advances and future perspectives. Bioengineering. 2023;10(2):165.
84. Lv Q, Zhou D, He Y, Xu T, Qiu X, Zeng J. Engineering functional electroconductive hydrogels for targeted therapy in myocardial infarction repair. Bioact Mater. 2025;(49):172–92.
85. Wei X, Chen S, Xie T, Chen H, Jin X, Yang J, Sahar S, Huang H, Zhu S, Liu N, Yu C. An MMP-degradable and conductive hydrogel to stabilize HIF-1α for recovering cardiac functions. Theranostics. 2022;12(1):127.
86. Zhang F, Zhang Y, Qian S, Qian X, Jiao J, Ma B, Chen J, Cheng H, Li X, Lin Y, Li H. Injectable and conductive nanomicelle hydrogel with α-tocopherol encapsulation for enhanced myocardial infarction repair. ACS Nano. 2024;18(14):10216–29.
87. Xiao L, Hui F, Tian T, Yan R, Xin J, Zhao X, Jiang Y, Zhang Z, Kuang Y, Li N, Zhao Y. A novel conductive antibacterial nanocomposite hydrogel dressing for healing of severely infected wounds. Front Chem. 2021;24(9):787886.
88. Schuh JC, Funk KA. Compilation of international standards and regulatory guidance documents for evaluation of biomaterials, medical devices, and 3-D printed and regenerative medicine products. Toxicol Pathol. 2019;47(3):344–57.
89. Li W, Zhou J, Xu Y. Study of the in vitro cytotoxicity testing of medical devices. Biomed Rep. 2015;3(5):617–20.
90. Tropp J, Collins CP, Xie X, Daso RE, Mehta AS, Patel SP, Reddy MM, Levin SE, Sun C, Rivnay J. Conducting polymer nanoparticles with intrinsic aqueous dispersibility for conductive hydrogels. Adv Mater. 2024 Jan;36(1):2306691.
91. Ashrafi M, Alonso-Rasgado T, Baguneid M, Bayat A. The efficacy of electrical stimulation in lower extremity cutaneous wound healing: a systematic review. Exp Dermatol. 2017;26(2):171–8.
92. Yang Y, Yao C, Huang WY, Liu CL, Zhang Y. Wearable sensor based on a tough conductive gel for real-time and remote human motion monitoring. ACS Appl Mater Interfaces. 2024;16(9):11957–72.
93. Frigoli M, Krupa MP, Hooyberghs G, Lowdon JW, Cleij TJ, Diliën H, Eersels K, van Grinsven B. Electrochemical sensors for antibiotic detection: a focused review with a brief overview of commercial technologies. Sensors. 2024;24(17):5576.
94. Aftab S, Bilge S. MXene-based electrochemical (bio) sensors for biomolecule detection: a review of recent advances and future perspective. Mater Today Chem. 2025;48:102947.
95. Wang Q, Jiao C, Wang X, Wang Y, Sun K, Li L, Fan Y, Hu L. A hydrogel-based biosensor for stable detection of glucose. Biosens Bioelectron. 2023;221:114908.
96. Seleka WM, Phasha MM, Kganyakgo LK, Makwakwa D, Makhado E. Preparation of a conductive hydrogel based on carrageenan, polyvinyl alcohol, and polypyrrole as a potential room temperature electrochemical hydrogen gas sensor. Microchem J. 2025;29:114091.

97. Zhang D, Ren B, Zhang Y, Xu L, Huang Q, He Y, Li X, Wu J, Yang J, Chen Q, Chang Y. From design to applications of stimuli-responsive hydrogel strain sensors. J Mater Chem B. 2020;8(16):3171–91.
98. Ding Y, Yan Y, Peng Q, Wang B, Xing Y, Hua Z, Wang Z. Multiple stimuli-responsive cellulose hydrogels with tunable LCST and UCST as smart windows. ACS Appl Polym Mater. 2020;2(8):3259–66.
99. Elella MH, Abdallah HM, Ali EA, Makhado E, Abd El-Ghany NA. Recent developments in conductive polysaccharide adsorbent formulations for environmental remediation: a review. Int J Biol Macromol. 2025;11:140915.
100. Liang J, Li X, Wu M, Chen C, Hu Z, Zhao M, Xue Y. MXene/polyaniline/sodium alginate composite gel: adsorption and regeneration studies and application in cu (II) and hg (II) removal. Sep Purif Technol. 2025;19(353):128298.

Polysaccharide-Based 3D Hydrogels for Advanced Food Applications

Mahmood Ahmed, Ali Abbas Aslam, Ahmad Saeed, and Awais Ali Aslam

Abstract Polysaccharide-based 3D hydrogels have emerged as innovative materials in advanced food applications due to their biocompatibility, biodegradability, and unique physicochemical properties. These hydrogels can retain significant amounts of water while maintaining structural integrity, making them suitable for applications such as food packaging, nutrient delivery, and texture modification. This chapter examines the synthesis, characterization, and application of polysaccharide-based 3D hydrogels, highlighting their role in improving food quality, shelf life, and safety. Additionally, the chapter discusses recent advancements in hydrogel engineering, including the integration of bioactive compounds, stimuli-responsive behavior, and smart packaging systems. Future perspectives on scaling up production, regulatory challenges, and commercialization opportunities are also addressed.

Keywords Polysaccharide hydrogels · 3D hydrogels · Food applications · Smart packaging · Nutrient delivery · Biodegradable materials · Sustainable food system

Abbreviations

FDA	Food and Drug Administration
PBHs	Polysaccharide-based 3D Hydrogels
PEG	Polyethylene glycol
TGase	Transglutaminase
GG	Guar Gum
PLA	Polylactic acid
LOM	Laminated Object Manufacturing

M. Ahmed (✉) · A. A. Aslam · A. Saeed
Department of Chemistry, Division of Science and Technology, University of Education, Lahore, Pakistan

A. A. Aslam
Department of Chemical Organic Technology and Petrochemistry, Silesian University of Technology, Gliwice, Poland

M. H. Abu Elella et al. (eds.), *3D Engineered Polysaccharides*,
https://doi.org/10.1007/978-3-032-26154-0_10

PAM	Pressure-assisted Microsyringe
CAD	Computer-aided design
CsNb	Norbornene-functionalized chitosan
KGM	Konjac glucomannan
APu	Aldehyde pullulan
GRAS	Generally recognized as Safe
E. Coli	*Escherichia Coli*
KC	κ-carrageenan
GDL	Glucono-δ-lactone
SDL	Selective deposition lamination
CMC	Carboxymethyl cellulose
PAAc	Poly(acrylic acid)
SMGHs	Soy meal-derived glycine-rich hydrolysates
LCST	Lower critical solution temperature
ADSCs	Adipose mesenchymal stem cells
LCA	Life cycle assessment

1 Introduction

As biodegradable, biocompatible, and high-quality polymers, edible polymers labeled as generally recognized as safe (GRAS) by the U.S. Food and Drug Administration (FDA) are an appropriate substitute for non-biodegradable and low-grade materials in various industries. Specifically, the applications of these polymers in food have been improved and modified due to vast studies on the optimization and customization of these polymers. Polymers form the basis of gels, which may absorb air, water, or oil in their 3D structures, spawning other variants like aerogels, hydrogels, and oleogels [1]. For several decades, these gels have been utilized in the biomedical field, particularly in wound healing [2]. Naturally derived gels should be affordable, renewable, biodegradable, and readily available. However, the increasing exploitation of natural resources has highlighted the need to incorporate new attributes, such as sustainability, structural variety, availability, and compatibility with biological systems, into natural polymers. This is often achieved through diverse techniques aimed at meeting the evolving requirements of the food, agricultural, and pharmaceutical industries [3, 4]. Furthermore, natural polysaccharides present certain drawbacks, including reduced viscosity over time, susceptibility to microbial growth, and an uncontrolled hydration rate. These issues have prompted the use of chemically modified synthetic alternatives, leading to the development of polysaccharide-based hydrogels [5].

Hydrogels are polymeric structures that are characterized by a hydrophilic nature and 3D networks that allow the uptake and storage of significant amounts of liquid. They consist of hydrophilic polymers, which can be natural, semi-synthetic, or synthetic, and are available in various forms and sizes [6]. These materials are formed through cross-linking [7], giving hydrogels the properties that make them viable in encapsulation, targeting, and delivery of therapeutic agents, such as hydrophobic molecules, without affecting the structural integrity. They are therefore often referred to as insoluble swollen solids. Furthermore, hydrogels exhibit antimicrobial qualities, and they have been useful in addressing antimicrobial resistance [8, 9].

Locust bean gum, gum ghatti, guar gum, selecan, xylan, starch, CS, SA, and pectin are examples of naturally occurring polysaccharides that have been developed and modified in countless ways to enhance the hydrogel behavior in a wide range of applications [10, 11]. In the food industry, hydrogels are used as food packaging materials, especially in packing high-moisture foods where humidity control is highly important. They are also useful in food-safety surveillance, food quality improvement, caloric intake control, and managing nutritional content [12]. The Tunable mechanical strength of a hydrogel is usually required by the application of a certain gel, which can be done by the use of physical, chemical, and enzymatic cross-linking methods, and other chemical transformations [13].

Animal, plant, microbial, and algae-derived polysaccharides possess a number of unique characteristics, including biodegradability, biocompatibility, and hydrophilic nature. The fact that they are renewable sources is a concern for the environment and the economy. In their swollen form, polysaccharide-based 3D hydrogels (PBHs) can hold over 90% of water in its free, bound and semi-bound states. Their useful application in various fields, such as pharmaceuticals, agriculture, cosmetics, and food industry, provoked the active research focus of the last several decades, which led to the creation of various structures of PBH [4, 13, 14].

PBHs are particularly useful in controlled drug delivery, toxic compounds and dyes adsorption, and life cell encapsulation scaffolds [15]. Due to the availability of hydrophilic functional groups (–OH, –CONH, $-CONH_2$, $-SO_3H$), these hydrogel formulations also have superior water-retention properties, which makes them a great option as wound-healing and biosensing integration platforms [16]. Their effects cut across a wide range of spheres, including medical to food and agriculture sectors, where their functional applications have been well identified. Their 3D structures promote encapsulation and modulation of texture in food systems, and also play a role in physical crosslinking, which influences moisture control in their 3D frameworks [17].

These 3D hydrogels are hydrophobic, hydrogen-bonded, and ionically cross-linked at the molecular level. It can be observed that PBHs can have varying textures, namely firm or hard gels (dense networks), soft or weak gels (more flexible networks), and fluid-like gels (adaptable, moldable textures), depending on the treatment and formulation [18, 19]. PBHs have not been widely used in the commercial sector, although they have tremendous potential. This is due in part to difficulties in precisely regulating the structure of gels and avoiding the creation of gels of high modulus, which can restrict flexibility. Also, issues related to immunogenicity and stability upon degradation have hindered its wider use [7].

Recent studies have indicated that polysaccharide 3D hydrogels have good control of the texture of the gel than individual polysaccharides. This advantage comes due to their wider range of phase behaviors such as soluble complex formation, coacervation, precipitation, co-solubility, and phase separation. The long-range and short-range molecular forces between proteins/peptides and polysaccharides and variations in their aqueous affinity are the rules of these interactions [20]. The combination of the functional characteristics of peptides and, more especially, their nutritional effect, and the outstanding rheological properties of polysaccharides enables the production of multifunctional composite hydrogels. These hydrogels can be designed to assume various shapes with superior nutritional properties, and thus, they can be considered effective alternatives to be used as functional food gels [21].

This chapter brings together and reviews an enormous number of works and technical literature on polysaccharide 3D hydrogels, making them look like an emergent field of research in itself. It starts with the introduction of the different polysaccharides that are commonly used in the process of forming of hydrogel. It then examines the fabrication methods and physical characteristics of these hydrogels with emphasis on the influence of different intrinsic and extrinsic conditions on these characteristics. The chapter also discusses the gelation processes and the use of polysaccharide 3D hydrogel in the food system. Lastly, it outlines existing issues and suggests possible research avenues in the future.

2 Structural and Functional Properties of Polysaccharide-Based Hydrogels

Polysaccharide-based 3D hydrogels have significant applications in the food and biomedical sectors due to their favorable attributes, such as biodegradability, biocompatibility, and non-toxic behavior at the site of application. In addition to these, hydrogels exhibit several notable characteristics, including their swelling capacity, elasticity, porosity, permeability, mechanical strength in nano-hydrogel form, biological compatibility, bioactivity, structural inhomogeneity, and exceptional absorption capacity under pressure [22]. According to Pushpamalar et al., evaluating mechanical strength and conducting toxicity assessments are crucial steps before hydrogels can be adopted for industrial use [23]. 3D hydrogels can retain approximately 90–99% water within their structure. This high-water content not only contributes to their hydrophilic nature but also allows for the incorporation of free-floating proteins and cells without promoting aggregation. Additionally, the interaction between the water and the polymer matrix, along with internal polymer movements, imparts viscoelastic properties to hydrogels [24].

Hydrogel formation is driven by both physical (weaker, non-covalent interactions) and chemical (strong, covalent interactions) cross-links between polymer chains. These types of bonding significantly influence the gel's properties, particularly how the material deforms after polymer relaxation (Fig. 1). For hydrogels to be effective across different applications, they should exhibit the following key characteristics [25, 26]:

(i) Exceptional absorption capacity
(ii) Appropriate rate of absorption depending on application requirements
(iii) Effective absorption performance under pressure
(iv) Controlled and sustained release of active agents, such as nutrients or drugs
(v) Low levels of unreacted monomers remaining in the product
(vi) Cost-effectiveness

Advanced or "smart" hydrogels incorporate chemical and structural components that allow them to respond to external stimuli such as temperature changes, pH variations, ionic strength shifts, light, magnetic or electric fields, and specific chemicals. These responsive materials are often developed by integrating multiple stimulus-sensitive systems (Fig. 2). When exposed to such triggers, intelligent hydrogels can undergo structural transitions or phase changes, making them suitable for use in various cutting-edge scientific and technological applications [28].

Polysaccharides are widely regarded as safe for food-related applications because of their non-toxic nature. Similarly, biodegradability is a fundamental requirement for hydrogels used in the biomedical field. A hydrogel is considered biodegradable when biological systems can break it down into non-toxic end products. The degradation characteristics are influenced by both the composition and the method of synthesis. Processes like hydrolysis and enzymatic solubilization contribute to the breakdown of hydrogel components. Findings by Ahmad et al. indicate that mechanisms such as

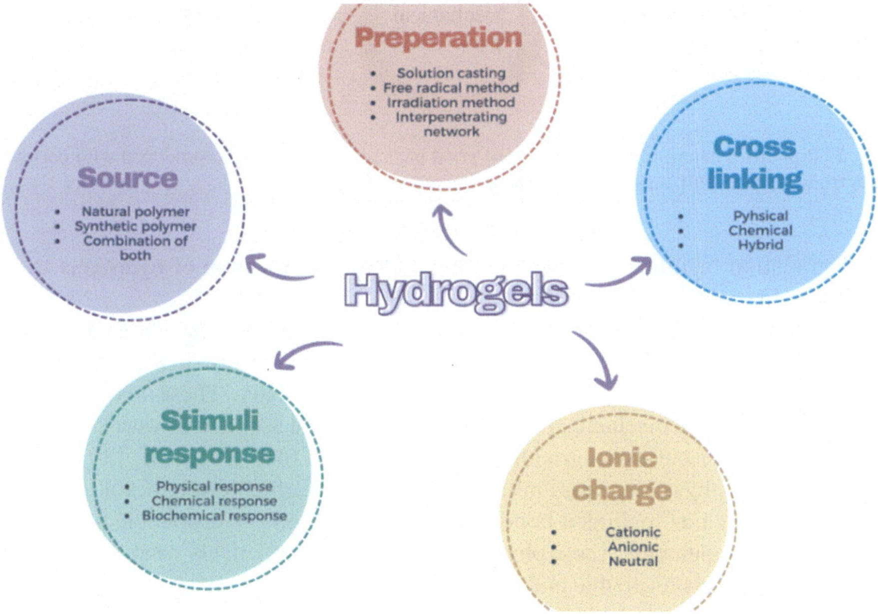

Fig. 1 Preparation strategies of hydrogels. Reproduced with permission from the Ref. [27], Elsevier, copyright 2022

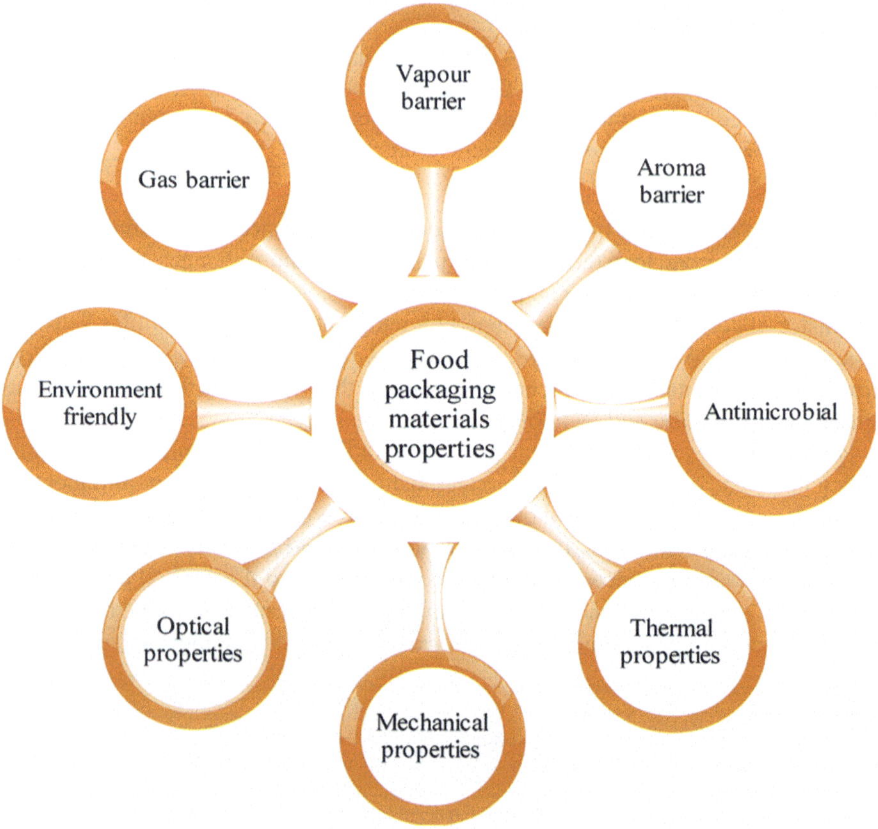

Fig. 2 Properties determining the quality of good packaging materials. Reproduced with permission from the Ref. [29], American Chemical Society, copyright 2018

biodegradation facilitate the disintegration and safe elimination of hydrogels from the body [5].

Chemically crosslinked hydrogels are formed through covalent bonds, typically transforming from a liquid state to a solid through mechanisms such as photopolymerization, enzyme-catalyzed reactions, or click chemistry. These hydrogels are known for their high mechanical strength and have found uses across various sectors, including pharmaceuticals, agriculture, cosmetics, and food processing [30].

In contrast, physically crosslinked hydrogels are formed through non-covalent interactions, such as hydrogen bonding, ionic interactions, hydrophobic effects, or molecular entanglements. These physical interactions are relatively weak, making the resulting hydrogels reversible and more prone to degradation. Importantly, physical hydrogels are synthesized without the use of chemical crosslinking agents. According to Parhi et al., physical hydrogels often display a more homogeneous structure compared to chemically crosslinked ones and show great potential for encapsulating and delivering bioactive compounds [31].

2.1 Swelling, Transparency, and Strength

The crosslinked polysaccharide polymers can form hydrophilic 3D frameworks due to abundant crosslinking points, and such insoluble networks exhibit good mechanical strengths [32]. The swelling process involves adsorbing either water or another solvent, and the solvent molecules diffuse into the 3D network due to the presence of hydrophilic groups, leading to the extrusion of polymeric chains and balanced via a thermodynamic swelling force of the polymeric chains. However, a higher number of crosslinking junction points causes a lower degree of swelling. A suitable crosslinking control is necessary during the fabrication of chemical gels when they are applied in biomedicine [33].

The swelling of hydrogels increases the average distance among junction points, and thus, the electrostatic repulsions of the amphiprotic groups in the polymeric chains are enhanced. This would accelerate the diffusion of the solvent molecules into 3D networks. The framework of hydrogels swells only perpendicularly toward the substrate. The swelling property of a hydrogel is essential for drug release applications. The swelling degree (SD) and the swelling ratio (SR) calculated for hydrogels [34] are expressed as.

$$SD = SR = \frac{m_{swollen} - m_{dry}}{m_{dry}} \tag{1}$$

Where $m_{swollen}$ and m_{dry} represent the weights of swollen and dried hydrogels, respectively.

The transparency of hydrogels depends on the light transmission through the gels. The openness of hydrogels varies with the treating temperature and coagulation bath during the gel process. The transparency is affected by the degree of phase separation (or spinodal decomposition), which involves a liquid or solid changing from one thermodynamic phase to two coexisting phases. In addition, the mechanical property of hydrogels can be enhanced by implementing a pre-gelation operation before the coagulation bath. Many hydrogels that exhibit both elastic and viscous properties are commonly referred to as viscoelastic hydrogels [35].

2.2 Biocompatible, Degradable, Injectable, and Environmentally Sensitive

Ideally, the hydrogels should be composed of scaffolds possessing a porous network, be nontoxic and biocompatible, and capable of enhancing cell culture and new tissue formation. Polysaccharides are a type of macromolecule used in the formation of hydrogels. They possess numerous compositions and abilities that may permit functionally tailored chemical modification. Depending on the water weight ratio, there exist homogeneous hydrogels, hydrogels with micropores, and hydrogels with macro-pores. The structure of the biodegradable segment should match the activity of a specific enzyme to enable a hydrogel to degrade [35]. Some other elements, such as low pH or elevated temperatures, are identified in vivo. Therefore, either pH and/or temperature-sensitive hydrogels may be applied for controllable drug delivery. Hydrogels that are responsive to specific molecules, such as glucose or antigens, or to light, pressure, and electrical signals, are also utilized in drug delivery and bioseparation [36].

2.3 *Hydrophilicity and Water Retention Capacity*

Hydrogels contain crosslinking between their molecules, which is primarily influenced by ionic functional groups within the polymer matrix, and plays a critical role in facilitating nutrient transport and enhancing the efficiency of drug release systems [37]. Research by Suflet et al. demonstrated that combining covalent and physical crosslinking strategies can produce hydrogels with rapid swelling rates and low elastic moduli [38]. The thermostability of an innovative hydrogel polymer is illustrated in Fig. 3. It swells when interacting with moisture (Aqueous water) and shrinks at a temperature of 50 °C, releasing liquid water [39].

The swelling behavior of hydrogels can be influenced by the type of cross-linking agent used, whether physical or chemical. For instance, chemically cross-linked

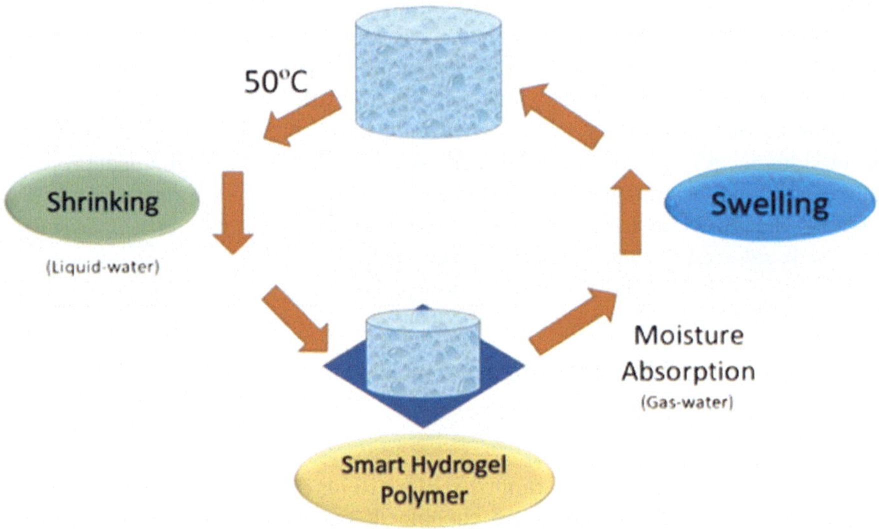

Fig. 3 Schematic representation of the thermosensitive mechanism of hydrogels. Reproduced with permission from the Ref. [27], Elsevier, copyright 2022

hydrogels have been synthesized using N, N'-methylene bisacrylamide as the cross-linker. In contrast, physically cross-linked hydrogels have been prepared via precipitation polymerization using dioxane as a solvent. This approach was applied to hydrogels made from poly(acrylamide-co-methyl methacrylate) (poly(AAm-co-MMA)) in varying compositions [40]. A study conducted several decades ago demonstrated this effect using poly(ethylene glycol) (PEG) with molecular weights of 860, 3900, and 9300 to create oligo(poly(ethylene glycol) fumarate) (OPF) hydrogels. The results showed that each PEG variant produced a hydrogel with a distinct water retention percentage, highlighting the critical role of molecular weight in determining the hydrogel's behavior [41].

2.4 Mechanical Properties and Stability

The mechanical characteristics of hydrogels are primarily influenced by the extent of cross-linking within their polymer network. A high density of cross-links generally results in stiffer hydrogels, whereas a lower degree of cross-linking produces softer, more flexible materials. This structural property significantly affects the functional capabilities of hydrogels, making them suitable for applications such as tendon and ligament repair, wound healing, tissue engineering, drug delivery systems, and cartilage replacement [42]. For instance, Singh et al. reported the development of a robust hydrogel based on Acacia gum polysaccharide, designed for use in wound dressings with enhanced mechanical strength [43].

Reactive functional groups such as -COOH, amino -NH^{3+}, and -OH are the main cause of biocompatibility and bioactivity of polysaccharide-based hydrogels. The groups allow chemical modification, which increases the level of interaction between the hydrogel and biological tissues. To be applicable in biomedical contexts, hydrogels should not only satisfy the requirements of biosafety but also respond to the biological environment, adjusting to the host environment and promoting cell activities [44]. For instance, Ali et al. have designed a pH-sensitive, non-toxic and biocompatible hydrogel made of citric acid crosslinked polysaccharides extracted from plant *Salvia spinosa L.* with a specific purpose of controlled drug delivery [45].

3 Stimuli-Responsive Polysaccharide Hydrogels

3.1 pH-sensitive Hydrogels

3D hydrogels that are pH-sensitive are developed to react to the local pH changes. Such hydrogels are generally either expanded or contracted in response to various chemical conditions and can be prepared by in-situ polymerization, making them especially well adapted to being incorporated into microfluidic systems. Among their major uses is the development of pH-responsive valves and systems that discharge substances as a result of pH changes [46].

The Human body shows significant differences in pH of various organs, tissues, and physiological states that provide useful prospects in developing site-specific drug delivery systems. They are functional based on their ability to experience structural/ phase changes in response to unique pH conditions. The pH of these transitions depends on several factors, and especially the chemical properties of the ionizable side chains of the hydrogel (Fig. 4) [47].

These smart biomaterials have special physical and chemical properties at a particular pH because of the availability of acidic or basic functional groups that are attached to the polymer chains [48]. The release of drugs through such hydrogels may take place through three main processes, i.e., diffusion, swelling, and chemical reactions. The most commonly known and used approach is the diffusion-based approach, which is based on the law of Fick, whereby the movement of drug molecules through the hydrogel is affected by the relative size of the drug particles and the pore structure of the hydrogel. When the drug molecules are considerably smaller than the pores, their diffusion is determined by the diffusion coefficient, but when the size of the

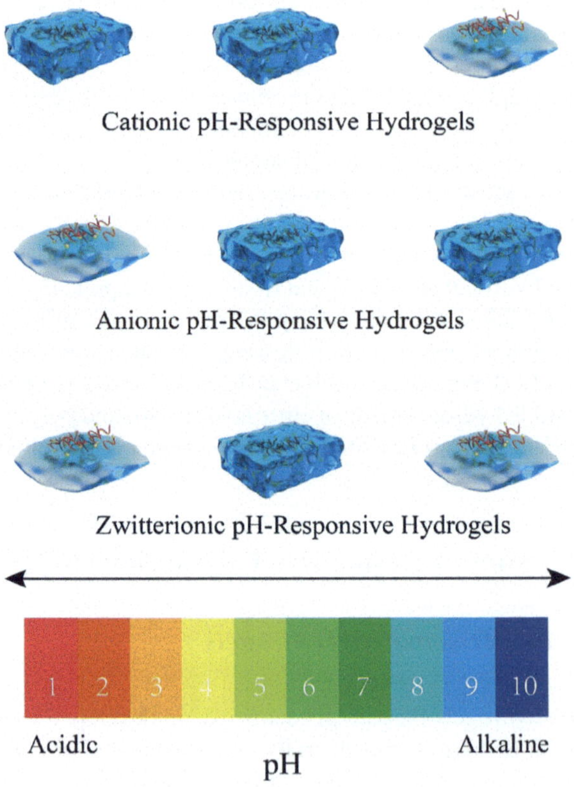

Fig. 4 Illustration shows the classification of pH-responsive hydrogels depending upon the phase change behaviors at varying pH levels. Reproduced from the Ref. [47], under Creative Commons Attribution (CC BY) license, 2023

drug is similar to the size of the pore, the network of tightly crosslinked hydrogel hinders its release [49].

In some cases, where the rate of hydrogel swelling is greater than the rate of drug release, the swelling process is the most important release process. This normally includes the absorption of water by the hydrogel and the subsequent desorption of the drug. These polymers change shape and volume considerably when hydrated in their dry or glassy form, and this has an influence on both drug release rates and structural properties, including crosslink density. 3D hydrogels can be permeated by the surface owing to their free intermolecular connections, which enable water or other bodily fluids to pass through the surface. After the drug is released, the hydrogel returns to its initial form once the absorbed water is expelled [49].

3.2 Temperature-Responsive Hydrogels

It is possible to have reversible gel formation in thermoresponsive polymers in solution above or below their critical solution temperature. 3D hydrogels based on polysaccharides with a lower critical solution temperature (LCST) of close physiological temperatures are of specific interest in wound healing because of their ability to create gels in vivo [50]. Multistimuli hydrogels, in particular those with dual responsiveness to different stimuli, are of great interest since their behavior is more dynamic, like biological systems.

In the context of diabetic wound healing, Wang et al. synthesized a temperature and pH-responsive hydrogel system that provided UV protection as well [51]. The synthesis of the system involved the oxidation of pullulan to produce aldehyde pullulan (APu) that was further crosslinked with Pluronic127-grafted polyethylenimine (PEI) in Schiff base reactions. The thermogelling properties were evaluated at 4, 25, and 37 °C at various concentrations of APu, and it was found that the sol-gel transition occurred above 25 °C [51]. Additionally, exosomes derived from adipose mesenchymal stem cells (ADSCs) were incorporated into the hydrogel via electrostatic interactions. Under acidic conditions (pH 5.5), cleavage of acid-sensitive imine bonds enhanced the release of exosomes, thereby promoting efficient wound healing.

Also, adipose mesenchymal stem cells (ADSCs) exosomes were introduced into the hydrogel through electrostatic interactions. In acidic environments (pH 5.5) and thus, the exosomes were liberated more by the cleavage of acid-sensitive imine bonds, which facilitated effective wound healing. Research has indicated that the tumor microenvironment, ischemic areas, and wound healing sites have a low pH. This has led to the invention of drug delivery systems that are selective to such acidic environments. One method is through the application of hydrogels that are both pH and temperature-sensitive. The thermosensitive hydrogels are categorized by the behavior of changing their phases upon different temperatures (Fig. 5) [47]. This is due to the fact that the thermal responsiveness of these hydrogels is due to the balance of hydrophobic and hydrophilic components in the polymer structure, which consists of both functional groups [52].

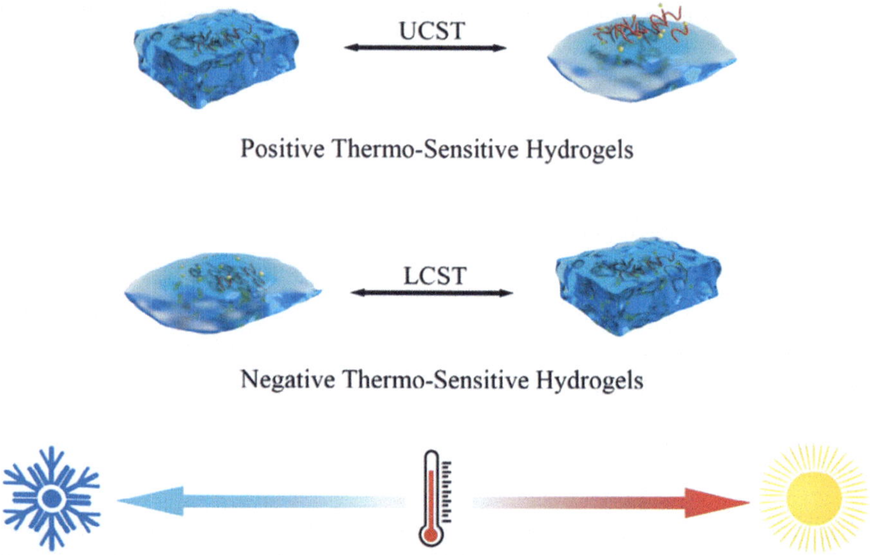

Fig. 5 Thermally sensitive hydrogels are classified according to their phase change behaviors at varying temperatures. Reproduced from the Ref. [47], under Creative Commons Attribution (CC BY) license, 2023

When the temperature varies, the interactions of the hydrophilic and hydrophobic parts of the polymer with water also vary, making changes in the dissolution of the crosslinked network and sol-gel transitions. The gel phase can stay relatively stationary, whereas the sol phase is free to move. The ability of the polymer to change into a gel is determined by the balance between the hydrophilic and hydrophobic properties. At the microscopic level, the polymer has temperature-sensitive units such that the hydrogel can gel at certain critical temperatures, either at the lower critical solution temperature (LCST) or the upper critical solution temperature (UCST). Once the system gets into the LCST, the polymer becomes less soluble and becomes hydrophobic, collapsing into a gel form. On the other hand, in hydrogels with a UCST, gelation takes place when the polymer cools, when the coil structure of the polymer becomes insoluble in the form of an aggregated structure [53].

The LCST is primarily influenced by the proportion of hydrophilic to hydrophobic groups in the polymer. Hydrogels that provide temperature-regulated, continuous drug release have been increasingly explored in recent years. These thermoresponsive systems present multiple advantages for drug delivery. Unlike traditional hydrogels, which often require surgical placement, temperature-sensitive hydrogels can be injected, thereby bypassing first-pass metabolism. Moreover, their sol–gel transition occurs naturally at body temperature, without the need for toxic cross-linking agents, making them safe for in vivo use [52].

When administered in liquid form, these hydrogels facilitate the uniform distribution of therapeutic agents throughout the matrix. Upon reaching body temperature, they rapidly transition into a gel, preventing an initial burst release and supporting a more controlled drug release profile. Additionally, this transition provides the hydrogel with structural stability post-injection [52].

4 Synthesis and Fabrication Techniques

4.1 Physical and Chemical Techniques

Proteins and polysaccharides of various natural sources have been explored as a wide range of techniques to produce food-grade hydrogels. The heat, cooling, enzymatic reactions, ionic interactions, and pH changes are common conventional methods that are often used for the formation of a hydrogel. More recently, novel technologies such as high-pressure and ultrasound treatment have broadened the range of options for customization of the physicochemical characteristics and performance of food-derived hydrogel.

Thermal crosslinking is recognized to be one of the simplest and cost-effective approaches to the environmentally friendly setting. During this procedure, protein hydrogels, especially those that consist of whey or egg proteins, are subjected to thermal denaturation, leading to the unfolding of proteins and their aggregation, which follows [17]. Composite hydrogels can be successfully prepared when the mixtures of protein and polysaccharides are heated at temperatures higher than the protein denaturation temperature, and in the presence of pH conditions in which the electrostatic interaction can occur effectively.

In some hydrogel systems, the thermal pre-treatment is required to achieve full dispersion and hydration of biopolymers, and then cooling should be carried out to induce gelation. Some of these are κ-carrageenan (KC) and agar, which gel when heated to 70–90 °C and allowed to cool to a lower temperature, thus inducing the formation of a structural network [54]. Globular proteins, including soy and whey, need to be heated (90–95 °C) in comparison to linear proteins, like gelatin (40–60 °C), which are required to be exposed to the heat to reveal their internal hydrophobic regions, thus allowing them to be gelated. The addition of co-agents, e.g., salts, acids, or secondary polymers, is usually advantageous to the cooling-induced gelation.

The gelation by enzyme offers a gentle and specific method for the formation of hydrogels. The most frequently used enzymes are transglutaminase (TGase), laccase, and horseradish peroxidase, all of which allow particular covalent reactions to be carried out in mild conditions [55, 56]. TGase is used to catalyze the formation of bonds between glutamine and lysine residues [57], laccase and horseradish peroxidase are utilized to encourage the formation of phenolic bonds, especially in polysaccharides. One-step techniques include the direct addition of enzymes to the biopolymer mixture and cooling, and two-step techniques include consecutive reactions of the enzymes to form layered gel networks [58, 59].

Electrostatic forces play a key role in the ionic gelation, in which the negatively charged functional groups (carboxyl and sulfonic acids) of biopolymers react with positively charged amines. Such ions as Fe^{3+}, Ca^{2+}, Na^{+}, and K^{+} are crosslinkers, and the ionic interactions are generally stronger between trivalent and divalent and weaker between trivalent and monovalent [4]. Two major methods are available, one of which involves placing ions directly into the gel-forming mixture, and the other method is a procedure by which the pre-existing gels are placed in ionic solutions. The former is optimal in the production of hydrogel beads (e.g., alginate-Ca^{2+}), and the latter can be used to form more homogenous and thermostable gel networks [60, 61].

Solubility, structural conformation, and charge characteristics of proteins and polysaccharides can be altered by adjusting the pH, and hence affect the gel formation. Reduction in pH can cause protein denaturation and aggregation close to the isoelectric point that facilitates the formation of a gel network [62]. Acidulants, such as glucono-δ-lactone (GDL), gradually hydrolyze into gluconic acid, slowly lowering the pH to optimize the gel texture. For example, fish protein hydrogels developed with 0.2% GDL exhibited enhanced rheological properties and network integrity. Incorporating gellan gum into such systems significantly amplified the gel's structural strength and improved sustained curcumin release [63].

High-pressure treatment (100–1000 MPa) is a modern non-thermal technique that enables the production of hydrogels while preserving the nutritional and sensory properties of food. It induces protein denaturation by disrupting electrostatic and hydrophobic interactions, rather than breaking hydrogen bonds like heat treatment does [64, 65]. Key outcomes include protein unfolding, acid group ionization, and aggregation into gel matrices or precipitates. This approach is efficient, safe, and minimizes energy usage and environmental impact.

Ultrasound (20–1000 kHz, 10–1000 W/cm^2) has emerged as a green, cost-effective strategy for enhancing hydrogel formation [66]. The cavitation phenomenon generates physical (shock waves, shear forces), chemical (free radicals), and thermal effects that modify the structure of proteins and polysaccharides, influencing their gelation behavior [67]. However, excessive exposure may lead to the overproduction of free radicals, thereby accelerating the oxidation process. Ultrasound is frequently used as a supportive tool in combination with other processing methods to optimize gelation and functionality in food and biotechnological products [68].

4.2 3D Printing Technologies

Advancements in 3D printing technology have enabled the use of diverse materials such as polymers, metals, and ceramics to fabricate objects layer by layer in a wide range of geometries [69]. Currently, a variety of 3D printing techniques are available, each suited to specific applications. These techniques can be broadly classified into binder jetting, sheet lamination, powder bed fusion, and material extrusion (Fig. 6) [70].

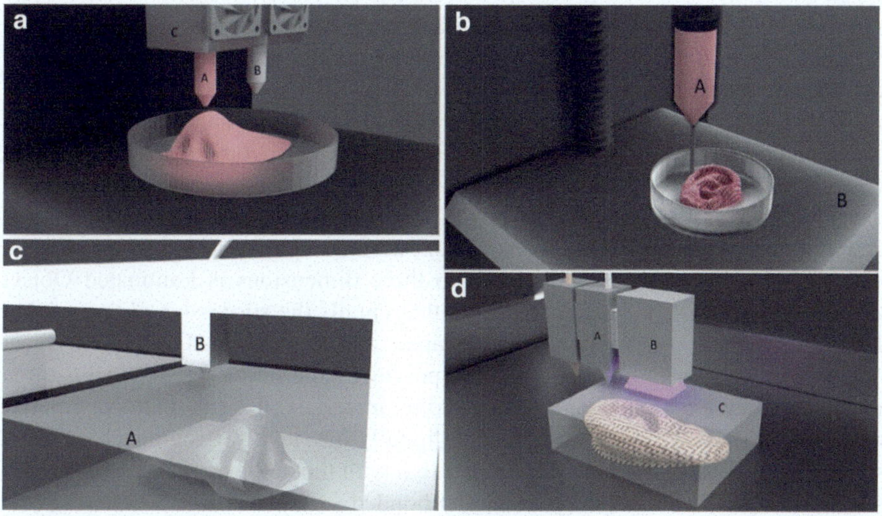

Fig. 6 **a** Dual-head FDM 3D printer. (A) Building material; (B) supporting material; (C) print heads. **b** Extrusion-based bioprinting. (A) Bioink; (B) build platform. **c** Inkjet 3D printing. (A) Powdered bed; (B) Binding liquid spraying nozzle. **d** PolyJet 3D printer. (A) Nozzle spraying photopolymer; (B) UV source; (C) supporting material. Reproduced from the Ref. [71], under Creative Commons Attribution (CC BY) license, 2018.

Fused Deposition Modeling (FDM) is a 3D printing technique that constructs objects by extruding a continuous filament of thermoplastic material (Fig. 6a) [72, 73]. One of its key advantages is its ability to produce complex shapes quickly and at a lower cost compared to conventional manufacturing techniques [73]. Invented by S. Scott Crump in the late 1980s, FDM was commercialized in the early 1990s by Stratasys Inc., headquartered in Edina, Minnesota, USA [74]. The standard FDM setup includes a printhead that moves along the X and Y axes, a build platform that shifts vertically (Z-axis), and a thermoplastic filament in cylindrical form as the printing material. Guided by a computer-aided design (CAD) model, the heated nozzle of the printhead melts or softens the filament, depositing it layer by layer onto the platform, where it solidifies upon cooling [75].

The software on the printer determines the motion of the printhead and feed of filament, allowing the object to be deposited in each new layer over the last layer one after the other until it is complete from bottom to top. The ultimate properties of FDM-printed scaffolds largely depend on three parameters, namely, shell arrangement, infill pattern, and layer thickness. Dimensional accuracy and surface finish are two important factors used to assess the outputs of FDM. The glass transition temperature of the material can have a significant effect on dimensional accuracy in that the layers printed solidify, undergoing shrinkage and contraction. Contrarily, enhanced accuracy is possible whereby the material cools and solidifies rapidly upon the extrusion [76]. FDM typically uses thermoplastic materials that melt at high temperatures and solidify rapidly when cooled. Among these, polylactic acid (PLA) is commonly used in the pharmaceutical and medical fields. However, other polymers, including cellulose-based derivatives and Eudragit®, have also been explored for such applications in recent years [77, 78].

FDM normally employs thermoplastic substances that melt at high temperatures and harden faster when cooled. Polylactic acid (PLA) is among them, and it is typically utilized in the pharmaceutical industry and in medicine. Other polymers, however, such as CEL-based derivatives and Eudragit®, have also been considered for such applications in recent years. Another additive technique applied in bioprinting to form tissue constructs into three dimensions is Laminated Object Manufacturing (LOM). This is a process that entails the successive gluing of thin sheets of material, usually polymers or paper, which are bonded together by heat or adhesive. In bioprinting, biodegradable polymer or cellulose-based paper is often selected as they are biocompatible scaffolding materials. It starts with a computer-aided design (CAD) model, which is cut digitally into layers. Each matching sheet is then cut, typically with a laser, so as to align with the outlines of every layer. This is so because these sheets are stacked and bonded together in a sequence to create the final 3D structure [79].

There are several benefits associated with LOM, such as speed of fabrication, low cost of production, and the capability of making constructive items on a large scale. The recent studies in LOM-based bioprinting dwell on the aspect of enhancing the levels of cellular integration and survival, fine-tuning the characteristics of employed material, and optimizing the state of the process to enhance the functionality of the scaffold. Interestingly, recent manufacturing has been undertaken to reproduce natural tissue surroundings by developing a scaffold with regulated permeability and hierarchical frameworks [80]. Multi-material printing, sacrificial materials are also being investigated, whereby more complex and functional tissue constructs can be fabricated to be used in regenerative medicine and drug testing procedures [81].

It begins with a computer representation of the object of interest, which is cut into very thin cross-sectional layers. The plates are then cut at a very fine level with a stamping or cutting tool and placed on a build platform. The layers are attached to the other layers by heat, pressure, or glue, based on the material properties. This process is carried out, add after add, till the entire object is created. SDL offers significant benefits to tissue engineering, including cases of building complicated geometries with high resolution and well-defined mechanical properties. The current advances in the SDL-based bioprinting aim to improve the cell viability and functionality of this process, fine-tune the process parameters, and design novel biomaterial compositions that are specific to tissue regeneration [81].

Extrusion-based printing, also called semi-solid extrusion printing, pressure-assisted Microsyringe (PAM) printing, or direct ink writing, is based on the deposition of a semi-solid material, including a gel or paste, via a nozzle or needle through controlled pressure (Fig. 6b) [82–84]. This process is like the fused deposition modeling (FDM), except that the nozzle is actually in motion (or the platform is actually in motion) so that there is no longer a need to have the material extrusion directed vertically to create a layer (or to have the platform drive horizontally to create a layer). The printing characteristics and the nature of the materials that will be used are the key factors that determine the properties of the printed object, which is popularly known as inks [85]. Material viscosity and its gelling properties are critical in this technique [86]. The ideal inks usually have shear-thinning characteristics and rapid recoveries [87]. In other words, they should be viscous to be structurally stable through printing, but not too viscous as to plug the nozzle or to deform uniform strands. After printing, rapid gelation is essential for the preservation of the shape of the printed structure [88, 89]. The process of solidification may take place in many ways, but it may include chemical crosslinking, crystallization, rearrangement of the molecules, and the establishment of hydrogen bonds or other non-covalent forces. This extrusion technique can tolerate a wide variety of printable materials, such as hydrogels, which have the added advantage of being tunable in viscosity, but also molten polymers, paste, and polymer solutions [90].

The 3D inkjet printing has become an influential method of producing polymeric, polysaccharide-based structures with great accuracy and fine control over the deposition of materials. Here, microscopic-sized droplets of a polysaccharide ink are ejected into microscale nozzles selectively to create complex structures in layers. The process is specifically applicable to polysaccharides since they can be formulated into shear-thinning inks with low viscosity that can be easily moved by the printing process but quickly solidify once deposited. Inkjet printing has very high spatial resolution and can make complex designs, gradients, and multi-material constructs, which are not achievable using conventional hydrogel fabrication processes. It also enables accurate localization of bioactive molecules, cells, enzymes, or nanoparticles in the printed structure, and this is very appealing in biomedical uses. Inkjet printing has also been used to produce wound dressing designs, drug delivery carriers, food package designs and tissue engineering scaffold designs. A digital customization that enables each structure to be customized to correspond to patient-specific or application-specific needs is also supported by the technique without changing the overall production process (Fig. 6c). Its non-contact nature is one of its largest merits as it makes it less contaminated and able to be used with sensitive biological parts. There are, however, still problems of optimizing ink stability, avoiding nozzle clogging, and increasing post-print crosslinking to achieve sufficient mechanical strength. In spite of these, the devolution of printer hardware, algorithmic droplet controls, and strategies of modifying polysaccharides are quickly enhancing print fidelity [71, 86].

Recent 3D PolyJet printing is a high-tech additive manufacturing strategy, which enables the manufacturing of polysaccharide-based high-resolution structures through simultaneous jetting of droplets of photopolymers. In contrast to traditional inkjet systems, PolyJet machines drop layers of liquid on a build platform and then cure them immediately by UV light, creating smooth surfaces and geometries of unsurpassed detail. This method is especially beneficial to polysaccharide-based formulations that may be either chemically modified or mixed with photo-curable compounds, hence attaining the needed viscosity and crosslinking properties needed to process PolyJet. The co-deposition of various compounds would allow researchers to combine the polysaccharide structure of various polysaccharides in a single construct to have gradients in mechanical strength, porosity, or bioactivity.

PolyJet printing finds extensive application in the manufacture of advanced biomedical prostitutes, microfluidic systems, tissue-engineering scaffolds, and controlled-release system prototypes. Its impressive capabilities in printing specific designs and structures make it adequately applicable in the production of patient-specific designs and structures with sophisticated internal channels. In addition, the thickness of the layer that could be achieved using the PolyJet printing process could be as thin as 16–30 microns, thereby allowing exceptionally smooth surfaces compared with the other 3D modalities of printing. Although the technique provides extensive design freedom, there are still issues associated with the creation of a completely biocompatible, degradable, and photo-crosslinkable polysaccharide ink. Even post-processing operations, like support removal and UV curing, have to be optimized to maintain the integrity of the fine hydrogel structures (Fig. 6d) [71, 91].

5 Applications in Food Systems

This is where the product is used in the fabrication of food systems. The 3D hydrogels based on polysaccharides have been utilized in innovative food applications due to their biocompatibility, edibility, and controllable functional features. They find wide application in active and intelligent packaging of foods and are used as delivery vehicles of antimicrobial and antioxidant food ingredients, and also as freshness indicators. These hydrogels ensure the protection of sensitive constituents like probiotics, vitamins, and flavor compounds in nutrient and bioactive delivery, which guarantees controlled release and enhanced stability. They can be designed to have customizable textures and rheological properties. Moreover, the 3D printing of these hydrogels will provide opportunities to personalize nutrition and functional foods, making it a sustainable and innovative aid in modern food technology.

5.1 Food Packaging and Preservation

Conventionally, food packaging is used as a shield that prevents food products from exterior factors like moisture, light, oxygen, microbial infection, mechanical stress, and dust. These environmental factors have the capacity to increase the rate of spoilage of food. The ideal packaging medium must be free of chemicals, must have a superior barrier quality, and be recyclable. The food packaging, therefore, should have three main roles to play: (i) mechanical protection during storage and transportation; (ii) barrier to permeation of gaseous substances, e.g., oxygen and water vapor; and (iii) resistance to microbial contamination. Its packaging also needs to be strong enough to maintain the quality of the food during a long period of time, but it should be resistant to various environmental factors. Since the food packaging market is a very competitive market, there is a high need to find low-cost material that yields high performance [92].

Traditionally, packaging materials such as plastics, metals, glass, and paperboard, which are derivatives of petroleum, have been in wide use because they are efficient and cost-effective. Among those, paper and cardboard are the only biodegradable; the rest are made of non-renewable resources. Such a situation has triggered an increase in the need to use more sustainable sources of packaging, which have renewable backplots like naturally occurring polysaccharides [93]. As consumer needs for food products that are of high quality and eco-friendliness increase, there is now interest in developing biodegradable and bioactive food packaging materials. These alternatives not only assist in maintaining the quality of the foods that are consumed but also address the environmental issue, such as waste reduction. Plant and animal-derived biopolymers (proteins, lipids, and their combinations) are being actively pursued as alternative materials to replace the traditional petrochemical-based packaging.

Recently, bio-based 3D hydrogels have been suggested to be used in the packaging of food. However, there has been little integration of superabsorbent polymers into the food sector, which includes the incorporation of hydrogels. The idea of hydrogel-based packaging, in turn, is in line with the developments toward environmentally-friendly and sustainable materials, but its real-life implementation is not a commonplace practice [94]. CMC is a polymer that is hydrophilic and biodegradable, which is widely used in the food, textile, and paper industry [95]. Fernandez et al. [96] described that biopolymer coating applied in direct contact with food must be applied similarly to so-called standalone biopolymer coatings.

Hydrogels have also been used in active packaging systems with primary uses of controlling the humidity level in the package and preventing microbial growth. These functions increase the quality of the products, stability, and shelf life [97]. Hydrogels that are specifically designed to generate moisture control are engineered using a divergent array of hydrogels [98]. In the food preservation environment, hydrogels would be used to absorb moisture or exudates that are excreted by produce in the process of natural respiration or as a result of physicochemical transitions in the packaged products. This especially applies where the surrounding setting determines the levels of accumulation of the water vapor, i.e., in the storage of fresh produce.

Fruits and vegetables are highly perishable products, and application of hydrogel-based packages is advantageous. They lower water activity, retard the activity of the spoilage organisms like mould and yeast, and maintain the integrity of perishable goods like fresh-cut produce and moisture-sensitive products (e.g, powdered food). Hydrogels can also contribute to the preservation of the texture of either crisp or crunchy products, including cookies or fried snacks [99].

Hydrogels are excellent absorbents, and their great potential can be utilized to create efficient food wrappings that are sustainable. Their properties on the box are adding durations to shelf life, better biodegradability, and mechanical strength [100]. Some packaging films have been made out of hydrogel matrices made of alginate, agar, collagen and antimicrobial compounds such as silver nanoparticles and grapefruit seed extract and hence makes foods such as potatoes remain longer on the shelves [101].

The use of edible biopolymers as both food ingredients and packaging materials is one of the benefits that help to maintain desirable properties, including a high barrier property, antimicrobial resistance, affordability, non-toxicity, and environmental sustainability [102]. Hydrogels are also used as a freshness detector, changing color in the presence of emitted food metabolites, so that they can give a visual representation of food quality [103]. As an example, baked goods that added hydrogels had a lower migration of moisture and maintained a higher level of softness at ambient temperature [104]. In a similar research, the antimicrobial-impregnated agar-based hydrogels could lengthen the shelf life of dairy products without affecting the sensory characteristics [105]. AgNPs and grapefruit seed extract-based hydrogel coatings were also developed successfully to be applied to the packaging of fresh potatoes.

3D printing enables the additive manufacturing of products according to computer-generated designs based on layer-by-layer manufacturing, giving the product a high level of personalization in both look and nutritional content. Nevertheless, not every food material has the mechanical properties needed to be successfully 3D printed, and the assurance of the appropriate material printability has to be ensured. The best materials to use in 3D printing are those colloidal materials that possess enough rigidity to maintain their shape after printing and, at the same time, possess enough flexibility to be easily extruded through small nozzles [106].

The critical parameters affecting printability are self-supporting capability, extrudability, and stability of structure, which are regulated by the storage modulus (G'), the loss modulus (G"), and the damping factor (tan δ). A reduced tan δ value is the sign of a better shape retention after extrusion [107]. Leelapunnawut et al. [107] used 3D printing to make plant-based meat analogs with pea protein isolate and alginate gel using transglutaminase (TGase) or KC in different concentrations (0.3–0.9%). Sample materials treated with KC recorded low tan 0 values relative to the controls, and the sample with 0.9% KC recorded the highest hardness (37.14 N) due to the loss of more moisture during cooking. The textural analysis revealed that TGase and KC-treated samples were very close to the texture of raw and cooked salmon, respectively.

One of the studies assessed the accuracy of 3D printing against a reference design that was a star, and the measurements included corner definition, line continuity, extrusion smoothness, and shape retention after printing. Milk-starch hydrogels with different concentrations of KC (0.3–1.5%) had good print fidelity. The KC-improved hydrogels were also especially appropriate in patients with dysphagia because of their adjustable texture. Moreover, these changes did not affect the bio accessibility of minerals, which still exceeded 80% [108].

Liu et al. [109] synthesized a 3D-printable pickering emulsion gel, which entraps Cur and is stabilized by pea protein isolate and KC. KC addition improved the viscosity and mechanical strength of the gel, thus leading to a better extrusion quality and shape stability. The printed cylindrical structures were found to be more shape-intelligent and harder with a higher concentration of KC of 0.1–0.5. The gels also exhibited the enhanced encapsulation efficiency and stability of curcumin, which remained highly bio-accessible during in-vitro digestion.

Microbial contamination in the food industry is one of the major issues in food preservation, especially in extending the shelf life and safety of perishable food.

A recent paper investigated the use of an antibacterial hydrogel that was prepared by electrospinning nanofibers made of gelatin, chitosan, and 3-phenyl lactic acid (PLA) to improve antimicrobial activity and increase the shelf life of refrigerated chicken. The hydrogel, with regard to its structure, water holding capacity, antibacterial activity, and preservation capacity, was strictly analyzed. The existence of amide bonds between the carboxyl groups of PLA and the amino groups of chitosan under acidic conditions was proven by Fourier-transform infrared (FTIR) spectroscopy and X-ray photoelectron spectroscopy (XPS), and helped to enhance the water retention and structural stability. The analysis of low-field nuclear magnetic resonance (LF-NMR) showed that PLA changed the internal structure of the hydrogel and distribution of water states within the hydrogel. The hydration of the hydrogel was observed under scanning electron microscopy, revealing a clear porous sponge-like morphology. The hydrogel was functionally effective in inhibiting the growth of typical foodborne pathogens, Staphylococcus aureus and Escherichia coli, to prolong chilled chicken meat shelf life to a maximum of four days [110].

Together, these results justify the use of multifunctional polysaccharide 3D hydrogel as a promising material to use in 3D food printing. This allows them to be tuned to achieve reliable and predictable printing results and allows new possibilities in the production of personalized, nutritionally functional food products by utilizing the latest additive manufacturing technologies.

5.2 *Nutrient and Bioactive Compound Delivery*

Food processing has led to the development of the delivery systems that protect sensitive chemical compounds, like enzymes and nutraceuticals, in the food processing process as well as adverse environmental factors, including changes in temperature, pH, oxidation, and light exposure. They are also useful in targeted gastrointestinal delivery, as well as regulated release of hydrophilic and hydrophobic active agents [111].

Protein-based hydrogels are highly appreciated in various fields due to their nutritional attributes, functional flexibility, amphiphilic characteristics, biocompatibility, and biodegradability [114]. Hydrogels are polymeric networks that cannot be dissolved in water but can absorb and retain it due to their crosslinked structure, as such a structure ensures mechanical integrity. Various crosslinking reagents, such as epoxy resin, glutaraldehyde, and formaldehyde, could be used, and physical or chemical methods, such as UV radiation, gamma irradiation, or dehydrothermal treatment, could be used to strengthen the hydrogel structure [115].

Recently, biopolymer-based hydrogels have been utilized as a delivery system in which targeted delivery of nutraceuticals and pharmaceuticals can be achieved. Carbohydrates and proteins are especially good materials because of their gelling property, modifiability, good interactions, biodegradability, and biocompatibility [116]. Buccal delivery is one of them, which takes advantage of the swelling characteristics of the hydrogel and the bioadhesive characteristics in contact with the body fluids to target the drug release in certain locations, e.g., the oral cavity, the stomach, or the intestines. As an example, Aftach® is a buccal hydrogel-based system that is used in the treatment of aphthous ulcers. It is made up of a two-layered system: an adhesive layer of hydroxypropyl cellulose and PLA, and a backing film of non-adhesive triamcinolone acetonide [117].

In the case of gastrointestinal usage, hydrogels must satisfy certain performance requirements. They are very effective in oral drug delivery due to their flexibility and capability of delivering the drug to the site sustainably and in a site-specific manner [118]. Hydrogel beads are engineered to be stable in certain regions of the digestive tract, but they are designed to release their contents when exposed to certain GI triggers [119]. Hoang et al. produced a hydrogel that was comprised of poly(acrylic acid) (PAAc) and norbornene-functionalized chitosan (CsNb) that were temperature and pH-responsive. This hydrogel showed almost full release (92%) in the presence of intestinal conditions (pH 7.4, 37 °C), and much restricted release in the presence of gastric conditions (pH 2.2), which implies that it can be used in colon-target delivery [120].

PBHs ensure a strong resistance to the external environmental factors, such as heat, oxygen, light, acidity, and enzymatic activity, which increases the integrity and functional lifespan of encapsulated bioactive compounds [121]. As an example, hydrogel delivery systems based on the curcumin (Cur) delivery through soy meal-derived glycine-rich hydrolysates (SMGHs) and k-carrageenan (KC) have been developed as a hydrogel delivery system formed mainly by electrostatic interactions. These SMGHs/KC hydrogels display phase-selective release under simulated gastric, intestinal, and colonic fluids (SGF, SIF, and SCF, respectively): minimal release (3.4%), moderate release (44.6%), and high release (about 90%) in the SGF phase, SIF phase, and SCF phase, respectively. These viscoelastic properties (G' and G") of the hydrogels are improved by the presence of KC in addition to a more uniform network structure that leads to a slower and selective release of Cur. Cryo-SEM imaging showed that there was significant degradation of the matrix and augmentation of the pore size during the SIF phase. Conversely, additional fragmentation on the SCF phase, which occurred with the assistance of gut microbiota obtained using the rat fecal matter, led to the release of Cur almost entirely [122].

The same method of cold gelation was used to prepare hydrogels using large yellow croaker (*Pseudosciaena crocea*) roe protein, gellan gum, and added with KC to encapsulate Cur. KC addition greatly enhanced the mechanical properties, doubling to tripling G' and G", and raising the water-holding capacity by 4–19%, and also causing a more-aggregated microstructure than the formulations containing no KC. Such improvements decreased the speed of erosion of the matrices, leading to a 70% decrease in Cur release in the course of the in vitro gastrointestinal digestion. The ability of Gellan gum to resist enzymatic breakdown prior to reaching the colon also helped to enhance this delayed release profile since microbial activity was required to break it down [123]. Moreover, Espinaco et al. [124] studied how chia oil-astaxanthin mixtures encapsulated in alginate gel beads were stored when they were in the presence of whey protein. Their results indicated high encapsulation efficiencies (>85% in all formulations), and whey protein was found to be much better in preserving astaxanthin and oxidative stability than beads prepared by using alginate only. Together, these studies suggest that PBHs can be effective nutraceutical delivery vehicles, which can protect and slowly release bioactive compounds in response to physiological conditions. This flexibility renders PBHs very promising in the application in the food, as well as biomedical, fields where sustained release is of interest.

5.3 Texture and Sensory Enhancement

3D polysaccharide-based multiphase hydrogels are used in food systems because of their ability to regulate rheological and textural values in multifaceted food systems. Textural characteristics of these systems are not only controlled by the macrospecific composition of the constituent components but also by the mesoscale microstructure patterns, which occur in processing. The manipulation of the internal microstructure of food matrices is therefore required to get desirable texture, mouthfeel, and eventually consumer satisfaction [125].

Jiang et al. [126] examined the impact of k-carrageenan (KC) on the rheological, textural, and microstructural properties of oyster protein gels (Fig. 7). Their results indicate that KC has a strong positive effect on the mechanical and structural characteristics of these gels, which is significant in the concentration range of 0.5–2%. It was found to increase storage modulus, hardness, cohesiveness, springiness, tackiness, and chewiness, along with a smaller and more complex network structure with a smaller mesh size and discrete sheet-like compartments. Both oyster protein and KC were also unfolded with the use of thermal treatment, revealing reactive functional groups. Further cooling led to the incorporation of KC in the protein matrix, which strengthened the structure. In addition, KC also facilitated the establishment of ionic, hydrogen, and disulfide bonds, which enhanced the construction of the β-sheet structures and reinforced the network crosslinking. At high concentrations of KC, there were visible flocculation events, which occurred due to the saturation of protein surfaces. Such findings support the hypothesis that a mechanistic comprehension of KC can produce new oyster-based products that have better structural and sensory characteristics.

The phase behavior of PBHs is one of the central concepts that should be thoroughly explained, as it forms the basis of controlling the textural characteristics of gelled food items and guides the creation of new food structures. Despite the fact that mixed polysaccharide protein hydrogels (MPPHS) are a potential field of study, the thermodynamic phase change of such mixtures at room temperature and high temperatures has not been sufficiently studied. The phase-separation phenomena can be effectively used in most cases to design processed foods with desired microstructural properties at various length scales.

In a recent study, the phase behavior of extracted grouper myofibrillar protein in combination with curdlan was investigated [127]. These ingredients are co-soluble at room temperature, and rheological evaluation showed that the optimum viscosity and

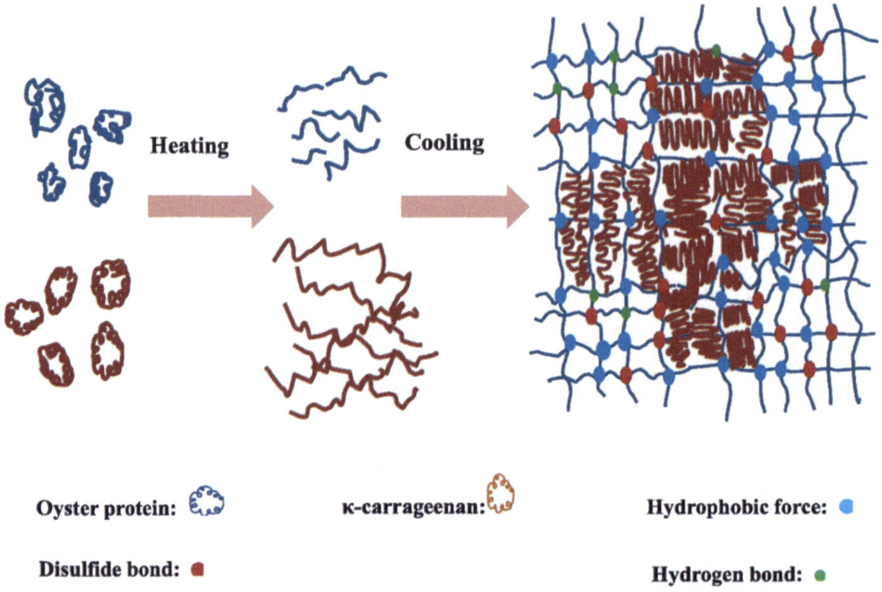

Fig. 7 Proposed mechanism of OP/κ-carrageenan gel. Reproduced with permission from the Ref. [126], Elsevier, copyright 2022

structural recovery were achieved at a curdlan concentration of 0.8%. On heating, the two biopolymers were thermodynamically incompatible, which phase-separated. The highest storage modulus (G') was obtained in a 0.6% curdlan solution, which produced a solid gel structure that was homogeneous with small and well-dispersed pores capable of holding water.

A study by Zhang et al. [128] examined the dynamic rheology of myofibrillar protein/konjac glucomannan (KGM) composite system. The phase transitions that occurred due to incremental changes in KGM concentration were irreversible and were classified into three distinct stages depending on relative changes in G' and G'' ($\alpha G'/\alpha G''$). At the bi-continuous phase ($\alpha G'/\alpha G'' > 1$), the system was homogeneous, and it enabled synergistic formation of networks between the protein and polysaccharide. The next related phase-separation step ($\alpha G'/\alpha G''$ approaches 1) was marked by the partial segregation but did not break the system's network integrity. The last stage was the segregate phase separation stage ($\alpha G'/\alpha G'' < 1$) in which the system faces deterioration because of complete thermodynamic incompatibility and KGM self-aggregation, resulting in the separation of phases and structural collapse. The authors have defined a direct relationship between phase behaviors and rheological parameters ($\alpha G'/\alpha G''$) as shown in Fig. 8.

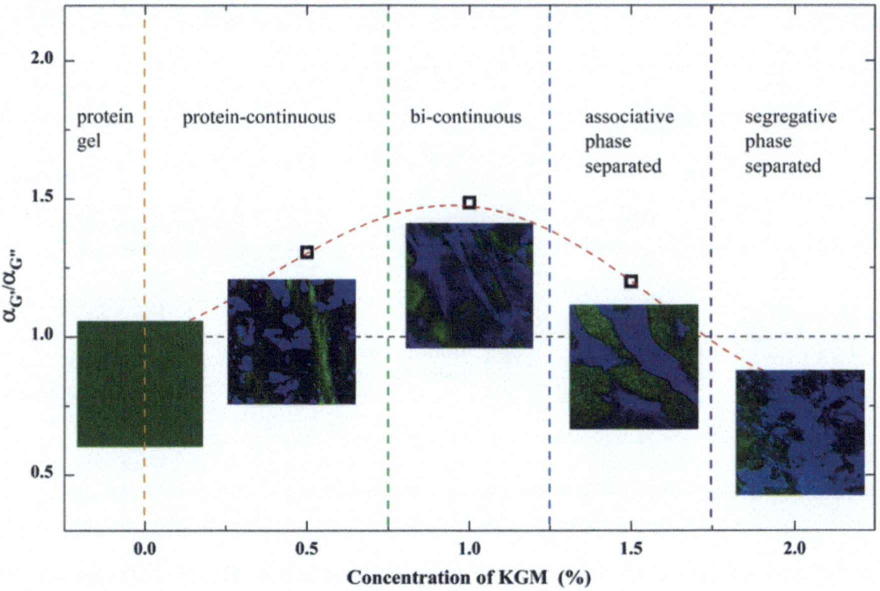

Fig. 8 Schematic representation of the correlation between phase behaviors and αG' and αG". Reproduced with permission from the Ref. [128], Elsevier, copyright 2023

5.4 *Smart Packaging Systems Using Polysaccharide Hydrogels*

Bio packaging based on renewable materials is central to reducing environmental pollution and also to curbing the energy issues that are related to petroleum-based plastics. Traditional paper packages are usually important because they do not provide adequate gas and oil resistance, thus requiring the creation of new technologies [129, 130].

Edible packaging in the form of polysaccharides is increasingly becoming popular due to characteristics associated with it, such as biodegradability, biocompatibility, safety, and easy accessibility by most people. The polysaccharide hydrogel films are normally transparent, very strong structurally, and they may have functional additives that may confer antioxidant and antibacterial characteristics. The tensile strength of such hydrogel films is similar to that of synthetic polymers, which increases the shelf life of ready-to-eat food, decreases the gas permeability, and decreases water-vapor transmission [131].

Sen et al. came up with a naturally antimicrobial hybrid hydrogel made of cationic starch and sodium alginate, which were bound together by electrostatic interactions. The ensuing product exhibited a significant thermal stability, good surface characteristics, and potent antimicrobial performance, which makes it appropriate for a broad range of food-packaging applications [132].

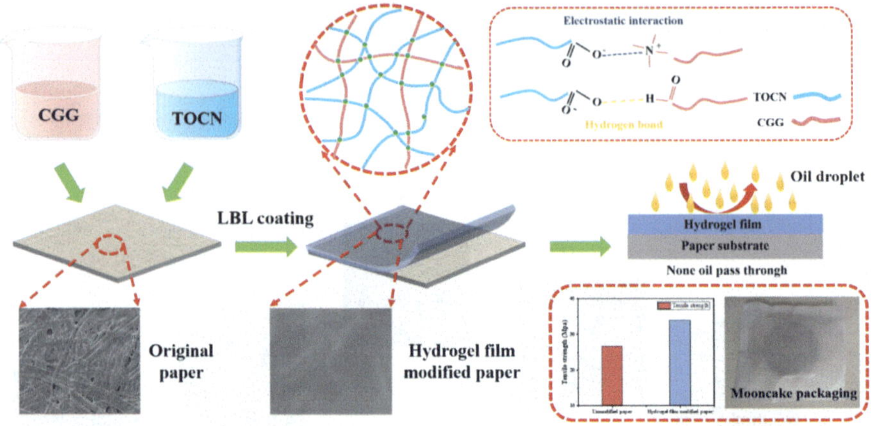

Fig. 9 A schematic diagram of the TOCN/CGG self-assembled hydrogel film modified paper for food packaging. Reproduced with permission from the Ref. [129], Elsevier, copyright 2021

Dai et al. studied a self-assembled hydrogel film made of TEMPO-oxidized CEL nanofibers and cationic GG as a paper-based packaging (Fig. 9). The deposition of the film in layers on paper substrates significantly improved the mechanical and barrier performance of the film. The resulting compound was oil-resistant and helped in maintaining the freshness of the food packed [129].

Benavides and collaborators created an alginate-based hydrogel film using mechanical stirring combined with ionic crosslinking. Alginates, when reacted with divalent or trivalent cations like calcium, form a three-dimensional network through interactions with 1,4-β-D-mannuronic acid and α-L-guluronic acid. This crosslinking process enhances water resistance, mechanical strength, cohesion, antibacterial properties, and rigidity, while also slowing the release of encapsulated substances. Moreover, the gradual release of calcium ions in acidic conditions promotes uniform matrix formation [133]. Wu et al. developed an edible hydrogel film using chitosan and agarose, incorporating anthocyanin as a pH-sensitive color indicator. The color change in anthocyanin, triggered by volatile nitrogenous compounds released during fish spoilage, enabled the hydrogel film to function as a freshness sensor for seafood [134]. Table 1 discusses the types and applications of hydrogels based on polysaccharides.

6 Safety, Regulatory, and Environmental Considerations

With the increasing emphasis on environmental responsibility and the sustainable use of natural resources, it has become crucial to evaluate the environmental impact of materials used in packaging. This is where Life Cycle Assessment (LCA) plays a vital role. LCA provides essential insights into the ecological consequences of

Table 1 Polysaccharide-based hydrogels and their key application in the food industry

Type of hydrogel	Materials used	Applications	References
Alginate hydrogels	Sodium alginate + Ca^{2+}	Moisture control, antimicrobial packaging, shelf-life extension, and lysozyme delivery	[18, 132, 133]
Agar-based hydrogels	Agar ± silver nanoparticles	Shelf-life extension of dairy, coatings for produce	[101, 105]
Chitosan-based hydrogels	Chitosan alone or with alginate	Antimicrobial packaging, lysozyme delivery	[11, 38]
CMC hydrogels	Carboxymethyl cellulose	Moisture control, biodegradable packaging	[94]
Protein-polysaccharide composite hydrogels	Protein + polysaccharide networks	Texture modification, encapsulation, 3D printing	[109]
κ-Carrageenan hydrogels	κ-carrageenan	3D printing of meat analogues; texture tuning	[109]
Konjac glucomannan hydrogels	KGM + proteins	Texture design, dysphagia foods, printing accuracy	[128]
Smart/ Stimuli-responsive hydrogels	pH-, temperature-, ion-, or light-responsive polymers	Freshness indicators, intelligent packaging	[39]
Nanocomposite hydrogels	Hydrogels with nanoparticles (AgNPs, CNF)	Strength, antimicrobial packaging	[135]
Electrostatic protein-polysaccharide hydrogels	Protein + polysaccharide complexes	Nutrient delivery, encapsulation	[20]
Gelatin-chitosan-PLA hydrogels	Gelatin + chitosan + PLA	Shelf-life extension of chicken, antimicrobial	[82]
Whey protein-alginate hydrogels	Whey protein aggregates + alginate	Encapsulation of astaxanthin-chia oil	[124]
Gellan gum-protein hydrogels	Gellan gum + roe protein isolate	Curcumin-loaded functional gels	[123]
Polyion-complex hydrogels	Oppositely charged polymers	Bio-based food packaging alternatives	[59]

various biopolymers used in sustainable packaging by analyzing the entire life cycle of a product from raw material extraction to design, manufacturing, usage, and end-of-life disposal or recycling processes [136]. Although LCA tools are increasingly being adopted in the packaging sector to assess the ecological footprint of packaging materials, there remains a need for more comprehensive data on both the materials themselves and the processes involved in converting them into functional packaging. Fundamentally, LCA aims to assess a product's total environmental burden by tracing its journey from resource extraction to disposal, offering a complete life cycle inventory and impact analysis [137].

Research on the life cycle impacts of biopolymer-based 3D hydrogels used in packaging remains limited. Existing data primarily focus on biopolymers derived from agricultural feedstocks. Previous LCA studies have often concentrated on impact categories such as global warming potential and fossil fuel depletion, as biopolymers are typically viewed as environmentally safer alternatives to conventional plastics [138]. However, some research also highlights potential drawbacks. While biopolymers are celebrated for being derived from renewable resources and for their biodegradability, the agricultural practices required to grow the raw materials can contribute to environmental degradation. Additionally, the energy demands associated with biopolymer production can sometimes surpass those of fossil-based polymers [139].

LCA research has been conducted on various types of plastics, ranging from fossil-based and biodegradable to non-biodegradable and bio-based plastics [140]. However, findings indicate that replacing conventional polymers with biopolymers does not automatically guarantee sustainability. In fact, effective waste management strategies, particularly recycling, often prove more beneficial than simple biodegradation approaches [139]. For instance, a Brazilian study assessed the energy and environmental impacts associated with producing cassava starch-based films. The results indicated that the majority of the ecological burden stemmed from cassava cultivation, the production of fossil-derived additives (such as glycerin and ethanol), and the energy consumed during the film manufacturing process using the casting technique [141]. Although hydrogels fulfill several green chemistry principles, including biodegradability, biocompatibility, and the use of renewable resources, and are generally regarded as safe and eco-friendly, there is still a noticeable gap in LCA data specific to hydrogel materials. While information is available on biopolymer extraction, casting techniques, additives used, and degradation profiles, comprehensive LCA studies on hydrogels remain sparse [139].

A review has been conducted recently on the use of life-cycle assessment (LCA) to measure the sustainability of thermoplastics, in comparison with thermoplastic materials and alternative forms of packaging solutions. The results have shown that, although bio-based and biodegradable packaging materials have the potential to alleviate the negative aspects of conventional plastics, they also raise serious environmental issues that can be associated with the agricultural feedstock involved in their manufacture. It is worth noting that even the emissions of greenhouse gases (GHG) during the biodegradation of these materials can contribute to the increased environmental problems. In addition, the construction of the appropriate infrastructure and the cost of composting further increase the economic cost. To overcome such concerns, the authors suggested that better agricultural methods should be adopted, including the use of renewable energy sources in cultivation, the use of low-energy, nutrient, and fertilizer feedstocks, and the use of modern nutrient-management methods in agriculture [139].

In a different study, Gujjala and Won [142] assessed the LCA of hydrogels made of lignin, which have potential uses in environmental remediation. They also included process design and techno-economic analysis in their study. It was discovered that the manufacturing of lignin hydrogel produced an average of 2.8 kg CO_2 equivalent and 1.1 kg oil equivalent per kilogram of hydrogel. Nevertheless, the implementation of heat recovery into the process provided significant energy savings of 78.9 and 83.4% heating and cooling, respectively, which points to the possibility of introducing the approach to the already available biorefinery systems to develop more sustainable alternatives [142].

Some of the considerations involved in the development of naturally derived hydrogels are the careful selection and synthesis of biopolymers, consumption of water and energy during film casting, use of solvents, cross-linking agents, and additives, e.g., crystalline nanofibers and nanocomposites. An example is the life cycle of cellulose-based hydrogels, which involves cellulose isolation in biomass through the separation of cellulose, lignin, and hemicellulose, the use of solvents and cross-linkers, and the control of the use of energy during the production and degradation of hydrogels. These factors should be taken into consideration in extensive LCA research of hydrogels [142].

Prior to the development of biopolymer-based hydrogels, a comprehensive and integrated evaluation of the sustainability of a particular product that considers not only the environmental footprint of the product but also its socio-economic contribution is needed. This holistic method will aid in the design of the materials that can successfully compete with traditional packaging systems, and at the same time be resources that are used responsibly and in the long term.

7 Future Trends and Challenges

Plastic packaging has been produced in the European continent to the extent of 92 million tons, and it is projected that the volume may increase to 1,124 million tons by the year 2050, assuming that the current trends remain the same [143]. The increase in the use of plastic packaging needs innovative approaches to make it biodegradable. Hydrogels with their various functional attributes have emerged as a potential solution, especially in food packaging. Use of polysaccharide-based hydrogels in food systems has a high regional difference, which can be explained by the fact that there are varying regulatory frameworks, agricultural practices, and local food supply needs. In the European Union, food-contact materials should be in line with EU Reg. No. 10/2011, which defines boundaries of compositions, particular migration limits, and approved additives to use on the biopolymer-based packaging [144]. FDA has acknowledged a number of polysaccharides, such as Alg, pectin, CS, and modified CEL, as Generally Recognized as Safe (GRAS), which makes their application as edible coatings, absorbents, and active packaging a possible use [145]. These authorizations have hastened the commercialization of hydrogel-based systems of moisture control and antimicrobial capabilities on packaged foods.

In resource-constrained societies like Sub-Saharan Africa and sections of Latin America, it is argued that it is better to capitalize on locally available agricultural wastes (e.g., cassava starch, plantain wastes, or seaweed-based alginates) to formulate cost-effective hydrogel packaging solutions. The Brazilian life-cycle assessment (LCA) study on films made of cassava starch revealed that the environmental burden is greatly affected due to the use of agricultural inputs and energy requirements of the film casting, and the strategies to optimize them regionally are necessary [141].

Nanotechnology is an interesting solution to these shortcomings. The use of nanoparticles in hydrogel matrices to form nanocomposite hydrogels has produced a lot of enhancements in mechanical, thermal, and water-resistant properties, which make them more applicable in food packaging [135]. As an example, the incorporation of CEL nanofibrils produced out of cotton into a CS grafted PAA matrix has been reported to improve the swelling behavior of the hydrogel [146]. Examples of nanomaterials that are currently used or are being investigated to be used include carbon nanotubes, CEL nanowhiskers, chitin whiskers, montmorillonite, micro fibrillated cellulose, and starch nanocrystals [13].

The effectiveness of nanotechnological improved packaging material mainly depends on the type of nanomaterials used in the composition of the film. Despite the issue of the cost of nanotechnology integration, research practices worldwide have adopted the strategies of coming up with new innovative approaches for the development of commercially viable nanocomposites of packaging materials. In addition to this, the creation of new packaging systems has the potential to minimize food waste. These systems integrate the capabilities of active and intelligent packaging, hence maintaining the quality of food, extending shelf life, presenting real-time information on the status of products, and being environmentally friendly in disposal [147].

Another promising field in the use of hydrogel is the use of perforated packaging, especially in the preservation of highly perishable food products [148]. This type of packaging includes micro- or macro-scale perforations, which allow the exchange of respiratory gases (e.g., oxygen and carbon dioxide) and hence, preserve freshness in the produce. The holes also increase the circulation of cold air in storage and transportation, which is really imperative to the quality of products [149].

The concept of using hydrogels has interesting potential in this regard because of their cooling characteristics, which can be used to enhance the efficiency of cold-chain horticultural produce. Low temperatures usually store fruits and vegetables in perforated containers to reduce the loss of moisture and shriveling (cucumbers, grapes, tomatoes, avocados, and berries). As an illustration, the freshness of grapes is usually maintained in perforated mono-PET or multi-layered plastic containers in bigger crates when transporting them [150].

Although perforated packaging made of plastic has been effective in increasing shelf life and reducing food waste, it has a high impact on environmental pollution due to the build-up of plastic waste. To overcome this, it can be proposed to use a dry or pre-hydrated hydrogel membrane in the packaging systems as a more environmentally friendly option. Their cross-linked, porous network structure is 3-dimensional and allows the exchange of gases and moisture to be controlled, which is why they are useful in keeping the optimal microenvironment surrounding perishable foods. Waste sorting is very important in maintaining sustainability in the environment. Taneepanichskul et al. [151] have exhaustively discussed the different end-of-life scenarios of biodegradable and compostable films. In the same way, in the case of biodegradable hydrogel-based packaging materials, sorting is necessary to increase the rate of composting and reduce contamination in the recycling streams. Before commercial implementation, relevant certifications, including TÜV AUSTRIA BELGIUM labeling or DIN CERTCO certification, are impossible to do without. According to Paul-Pont et al. [152], there is still a big gap in assessing the materials in the real-life environment because it is still unclear what happens to them in the environment, their decomposition rates, and toxicological issues.

Other toxicological issues, as posed by various studies [153, 154] on bio-based and biodegradable materials, also need to be incorporated into hydrogel-based films when produced at a commercial level. Without the existing laws that require certification, the best practice would be to use these materials along with food waste. Social organizations like BioPak New Zealand are already moving towards this reality. To make compostable packaging, such as hydrogel films, a part of responsible recycling and composting streams in the future will be necessary.

8 Conclusion

Hydrogels made of polysaccharides have become integrated, adaptable, and viable materials to apply in various activities in food preservation, active packaging, nutrient delivery, and quality control. The chapter highlighted the use of their biocompatibility, tunable physicochemical properties, and the ability to impart active agents, making them promising alternatives to traditional synthetic materials. A comparative analysis of hydrogel systems shows obvious benefits in moisture retention, mechanical stability, controlled release behavior, and incorporation of smart sensing properties, especially in combinations with new nanomaterials and bioactive compounds. Although these advantages are there, there are still a number of issues, such as low mechanical strength, inconsistency in quality of raw materials, high costs of production, and regulatory uncertainties in the international markets. To solve the problems, special optimisation of polymer blends, reinforcing agents, and standardisation of fabrication procedures need to be developed. In addition, local regulatory frameworks and supply-chain demands shall be taken into account during the initial stage of product development to make sure that the products are safe and scalable to the food systems.

Moving forward, future research should prioritize (i) designing multifunctional hydrogels capable of combining antimicrobial, antioxidant, and sensing properties; (ii) advancing green and energy-efficient fabrication techniques; (iii) conducting comprehensive life cycle and migration studies to support regulatory approval; and (iv) developing region-specific formulations that leverage local biopolymer resources. Interdisciplinary collaboration among material scientists, food technologists, and regulatory bodies will be essential to translate laboratory-scale hydrogel innovations into commercially viable solutions. By addressing these research and implementation gaps, polysaccharide-based hydrogels can play a transformative role in shaping the next generation of sustainable and intelligent food systems.

References

1. Cao Y, Mezzenga R. Design principles of food gels. Nature Food. 2020;1(2):106–18.
2. Klein M, Poverenov E. Natural biopolymer-based hydrogels for use in food and agriculture. J Sci Food Agric. 2020;100(6):2337–47.
3. Reddy MS, Ponnamma D, Choudhary R, Sadasivuni KK. A comparative review of natural and synthetic biopolymer composite scaffolds. Polym [Internet]. 2021;13(7):1105.
4. Ahmad S, Ahmad M, Manzoor K, Purwar R, Ikram S. A review on latest innovations in natural gums based hydrogels: Preparations & applications. Int J Biol Macromol. 2019;136:870–90.
5. Ahmad Z, Salman S, Khan SA, Amin A, Rahman ZU, Al-Ghamdi YO, et al. Versatility of hydrogels: from synthetic strategies, classification, and properties to biomedical applications. Gels [Internet]. 2022;8(3)
6. Mirzaei M, Okoro OV, Nie L, Petri DF, Shavandi A. Protein-based 3D biofabrication of biomaterials. Bioeng [Internet]. 2021;8(4)
7. Shi Z, Gao X, Ullah MW, Li S, Wang Q, Yang G. Electroconductive natural polymer-based hydrogels. Biomaterials. 2016;111:40–54.
8. Getachew AT, Jacobsen C, Holdt SL. Emerging technologies for the extraction of marine phenolics: opportunities and challenges. Mar Drugs [Internet]. 2020;18(8)
9. Lima CSAD, Balogh TS, Varca JPRO, Varca GHC, Lugão AB, Camacho-Cruz AL, et al. An updated review of macro, micro, and nanostructured hydrogels for biomedical and pharmaceutical applications. Pharm [Internet]. 2020;12(10)
10. Rahman MM, Netravali AN, Tiimob BJ, Apalangya V, Rangari VK. Bio-inspired "green" nanocomposite using hydroxyapatite synthesized from eggshell waste and soy protein. J Appl Polym Sci. 2016;133(22)
11. Arakawa C, Ng R, Tan S, Kim S, Wu B, Lee M. Photopolymerizable chitosan–collagen hydrogels for bone tissue engineering. J Tissue Eng Regenerative Med. 2017;11(1):164–74.
12. Li J, Xin J, Yin L. Hydrogel: diversity of structures and applications in food science. Food Rev Int. 2021;37(3):313–72.
13. Batista RA, Espitia PJP, Quintans JSS, Freitas MM, Cerqueira MÂ, Teixeira JA, et al. Hydrogel as an alternative structure for food packaging systems. Carbohydr Polym. 2019;205:106–16.
14. Nangia S, Sudhir W, Katyal D. A review on environmental applications of chitosan biopolymeric hydrogel based composites. J Macromol Sci Part A. 2018;55(11-12):747–63.
15. Reakasame S, Boccaccini AR. Oxidized alginate-based hydrogels for tissue engineering applications: a review. Biomacromolecules. 2018;19(1):3–21.
16. Hu X, Wang Y, Zhang L, Xu M. Construction of self-assembled polyelectrolyte complex hydrogel based on oppositely charged polysaccharides for sustained delivery of green tea polyphenols. Food Chem. 2020;306:125632.
17. Liu K, Chen Y-Y, Zha X-Q, Li Q-M, Pan L-H, Luo J-P. Research progress on polysaccharide/protein hydrogels: preparation method, functional property and application as delivery systems for bioactive ingredients. Food Res Int. 2021;147:110542.
18. Senturk Parreidt T, Müller K, Schmid M. Alginate-based edible films and coatings for food packaging applications. Foods [Internet]. 2018;7(10)
19. Wu T, Huang J, Jiang Y, Hu Y, Ye X, Liu D, et al. Formation of hydrogels based on chitosan/alginate for the delivery of lysozyme and their antibacterial activity. Food Chem. 2018;240:361–9.
20. Le XT, Rioux L-E, Turgeon SL. Formation and functional properties of protein–polysaccharide electrostatic hydrogels in comparison to protein or polysaccharide hydrogels. Adv Colloid Interface Sci. 2017;239:127–35.

21. Hosseini H, Jafari SM. Introducing nano/microencapsulated bioactive ingredients for extending the shelf-life of food products. Adv Colloid Interface Sci. 2020;282:102210.
22. Ravani L, Esposito E, Bories C, Moal VL-L, Loiseau PM, Djabourov M, et al. Clotrimazole-loaded nanostructured lipid carrier hydrogels: thermal analysis and in vitro studies. Int J Pharm. 2013;454(2):695–702.
23. Pushpamalar J, Meganathan P, Tan HL, Dahlan NA, Ooi L-T, Neerooa BN, et al. Development of a polysaccharide-based hydrogel drug delivery system (DDS): an update. Gels [Internet]. 2021;7(4)
24. Nam S, Hu KH, Butte MJ, Chaudhuri O. Strain-enhanced stress relaxation impacts nonlinear elasticity in collagen gels. Proceed Nat Acad Sci. 2016;113(20):5492–7.
25. Nair AV, Raman M, Doble M. Chapter 13—Polysaccharide-based hydrogels for targeted drug delivery. In: Holban A-M, Grumezescu AM, editors. Materials for biomedical engineering. Elsevier; 2019. p. 343–82.
26. Nagam SP, Jyothi AN, Poojitha J, Aruna S, Nadendla RR. A comprehensive review on hydrogels. Int J Curr Pharm Res. 2016;8(1):19–23.
27. Manzoor A, Dar AH, Pandey VK, Shams R, Khan S, Panesar PS, et al. Recent insights into polysaccharide-based hydrogels and their potential applications in food sector: a review. Int J Biol Macromol. 2022;213:987–1006.
28. Chirani N, Yahia L, Gritsch L, Motta FL, Chirani S, Farè S. History and applications of hydrogels. J Biomed Sci. 2015;4(02):1–23.
29. Ali A, Ahmed S. Recent advances in edible polymer based hydrogels as a sustainable alternative to conventional polymers. J Agric Food Chem. 2018;66(27):6940–67.
30. Cao H, Duan L, Zhang Y, Cao J, Zhang K. Current hydrogel advances in physicochemical and biological response-driven biomedical application diversity. Signal Transduction Targeted Therapy. 2021;6(1):426.
31. parhi R. Cross-linked hydrogel for pharmaceutical applications: a review. Adv Pharm Bull. 2017;7(4):515–30.
32. Bae J, Li Y, Zhang J, Zhou X, Zhao F, Shi Y, et al. A 3D nanostructured hydrogel-framework-derived high-performance composite polymer lithium-ion electrolyte. Angew Chem Int Edit. 2018;57(8):2096–100.
33. Liu W, Erol O, Gracias DH. 3D printing of an in situ grown MOF hydrogel with tunable mechanical properties. ACS Appl Mater Interfaces. 2020;12(29):33267–75.
34. Kang KS, Lee S-I, Hong JM, Lee JW, Cho HY, Son JH, et al. Hybrid scaffold composed of hydrogel/3D-framework and its application as a dopamine delivery system. J Controlled Release. 2014;175:10–6.
35. Song M, Wang J, He J, Kan D, Chen K, Lu J. Synthesis of hydrogels and their progress in environmental remediation and antimicrobial application. Gels. 2022;9(1):16.
36. Chen Z, Song S, Zeng H, Ge Z, Liu B, Fan Z. 3D printing MOF nanozyme hydrogel with dual enzymatic activities and visualized glucose monitoring for diabetic wound healing. Chem Eng J. 2023;471:144649.
37. Mayorova OA, Jolly BCN, Verkhovskii RA, Plastun VO, Sindeeva OA, Douglas TEL. pH-sensitive dairy-derived hydrogels with a prolonged drug release profile for cancer treatment. Mater [Internet]. 2021;14(4)
38. Suflet DM, Popescu I, Pelin IM, Ichim DL, Daraba OM, Constantin M, et al. Dual cross-linked chitosan/PVA hydrogels containing silver nanoparticles with antimicrobial properties. Pharm [Internet]. 2021;13(9)
39. El-Husseiny HM, Mady EA, Hamabe L, Abugomaa A, Shimada K, Yoshida T, et al. Smart/stimuli-responsive hydrogels: cutting-edge platforms for tissue engineering and other biomedical applications. Mater Today Bio. 2022;13:100186.
40. Begam T, Nagpal AK, Singhal R. A comparative study of swelling properties of hydrogels based on poly(acrylamide-co-methyl methacrylate) containing physical and chemical crosslinks. J Appl Polym Sci. 2003;89(3):779–86.
41. Temenoff JS, Athanasiou KA, Lebaron RG, Mikos AG. Effect of poly(ethylene glycol) molecular weight on tensile and swelling properties of oligo(poly(ethylene glycol) fumarate) hydrogels for cartilage tissue engineering. J Biomed Mater Res. 2002;59(3):429–37.

42. Niemczyk-Soczynska B, Zaszczyńska A, Zabielski K, Hydrogel SP. Electrospun and composite materials for bone/cartilage and neural tissue engineering. Mater [Internet]. 2021;14(22)
43. Singh B, Sharma S, Dhiman A. Acacia gum polysaccharide based hydrogel wound dressings: Synthesis, characterization, drug delivery and biomedical properties. Carbohydr Polym. 2017;165:294–303.
44. Weng T, Zhang W, Xia Y, Wu P, Yang M, Jin R, et al. 3D bioprinting for skin tissue engineering: current status and perspectives. J Tissue Eng. 2021;12:20417314211028574.
45. Ali A, Hussain MA, Haseeb MT, Abbas Bukhari SN, Tabassum T, Farid-ul-Haq M, et al. A pH-responsive, biocompatible, and non-toxic citric acid cross-linked polysaccharide-based hydrogel from Salvia spinosa L. offering zero-order drug release. J Drug Deliv Sci Technol. 2022;69:103144.
46. Naghieh S, Chen X. Printability—A key issue in extrusion-based bioprinting. J Pharm Anal. 2021;11(5):564–79.
47. Yu H, Gao R, Liu Y, Fu L, Zhou J, Li L. Stimulus-responsive hydrogels as drug delivery systems for inflammation targeted therapy. Adv Sci. 2024;11(1):2306152.
48. Zhuo S, Zhang F, Yu J, Zhang X, Yang G, Liu X. pH-sensitive biomaterials for drug delivery. Mol [Internet]. 2020;25(23)
49. Rizwan M, Yahya R, Hassan A, Yar M, Azzahari AD, Selvanathan V, et al. pH sensitive hydrogels in drug delivery: brief history, properties, swelling, and release mechanism. Mater Sel Appl Polym [Internet]. 2017;9(4)
50. Klouda L, Mikos AG. Thermoresponsive hydrogels in biomedical applications. Eur J Pharm Biopharm. 2008;68(1):34–45.
51. Wang M, Wang C, Chen M, Xi Y, Cheng W, Mao C, et al. Efficient angiogenesis-based diabetic wound healing/skin reconstruction through bioactive antibacterial adhesive ultraviolet shielding nanodressing with exosome release. ACS Nano. 2019;13(9):10279–93.
52. Huang H, Qi X, Chen Y, Wu Z. Thermo-sensitive hydrogels for delivering biotherapeutic molecules: a review. Saudi Pharm J. 2019;27(7):990–9.
53. Saha S, Banskota S, Roberts S, Kirmani N, Chilkoti A. Engineering the architecture of elastin-like polypeptides: from unimers to hierarchical self-assembly. Adv Ther. 2020;3(3):1900164.
54. Lu X, Chen J, Guo Z, Zheng Y, Rea MC, Su H, et al. Using polysaccharides for the enhancement of functionality of foods: a review. Trends Food Sci Technol. 2019;86:311–27.
55. Chen H, Gan J, Ji A, Song S, Yin L. Development of double network gels based on soy protein isolate and sugar beet pectin induced by thermal treatment and laccase catalysis. Food Chem. 2019;292:188–96.
56. Zhang Y, Fan Z, Xu C, Fang S, Liu X, Li X. Tough biohydrogels with interpenetrating network structure by bienzymatic crosslinking approach. Eur Polym J. 2015;72:717–25.
57. Gharibzahedi SMT, Roohinejad S, George S, Barba FJ, Greiner R, Barbosa-Cánovas GV, et al. Innovative food processing technologies on the transglutaminase functionality in protein-based food products: trends, opportunities and drawbacks. Trends Food Sci Technol. 2018;75:194–205.
58. Sorde KL, Ananthanarayan L. Effect of transglutaminase treatment on properties of coconut protein-guar gum composite film. LWT. 2019;115:108422.
59. Hou J-J, Guo J, Wang J-M, He X-T, Yuan Y, Yin S-W, et al. Edible double-network gels based on soy protein and sugar beet pectin with hierarchical microstructure. Food Hydrocolloids. 2015;50:94–101.
60. Anamizu M, Tabata Y. Design of injectable hydrogels of gelatin and alginate with ferric ions for cell transplantation. Acta Biomater. 2019;100:184–90.
61. Liu S, Wang X, Peng Y, Wang Z, Ran R. Highly stretchable, strain-sensitive, and antifreezing macromolecular microsphere composite starch-based hydrogel. Macromol Mater Eng. 2021;306(9):2100198.
62. Du Y-N, Xu S-Q, Wang Y-Q, Wang X-C, Wu H-T. Fabrication and characterization of cold-set large yellow croaker (Pseudosciaena crocea) roe protein isolate gels. LWT. 2022;158:113135.

63. Xu Y, Jiang Q, Xia W. Effect of glucono-δ-lactone acidification and heat treatment on the physicochemical properties of silver carp mince. LWT—Food Sci Technol. 2011;44(9):1952–7.
64. Li X, He X, Mao L, Gao Y, Yuan F. Modification of the structural and rheological properties of β-lactoglobulin/κ-carrageenan mixed gels induced by high pressure processing. J Food Eng. 2020;274:109851.
65. Su J, Wang L, Dong W, Wei J, Liu X, Yan J, Ren F, Yuan F, Wang P. Fabrication and characterization of ultra-high-pressure (UHP)-induced whey protein isolate/κ-carrageenan composite emulsion gels for the delivery of curcumin. Front Nutr. 2022;9:839761.
66. Zheng X, Zou B, Zhang J, Cai W, Na X, Du M, et al. Recent advances of ultrasound-assisted technology on aquatic protein processing: Extraction, modification, and freezing/thawing-induced oxidation. Trends Food Sci Technol. 2024;144:104309.
67. Xe R, Li C, Yang F, Huang Y, Huang C, Zhang K, et al. Comparison of hydrodynamic and ultrasonic cavitation effects on soy protein isolate functionality. J Food Eng. 2020;265:109697.
68. Wu D, Tu M, Wang Z, Wu C, Yu C, Battino M, et al. Biological and conventional food processing modifications on food proteins: Structure, functionality, and bioactivity. Biotechnol Adv. 2020;40:107491.
69. Mohammed A, Elshaer A, Sareh P, Elsayed M, Hassanin H. Additive manufacturing technologies for drug delivery applications. Int J Pharm. 2020;580:119245.
70. Trenfield SJ, Awad A, Goyanes A, Gaisford S, Basit AW. 3D printing pharmaceuticals: drug development to frontline care. Trends Pharmacol Sci. 2018;39(5):440–51.
71. Tappa K, Jammalamadaka U. Novel biomaterials used in medical 3D printing techniques. J Funct Biomater [Internet]. 2018;9(1)
72. Melocchi A, Briatico-Vangosa F, Uboldi M, Parietti F, Turchi M, von Zeppelin D, et al. Quality considerations on the pharmaceutical applications of fused deposition modeling 3D printing. Int J Pharm. 2021;592:119901.
73. Cailleaux S, Sanchez-Ballester NM, Gueche YA, Bataille B, Soulairol I. Fused deposition modeling (FDM), the new asset for the production of tailored medicines. J Controlled Release. 2021;330:821–41.
74. Cooper K. Rapid prototyping technology: selection and application. CRC press; 2001.
75. Walker JL, Santoro M. 9—Processing and production of bioresorbable polymer scaffolds for tissue engineering. In: Perale G, Hilborn J, editors. Bioresorbable polymers for biomedical applications. Woodhead Publishing; 2017. p. 181–203.
76. Bähr F, Westkämper E. Correlations between influencing parameters and quality properties of components produced by fused deposition modeling. Proced CIRP. 2018;72:1214–9.
77. Grigora M-E, Terzopoulou Z, Tsongas K, Klonos P, Kalafatakis N, Bikiaris DN, et al. Influence of reactive chain extension on the properties of 3D printed poly(lactic acid) constructs. Polym [Internet]. 2021;13(9)
78. Garzon-Hernandez S, Arias A, Garcia-Gonzalez D. A continuum constitutive model for FDM 3D printed thermoplastics. Compos Part B: Eng. 2020;201:108373.
79. Luo S, Zhang X. High-quality 3D printing of ethylene vinyl acetate with direct pellet-based FDM for medical applications: Mechanical analysis, energy absorption and recovery evaluation. J Mech Behav Biomed Mater. 2023;148:106231.
80. Poomathi N, Singh S, Prakash C, Subramanian A, Sahay R, Cinappan A, et al. 3D printing in tissue engineering: a state of the art review of technologies and biomaterials. Rapid Prototyping J. 2020;26(7):1313–34.
81. Ruiz-Cantu L, Gleadall A, Faris C, Segal J, Shakesheff K, Yang J. Multi-material 3D bioprinting of porous constructs for cartilage regeneration. Mater Sci Eng C. 2020;109:110578.
82. Wulle F, Gorke O, Schmidt S, Nistler M, Tovar GEM, Riedel O, et al. Multi-axis 3D printing of gelatin methacryloyl hydrogels on a non-planar surface obtained from magnetic resonance imaging. Add Manuf. 2022;50:102566.
83. Boularaoui S, Al Hussein G, Khan KA, Christoforou N, Stefanini C. An overview of extrusion-based bioprinting with a focus on induced shear stress and its effect on cell viability. Bioprinting. 2020;20:e00093.

84. Angelats Lobo D, Ginestra P. Cell bioprinting: the 3D-Bioplotter™ case. Mater [Internet]. 2019;12(23)
85. Ligon SC, Liska R, Stampfl J, Gurr M, Mülhaupt R. Polymers for 3D printing and customized additive manufacturing. Chem Rev. 2017;117(15):10212–90.
86. Hölzl K, Lin S, Tytgat L, Van Vlierberghe S, Gu L, Ovsianikov A. Bioink properties before, during and after 3D bioprinting. Biofabrication. 2016;8(3):032002.
87. Catoira MC, González-Payo J, Fusaro L, Ramella M, Boccafoschi F. Natural hydrogels R&D process: technical and regulatory aspects for industrial implementation. J Mater Sci: Mater Med. 2020;31(8):64.
88. Schwab A, Levato R, D'Este M, Piluso S, Eglin D, Malda J. Printability and shape fidelity of bioinks in 3D bioprinting. Chem Rev. 2020;120(19):11028–55.
89. Azad MA, Olawuni D, Kimbell G, Badruddoza AZ, Hossain MS, Sultana T. Polymers for extrusion-based 3D printing of pharmaceuticals: a holistic materials–process perspective. Pharm [Internet]. 2020;12(2):124.
90. Kyle S, Jessop ZM, Al-Sabah A, Whitaker IS. 'Printability' of candidate biomaterials for extrusion based 3D printing: state-of-the-art. Adv Healthc Mater. 2017;6(16):1700264.
91. Patpatiya P, Chaudhary K, Shastri A, Sharma S. A review on polyjet 3D printing of polymers and multi-material structures. Proceed Inst Mech Eng Part C: J Mech Eng Sci. 2022;236:095440622210795.
92. Puligundla P, Jung J, Ko S. Carbon dioxide sensors for intelligent food packaging applications. Food Control. 2012;25(1):328–33.
93. Regubalan B, Pandit P, Maiti S, Nadathur GT, Mallick A. Potential bio-based edible films, foams, and hydrogels for food packaging. In: Ahmed S, editor. Bio-based materials for food packaging: green and sustainable advanced packaging materials. Singapore: Springer Singapore; 2018. p. 105–23.
94. Roy N, Saha N, Kitano T, Saha P. Biodegradation of PVP–CMC hydrogel film: a useful food packaging material. Carbohydr Polym. 2012;89(2):346–53.
95. Anita BB, Thatheyus A, Ramya D. Biodegradation of carboxymethyl cellulose using Aspergillus flavus. Sci Int. 2013;1:85–91.
96. Fernández A, Cava D, Ocio MJ, Lagarón JM. Perspectives for biocatalysts in food packaging. Trends Food Sci Technol. 2008;19(4):198–206.
97. Shewan HM, Stokes JR. Review of techniques to manufacture micro-hydrogel particles for the food industry and their applications. J Food Eng. 2013;119(4):781–92.
98. Farris S, Schaich KM, Liu L, Piergiovanni L, Yam KL. Development of polyion-complex hydrogels as an alternative approach for the production of bio-based polymers for food packaging applications: a review. Trends Food Sci Technol. 2009;20(8):316–32.
99. de Azeredo HMC. Antimicrobial nanostructures in food packaging. Trends Food Sci Technol. 2013;30(1):56–69.
100. Celli A, Sabaa MW, Jyothi AN, Kalia S. Chitosan and starch-based hydrogels via graft copolymerization. In: Kalia S, editor. Polymeric hydrogels as smart biomaterials. Cham: Springer International Publishing; 2016. p. 189–234.
101. Wang L-F, Rhim J-W. Preparation and application of agar/alginate/collagen ternary blend functional food packaging films. Int J Biol Macromol. 2015;80:460–8.
102. Risch SJ. Food packaging history and innovations. J Agric Food Chem. 2009;57(18):8089–92.
103. Baek S, Kim D, Jeon SL, Seo J. Preparation and characterization of pH-responsive poly(N,N-dimethyl acrylamide-co-methacryloyl sulfadimethoxine) hydrogels for application as food freshness indicators. React Funct Polym. 2017;120:57–65.
104. Manzocco L, Anese M, Calligaris S, Quarta B, Nicoli MC. Use of monoglyceride hydrogel for the production of low fat short dough pastry. Food Chem. 2012;132(1):175–80.
105. Incoronato AL, Conte A, Buonocore GG, Del Nobile MA. Agar hydrogel with silver nanoparticles to prolong the shelf life of Fior di Latte cheese. J Dairy Sci. 2011;94(4):1697–704.
106. Godoi FC, Prakash S, Bhandari BR. 3d printing technologies applied for food design: status and prospects. J Food Eng. 2016;179:44–54.

107. Leelapunnawut S, Ngamwonglumlert L, Devahastin S, Derossi A, Caporizzi R, Chiewchan N. Effects of texture modifiers on physicochemical properties of 3D-printed meat mimics from pea protein isolate-alginate gel mixture. Foods [Internet]. 2022;11(24)
108. Bitencourt BS, Guedes JS, Saliba ASMC, Sartori AGO, Torres LCR, Amaral JEPG, et al. Mineral bioaccessibility in 3D printed gels based on milk/starch/κ-carrageenan for dysphagic people. Food Res Int. 2023;170:113010.
109. Liu Z, Ha S, Guo C, Xu D, Hu L, Li H, et al. 3D printing of curcumin enriched Pickering emulsion gel stablized by pea protein-carrageenan complexes. Food Hydrocolloids. 2024;146:109170.
110. Brownlee IA, Allen A, Pearson JP, Dettmar PW, Havler ME, Atherton MR, et al. Alginate as a source of dietary fiber. Crit Rev Food Sci Nutr. 2005;45(6):497–510.
111. Benito-Peña E, González-Vallejo V, Rico-Yuste A, Barbosa-Pereira L, Cruz JM, Bilbao A, et al. Molecularly imprinted hydrogels as functional active packaging materials. Food Chem. 2016;190:487–94.
112. Faridi Esfanjani A, Jafari SM. Biopolymer nano-particles and natural nano-carriers for nano-encapsulation of phenolic compounds. Colloids Surf B: Biointerfaces. 2016;146:532–43.
113. Jafari SM. 1—An overview of nanoencapsulation techniques and their classification. In: Jafari SM, editor. Nanoencapsulation technologies for the food and nutraceutical industries. Academic Press; 2017. p. 1–34.
114. Jafari SM. An introduction to nanoencapsulation techniques for the food bioactive ingredients. Nanoencapsulation Food Bioact Ingredients. 2017:1–62.
115. Snyders R, Shingel KI, Zabeida O, Roberge C, Faure M-P, Martinu L, et al. Mechanical and microstructural properties of hybrid poly(ethylene glycol)–soy protein hydrogels for wound dressing applications. J Biomed Mater Res Part A. 2007;83A(1):88–97.
116. Lima AC, Praveen S, Mano JF. Production methodologies of polymeric and hydrogel particles for drug delivery applications. Expert Opin Drug Deliv. 2012;9(2):231–48.
117. Vilela C, Pinto RJB, Pinto S, Marques P, Silvestre A, Barros CSRF. Polysaccharide based hybrid materials: metals and metal oxides, graphene and. Carbon Nanotubes: Springer; 2018.
118. Chinna Reddy P, Chaitanya KS, Madhusudan RY. A review on bioadhesive buccal drug delivery systems: current status of formulation and evaluation methods. Daru: J Fac Pharm Tehran Univ Med Sci. 2011;19(6):385–403.
119. Sharpe LA, Daily AM, Horava SD, Peppas NA. Therapeutic applications of hydrogels in oral drug delivery. Expert Opin Drug Deliv. 2014;11(6):901–15.
120. Zhang Z, Zhang R, Chen L, Tong Q, McClements DJ. Designing hydrogel particles for controlled or targeted release of lipophilic bioactive agents in the gastrointestinal tract. Eur Polym J. 2015;72:698–716.
121. Mahinroosta M, Jomeh Farsangi Z, Allahverdi A, Shakoori Z. Hydrogels as intelligent materials: A brief review of synthesis, properties and applications. Mater Today Chem. 2018;8:42–55.
122. Yan J-N, Cui X-F, Jiang X-Y, Li L, Sun W, Wu H-T. Complex characterization and formation mechanism of scallop (Patinopecten yessoensis) protein hydrolysates/κ-carrageenan/konjac gum composite gels. J Food Sci. 2022;87(7):2953–64.
123. Du Y-N, Yan J-N, Xu S-Q, Wang Y-Q, Wang X-C, Wu H-T. Formation and characteristics of curcumin-loaded binary gels formed from large yellow croaker (Pseudosciaena crocea) roe protein isolate and gellan gum. Food Chem. 2023;405:134759.
124. Espinaco BY, Niizawa I, Zorrilla SE, Sihufe GA. Whey protein aggregates-alginate composite gel for astaxanthin-chia oil encapsulation. Eur J Lipid Sci Technol. 2023;125(8):2300010.
125. Wijaya W, Patel AR, Setiowati AD, Van der Meeren P. Functional colloids from proteins and polysaccharides for food applications. Trends Food Sci Technol. 2017;68:56–69.
126. Jiang S, Ma Y, Wang Y, Wang R, Zeng M. Effect of κ-carrageenan on the gelation properties of oyster protein. Food Chem. 2022;382:132329.
127. Zhang C, Chen L, Teng H. Phase behavior of the gelation process of myofibrillar protein-curdlan blended system: discussion based on rheology and gel properties. Food Chem. 2024;437:137839.

128. Zhang T, Chen S, Xu X, Zhuang X, Chen Y, Xue Y, et al. Effects of konjac glucomannan on physical properties and microstructure of fish myofibrillar protein gel: phase behaviours involved. Food Hydrocolloids. 2023;134:108034.
129. Dai L, Xi X, Li X, Li W, Du Y, Lv Y, et al. Self-assembled all-polysaccharide hydrogel film for versatile paper-based food packaging. Carbohydr Polym. 2021;271:118425.
130. Wu S, Wang W, Yan K, Ding F, Shi X, Deng H, et al. Electrochemical writing on edible polysaccharide films for intelligent food packaging. Carbohydr Polym. 2018;186:236–42.
131. Cazón P, Velazquez G, Ramírez JA, Vázquez M. Polysaccharide-based films and coatings for food packaging: a review. Food Hydrocolloids. 2017;68:136–48.
132. Şen F, Uzunsoy İ, Baştürk E, Kahraman MV. Antimicrobial agent-free hybrid cationic starch/sodium alginate polyelectrolyte films for food packaging materials. Carbohydr Polym. 2017;170:264–70.
133. Benavides S, Villalobos-Carvajal R, Reyes JE. Physical, mechanical and antibacterial properties of alginate film: Effect of the crosslinking degree and oregano essential oil concentration. J Food Eng. 2012;110(2):232–9.
134. Wu S, Wang W, Yan K, Ding F, Shi X, Deng H, et al. Electrochemical writing on edible polysaccharide films for intelligent food packaging. Carbohydr Polym. 2018;186:236–42.
135. Vashist A, Kaushik A, Ghosal A, Bala J, Nikkhah-Moshaie R, Wani AW, et al. Nanocomposite hydrogels: advances in nanofillers used for nanomedicine. Gels [Internet]. 2018;4(3)
136. Enríquez-Martínez V, Niembro-García IJ, Marmolejo-Saucedo JA. A life cycle assessment (LCA) of antibacterial gel production. In: Computer science and engineering in health services. Cham: Springer International Publishing; 2021.
137. Yates MR, Barlow CY. Life cycle assessments of biodegradable, commercial biopolymers—A critical review. Resour Conserv Recycl. 2013;78:54–66.
138. Hottle TA, Bilec MM, Landis AE. Biopolymer production and end of life comparisons using life cycle assessment. Resour Conserv Recycl. 2017;122:295–306.
139. Banerjee R, Ray SS. Sustainability and life cycle assessment of thermoplastic polymers for packaging: a review on fundamental principles and applications. Macromol Mater Eng. 2022;307(6):2100794.
140. Semba T, Sakai Y, Sakanishi T, Inaba A. Greenhouse gas emissions of 100% bio-derived polyethylene terephthalate on its life cycle compared with petroleum-derived polyethylene terephthalate. J Cleaner Prod. 2018;195:932–8.
141. de Léis CM, Nogueira AR, Kulay L, Tadini CC. Environmental and energy analysis of biopolymer film based on cassava starch in Brazil. J Clean Prod. 2017;143:76–89.
142. Gujjala LKS, Won W. Process development, techno-economic analysis and life-cycle assessment for laccase catalyzed synthesis of lignin hydrogel. Bioresour Technol. 2022;364:128028.
143. Kim YH, Priyadarshi R, Kim J-W, Kim J, Alekseev DG, Rhim J-W. 3D-printed pectin/carboxymethyl cellulose/ZnO bio-inks: comparative analysis with the solution casting method. Polym [Internet]. 2022;14(21)
144. Union E. Commission Regulation (EU) No 10/2011 of 14 January 2011 on plastic materials and articles intended to come into contact with food 211 [Available from: https://eur-lex.europa.eu/eli/reg/2011/10/oj/eng.
145. Administration UFaD. Inventory of Food Contact Substances Listed in 21 CFR [Available from: https://www.fda.gov/food/packaging-food-contact-substances-fcs/inventory-food-contact-substances-listed-21-cfr
146. Spagnol C, Rodrigues FHA, Neto AGVC, Pereira AGB, Fajardo AR, Radovanovic E, et al. Nanocomposites based on poly(acrylamide-co-acrylate) and cellulose nanowhiskers. Eur Polym J. 2012;48(3):454–63.
147. Chen S, Brahma S, Mackay J, Cao C, Aliakbarian B. The role of smart packaging system in food supply chain. J Food Sci. 2020;85(3):517–25.
148. Rodov V, Porat R, Sabag A, Kochanek B, Friedman H. Microperforated compostable packaging extends shelf life of ethylene-treated banana fruit. Foods [Internet]. 2022;11(8)
149. Foudazi R, Zowada R, Manas-Zloczower I, Feke DL. Porous hydrogels: present challenges and future opportunities. Langmuir. 2023;39(6):2092–111.

150. Saha N, Zaandra O, Bandyopadhyay S, Saha P. Bacterial cellulose based hydrogel film for sustainable food packaging. In: Katiyar V, Gupta R, Ghosh T, editors. Advances in sustainable polymers: processing and applications. Singapore: Springer Singapore; 2019. p. 237–45.
151. Taneepanichskul N, Hailes HC, Miodownik M. Automatic identification and classification of compostable and biodegradable plastics using hyperspectral imaging. Front Sustain. 2023;4:2023.
152. Paul-Pont I, Tallec K, Gonzalez-Fernandez C, Lambert C, Vincent D, Mazurais D, et al. Constraints and priorities for conducting experimental exposures of marine organisms to microplastics. Front Mar Sci. 2018;5:2018.
153. Capolupo M, Rafiq A, Coralli I, Alessandro T, Valbonesi P, Fabbri D, et al. Bioplastic leachates characterization and impacts on early larval stages and adult mussel cellular, biochemical and physiological responses. Environ Pollut. 2023;319:120951.
154. Zimmermann L, Dombrowski A, Völker C, Wagner M. Are bioplastics and plant-based materials safer than conventional plastics? In vitro toxicity and chemical composition. Environ Int. 2020;145:106066.

Polysaccharide-Based 3D-Printed Nanocomposites: Fabrication and Applications

Stephen Sunday Emmanuel and Ademidun Adeola Adesibikan

Abstract Polysaccharides are intricate biopolymers abundantly found in nature and hold astounding potential in fostering sustainability initiatives by providing eco-friendly solutions for waste management, pollutant remediation, and the development of recyclable materials, which aligns with the global agenda for a sustainable future. Notably, advancements in 3-D technology have revolutionized the world of material synthesis, including the polysaccharide fabrication landscape by enabling layer-by-layer architectural designing of materials with improved properties such as tunability, biocompatibility, recyclability, specificity, selectivity, and adsorptivity, which are important in various applications. Thus, this book chapter focuses on 3D-engineered polysaccharide nanocomposites (TDPSNC) for different applications. Firstly, this work presents various 3-D strategies for the synthesis of TDPSNC to set the tone of the study. Thereafter, recent advances in environmental, energy, and medical applications of various TDPSNCs, especially in environmental remediation, were pragmatically discussed with a specific focus on their performance. Furthermore, their recyclability/biodegradability potential was analyzed to provide empirical clues about their industrial applicability, cost-effectiveness, and circular economy value. In the end, challenges, frontiers, and future prospects in the areas of artificial intelligence, machine learning, and neural networks for TDPSM properties manipulation were highlighted to advance the field.

Keywords Chitosan · Cellulose · Biopolymer · Supercapacitor · Dye pollutant · Wound-healing · Tissue engineering

List of Abbreviations

TDPSNC	3D-engineered polysaccharide nanocomposites
NPs	Nanoparticles

S. S. Emmanuel (✉) · A. A. Adesibikan
Department of Industrial Chemistry, Faculty of Physical Sciences, University of Ilorin, Ilorin, Nigeria
e-mail: stephenemmanuel6011@gmail.com

M. H. Abu Elella et al. (eds.), *3D Engineered Polysaccharides*,
https://doi.org/10.1007/978-3-032-26154-0_11

MOFs	Metal organic frameworks
COFs	Covalent organic frameworks
XRD	X-ray Diffraction
UV	Ultraviolet
BC	Biochar
PFAS	Per- and Polyfluoroalkyl Substances
MB	Methylene blue
FI	Freundlich isotherm
PFO	Pseudo-first order
GO	Graphene Oxide
ZIF	Zeolitic Imidazolate Framework
EDX	Energy-Dispersive X-ray
ESI	Electrostatic interaction
PFOS	Perfluorooctane Sulfonate
PDGF	Platelet-Derived Growth Factor
MIC	Minimum Inhibitory Concentration
SEM	Scanning Electron Microscopy
TGF-β	Transforming Growth Factor Beta
HEPG2	Human Hepatocellular Carcinoma Cell Line 2

1 Introduction

Recently, a class of novel materials represented by nanocomposites (material that consists of several phases where at least one, two, or three dimensions are in the nanometer range) has proven to be a promising option in the domain of nanotechnology [1–3]. Specifically, in search of eco-friendly nanocomposite materials with hierarchical organization and better performance, biopolymer nanocomposites have emerged as a paradigm that has improved physicochemical properties, which are relevant for the field of modern science and technology [4–7]. More specifically, among the emerging biopolymer nanocomposites, polysaccharide (multiple saccharide units joined to one another through glycosidic linkages) nanocomposites have come to the forefront owing to their abundance and multifunctionality [8–10]. Polysaccharide-based nanocomposites represent a class of auspicious materials with enhanced characteristics that are essential for the recent and future demands of nanotechnology advances owing to the fact that they synergistically combine the strength and functionality of integral materials to generate fresh nanomaterials with improved values (brilliant mechanical, efficacious antibacterial properties, optical, and electrical properties along with potential biocompatibility, sustainability, eco-friendly processing, and low toxicity) [7, 11]. This collection of distinct and often complementary qualities presents chances to develop strong, flexible, stretchable, compressible, and multifunctional devices with additional capabilities in chemical sensing, light absorption/response, and smart-shielding of gas [12, 13].

Notably, advancements in 3-D technology have revolutionized the world of material synthesis [14], including the polysaccharide-based nanocomposite fabrication landscape by enabling layer-by-layer architectural designing of materials with improved properties such as tunability, tailored morphologies and structures, biocompatibility, recyclability, specificity, flexibility, availability/accessibility of multiple reactive spots for introducing more functionalities, selectivity, and absorptivity, which are important in various applications [15, 16]. This recent innovation in the field of TDPSNC leads to diverse applications, as presented in Fig. 1, including drug delivery [17], food packaging [18], photo(catalysis) [19], environmental remediation [20], biomedical applications [15], and energy storage [21].

Interestingly, a thorough literature survey shows that several original research studies have been conducted on the use of TDPSNC in the various fields mentioned above. Thus, there is a need for an update on the body of knowledge on this subject. To establish the tone for the investigation, this paper first offers a variety of 3-D TDPSNC

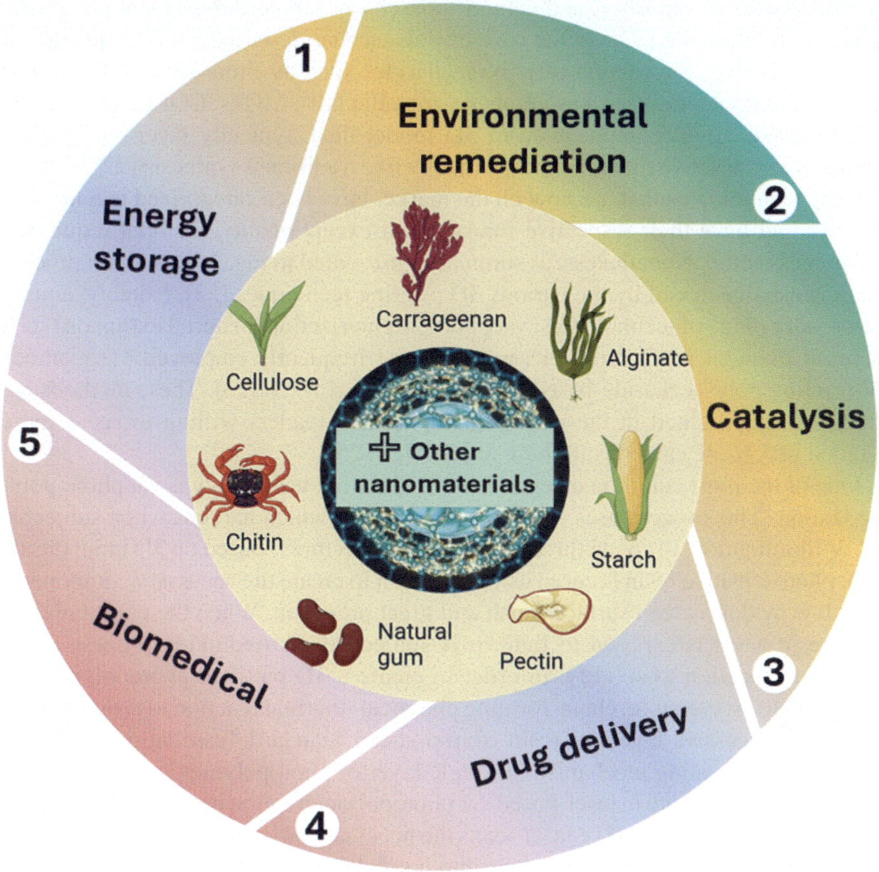

Fig. 1 Various applications of polysaccharide nanocomposites

synthesis techniques. Following that, a practical discussion of current developments in the environmental, energy, and medical applications of different TDPSNCs, particularly in environmental remediation, with an emphasis on their performance, was held. In order to offer empirical hints on their industrial usability, cost-effectiveness, and circular economy worth, their capacity for recycling and biodegradation was also examined. At the end, salient dilemmas, challenges, frontiers, and intriguing future research hotspots were underscored.

2 Overview of 3D Printing Technology for Polysaccharide-Based Nanocomposite Fabrication

3D printing technology is usually employed to engineer/develop three-dimensional architectures in a layer-by-layer fashion using various materials such as polymers, metallic pastes, composites, and ceramics based on a three-dimensional computer-aided digital data file [22]. In the case of polysaccharide-based nanocomposites, it typically involves the joining of polysaccharides such as chitosan, cellulose, alginate, etc., and materials such as graphene, metallic NPs, MOFs, COFs, carbon dots, etc., to make a nanocomposite from 3D model data, typically layer-by-layer, in contrast to subtractive manufacturing strategy like traditional synthesis [22]. The 3D printing technologies that are now on the market have been categorized in a number of ways and have their respective fundamental exceptionality, material requisites, and process setup. Nonetheless, as summarily presented in Fig. 2, a common process chain is usually shared by the various 3D printing techniques [23]. Notably, among the broader range of technologies, vat photopolymerization, material extrusion (such as fused filament fabrication) etc. are the most frequently employed 3D methods for printing polysaccharide-based nanocomposites (Fig. 3) [23]. These methods are succinctly highlighted in the next paragraphs, and readers willing to see a more detailed procedure can consult these articles [22–25].

One of the most common categories of 3D printing technology is vat photo polymerization. This process uses resin (usually liquid), which hardens when subjected to UV illumination, to create three-dimensional structures. Based on 3D input digital data, photosensitive resin is cured using UV light to create the necessary components [23]. It provides a clean surface finish and great precision. When the photopolymer mixture in a vat is exposed to light, it is subject to a crosslinking process at the irradiation location to solidify. In order to create a 3D part, vat photopolymerization uses this irreversible, chain-forming chemical–thermal reaction to progressively construct successive layers of resin from a sliced Standard Tessellation Language file. Using a recoating mechanism, a single layer of photopolymer is first deposited onto the build substrate in laser-based vat photopolymerization procedures like stereolithography. After that, a UV laser scans the necessary structure on the resin surface, crosslinking the liquid photopolymer into a solid component. To construct further layers of the part, the same procedures are repeated by recoating new resin layers and

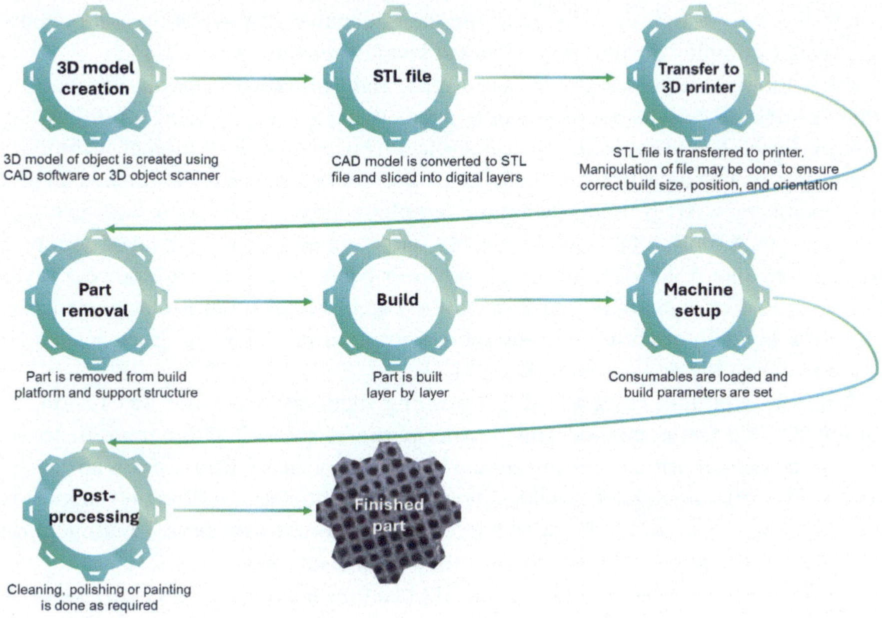

Fig. 2 Generic overview of the 3D printing process

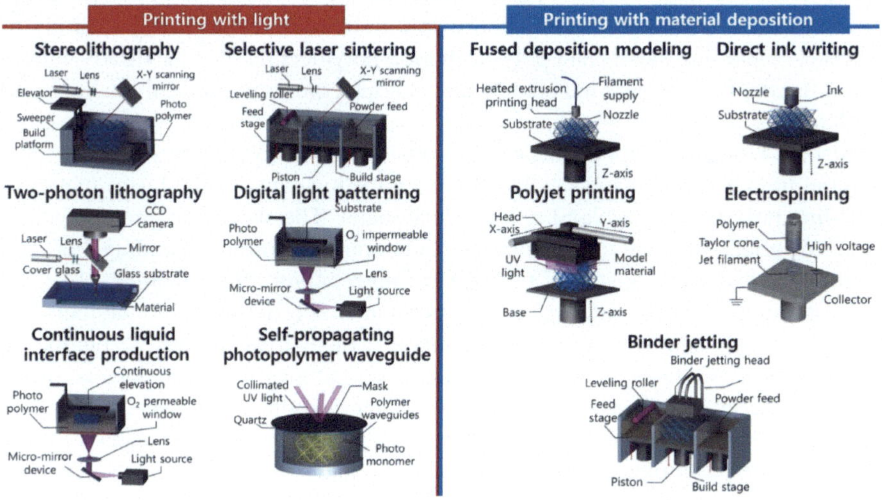

Fig. 3 Commonly used 3D printing techniques [26]. Reused with authorization from Springer Nature (License code: 6085380976311)

using a laser to design them. This technique is capable of fabricating components with overly complex designs at the highest level of resolution [23, 24].

Interestingly, stereolithography and digital light processing are similar, but the latter is faster since it uses a projector beam as the UV source, curing a full layer in a single flash. The laminated object manufacturing approach is another technique. According to the data from each layer of the slicing software, laminated object manufacturing, a 3D printing technique, employs knives or lasers to cut sheets of polymeric materials. A heated roller stacks and compacts each fresh layer on top of the previous one. Likewise, a high-power laser beam is directed at a powder layer that is applied over the build platform in the Direct Metal Laser Sintering process. Using the layer information that the dicing program provides, the laser beam will heat and merge the powder particles [25].

Another important category of 3D printing methods is powder bed fusion. In the powder bed fusion method, tiny powder particles are melted and fused to create three-dimensional structures using either heat energy or an electron beam as an energy carrier. The leftover powder provides the necessary support; therefore, no support is needed while constructing the overhang and unsupported structures. Contingent on the material, the finished 3D pieces are often porous and rough [22, 23].

Binder jetting is also a very crucial 3D printing technique, especially when it involves the use of biopolymers, including polysaccharides. This method creates three-dimensional things layer by layer using two materials: liquid and powder. To construct a solid three-dimensional structure, the liquid substance binds the layers together like an adhesive. Thus, it is possible to build 3D items using a variety of materials, including metals and polysaccharide polymers. Part of the disadvantages is that the final 3D printed nanocomposite can have poor mechanical qualities and brittleness [22].

Also, 3D material components can be constructed by bonding tiny sheets of raw material together in a process called sheet lamination. This method uses materials including polysaccharide-based polymer and metal. The sheets are joined together using a variety of techniques, including clamping, ultrasonic welding, thermal bonding, and adhesive bonding. It is a quick and economical procedure [22].

Furthermore, for 3D printed nanocomposites involving the use of biopolymers like polysaccharides, another very suitable method is the extrusion-based technique. In extrusion-based systems, sufficient pressure is used to forcefully eject the raw materials from the hot extrusion orifice. The molten material is then successively deposited onto the substrate from the bottom up to create a three-dimensional nanocomposite. The technique has limited layer thickness accuracy and is not that fast when compared to other 3D printing techniques [22].

In another method known as material jetting, to create the finished 3D nanocomposite, inkjet printheads are utilized in this process to distribute raw materials onto the construction platform. A 3D nanocomposite made of a variety of materials may be constructed using many printheads. One of the main advantages is that this technique makes it possible to fabricate 3D printed polysaccharide-based nanocomposite materials with a high degree of precision, exquisite finishing, and a wide range of colors [22, 23].

Also, in the directed energy deposition 3D printing method, to create a 3D material, the powder is simultaneously deposited and fused using a plasma arc, electron beam, or laser. This method is usually utilized to install new components or fix existing ones. A vacuum or inert environment is usually used for the whole printing process. This technology may be used to produce fully dense 3D printed nanocomposite materials with significantly tunable microstructural characteristics [22].

Fused filament fabrication, also referred to by its trademark, fused deposition modeling, is also among the frequently employed 3D printing technologies [23]. This technology is widely used because it can be used to create high-quality, tailored materials at a reasonably affordable price by both industry and non-specialists. According to premade 3D computer-aided digital data models, the procedure entails feeding polymer filaments or melting them into a heated jet, through which strands of the polymer are extruded to create the necessary components layer by layer. Unlike traditional fused filament fabrication processes, recent advances have led to the development of screw extrusion-based systems that facilitate the utilization of pellet feedstock and enable the processing of a wider variety of thermoplastics, including those of polysaccharide origin [22, 23]. Another very common 3D printing technique akin to fused deposition modeling is direct ink writing. Recent advances show that several direct ink writing strategies and ink designs are available, including robotic deposition and inkjet printing techniques, and polymer melts of polysaccharide origin and colloidal gel ink designs. These advances have made direct ink writing more auspicious for multi-material (including polysaccharide-based nanocomposite) 3D printing [23].

Overall, the size/morphology of the object being printed, the nozzle size, the height of the extruded layer, scanning speed, laser power, polysaccharide composition, and the printer's default speed all affect how long a production takes. Thus, it is crucial to maintain a healthy balance between printing time and quality [25, 27].

3 Application in Environmental Remediation

Environmental remediation remains one of the key aspects of environmental sustainability, and the removal of various pollutants is a major effort that has been put forward to achieve this [28, 29]. Various TDPSNC have been employed for the removal of these pollutants, including pesticides and agrochemicals, heavy metals, radio-contaminants, PFAS, pharmaceuticals, dyes, etc. This section deal with the performance of these TDPSNC for the remediation of these environmental pollutants and was summarized in Table 1.

Table 1 Summary of performance of 3D printed polysaccharide nanocomposites for the remediation of various environmental pollutants

Polysaccharide-based 3D printed nanocomposites	Pollutant removed	Performance/ removal efficiency (%)	Time taken (minutes)	References
Cellulose-MOFs@ZnO	Methyl blue	~ 100	10	[36]
Cellulose-MOFs@ZnO	Rhodamine B	85	10	[36]
Phosphorylated cellulose gel@MoS_2	Methylene blue	99	40	[37]
κ-carrageenan hydrogels@graphene oxide	Methyl violet	> 90	–	[34]
κ-carrageenan hydrogels@graphene oxide	Methylene blue	> 90	–	[34]
Chitosan@nano-TiO_2	Methylene blue	11.148[a]	–	[57]
Calcium alginate-gelatin@MOF	Methylene blue	99.8	20	[41]
Chitosan@graphene oxide@silicon oxycarbide polymer-derived ceramic monolithic	Methyl orange	91.16	160	[38]
Alginate@biochar	Methylene blue	214.6[a]	120	[35]
Chitin/cellulose/TiO_2	Methylene blue	~ 80	120	[40]
Cellulose@alginate monolithic hydrogel	Methylene blue	96	120	[39]
Alginate@biochar	Ibuprofen	111.4[a]	120	[35]
Chitosan@TiO_2	Amoxicillin	100	> 180	[45]
Chitosan@Ag@TiO_2	Amoxicillin	100	120	[44]
Chitosan@Ag@TiO_2	Paracetamol	100	120	[44]
Sodium alginate@bentonite clay	Paraquat	2.29[a]	–	[47]
Cellulose nanofibrils@sodium alginate@COF	PFOS	118.5–410.8[a]	–	[51]
Chitosan@ZIF-67	Pb^{2+}	5.4[a]	–	[50]
Calcium alginate@GO	Uranium	740.74	–	[49]
Cellulose nanofibrils@sodium alginate@COF	Al^{3+}, Cu^{2+}, Ca^{2+}, Na^{+}, Zn^{2+}, Fe^{3+}, and Co^{2+}	118.5–410.8[a]	180	[51]
Cellulose-MOFs@ZnO	Co^{2+}	33.5[a]	-	[36]
Cellulose-MOFs@ZnO	Cu^{2+}	101.2[a]	-	[36]
Calcium alginate@GO	Uranium	200.1[a]	-	[54]

(continued)

Table 1 (continued)

Polysaccharide-based 3D printed nanocomposites	Pollutant removed	Performance/ removal efficiency (%)	Time taken (minutes)	References
Alginate@nanoclay	Pb^{2+}	532[a]	60	[53]
Cellulose@leaf-like ZIF	Cu^{2+}	554.8[a]	–	[52]
Cellulose-MOFs@ZnO	CO_2	27.7[a]	–	[36]
Cellulose nanofibrils@sodium alginate@COF	CO_2	19.9[a]	–	[51]
Cellulose@leaf-like ZIF	CO_2	1.15[b]	–	[52]

[a] = mg/g, [b] = mmol/g

3.1 Dye Removal

Dyes are colouring substances that are used in virtually all essential industries (including food, textile, pulp, paper, leather, clothing, and pharmaceutical industries) to impart colour or polish various products [30, 31]. Unfortunately, from these aforementioned industries, dye effluent usually escapes to the environment without proper treatment, thus making dye pollutants one of the prominent hazardous pollutants within the ecosystem [32, 33]. Interesting, polysaccharide-based 3D nanocomposites are being explored as possible sustainable materials for the cleanup of these dye pollutants.

For instance, Majooni et al. [34] fabricated 3D printed scaffolds nanocomposite of κ-carrageenan hydrogels@graphene oxide and employed it for the removal of four organic dye pollutants. The nanocomposite was found to perform better than bare κ-carrageenan. Specifically, as shown in Table 1, over 90% of methylene blue and methyl violet were removed. According to the authors, the high removal efficiency is due to the fact that these dye pollutants not only interact with the hydrogel via electrostatic forces between the cationic dyes and the anionic hydrogels but also profit from metachromatic interactions with carrageenan. Furthermore, it was observed that the dye pollutant remediation by the nanocomposite is better described by the Langmuir model, which suggests a monolayer physisorption process. In addition, the nanocomposite also exhibits good reusability potential over six cycles. This improved performance of the 3D printed nanocomposite can be ascribed to enhanced chemical stability, reduced swelling ratio, and increased storage modulus, all signifying an improved crosslinking [34]. In a comparable study [35], 3D-printed nanocomposite hydrogels consisting of alginate and rice husk BC were employed as an adsorbent for the adsorptive elimination of methylene blue from tap water. The 3D printed nanocomposite demonstrated an adsorption capacity of 214.6 mg/g, which is 58% > that of traditional hydrogel without BC, and this showcases the advantage of biochar incorporation. And this can be attributed to the fact that the 3D-printed alginate@biochar nanocomposite displayed a maximum swelling rate that is

2.2X more than the bare hydrogel, which makes the polymer network more expandable. The remediation process obeys the Freundlich isotherm (FI) and pseudo-first order (PFO) model, which shows that the dye remediation mechanism is multilayer sorption (largely via ESIs between the negatively charged surface of 3D printed polysaccharide nanocomposite and the cationic dye pollutant molecules). Moreover, the presence of OH functional groups within the 3D printed nanocomposite infers the possibility of H-bonding interactions with the dye pollutant molecules coupled with the participation of π–π interactions enhanced by the carbonaceous nature of biochar. In addition, the 3D printed nanocomposite can also be reused 20 times without strongly losing its adsorption capacity (>80% adsorption efficiency at 20th cycle). The cost analysis (from the rice husk biochar synthesis, polymer solutions, and energy consumption) also revealed that the estimated cost of fabricating 1 g of the 3D-printed alginate@biochar nanocomposite is about $ 1.06, which is competitive with other sorbent materials. All of these suggest the 3D-printed alginate@biochar nanocomposite stability and real-world applicability for the treatment of industrial wastewater. [35].

Similarly, 3D printed cellulose-MOFs@ZnO was applied as a catalytic adsorbent for the remediation of methyl blue and rhodamine B, and in 10 min, ~ 100% and 85% of the organic dye pollutant was eliminated [36]. In like manner, a 3D printed phosphorylated cellulose gel@MoS_2 nanocomposite was synthesized and employed for the photocatalytic degradation of methylene blue dye pollutant. Remarkably, in 40 min, 99% degradation efficiency, which is equivalent to 202.97 mg/g degradation capacity, was accomplished at pH 9 and a dye concentration of 100 mg/L. The nanocomposite also demonstrates excellent reusability potential in that there was only ~ 5.3% variation in degradation output following five successive rounds of recyclability experiment [37]. The adsorption performance and recyclability efficiency are comparable to what was achieved for Methyl orange in 160 min using 3D printed chitosan@graphene oxide@silicon oxycarbide polymer-derived ceramic monolithic at pH 3.0 and temperature of 30 °C [38]. In another study, 66.31 mg/g adsorption capacity and 96.8% efficiency at 8 reusability cycles were achieved by 3D printed cellulose@alginate monolithic hydrogel for 5.0 mg/L of methylene blue dye pollutant remediation at 298 K. The isotherm, kinetics, and thermodynamic modeling revealed that the adsorption operation is spontaneous, exothermic (suggesting a low temperature advantage), chemisorption, and a monolayer with intraparticle diffusion playing a crucial role as the rate-determining step. The mechanism of adsorption was opined to be hydrogen bonding, ESIs via COO^- and OH of the 3D nanocomposite material, and the nitrogen group of the dye pollutant. In addition, it was revealed that as the solution pH moves from 4 to 1, the adsorption performance of the 3D printed nanocomposite drastically diminished. This was attributed to the fact that the surface zeta potential of 3D printed cellulose@alginate monolithic hydrogel clearly surged, thus triggering weaker electrostatic attraction and strong electrostatic repulsion to the positively charged methylene blue dye pollutant molecule. [39].

Also, Lei's research group used a 3D-printed TiO_2/chitin/cellulose nanocomposite as a catalyst for the degradation of MB under 300 W mercury lamp irradiation. After 2 h of light irradiation, roughly 80% of the dye pollutant was degraded by the 3D printed nanocomposite compared to bare TiO_2, cellulose, and TiO_2/cellulose, which degrades less than 40%. In addition, in the absence of light irradiation, the removal capacity of TiO_2/chitin/cellulose nanocomposite was also not good, and this can be attributed to the fact that without light irradiation, there was poor production of •OH and $•O^{-2}$ radicals that facilitate the degradation operation. It was observed that as the chitin concentration increases in the 3D nanocomposite, the photocatalytic degradation performance also increases, and this points to the reality that chitin/cellulose nano/microporous structure protects the structure of TiO2 NPs in the 3D printed polysaccharide nanocomposite. Furthermore, more adsorption sites were imparted on the nanocomposite as more chitin components were introduced on the 3D printed polysaccharide nanocomposite, which enhances the overall methylene blue photocatalytic degradation output. Additionally, the 3D-printed TiO_2/chitin/cellulose nanocomposite exhibited improved reusability performance over five consecutive cycles, and this suggests that it can be employed in real-life dye pollutant treatment [40].

In another experiment, 3D scaffold printed MOF@calcium alginate-gelatin was synthesized, and its capacity for the remediation of methylene blue was assessed. Notably, the 3D nanocomposite had 99.8% adsorption effectiveness at 20 min, whereas the bare calcium alginate-gelatin scaffolds without MOF only had an adsorption efficiency of 43.6% under the same experimental conditions. The superior adsorption capabilities towards the organic dye pollutants and their mixes were facilitated by the meso/macroporous structures of the TDPSNC, as well as the intrinsic microporous structure of MOFs. Additionally, it is simple to regenerate the 3D-printed nanocomposite sorbents with 0.1 M HCl, which can be seen in Fig. 4a by the gradual return of the 3D nanocomposite original color within 1 h. The regeneration allows for up to 10 rounds of usage without compromising adsorption performance (82% efficiency at the 10th cycle). In contrast, the MOF without the polysaccharides can only be reused twice at most. Furthermore, the characteristic peaks observed for Cu-BTC in the XRD analysis (Fig. 4b) of the spent nanocomposite revealed that the MOF component remains intact. However, the BET surface area of the spent 3D nanocomposite, although still higher than that of calcium alginate-gelatin, was slightly reduced, and this can be attributed to partial collapsing and leakage of material during the dye uptake process and regeneration phase. Interestingly, as displayed in the SEM pictures of the spent 3D-printed nanocomposite in Fig. 4d, it can be plainly noticed that the pores and walls maintain their integrity without cracking. Thus, the finding revealed that crystalline and chemical structures of the material remained unchanged during the regeneration process, signifying a satisfactory recyclability of the TDPSNC. This further points to industrial-scale potential and improved stability of the polysaccharide-based 3D printed nanocomposites compared to individual materials [41].

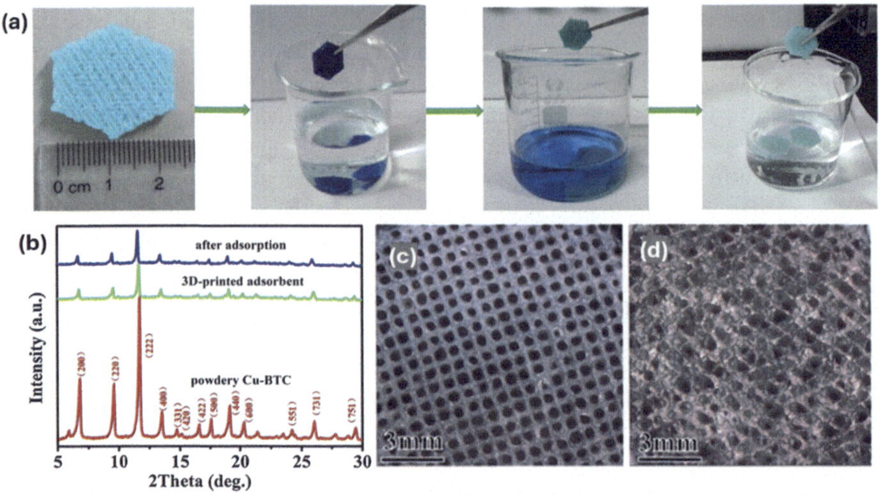

Fig. 4 **a** The regeneration process (with colour changes) of 3D-printed calcium alginate-gelatin@MOF by 0.1 M HCl, **b** XRD analysis of Cu-BTC MOFs, the fresh and expended 3D-printed calcium alginate-gelatin@MOF, SEM images of **c** fresh and **d** expended 3D-printed calcium alginate-gelatin@MOF [41]. Reused with authorization from Elsevier (License code: 6084531240613)

3.2 Pharmaceutical Pollutants Remediation

Pharmaceutical pollutants are another type of common pollutant, especially in aquatic bodies [42, 43]. Interestingly, advanced nanomaterials like 3D printed polysaccharide nanocomposites are being explored for the sustainable remediation of these pollutants.

For example, in a particular research [35], 3D-printed nanocomposite hydrogels consisting of alginate and rice husk biochar were employed as an adsorbent for the adsorptive elimination of ibuprofen from tap water. As shown in Table 1, the 3D printed nanocomposite demonstrated an adsorption capacity of 111.4 mg/g, which is 48% > that of traditional hydrogel without biochar, and this showcases the advantage of biochar incorporation, and this can be attributed to the fact that the 3D-printed alginate@biochar nanocomposite displayed a maximum swelling rate that is 2.2× greater than the bare hydrogel, which makes the polymer network more expandable. The remediation process obeys the FI and PFO model, which shows that the pharmaceutical remediation mechanism is multilayer sorption (largely via ESIs between the negatively charged surface of 3D printed polysaccharide nanocomposite and the positive molecules of the pharmaceutical pollutant). Moreover, the presence of OH functional groups within the 3D printed nanocomposite infers the possibility of H-bonding interactions with the pharmaceutical pollutant molecules, coupled with the participation of van der Waals and π–π interactions enhanced by the carbonaceous nature of biochar. In addition, the 3D printed nanocomposite can also be reused 20 times without strongly losing its adsorption capacity (>80% adsorption efficiency at the 20th cycle). The cost analysis (from the rice husk biochar synthesis, polymer solutions, and energy consumption) also revealed that the estimated cost of fabricating 1 g of the 3D-printed alginate@biochar nanocomposite is about $ 1.06, which is competitive with other sorbent materials. All of these suggest the 3D-printed alginate@biochar nanocomposite stability and real-world applicability for the treatment of industrial wastewater. [35].

In another study, a chitosan/Ag/TiO_2 TDPSNC was utilized for the photocatalytic degradation of amoxicillin and paracetamol under light irradiation by a 125 W Hg-vapor lamp. Interestingly, in just 120 min, a 100% degradation was recorded for both pharmaceutical pollutants. This efficiency is far higher than that of bare chitosan and TiO_2 individually, and this points to the synergistic impact of the 3D printed nanocomposite. Specifically, the presence of chitosan with an array of oxygenated functional groups can boost the production of OH radicals needed for the degradation process and enhance surface contact between the 3D printed nanocomposite and the pharmaceutical pollutant molecules [44]. In parallel, 100% degradation efficiency was also reported for the removal of amoxicillin under 125 W Hg-vapor lamp irradiation using 3D printed chitosan@TiO_2 scaffolds. This 3D printed nanocomposite also demonstrates excellent reusability potential of 80% over 3 cycles, and this can be ascribed to its good mechanical strength [45].

3.3 *Pesticides and Agrochemical Remediation*

Pesticides and agrochemical residues (including organochlorine and organophosphorus) are another class of hazardous pollutants that have garnered the attention of the research community due to their widespread persistence as well as ecotoxicity [46] and 3D printed nanocomposite materials are part of the emerging materials being tried to clean up these pollutants. However, in the use of 3D printed polysaccharide nanocomposites, not so much has been done. Amazingly, the few studies available suggest that this type of 3D nanomaterial has potential.

For example, in a particular study [47], 3D printed sodium alginate and bentonite clay nanocomposite hydrogels were fabricated and applied for the adsorptive remediation of paraquat pesticide. Notably, a 2.29 mg/g adsorption capacity was spontaneously achieved at ~ 7.0 pH. Above pH 7.0, the adsorption capacity dropped. This effect may result from differences in the 3D nanocomposite surface charges, which have a direct impact on how well they interact with paraquat pesticide molecules. The 3D nanocomposite majorly exhibited negative surface charges at neutral pH, where the removal percentages were highest, favoring interaction with the positively charged paraquat pesticide molecules. Moreover, as the nanocomposite dose increases, the pesticide adsorption output also increases due to an upsurge in the available active binding sites. In addition, as seen in the SEM image presented in Fig. 5, the paraquat pesticide molecules were situated on the surface of the 3D nanocomposite. Specifically, the paraquat pesticide molecules were sorbed on the

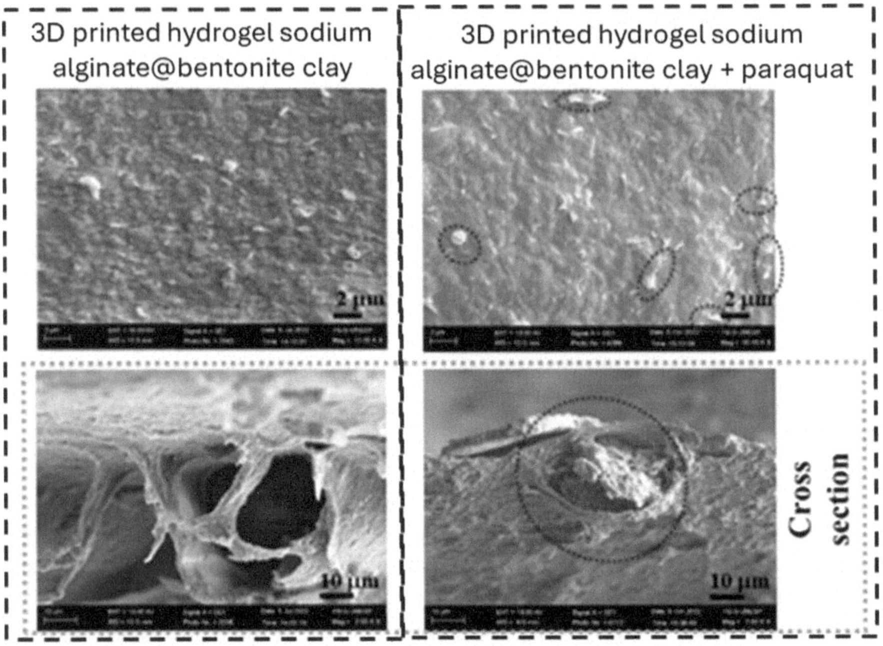

Fig. 5 SEM images of the 3D printed hydrogel sodium alginate@bentonite clay before and after paraquat adsorption [47]. Reused with authorization from Elsevier (License code: 6084541056983)

external surface and in the interior of the 3D nanocomposite, which was visibly noticed inside the nanocomposite pores as well, confirming that robust adsorption has taken place. The remediation process was best fitted to the PFO and FI model, which suggests that the pesticide elimination process is a multilayer physisorption operation via ESIs. Furthermore, the 3D printed nanocomposite can be reused for five rounds while still delivering around 70% of its original adsorption capacity. This shows that the nanocomposite is eco-friendly and cost-effective, and can also contribute to the advancement of the circular economy while guaranteeing water security. Nevertheless, future researchers should look into how to improve the salt resistance of the 3D printed nanocomposite, as salts like NaCl, KCl, $MgSO_4$, and $CaCl_2$ were observed to affect the adsorption (50% reduction) of the material for the paraquat pesticide [47].

3.4 Remediation of Heavy Metals and Radio-Contaminants

Heavy metals and radio-contaminants are a class of dangerous inorganic contaminants that are dominant in the environment and are inevitable because they typically emanate from essential human day-to-day activities, which cannot be truncated [48].

3D printed polysaccharide nanocomposites represent a class of classic functional materials that can be used to eliminate these hazardous inorganic pollutants.

For example, in a particular study, a 3D printed calcium alginate@GO nanocomposite was employed for the removal of uranium from aqueous media. Notably, the optimized 3D printed calcium alginate@GO nanocomposite was found to have hierarchical porosity and delivered uranium adsorption capacity of 740.74 mg/g at pH 5.0, which is about 2 to 3 times greater than that of bare traditional alginate. Interestingly, the material also demonstrates excellent stability and reusability by maintaining 97.7% efficiency after 9 regeneration rounds, exceeding that of existing adsorbents. This shows that the 3D printed calcium alginate@GO nanocomposite can be applied for nuclear effluent treatment [49]. In another research, 3D printed cellulose-MOFs@ZnO was applied as an adsorbent for the remediation of Co^{2+} and Cu^{2+}, with adsorption capacities of 33.5 and 101.2 mg/g, respectively. Furthermore, the adsorbed heavy metal ions were observed to be distributed on the surface of 3D nanocomposite homogenously, and this was evident in the change in the nanocomposite color from white to violet and turquoise color following the uptake of Co^{2+} and Cu^{2+}, respectively. According to the authors, in the case of Co^{2+}, the change in colour was a result of the formation of ZIF-67, a Co-based ZIF MOF with isostructural to ZIF-8. Moreover, the nanocomposite material was found to maintain its 3D structural integrity post adsorption, and this suggests that this material has sustainable capacity for industrial-scale applications [36]. Similarly, in another experiment, 3D printed chitosan@ZIF-67 was employed for the elimination of Pb(II). An adsorption capacity of 5.4 mg/g was achieved through H-bonding, complexation, and ion exchange involving C=O, OH, NH, and NH2 functional groups of the 3D printed nanocomposite, and this was supported by the fitness of the adsorption data into Langmuir isotherm and pseudo-second-order kinetic models. Additionally, in a multipollutant system containing Cu(II) and Cd(II) and a fixed-bed column adsorption system, the 3D printed chitosan@ZIF-67 also performed excellently. This suggests industrial applicability potential, including remediation of acid mine effluent [50].

In another research, 3D printed Cellulose nanofibrils@sodium alginate@COF demonstrated a remarkable remediation capacity of 410.8 mg/g for 100 mg/L of Cu^{2+}. Interestingly, when the same 3D printed nanocomposite material was tested for its selectivity in a multipollutant system, it showed higher selectivity towards Fe^{2+} compared to Cu^{2+} and other metals (Na^{+}, Co^{2+}, Al^{3+}, and Zn^{2+}). Furthermore, the 3D printed nanocomposite can be reused for 3 rounds while still delivering around 80% of its original Cu^{2+} adsorption capacity. In addition, the SEM analysis revealed that no structural or morphological damage was done to the 3D printed nanocomposite material during the heavy metal remediation process, suggesting high stability and potential for industrial wastewater treatment [51]. Abdelhamid et al. also recorded high adsorption selectivity toward Fe^{3+} and Co^{2+} and Cu^{2+} by 3D printed cellulose@leaf-like ZIF MOFs in contrast to Na^{+}, Al^{3+}, and Ca^{2+} via an ion exchange mechanism. It was stated that hard heavy metals like Fe^{3+} tend to interact better with the OH groups of the 3D printed cellulose@leaf-like ZIF MOFs [52]. Shahbazi's research group did not differ either, as 3D printed alginate@nanoclay displayed good selectivity towards Fe^{3+} and Cr^{3+} due to their higher charge radius ratio, which enabled the

3D printed nanocomposite to capture them effortlessly. Although when employed for the adsorption of 500 mg/L Pb(II), the nanocomposite exhibited excellent remediation capacity of 532 mg/g in 60 min at 298 K [53]. This suggests that the 3D printed alginate@nanoclay can be employed for the cleanup of acid mine drainage. Similarly, when a 3D printed calcium alginate@GO nanocomposite was used for the adsorption of uranium radio-contaminant, an adsorption capacity of 200.1 mg/g was recorded at pH 2.8. Additionally, the 3D printed nanocomposite material (regenerated using 1 mol/L HCl) maintained 99.7% of its original efficiency and original structural morphology after 8 consecutive rounds of reuse. Moreover, in the midst of other radio-contaminants such as As^{3+}, Cd^{2+}, Co^{2+}, Cr^{3+}, Cu^{2+}, Hg^{2+}, Ni^{2+}, the 3D printed polysaccharide nanocomposite still shows strong selectivity (99.7%.) towards uranium. This activity was ascribed to the robust affinity between the oxygenated functional groups on 3D printed calcium alginate@GO nanocomposite and uranium [54]. It can be inferred that this 3D printed material can be deployed on an industrial scale for real radio-contaminated water or nuclear effluent treatment.

3.5 Carbon Capture/sequestration

Carbon capture using advanced functional materials, including 3D printed polysaccharide nanocomposites, remains part of an emerging strategy to mitigate CO_2 emissions and move a step closer to a zero-carbon economy [51, 55, 56]. For instance, in a particular research, 3D printed cellulose-MOFs@ZnO was used as an adsorbent for the remediation of CO_2 with an adsorption capacity of 27.7 mg/g. Interestingly, it was observed that this 3D nanocomposite can be reused six times without losing its adsorption capacity, and this points to a strong stability profile and better economic value, which are crucial for pilot-scale deployment [36]. Similarly, Abdelhamid et al. fabricated a 3D printed polysaccharide nanocomposite comprising cellulose nanofibrils, sodium alginate, and COF. This 3D nanocomposite was employed for the adsorption of CO_2. The research findings show that the nanocomposite materials have abundant active sites and deliver 19.9 mg/g adsorption capacity for CO_2. When the 3D printed nanocomposite was reused 3 times, the adsorption capacity only reduced slightly to 16.2 mg/g, which is still well above 80% of the original adsorption performance [51].

In another study, 3D printed cellulose@leaf-like ZIF MOFs were explored for the capturing of CO_2. At 1 bar, a sequestration capacity of 1.15 mmol/g through an open-gate mechanism. Remarkably, the adsorption capacity displayed by the 3D printed polysaccharide nanocomposite was far higher than that of ordinary ZIF-8 powder. It was noted that the high performance of the 3D nanocomposite is due to the improved CO_2 diffusion within the large pores of the material. In addition, the 3D printed nanocomposite also demonstrated good performance over 4 adsorption–desorption cycles [52].

3.6 Other Pollutant Remediation

3D printed polysaccharide nanocomposites have also been employed to remove other dangerous pollutants like PFAS. In a particular study, Abdelhamid and coworkers fabricated a 3D printed polysaccharide nanocomposite comprising cellulose nanofibrils, sodium alginate, and COF. This 3D nanocomposite was employed for the adsorption of PFOS. The research findings show that the nanocomposite materials have abundant active sites and deliver 118.5–410.8 mg/g adsorption capacity for PFOS. The uptake of PFOS by the 3D printed nanocomposite was established using SEM analysis (Fig. 6a) and EDX mapping (Fig. 6b, c). From the two spectroscopic analyses, a precipitate of PFOS was noticed on the surface of the 3D printed polysaccharide nanocomposite, which confirms that adsorption has taken place. Furthermore, EDX analysis suggests the homogeneous distribution of PFOS on the 3D printed polysaccharide nanocomposite surface. Specifically, EDX results presented in Fig. 6b, c established the existence of F and S elements belonging to the PFOS pollutant molecule and thus confirm excellent remediation of PFOS by the 3D printed polysaccharide nanocomposite. Furthermore, the 3D nanocomposite also demonstrates high adsorptive selectivity toward PFOS in a multipollutant system containing various heavy metals [51].

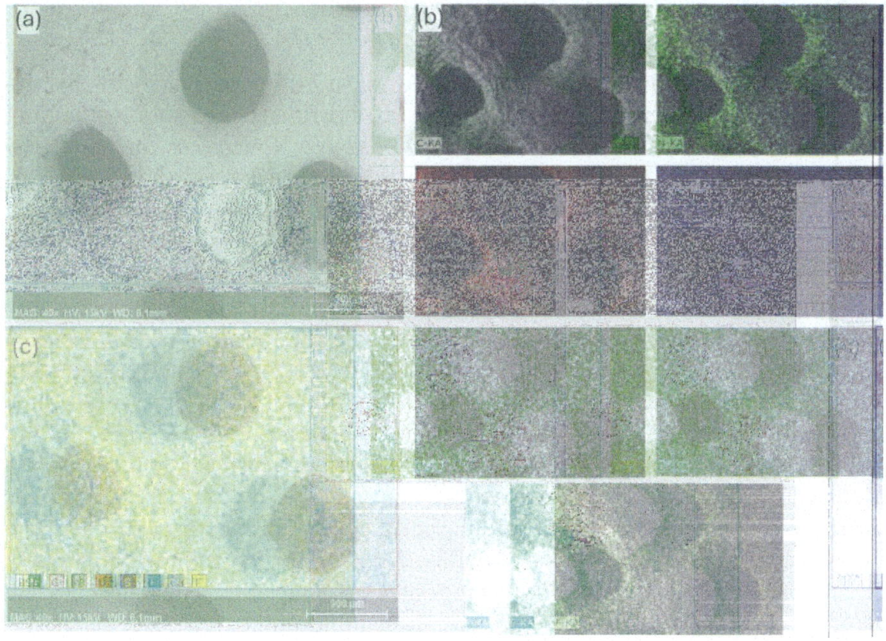

Fig. 6 **a** SEM and **b**, **c** EDX analysis/mapping of 3D printed Cellulose nanofibrils@sodium alginate@COF after PFOS adsorptive remediation [51]. ACS copyright 2023 (Open access, which permits unrestricted use)

Summarily, the findings here revealed that polysaccharide-based 3D printed nanocomposites are good materials for the cleanup of numerous pollutants from various environmental compartments, delivering over 80% efficiency with good reusability potential up to 8 cycles while maintaining structural integrity. Furthermore, various studies showed that the adsorption efficiency of polysaccharide-based 3D printed nanocomposites is largely governed by their intrinsic structural features. Functional groups such as hydroxyl, carboxyl, and amino moieties provide active binding sites that interact with pollutants through hydrogen bonding, electrostatic attraction, ion exchange, or coordination. For instance, hydroxyl and carboxyl groups favor complexation with heavy metal ions and hydrogen bonding with anionic dyes, while amino groups enhance affinity toward negatively charged contaminants. In addition, the degree of branching and crystallinity influences the accessibility of these functional groups, modulating swelling behavior and diffusion pathways for adsorbates. Chemical modifications such as cross-linking further enhance porosity and surface area, thereby improving active site exposure and selectivity across diverse classes of pollutants such as dyes, metals, and pharmaceuticals. These mechanistic insights highlight the pivotal role of polysaccharide structure in dictating adsorption performance beyond the predictions of classical isotherm models.

4 Medical and Drug Delivery Application

Over the years, scientists have continued to make efforts in search of alternative solutions and materials to tackle various medical issues and sustainable drug delivery. Interestingly, in recent years, biopolymer-based materials, including 3D printed polysaccharide-based nanocomposites, have emerged as low-cost biodegradable nanomaterials with low toxicity [58–61]. This section focuses on the medical application and drug delivery potential of various 3D printed polysaccharide-based nanocomposites.

For instance, in research conducted by Biranje et al., a 3D bioprinted nanocellulose composite (cellulose nanofibrils@Epsilon-poly-L-lysine) scaffold was fabricated and evaluated for its wound healing potential and antibacterial capacity. According to research findings, the 3D composite scaffold has a large specific surface area and a mesoporous structure, which guarantees a high adsorption capacity of red blood cells in wound healing. Furthermore, the three-dimensional composite scaffold has antimicrobial efficacy against bacteria that cause wound contaminations [62]. In a comparable study, 3D scaffold printed cellulose nanocrystals@polylactic acid was assessed for its cytocompatibility. Notably, the cytocompatibility study showed that 3D printed scaffold nanocomposites are compatible with bone cells and non-toxic. This finding indicates that the 3D scaffold printed cellulose nanocrystals@polylactic acid nanocomposite is a good biomaterial that can be employed in regenerative medicine and tissue repairing/engineering [63]. In another experiment, a 3D printed and electrospun bilayer pharmaceutical membrane containing chitosan, polycaprolactone, diltiazem, and polyvinyl alcohol was evaluated for its medicinal properties. Interestingly, the 3D printed nanocomposite was reported to show improved antibacterial activity, and the addition of diltiazem boosts the proliferation, viability, and migration of fibroblasts and adipose-derived stem cells. In addition, the rat model experiment also showed that the 3D printed material possesses good anti-inflammatory and wound healing [64]. This research finding is similar to what was reported for a TDPSNC scaffold comprising alginate, chitosan, and Barijeh-loaded niosomes [65]. The nanocomposite greatly showed antibacterial activity against Staphylococcus aureus and Pseudomonas aeruginosa while successfully suppressing biofilm formation by 74–80%. Amazingly, within 10 days, the 3D printed nanocomposite also delivered 90% wound healing with no sign of cytotoxicity against fibroblast cells. Moreover, TGF-β expression doubled while IL-6 expression dropped to almost baseline levels, indicating tissue integration and reduced inflammation [65]. The histopathological examination also showed much more collagen deposition, a crucial sign of successful healing [65–67]. Similarly, a 3D printed cellulose nanofiber@chitosan hydrogel was prepared, and excellent cytocompatibility was demonstrated by the nanocomposite, which promoted fibroblast cell adhesion, growth, and proliferation. In the end, the simultaneous proliferation of cells and biodegradation of cellulose nanofiber within the 3D printed cellulose nanofiber@chitosan scaffolds demonstrates the exceptional potential of 3D printed nanocomposite materials in regenerative medicine and wound healing [68].

In another experiment, a 3D printed polysaccharide-based nanocomposite of alginate@cellulose@Ag NPs was fabricated and assessed for its cytotoxicity and antimicrobial activity. The antibacterial test conducted against S. Aureus and P. aeruginosa revealed that the composite biomaterial has significant antimicrobial action at a low NP dosage (MIC 10 μg/mL). Because of their poor mechanical features and solubility, silver NPs embedded in alginate alone quickly spread throughout the culture medium when HEPG2 cells are present, according to the results. The authors were able to get around this problem and create scaffolds with a better mechanical profile against HEPG2 cells by including cellulose into the alginate@Ag nanocomposite. This increased the effectiveness of silver NP cytotoxicity over time. Therefore, using this 3D printed polysaccharide-based nanocomposite may be a good way to develop novel active ingredients for medical and clinical purposes [69]. In a similar study, 3D-printed alginate@cellulose hydrogels with in situ-fabricated Cu nanostructure composite were tested for their antimicrobial behavior. The study shows that the 3D printed polysaccharide-based nanocomposite was able to work well against Staphylococcus aureus and Escherichia coli strains. This finding suggests that the 3D printed nanocomposite can be employed in regenerative medicine and tissue engineering [70]. In a similar study, a 3D printed cellulose nanocrystals@polycaprolactone@Ag nanocomposite scaffold was fabricated for bone tissue engineering. Notably, the 3D printed nanocomposite scaffold was reported to have enhanced hydroxyapatite nucleation, hydrophilicity, and biodegradability, all of which are vital for osteoconductivity. The scaffolds showed cytocompatibility with mesenchymal stem cells, mechanical features appropriate for bone repair, and strong antibacterial activity against Escherichia coli. These results demonstrate how 3D printed nanocomposite scaffolds provide improved resorbability, antibacterial protection, and structural flexibility, and may be used as cutting-edge 3D biomaterials for bone tissue engineering [71].

Overall, findings from this discussion here revealed that 3D printed polysaccharide-based nanocomposites possess strong antibacterial, anti-inflammatory, and antioxidant properties (>70% efficacy) and are excellent sustainable materials for wound healing, tissue engineering, and skin regeneration with low cytotoxicity. However, serious clinical trials are still required for full real-life application and approval.

5 Application in Energy Storage

Energy remains the key anchor leg of the world economy and industries. However, the global economy's fast expansion has made the impending crisis of fossil fuels' rapid depletion worse because they are widely used for transportation, heating, and electricity generation [72, 73]. In fact, by 2050, world primary energy consumption is expected to surge by 3 folds [74]. Thus, efforts are being directed to meet the expected energy demand while working toward zero carbon emission and clean energy technology [75]. Interestingly, one of the key efforts to achieve this is the integration of energy storage devices with renewable energy resources. This is due to the fact that the development of strong and sustainable next-generation energy storage devices depends largely on the materials in order to boost the overall performance [72, 75]. Recently, 3D printed polysaccharide-based nanocomposites have emerged as sustainable energy storage materials, and, in this section, their performance as seen in open literature was discussed.

In a particular study, Zhou's research group developed a 3D printed cellulose@$Ti_3C_2T_x$ MXene nanofiber to evaluate its electrochemical performance for a symmetrical solid-state supercapacitor. Notably, the 3D printed cellulose@$Ti_3C_2T_x$ MXene nanofiber supercapacitor was able to deliver an energy density of 101 μWh cm^{-2} at a power density of 0.299 mW cm^{-2}. In addition, following 5000 rounds, the 3D printed nanocomposite material was able to sustain a capacitance retention rate of 85%. The excellent performance of the 3D nanocomposite material can be attributed to the hierarchically porous structures and enhanced surface area accessibility, which improves ion transport efficiency and capacitive performance. This suggests that a 3D printed cellulose-based polysaccharide nanocomposite is a potential next generation of electrochemical energy storage devices [76]. In another experiment, 3D printed chitosan@AgNPs@pencil graphite electrodes were developed into an electrode material for a supercapacitor. Remarkably, the 3D printed nanocomposite supercapacitor device demonstrated good cyclic stability, with a storage capacity of 367.16 mF cm^{-2} at a current density of 1 mA cm^{-2}. The study outcome indicates that the fabricated 3D printed polysaccharide-based nanocomposite hybrid electrode material exhibited brilliant electrochemical activity for the supercapacitor, which can be further evaluated for other important real-world electronic applications [77].

In another study, a 3D-printable polysaccharide nanocomposite consisting of cellulose fibers and poly(3,4-ethylene dioxythiophene) poly(styrene sulfonate) was assessed for its potential utilization as wearable supercapacitors. From the research findings, the 3D printed nanocomposite was able to deliver 211 F/g specific capacitances. This shows that the 3D printed nanocomposite has the potential to be used as cheap, eco-friendly wearable electronics [78].

Summarily, it can be inferred that 3D printed polysaccharide-based nanocomposite has great potential as a next-generation energy storage device, which can be used in electronic devices. However, the available research studies are still scanty, and thus, future researchers should put in more effort on this subject.

6 Challenges and Future Research Directions

Notably, in spite of advances and the remarkable performance of polysaccharide-based 3D nanocomposites revealed in this chapter, some challenges highlighted below remain (Fig. 7). These include scalability problem while fabricating and applying the nanocomposites for use in industry, poor nanoparticle dispersion within the biopolymer matrix may have an impact on performance and homogeneity, inconsistent performance across a range of environmental conditions, such as temperature, pH, and humidity, limited structural integrity originating from weak interfacial contacts between hydrophilic polysaccharides and hydrophobic nanoparticles and limited stability over the long term, particularly for biomedical and water treatment applications.

To address the various challenges underscored above, future researchers should concentrate on devising advanced functionalization methods to enhance the interfacial bonding and compatibility of nanofillers with polysaccharide matrices, investigating hybrid nanocomposites, which combine several nanomaterial types to provide complementary effects. Developing smart or stimuli-responsive nanocomposites for selective adsorption, targeted distribution, or controlled release, optimizing green synthesis pathways (using biogenic entities) that is economical, scalable, and environmentally safe [42, 46], perform techno-economic analysis to holistically have grounded knowledge of the entire process cost, from the nanocomposite production to application, integrate the use of artificial intelligence based on existing data set to simulate or predict the best match of polysaccharide/nanoparticles, and polysaccharide-based 3D nanocomposites/application for enhanced performance, time, cost and energy saving and assessing lifecycle performance to better meet the objectives of the circular economy and sustainability. In addition, there is a need for future researchers to focus on field and clinical trials for commercialization purposes.

Fig. 7 Overview of challenges and future study outlook

7 Conclusion

Polysaccharide-based 3D nanocomposites represent a groundbreaking, sustainable material with immense potential across a wide range of applications. By and large, this chapter reveals that polysaccharide nanocomposites offer a green, multifunctional platform that bridges the gap between high-performance materials and sustainable development goals, positioning them as key players in next-generation material science. In particular, their synthesis, which combines naturally abundant biopolymers with functional nanomaterials, enables the fabrication of eco-friendly composites with improved mechanical, optical, and biological properties. Advances in synthesis methods, like in situ nanoparticle generation, electrospinning, and solution casting, have enabled fine control over nanocomposite structure and functionality, making them suitable for specific performance-driven uses. Significant applications in industries including biomedicine, drug delivery, environmental remediation (e.g., dye removal, heavy metal adsorption, etc.), photocatalysis, and energy storage have been made possible by the incorporation of nanoparticles into polysaccharide matrices. Various 3D printed nanocomposite materials displayed excellent pollutant removal capacity, high capacitance, conductivity, and energy density, strong antibacterial, antioxidant, and anti-inflammatory properties, an improved drug delivery system, remarkable wound healing, and tissue engineering potential. Specifically, this improved performance of the 3D printed nanocomposite can be ascribed to enhanced chemical and mechanical stability, biodegradability, low toxicity, reduced swelling ratio, and increased storage modulus. Notably, they are especially appealing for biomedical and environmental applications where material safety and sustainability are crucial due to their biocompatibility and biodegradability. Despite its impressive qualities, issues with long-term stability, property consistency, and large-scale production still exist. In order to optimize their performance for specific applications, future research should concentrate on refining synthesis techniques, enhancing nanoparticle dispersion, and investigating novel functionalization pathways.

References

1. Tang XZ, Kumar P, Alavi S, Sandeep KP. Recent advances in biopolymers and biopolymer-based nanocomposites for food packaging materials. Crit Rev Food Sci Nutr. 2012;52:426–42.
2. Omanović-Mikličanin E, Badnjević A, Kazlagić A, Hajlovac M. Nanocomposites: a brief review. Health Technol (Berl). 2020;10:51–9.
3. Qadir A, Le TK, Malik M, Min-Dianey KAA, Saeed I, Yu Y, et al. Representative 2D-material-based nanocomposites and their emerging applications: a review. RSC Adv. 2021;11:23860–80.
4. Sharma B, Malik P, Jain P. Biopolymer reinforced nanocomposites: A comprehensive review. Mater Today Commun. 2018;16:353–63.
5. Hassan T, Salam A, Khan A, Khan SU, Khanzada H, Wasim M, et al. Functional nanocomposites and their potential applications: a review. J Polym Res. 2021;28:36.
6. Bitinis N, Hernández M, Verdejo R, Kenny JM, Lopez-Manchado MA. Recent advances in clay/polymer nanocomposites. Adv Mater. 2011;23:5229–36.
7. Zienkiewicz-Strzałka M, Deryło-Marczewska A. Small AgNP in the biopolymer nanocomposite system. Int J Mol Sci. 2020;21:9388.
8. Zheng Y, Monty J, Linhardt RJ. Polysaccharide-based nanocomposites and their applications. Carbohydr Res. 2015;405:23–32. https://doi.org/10.1016/j.carres.2014.07.016.
9. Liu Q, Duan B, Xu X, Zhang L. Progress in rigid polysaccharide-based nanocomposites with therapeutic functions. J Mater Chem B. 2017;5:5690–713.
10. Dufresne A, Castaño J. Polysaccharide nanomaterial reinforced starch nanocomposites: a review. Starch-Stärke. 2017;69:1500307.
11. Ghazzy A, Naik RR, Shakya AK. Metal–polymer nanocomposites: a promising approach to antibacterial materials. Polymers (Basel) 2023;15. https://doi.org/10.3390/polym15092167.
12. Yang W, Zhao Z, Wu K, Huang R, Liu T, Jiang H, et al. Ultrathin flexible reduced graphene oxide/cellulose nanofiber composite films with strongly anisotropic thermal conductivity and efficient electromagnetic interference shielding. J Mater Chem C. 2017;5:3748–56.
13. Xiong R, Grant AM, Ma R, Zhang S, Tsukruk VV. Naturally-derived biopolymer nanocomposites: interfacial design, properties and emerging applications. Mater Sci Eng R Reports. 2018;125:1–41. https://doi.org/10.1016/j.mser.2018.01.002.
14. Shahbazi M, Jäger H. Current status in the utilization of biobased polymers for 3D printing process: a systematic review of the materials, processes, and challenges. ACS Appl Bio Mater. 2020;4:325–69.
15. Fabregat-Safont D, Gracia-Marín E, Ibáñez M, Pitarch E, Hernández F. Analytical key issues and challenges in the LC-MS/MS determination of antibiotics in wastewater. Anal Chim Acta. 2023;1239: 340739.
16. Mallakpour S, Ramezanzade V, Hussain CM. 11 - Novel biomaterials based on diverse polysaccharide gums with gelatin used in three-dimensional printing techniques. In: Mallakpour S, Mustansar Hussain CBT-MAM, editors. Addit Manuf Mater Technol. Elsevier; 2024, p. 281–98. https://doi.org/10.1016/B978-0-323-95383-2.00009-3.

17. Teymourlouei AS, Naghib SM, Mozafari MR. Stimuli-responsive graphene-polysaccharide nanocomposites for drug delivery and tissue engineering. Curr Org Synth. 2025;22:211–33.
18. Deng J, Zhu E-Q, Xu G-F, Naik N, Murugadoss V, Ma M-G, et al. Overview of renewable polysaccharide-based composites for biodegradable food packaging applications. Green Chem. 2022;24:480–92.
19. Rodrigues VG, Henrique G, Sousa-Vidal ÉK, de Souza RMM, Tavares EFC, Mezzalira N, et al. Thyroid under attack: the adverse impact of plasticizers, pesticides, and PFASs on thyroid function. Endocrines. 2024;5:430–53.
20. Jaymand M. Biosorptive removal of cationic dyes from ternary system using a magnetic nanocomposite hydrogel based on modified tragacanth gum. Carbohydr Polym Technol Appl. 2024;7: 100403.
21. Jalali A, Gupta T, Pakharenko V, Rejeb Z Ben, Kheradmandkeysomi M, Sain M, et al. Lightweight 3D-printed chitosan/MXene aerogels for advanced electromagnetic shielding, energy harvesting, and thermal management. Carbohydr Polym 2025;352:123252.
22. Parupelli SK, Desai S. The 3D printing of nanocomposites for wearable biosensors: recent advances, challenges, and prospects. Bioengineering 2024;11. https://doi.org/10.3390/bioengineering11010032.
23. Alo OA, Mauchline D, Otunniyi IO. 3D-printed functional polymers and nanocomposites: defects characterization and product quality improvement. Adv Eng Mater. 2022;24:2101219. https://doi.org/10.1002/adem.202101219.
24. Silva M, Pinho IS, Covas JA, Alves NM, Paiva MC. 3D printing of graphene-based polymeric nanocomposites for biomedical applications. Funct Compos Mater. 2021;2:8.
25. Jafferson JM, Pattanashetti S. Use of 3D printing in production of personal protective equipment (PPE)—a review. Mater Today Proc. 2021;46:1247–60. https://doi.org/10.1016/j.matpr.2021.02.072.
26. Kim S, Kim DH, Kim W, Cho YT, Fang NX. Additive manufacturing of functional microarchitected reactors for energy, environmental, and biological applications. Int J Precis Eng Manuf Technol. 2021;8:303–26.
27. Ford J, Goldstein T, Trahan S, Neuwirth A, Tatoris K, Decker S. A 3D-printed nasopharyngeal swab for COVID-19 diagnostic testing. 3D Print Med 2020;6:21.
28. Emmanuel SS, Olawoyin CO, Adesibikan AA, Opatola EA. A pragmatic review on biopolymerized metallic nano-architecture for photocatalytic degradation of recalcitrant dye pollutants. J Polym Environ. 2024;32:1–30.
29. Bayode AA, Emmanuel SS, Akinyemi AO, Ore OT, Akpotu SO, Koko DT, et al. Innovative techniques for combating a common enemy forever chemicals: a comprehensive approach to mitigating per- and polyfluoroalkyl substances (PFAS) contamination. Environ Res. 2024;261: 119719. https://doi.org/10.1016/j.envres.2024.119719.
30. Emmanuel SS, Adesibikan AA, Saliu OD, Opatola EA. Greenly biosynthesized bimetallic nanoparticles for ecofriendly degradation of notorious dye pollutants: a review. Plant Nano Biol. 2023;3: 100024.
31. Ahmaruzzaman M, Roy S, Singha A, Rtimi S, Aminabhavi TM. Emerging nanotechnologies in adsorption of dyes: a comprehensive review of carbon and metal oxide-based nanomaterials. Adsorption. 2025;31:34. https://doi.org/10.1007/s10450-024-00588-y.
32. Emmanuel SS, Adesibikan AA, Opatola EA, Olawoyin CO. A pragmatic review on photocatalytic degradation of methyl orange dye pollutant using greenly biofunctionalized nanometallic materials: a focus on aquatic body. Appl Organomet Chem. 2023;3: e7108. https://doi.org/10.1002/aoc.7108.
33. Dhila H, Bhapkar A, Bhame S. Metal oxide/biochar hybrid nanocomposites for adsorption and photocatalytic degradation of textile dye effluents: a review. Desalin Water Treat. 2025;321: 101004. https://doi.org/10.1016/j.dwt.2025.101004.
34. Majooni Y, Fayazbakhsh K, Yousefi N. Toward mechanically robust and highly recyclable adsorbents using 3D printed scaffolds: a case study of encapsulated carrageenan hydrogel. Chem Eng J. 2024;494: 152672. https://doi.org/10.1016/j.cej.2024.152672.

35. Silva EC, Soares VR, Nörnberg AB, Fajardo AR. Recyclable 3D-printed composite hydrogel containing rice husk biochar for organic contaminants adsorption in tap water. ACS Appl Polym Mater. 2023;5:8415–29. https://doi.org/10.1021/acsapm.3c01534.
36. Nasser Abdelhamid H, Sultan S, Mathew AP. Binder-free three-dimensional (3D) printing of cellulose-ZIF8 (CelloZIF-8) for water treatment and carbon dioxide (CO_2) adsorption. Chem Eng J. 2023;468: 143567. https://doi.org/10.1016/j.cej.2023.143567.
37. Ranjan R, Rai R, Naik K, Parmar AS, Dhar P. In-situ growth of MoS2 on 3D printed phosphorylated cellulose for simultaneous dye degradation and microbial inactivation. Cellulose. 2025;32:1959–78.
38. Xin Yong GT, Koh W Bin, Binti Mohd Yusoff NH, Norman A, Chong CH, Cheah KH, et al. 3D-printed SiOC ceramics grafted with chitosan-graphene oxide composite for enhanced dye adsorption. Sep Purif Technol 2025;362:131669. https://doi.org/10.1016/j.seppur.2025.131669.
39. Yuan J, Yi C, Jiang H, Liu F, Cheng GJ. Direct ink writing of hierarchically porous cellulose/alginate monolithic hydrogel as a highly effective adsorbent for environmental applications. ACS Appl Polym Mater. 2021;3:699–709. https://doi.org/10.1021/acsapm.0c01002.
40. Li L, Li J, Luo H, Li S, Yang J. Physicochemical and photocatalytic properties of 3D-printed TiO_2/chitin/cellulose composite with ordered porous structures. Polymers (Basel) 2022;14. https://doi.org/10.3390/polym14245435.
41. Pei R, Fan L, Zhao F, Xiao J, Yang Y, Lai A, et al. 3D-Printed metal-organic frameworks within biocompatible polymers as excellent adsorbents for organic dyes removal. J Hazard Mater. 2020;384: 121418. https://doi.org/10.1016/j.jhazmat.2019.121418.
42. Emmanuel SS, Esan AO, Afigo FSO, Adesibikan AA, Idris MO. A review on eco-sustainable photocatalytic degradation of pharmaceutical pollutants using biosynthesized nanoparticles. J Chinese Chem Soc. 2024;71:1332–57. https://doi.org/10.1002/jccs.202400250.
43. Bayode AA, Olisah C, Emmanuel SS, Adesina MO, Koko DT. Sequestration of steroidal estrogen in aqueous samples using an adsorption mechanism: a systemic scientometric review. RSC Adv. 2023;13:22675–97.
44. Bergamonti L, Graiff C, Bergonzi C, Potenza M, Reverberi C, Ossiprandi MC, et al. Photodegradation of pharmaceutical pollutants: new photocatalytic systems based on 3D printed scaffold-supported Ag/TiO_2 nanocomposite. Catalysts 2022;12. https://doi.org/10.3390/catal12060580.
45. Bergamonti L, Bergonzi C, Graiff C, Lottici PP, Bettini R, Elviri L. 3D printed chitosan scaffolds: A new TiO_2 support for the photocatalytic degradation of amoxicillin in water. Water Res 2019;163. https://doi.org/10.1016/j.watres.2019.07.008.
46. Emmanuel SS, Idris MO, Olawoyin CO, Adesibikan AA, Aliyu AA, Suleiman AI. Biosynthesized metallic nanoarchitecture for photocatalytic degradation of emerging organochlorine and organophosphate pollutants: a review. ChemistrySelect. 2024;9: e202304956. https://doi.org/10.1002/slct.202304956.
47. Baigorria E, Souza dos Santos S, de Moura MR, Fraceto LF. Nanocomposite hydrogels 3D printed for application in water remediation. Mater Today Chem 2023;30:101559. https://doi.org/10.1016/j.mtchem.2023.101559.
48. Kashaudhan K, Pande PP, Sharma J, Chaurasiya A, Shankar R, Nath A. Facile synthesis of XGX-based hydrogel and its application for the removal of heavy metal ions from aqueous solutions. J Porous Mater. 2025;32:915–33. https://doi.org/10.1007/s10934-025-01749-2.
49. Wang N, Xu Y, Wei Y, Chen Q, Song F, Liu B. Proportion-optimized 3D-printed calcium alginate/graphene oxide adsorbent for efficient uranium separation from aqueous solutions. Sep Purif Technol. 2025;377: 134199. https://doi.org/10.1016/j.seppur.2025.134199.
50. Wu T, Karimi-Maleh H, Li Y, Zhang D, Zhang Z, Zhong N, et al. 3D printed porous chitosan/metal–organic framework composites as effective adsorbents to remove heavy metals from wastewater. Chem Eng J. 2024;493: 152780.
51. Abdelhamid HN, Sultan S, Mathew AP. Three-dimensional printing of cellulose/covalent organic frameworks (CelloCOFs) for CO_2 adsorption and water treatment. ACS Appl Mater Interfaces. 2023;15:59795–805. https://doi.org/10.1021/acsami.3c13966.

52. Abdelhamid HN, Sultan S, Mathew AP. 3D printing of cellulose/leaf-like zeolitic imidazolate frameworks (CelloZIF-L) for adsorption of carbon dioxide (CO_2) and heavy metal ions. Dalt Trans. 2023;52:2988–98.
53. Shahbazi M, Jäger H, Ahmadi SJ, Lacroix M. Electron beam crosslinking of alginate/nanoclay ink to improve functional properties of 3D printed hydrogel for removing heavy metal ions. Carbohydr Polym. 2020;240: 116211. https://doi.org/10.1016/j.carbpol.2020.116211.
54. Song F, Niu Y, Wang N, Chen W, Xie W, Liu B. Fabrication of three-dimensional printed hybrid graphene oxide/calcium alginate network by 3D printing technology with outstanding U(VI) recycling capacity in acidic solution. Chem Eng J. 2023;457: 141245. https://doi.org/10.1016/j.cej.2022.141245.
55. Yoon H-J, Atsbha TA, Yoon T, Shin D, An J, Zarei M, et al. Sustainable and feasible carbon capture and utilization pathways towards net-zero. Renew Sustain Energy Rev. 2025;211: 115331. https://doi.org/10.1016/j.rser.2025.115331.
56. Määttä S, de Gooyert V. A critical review of social scientific research on carbon capture and storage. Renew Sustain Energy Rev. 2025;208: 115063. https://doi.org/10.1016/j.rser.2024.115063.
57. Dung HT, Chinh NT, Lu LT, Hoang T. Methylene blue adsorption of chitosan/nano-TiO_2 films prepared by 3D printing method. Vietnam J Chem. 2021;59:319–25.
58. Emmanuel SS, Adesibikan AA, Olusola Olawoyin C, Bayode AA. A Review of antithrombolytic, anticoagulant, and antiplatelet activities of biosynthesized metallic nanostructured multifunctional materials. ChemistrySelect 2023;8:e202302712. https://doi.org/10.1002/slct.202302712.
59. Saranya S, Radha KV. Review of nanobiopolymers for controlled drug delivery. Polym Plast Technol Eng. 2014;53:1636–46.
60. Varma K, Gopi S. Biopolymers and their role in medicinal and pharmaceutical applications. Biopolym Ind Appl. Elsevier; 2021, p. 175–91.
61. Biswas MC, Jony B, Nandy PK, Chowdhury RA, Halder S, Kumar D, et al. Recent advancement of biopolymers and their potential biomedical applications. J Polym Environ. 2022;30:51–74.
62. Biranje SS, Shi Y, Sun J, Cheng L, Jiao H, Lu X, et al. Cellulose nanofibril/polylysine-based 3D composite antibacterial scaffold for wound healing applications. Cellulose. 2023;30:5289–306.
63. N'gatta KM, Belaid H, El Hayek J, Assanvo EF, Kajdan M, Masquelez N, et al. 3D printing of cellulose nanocrystals based composites to build robust biomimetic scaffolds for bone tissue engineering. Sci Rep 2022;12:21244.
64. Mehdikhani M, Yilgör P, Poursamar SA, Etemadi N, Gokyer S, Navid S, et al. A hybrid 3D-printed and electrospun bilayer pharmaceutical membrane based on polycaprolactone/chitosan/polyvinyl alcohol for wound healing applications. Int J Biol Macromol. 2024;282: 136692. https://doi.org/10.1016/j.ijbiomac.2024.136692.
65. Bidaki A, Rezaei N, Kazemi S, Ali SN, Ziaei S, Moeinzadeh A, et al. 3D printed bioengineered scaffold containing chitosan, alginate, and Barijeh-loaded niosomes enabled efficient antibiofilm activity and wound healing. Int J Biol Macromol. 2025;311: 143743. https://doi.org/10.1016/j.ijbiomac.2025.143743.
66. Oluwaniyi OO, Adesibikan AA, Emmanuel SS. Evaluation of wound-healing activity of securidaca longepedunculata root extract in Male Wistar Rats. ChemistrySelect. 2022;7: e202200711.
67. Awolola GV, Emmanuel SS, Adesibikan AA. Phytomedicine plus evaluation of phytoconstituent and wound-healing potential of methanolic waste shell extract of Elaeis guineensis Jacquin in female rats. Phytomedicine Plus. 2021;1: 100126. https://doi.org/10.1016/j.phyplu.2021.100126.
68. Tamo AK, Tran TA, Doench I, Jahangir S, Lall A, David L, et al. 3D Printing of cellulase-laden cellulose nanofiber/chitosan hydrogel composites: towards tissue engineering functional biomaterials with enzyme-mediated biodegradation. Materials (Basel) 2022;15. https://doi.org/10.3390/ma15176039.
69. Bergonzi C, Remaggi G, Graiff C, Bergamonti L, Potenza M, Ossiprandi MC, et al. Three-dimensional (3D) printed silver nanoparticles/alginate/nanocrystalline cellulose hydrogels:

study of the antimicrobial and cytotoxicity efficacy. Nanomaterials 2020;10. https://doi.org/10.3390/nano10050844.
70. Gutierrez E, Burdiles PA, Quero F, Palma P, Olate-Moya F, Palza H. 3D printing of antimicrobial alginate/bacterial-cellulose composite hydrogels by incorporating copper nanostructures. ACS Biomater Sci Eng. 2019;5:6290–9. https://doi.org/10.1021/acsbiomaterials.9b01048.
71. N'Gatta KM, Assanvo EF, El Hayek J, Masquelez N, Kamgang Syapnjeu P, Deabate S, et al. 3D-printed polycaprolactone scaffolds reinforced with cellulose nanocrystals and silver nanoparticles for bone tissue engineering. ACS Appl Mater Interfaces 2025;17:41028–43. https://doi.org/10.1021/acsami.5c06504.
72. Meena D, Kumar R, Gupta S, Khan O, Gupta D, Singh M. Energy storage in the 21st century: a comprehensive review on factors enhancing the next-generation supercapacitor mechanisms. J Energy Storage. 2023;72: 109323. https://doi.org/10.1016/j.est.2023.109323.
73. Shaheen I, Hussain I, Zahra T, Javed MS, Shah SSA, Khan K, et al. Recent advancements in metal oxides for energy storage materials: design, classification, and electrodes configuration of supercapacitor. J Energy Storage. 2023;72: 108719. https://doi.org/10.1016/j.est.2023.108719.
74. Bauquis PR. A reappraisal of energy supply and demand in 2050. Oil Gas Sci Technol Rev IFP. 2001;56:389–402.
75. Dar MA, Majid SR, Satgunam M, Siva C, Ansari S, Arularasan P, et al. Advancements in supercapacitor electrodes and perspectives for future energy storage technologies. Int J Hydrogen Energy. 2024;70:10–28. https://doi.org/10.1016/j.ijhydene.2024.05.191.
76. Zhou G, Li M-C, Liu C, Wu Q, Mei C. 3D printed Ti3C2Tx MXene/cellulose nanofiber architectures for solid-state supercapacitors: ink rheology, 3D printability, and electrochemical performance. Adv Funct Mater. 2022;32:2109593. https://doi.org/10.1002/adfm.202109593.
77. Salve M, Mandal A, Amreen K, Pattnaik PK, Goel S. Greenly synthesized silver nanoparticles for supercapacitor and electrochemical sensing applications in a 3D printed microfluidic platform. Microchem J. 2020;157: 104973. https://doi.org/10.1016/j.microc.2020.104973.
78. Jain K, Wang Z, Garma LD, Engel E, Ciftci GC, Fager C, et al. 3D printable composites of modified cellulose fibers and conductive polymers and their use in wearable electronics. Appl Mater Today. 2023;30: 101703. https://doi.org/10.1016/j.apmt.2022.101703.

4D Printing of Polysaccharides: Fabrication and Applications

Muhammad Humais Yamin, Qudsia Kanwal, Muhammad Mubashar Rasool, Rimsha Shahzadi, and Muhammad Hamza

Abstract This chapter explores the innovative 4D printing of polysaccharides, emphasizing its transformational capability for producing dynamic and stimuli-responsive materials. By integrating time as the fourth dimension, 4D printing enhances the functionalities of traditional 3D printing to create structures that can adapt, transform, or modify their properties in response to environmental stimuli such as temperature, pH, moisture, or light. Due to their natural abundance, biocompatibility, and simplicity of chemical modification, polysaccharides are exemplary prospects for designing 4D-printed materials. The chapter examines several polysaccharide additive manufacturing processes, including advanced printing technologies and post-printing functionalization approaches. It offers a comprehensive analysis of the stimuli-responsive behaviours shown by polysaccharide-based materials and their mechanism of action. Moreover, it highlights the diverse applications of 4D-printed polysaccharides in domains like drug delivery, tissue engineering, soft robotics, and environmental remediation. Challenges such the precise control over material or shape transformations, scalability, and material stability are also addressed. The chapter anticipates future developments in polysaccharide-based 4D printing, enabling the creation of more innovative, sustainable, and versatile materials for advanced applications.

Keywords 4D printing · Polysaccharides · Chemical and physical modifications · Smart polysaccharides · Biomedical applications

---

Muhammad Humais Yamin and Qudsia Kanwal are contributed equally.

M. H. Yamin
Department of Physics, University of Gujrat, Gujrat, Pakistan

Q. Kanwal · M. M. Rasool · R. Shahzadi
Department of Chemistry, The University of Lahore, Lahore, Pakistan

M. Hamza (✉)
Additive Manufacturing Institute, Shenzhen University, Shenzhen, China
e-mail: muhammadhamza7596173@gmail.com

M. H. Abu Elella et al. (eds.), *3D Engineered Polysaccharides*,
https://doi.org/10.1007/978-3-032-26154-0_12

Abbreviations

3D	Three Dimensional
4D	Four Dimensional
Ba^{2+}	Barium Ion
BCS	Breast-Conserving Surgery
Ca^{2+}	Calcium Ion
$CaCO_3$	NPs-GDL Calcium Carbonate Nanoparticles–Gluconic Delta-Lactone
CAD	Computer-Aided Design
Cap	Calcium Phosphate
CMC	Cellulose Sodium Salt
CNC	Cellulose Nanocrystals
–COOH	Carboxyl Group
DIW	Direct Ink Writing
DSC	Differential Scanning Calorimetry
EMA	European Medicines Agency
FDA	Food and Drug Administration
Fe^{2+}	Ferrous Ion
Fe_2O_3	Iron (III) oxide
Fe^{3+}	Ferric Ion
Fe_3O_4	Magnetite
FEM	Finite Element Model
G	α-L-guluronic Acid
GelMA	Gelatin Methacryloyl
GG	Gellan Gum
HA	Hydroxyapatite
HAMA	Hyaluronic Acid Methacryloyl
HA	Hyaluronic acid
K^+	Potassium Ion
KGM	Konjac Glucomannan
LCST	Lower Critical Solution Temperature
M	β-D-mannuronic acid
MC	Methylcellulose
ML	Machine Learning
Na^+	Sodium Ion
NaOH	Sodium Hydroxide
$NH_3{\cdot}H_2O$	Ammonium hydroxide
$-NHCOCH_3$	Acetanilide Group
–OH	Hydroxide Group
PCL	Polycaprolactone
PEG	Polyethylene glycol
pH	Potential of Hydrogen
RH	Relative Humidity

ROS	Reactive Oxygen Species
SA	Sodium Alginate
SA-MA	Methacrylated alginate
SPAs	Soft Pneumatic Actuators
Sr^{2+}	Strontium Ion
TPU	Thermoplastic Polyurethane
UCST	Upper Critical Solution Temperature
UV	Ultra Violet

1 Introduction

4D printing technology has revolutionized the additive manufacturing field by introducing a fourth dimension, which is time, into the fabrication process. This innovative technology enables the formation of responsive structures that can alter shape, function, or features in response to environmental stimuli, e.g., temperature, pH, moisture, or light [1–3]. Different materials are explored for 4D printing, polysaccharides are commonly used because of their natural abundance, biocompatibility, and chemical modification feature [2]. These biopolymers offer a unique combination of properties that make them highly suitable for designing 4D-printed materials with applications in drug delivery, tissue engineering, soft robotics, and environmental remediation.

These are natural polymers like alginate [4], cellulose [5], chitin [6], chitosan [7], gellan gum [8], hyaluronic acid [9], konjac glucomannan [10], pullulan, and starch. All these polysaccharides will have distinct characteristics, which can be utilized in particular uses in 4D printing. Alginate has great biocompatibility and gelation characteristics that render it appropriate in drug delivery and tissue engineering [4]. Cellulose, the most widespread biopolymer on the planet, is high in mechanical strength and biodegradability, and is appropriate to build strong and sustainable materials. Chitosan is a derivative of chitin that is antimicrobial, and is biocompatible; it has applications in wound healing and bioactive delivery systems [5]. The unique properties of polysaccharides, i. e. chemical modifiability and stimuli-responsive behavior, render them the ideal candidates to use in 4D printing [11]. Printability, mechanical and responsive properties of chemically modified polysaccharide are improved compared to non-modified [12]. The addition of polysaccharides to other biopolymers or the addition of functional groups dramatically changes their characteristics. Alginate and methylcellulose mixing to enhance their shape-morphing property [13] and chitosan and polyethylene glycol mixing to enhance their extrusion property [14]. Besides, the properties of these biopolymers are also further personalized by the introduction of photo-crosslinkable units or oxidation reactions to particular uses [15]. One of the factors that render polysaccharides applicable in 4D printing is the stimuli-responsive nature of these polymers [16]. These materials are sensitive to shape and properties depending on the different physical, chemical, and biological stimuli [17]. Alginate and chitosan are polysaccharides with a pH-sensitive swelling behavior and are thus suitable in targeted drug delivery [18]. Likewise, temperature-dependent polysaccharides such as agarose and sodium alginate may have sol–gel transitions, allowing the formation of dynamic structures that respond to changes in environment [19]. Advanced printing technology and post-printing functionalization methods are fabrication methods of the 4D printing of polysaccharides. These methods permit making sophisticated structures that can have a controlled shape and characteristics. As an illustration, 3D programmed structures are manufactured using extrusion-based printing techniques in order to alter shape as time passes [19]. Functionalization methods post-printing, e.g. chemically grafting and crosslinking, enhances material properties further, and is applicable in multiple applications. Regardless of the many benefits of 4D-printed polysaccharides, there are a number of challenges that must be overcome to achieve the wide implementation [20]. These are material constraints, accuracy in the transformations caused by stimulus, scalability, and regulatory issues. Also, accurate control over stimuli-initiated changes and reliability between large-scale manufacturing processes are important in the successful realization of these materials.

This chapter gives a detailed description of the new 4D printed polysaccharides and its process, properties and various application. It starts with shown polysaccharides as practical 4D printable materials, such as alginate, cellulose, chitin, chitosan, gellan glucomannan, pullulan, and starch. Their distinctive features in terms of chemical modifiability and stimuli-responsiveness are outlined, which are likely to enable the generation of dynamic and adaptive materials. The chapter then explores the fabrication methods applied in polysaccharides 4D printing, the comparison between 3 and 4D printing processes and the advanced printing technology and functionalizing after printing. It explores stimuli responsive behaviour of polysaccharides such as thermal behaviour, pH sensitive behaviour, moisture responsive behaviour, light responsive behaviour and multi-stimulus responsive behaviour. Applications of the 4D-printed polysaccharides are widely distributed, including biomedical uses as dynamic drug delivery systems and tissue engineering scaffolds, soft robotics and actuators, environmental uses such as smart filtration systems, and consumer uses such as adaptive textiles and customized packaging systems. In this chapter, the challenges of the 4D printing of polysaccharides are discussed, such as material constraints, sensitivity of stimuli-induced changes, scalability and regulation issues of biomedical applications. Lastly, it provides future insights into the 4D printed polysaccharide evolution, to create more innovative, sustainable, and adaptation materials to be used in superior applications.

2 Polysaccharides as Functional Materials for 4D Printing

4D printing is a progressive development of 3D printing that aims at creating dynamic structures that have the potential to modify shape, functionality, or properties throughout time in reaction to external influences. Natural and biocompatible polysaccharides are a promising choice of materials to 4D print because of their controllable characteristics and the ability to react to stimuli. This section discusses different kinds of polysaccharides that are used in 4D printing.

2.1 Types of Polysaccharides Used for 4D Printing

Alginate is a naturally occurring polyanionic polysaccharide composed of β-D-mannuronic acid (M) and α-L-guluronic acid (G) units linked by 1 → 4 glycosidic bonds, valued for its excellent biocompatibility, non-toxicity, and biodegradability as both a polymer and nanocarrier [21]. The formation of alginate occurs through an ionotropic gelation process, where sodium ions in sodium alginate are replaced by divalent cations (Ca^{2+}, Sr^{2+}, Ba^{2+}, etc.) that specifically bind to α-L-guluronic acid blocks, creating cross-links between polymer chains to form a three-dimensional network [22, 23]. This versatile biopolymer combines ease of handling with tunable properties and chemical modifiability, making it particularly suitable for developing drug delivery system [4].

Cellulose, the most abundant biopolymer on Earth, is a linear polysaccharide consisting of β-(1 → 4)-linked D-glucose units primarily derived from wood pulp (85–88%) and cotton lint (12–15%) [24]. As the main component of terrestrial biomass, it represents a highly renewable and cost-effective natural resource [5]. Various cellulose derivatives, including carboxymethyl cellulose and hydroxypropyl methyl cellulose, are commonly employed as polysaccharide scaffold materials [25]. These modified cellulose derivatives can form three-dimensional networks using different fabrication methods, and when combined with stimuli-responsive materials, they can be engineered into intelligent drug delivery systems capable of responding to environmental changes [26, 27].

Chitin, a linear β-(1–4)-N-acetyl-d-glucosamine polymer found in arthropod exoskeletons and fungal cell walls, serves as a structural component in nature, typically forming crystalline microfibrils [28]. Its derivative, chitosan, is produced through partial deacetylation, resulting in a linear cationic heteropolysaccharide composed of β-1,4-linked glucosamine and N-acetylglucosamine residues [29]. Renowned for its biodegradability, biocompatibility, nontoxicity, antibacterial properties, and film-forming capabilities, chitosan has broad applications in wound healing and bioactive delivery systems. Chitosan exhibits pH-sensitive swelling behavior, dissolving in acidic (gastric) conditions due to protonated amine groups causing electrostatic repulsion, while remaining insoluble in alkaline (intestinal) environments [30]. This property makes chitosan an ideal carrier for lipophilic bioactive compounds, enhancing their gastrointestinal absorption. Carboxymethyl chitosan, an amphoteric derivative, offers superior biocompatibility, biodegradability, and antibacterial activity compared to unmodified chitosan [31]. For instance, hydrochloride and carboxymethyl chitosan have been used to encapsulate quercetin, significantly improving its chemical stability, solubility, and bioavailability.

Gellan gum (GG) is a linear, water-soluble anionic polysaccharide produced by Pseudomonas elodea, composed of repeating units of β-D-glucose, β-D-glucuronic acid, α-L-rhamnose, and β-D-glucose [32]. It is considered safe to consume and has properties of great gelling, biocompatibility, and biodegradability, thus is commonly used as a food additive stabilizer, thickener, and gelator. Beyond food applications, its low cytotoxicity has spurred interest in biomedical engineering, where it serves as a promising drug delivery agent and cell carrier [8]. The polysaccharide readily forms gels at low concentrations in the presence of mono- (Na^+, K^+) or divalent (Ca^{2+}) cations. Additionally, oxidative modification of it through carboxyl group introduction enhances its pH sensitivity, improving water solubility and reducing gelation temperature, key properties for developing responsive biomaterials [33].

Hyaluronic acid (HA) is a linear polysaccharide composed of alternating units of 2-acetamido-2-deoxy-D-glucose and D-glucuronic acid, possessing strong anionic properties that facilitate electrostatic interactions with other compounds [34]. While native HA exhibits excellent cytocompatibility, modified HA derivatives can be engineered to enhance mechanical and biological properties while acquiring novel functionalities [9]. To improve its stability, functional groups (–COOH, –OH, –$NHCOCH_3$) are often chemically modified. HA is particularly valued in biomedical applications for promoting tissue healing, cell proliferation/migration, angiogenesis, and inflammation modulation [35]. Current applications include cartilage regeneration, osteochondral defect treatment using sponge matrices, and controlled drug delivery systems, leveraging HA's unique biological activities and tunable material properties for therapeutic purposes [36].

Konjac glucomannan (KGM) is a natural polysaccharide composed of β-(1 → 4)-linked D-mannose and D-glucose residues in a 1.6:1 or 1.4:1 molar ratio, with approximately one acetyl group per 17–19 sugar units at the C-6 position that enhances its solubility and gelation properties [37]. As KGM is specifically degraded by colonic bacterial enzymes, it serves as an effective targeted carrier for controlled drug release in the colon [38]. Chemical modification through sodium periodate oxidation introduces reactive aldehyde groups, creating oxidized KGM with enhanced functionality. Sulfhydryl-oxidized KGM hydrogel microspheres effectively protect probiotics, significantly improving their survival in gastric conditions and prolonging intestinal residence time to increase probiotic abundance [10]. These unique properties make KGM-based materials particularly valuable for developing advanced delivery systems targeting gastrointestinal applications.

Pullulan is an extracellular microbial polysaccharide primarily produced by yeasts, composed of linear maltotriose repeating units where two α-(1 → 4)-linked glucopyranose rings are connected by α-(1 → 6) glycosidic bonds, with partial acid hydrolysis yielding panose and isopanose structures [39]. This unique biopolymer exhibits distinctive properties including low solution viscosity, water solubility, biodegradability, and neutral taste/odor, along with compatibility with dilute alkaline solutions. These characteristics have enabled its widespread utilization in biomedical, pharmaceutical, and food applications. Furthermore, chemical modifications such as oxidation allow pullulan to form pH-responsive through interactions with other polysaccharides, making it particularly valuable for developing advanced bioactive delivery systems [10]. The polymer's structural versatility and functional adaptability position it as a promising material for designing stimulus-responsive drug carriers.

Starch, a natural plant-based polysaccharide composed of linear amylose (20–25%) and branched amylopectin (75–80%), has gained significant attention in biomaterials due to its low cost, biocompatibility, and biodegradability [40]. However, its susceptibility to gastric degradation necessitates modification through physical (e.g., pregelatinization), chemical (e.g., oxidation), or enzymatic methods to create stable drug carriers. Modified starch-based hydrogels, particularly those incorporating carboxyl groups, exhibit pH-responsive behavior through swelling and controlled release mechanisms [41], enabling targeted delivery of hydrophobic bioactives like curcumin. Advanced composites such as oxidized starch/CuO hydrogels demonstrate pH-dependent release of drugs (e.g., ibuprofen, naproxen) with sustained colon-specific delivery potential, highlighting starch's promise as a smart biomaterial for pharmaceutical applications [42].

2.2 *Unique Properties of Polysaccharides for 4D Printing*

2.2.1 Chemical Modifiability

Polysaccharides are a class of biodegradable polymers that are highly sensitive to various physical, chemical, and biological stimuli, making them ideal candidates for 4D printing applications. Their ability to respond to stimuli allows for the creation of dynamic and adaptive structures that can change shape or properties over time. However, to fully harness their potential in 4D printing, polysaccharides often require chemical modification to enhance their printability, mechanical strength, and responsiveness to stimuli.

Blending with Other Polymers: One common approach to improving the printability and mechanical properties of polysaccharides is blending them with other biopolymers [43]. As an example, the issue of the poor shape-morphing characteristic of alginate may be addressed by incorporating it with methylcellulose [13] or dopamine [44]. This leads to hydrogels that are better rheologically and extrudability thus they can be used in 4D printing. In the same manner, chitosan, with low printability, can be added to polyethylene glycol (PEG), gelatin, or pectin to regulate its viscosity and improve its extrusion properties [45].

Functionalization with Responsive Groups: The addition of hydrophobic, acidic, basic or other chemical functional groups to the polysaccharide backbone can drastically change their characteristics and stimuli responsiveness [46]. Indicatively, polysaccharides such as pullulan can be modified with photo-crosslinkable methacrylic units to make them light-responsive and printable [47]. The use of oxidation reactions to add carboxyl groups to polysaccharides can also be performed to allow further crosslinking reactions to enhance mechanical strength and shape-morphing properties [48].

2.2.2 Stimuli-Responsive Behavior

The polysaccharides are sensitive to numerous physical (temperature, humidity, electric field, magnetic field, light), chemical (pH value and ion concentration) and biological (enzymes and peptides) stimuli [49], and as such are well suited to 4D printing uses. An example of such materials is the ability to change shape or properties over time with response to certain stimuli, allowing the formation of dynamic and adaptation structures. We are talking about the stimuli-responsive nature of polysaccharides and their use in 4D printing in this case (recent research results validate this point).

3 Fabrication Techniques in 4D Printing of Polysaccharides

3.1 Comparison of 3D and 4D Printing Processes

3D printing and 4D printing of polysaccharides both have unique characteristics and applications. 3D printing is a layer-by-layer fabrication process that creates static structures based on a designed model, using materials like alginate, chitosan, and hyaluronic acid. It is widely used for creating static scaffolds for tissue engineering and medical devices. On the other hand, 4D printing of polysaccharides relies on the material's ability to respond to specific stimuli, chemical, physical, or biological, enabling shape or property changes over time. Polysaccharides exhibit remarkable sensitivity to various triggers, making them ideal for 4D applications. For instance, they can react to physical stimuli such as temperature, light, electric or magnetic fields, and pressure, as well as chemical cues like reactive oxygen species, redox agents, glucose, enzymes, and ions. Chitosan, for example, responds to glucose, pH (particularly in acidic conditions due to its amine groups), and electric fields [50]. Similarly, agarose, sodium alginate, and hyaluronic acid are temperature-sensitive, while chitosan and agarose react to voltage changes [51]. Alginate-glycerin systems respond to pH shifts, and hyaluronic acid can deform under mechanical tension. Additionally, composite materials like cellulose-dextran-graphene blends demonstrate dual responsiveness to pH and near-infrared light [52, 53].

3.2 Printing Technologies for Polysaccharides

Since the introduction of 3D printing in 1986, the manufacturing industry has transformed, reducing time, energy, and waste through direct fabrication from CAD designs. In healthcare, 3D printing has enabled the creation of patient-specific scaffolds and hydrogels, offering precise control over shape and structure for tissue engineering and drug delivery. However, most traditional 3D-printed hydrogels are static and do not adapt to dynamic internal biological changes. 3D printing of

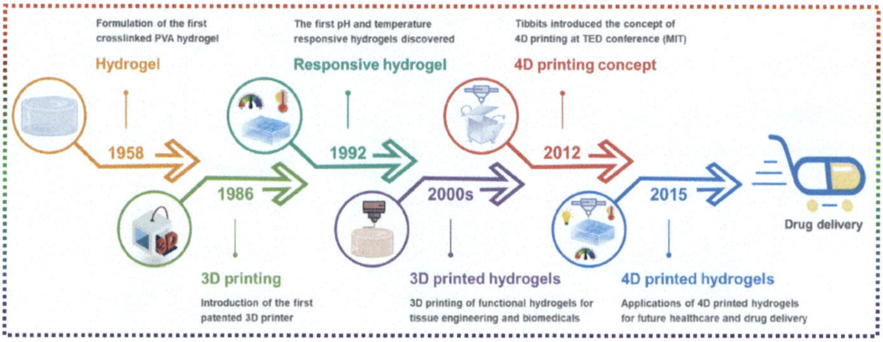

Fig. 1 A brief history of polysaccharides hydrogels from 3 to 4D for drug delivery applications. Adapted from [55]

polysaccharides utilizes various techniques to produce three-dimensional objects, including extrusion-based printing (direct ink writing, fused deposition modeling), stereolithography and digital light processing, selective laser sintering, and inkjet printing [54]. The development of dynamic materials has led to the emergence of 4D printing, where time is the fourth dimension, allowing objects to change shape in response to stimuli like light, heat, and magnetic fields. This advancement has generated significant interest in 4D-printed smart hydrogels, which can interact with their environment and exhibit various responses, enhancing their applications in healthcare [55], as shown in Fig. 1.

3.3 Post-Printing Functionalization and Crosslinking Techniques

Post-printing functionalization and crosslinking techniques play a crucial role in enhancing the properties of 4D printed polysaccharide hydrogels. After the initial printing process, various methods can be employed to modify the hydrogels' chemical and physical characteristics. Functionalization techniques, such as chemical grafting or surface modification, as shown in Fig. 2a, allow for the introduction of specific functional groups that can enhance biocompatibility, improve drug loading capacity, or enable responsiveness to environmental stimuli [56]. The hydrogel structure is stabilized by means of crosslinking, (physical, chemical, or enzymatic) to enhance mechanical strength, and to maintain shape stability through dynamic transformations as depicted in Fig. 2b. These methods do not only maximize the performance of the hydrogels but also increase their use in drug delivery, tissue engineering, and smart biomedical devices where specific control over their behavior is required [57]. A list of polysaccharides with applied modifiers, resulting functionalities, properties, and proposed applications in the Table 1.

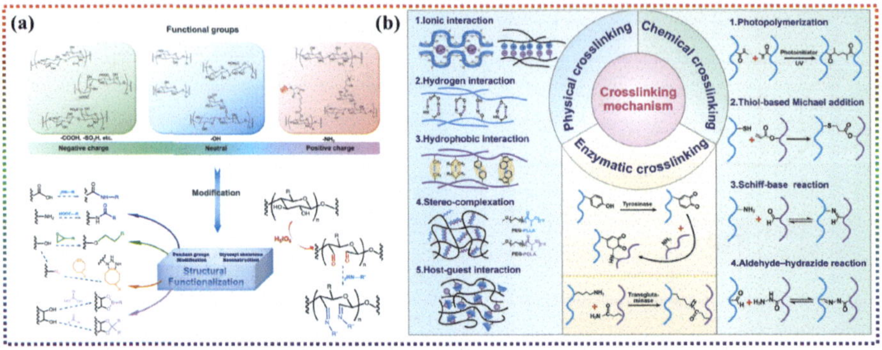

Fig. 2 An overview of **a** functionalization techniques, including chemically functional group modification and structural modification. Adapted from [56]. Crosslinking mechanisms including, physical, chemical, and enzymatic crosslinking. Adapted from [57]

Table 1 Polysaccharides with applied modifiers, resulting functionalities, properties, and proposed applications

Polysaccharide	Functionalize Material	Properties	Applications	References
Chitosan	Catechol	Thermosensitive	Wound dressing	[58]
Chitosan	Lauric acid	Physicochemical properties	Mucosal drug delivery	[59]
Chitosan	Gelatine	Adhesiveness	Tissue adhesive	[60]
Chitosan	Pyrogallol	Adhesiveness to wet surfaces	Haemostasis	[61]
Chitosan	Polyethylene glycol	Fast gelation, Haemostatic	Tissue adhesive	[62]
Sodium alginate	Polyacrylamide	Stretchability, stability	Wound dressing	[63]
Agarose	Polyethylene glycol	Haemostasis	Wound dressing	[64]
Dextran	Aldehyde group	Blood absorption, haemostasis	Haemostasis	[65]
Levan	Chitosan and alginate	Mechanical properties	Free-standing layer-by-layer film	[66]
Chitin	Chitosan and alginate	Haemostatic, Bio-adhesive	Surgical adhesives	[67]

4 Stimuli-Responsive Mechanisms in Polysaccharide-Based 4D Printing

4.1 Thermal Responsiveness

Thermal-responsive polysaccharides represent an important class of smart biomaterials that undergo reversible sol–gel transitions in response to temperature changes, making them ideal for biomedical applications. These polymers contain both hydrophilic and hydrophobic groups that enable distinct swelling-shrinking behavior at either the upper critical solution temperature (UCST) or lower critical solution temperature (LCST) [68]. When the temperature exceeds the LCST (typically near physiological range), increased hydrophobic interactions cause network contraction, while temperatures below LCST promote swelling through weakened hydrophobic effects. Various polysaccharides exhibit this behavior, chitosan forms LCST systems through its amine groups, while cellulose derivatives undergo helix-coil transitions during freeze–thaw cycles [69]. The thermosensitivity can be precisely tuned by incorporating hydrophobic/hydrophilic comonomers via free radical polymerization or blending with synthetic polymers like polyethylene glycol. Such programmable phase transitions at biologically relevant temperatures enable controlled drug release, where payloads remain protected below LCST and rapidly release above it. These intelligent polysaccharides demonstrate particular promise for injectable therapies, as they can be administered as low-viscosity solutions that spontaneously gel at body temperature to form localized drug depots.

4.2 pH-Sensitive Behaviors

The pH-sensitive properties of 4D-printed polysaccharide-based materials make them promising candidates for targeted drug delivery, leveraging physiological pH variations as a key stimulus. Tumor tissues and infected wounds often exhibit an acidic microenvironment (pH 6.0–7.0) due to anaerobic respiration of cancer cells and bacterial activity, contrasting with the near-neutral pH (7.2–7.4) of healthy tissues and blood. Acidic conditions also found in gastrointestinal tracts (pH 1–3). This pH gradient enables the design of smart drug carriers that selectively release therapeutics in diseased regions, enhancing treatment precision while minimizing off-target effects [70]. Polysaccharides including HA, chitosan, and alginate, exhibit pH responsiveness through their dissociable functional groups, where hydrogen bonding and intermolecular forces weaken with pH variations, triggering discontinuous swelling. HA's carboxyl groups ionize at pH 6–7, enabling pH-triggered protein release (e.g., 91% release at pH 5 vs. 18.4% at pH 7.4 over 24 h) [71]. Alginate's carboxylic acid groups confer both hydrophilicity and pH sensitivity, with protein-loaded complexes showing accelerated release at higher pH [72]. Cellulose-based hydrogels exhibit injectability and self-healing while enhancing doxorubicin's antitumor efficacy in osteosarcoma models.

4.3 Moisture-Driven Transformations

While not typically classified as "smart" materials, moisture-responsive polysaccharide hydrogels demonstrate remarkable utility in 4D printing systems. These hydrophilic polymers exhibit controlled swelling behaviors that can be precisely engineered to create dynamic, shape-changing structures. By strategically manipulating crosslinking density within printed constructs, either through gradient crosslinking patterns or composite material formulations, designers can program specific deformation responses to aqueous environments. This differential swelling capability enables complex morphing behaviors, including bending, twisting, and folding, when exposed to moisture gradients or solution changes [73]. The inherent hygroscopic nature of polysaccharides like alginate, chitosan, and cellulose derivatives makes them particularly suitable for such moisture-driven 4D transformations. Researchers have developed various approaches to control swelling mismatches, including:

- Spatially controlled crosslinking densities
- Multi-material printing with varying swell capacities
- Layered architectures combining responsive and non-responsive regions

These moisture-responsive systems open new possibilities for biomedical devices, soft robotics, and adaptive structures that can autonomously reconfigure in aqueous environments.

4.4 Light-Induced Changes

Light-responsive polysaccharides exhibit tunable properties through three primary mechanisms of photo stimulation: (1) Photosensitive group grafting, where modified polysaccharides undergo reversible phase transitions upon photon absorption; (2) Photoionization, where embedded photosensitive molecules generate ions under light exposure, altering osmotic pressure to trigger swelling/deswelling; and (3) Photochemical modulation, where light-responsive compounds undergo cleavage or isomerization to change hydrogel network properties. These mechanisms enable precise spatiotemporal control over drug release, degradation kinetics, and mechanical properties in biomedical applications such as wound dressings and targeted therapy. For instance, UV-responsive oxidized alginate achieves on-demand antibiotic release through photocleavage, while visible-light-sensitive cellulose derivatives allow non-invasive gelation for 3D cell culture. The natural polysaccharide backbone ensures biocompatibility, while the photo-responsive modifications introduce advanced functionality for smart therapeutic systems.

4.5 *Magnetic-Sensitive Behaviors*

Magnetic-responsive polysaccharide is engineered by incorporating magnetic nanoparticles (γ-Fe_2O_3, Fe_3O_4, or $CoFe_2O_4$) into the polymer network, with their properties determined by nanoparticle type, size, concentration, and distribution. These enable dual functionality: (1) magnetic targeting for precise localization of injectable carriers, and (2) remote-controlled drug release triggered by alternating magnetic fields that induce both mechanical deformation and thermal effects (via magnetic hyperthermia), simultaneously promoting deswelling and payload release. While synthetic polymers often utilize grafting methods, natural polysaccharides (chitosan, alginate, cellulose) primarily employ blending or in situ approaches due to limited active sites for covalent modification.

In **the blending method, pre-synthesized** magnetic nanoparticles are dispersed into a polysaccharide solution before cross-linking, physically entrapping the particles [74, 75]. The in-situ method involves diffusing Fe^{2+}/Fe^{3+} ions into a pre-formed polymer mixture, followed by alkaline precipitation ($NaOH/NH_3{\cdot}H_2O$) to generate nanoparticles within the network. These strategies yield unique advantages:

- Spatiotemporal control for targeted drug/cell delivery.
- Combined hyperthermia-chemo therapy capabilities.
- Easy retrieval post-application via magnetic separation.
- Enhanced MRI contrast: Iron oxide-loaded polysaccharides improve imaging resolution.

Such systems are particularly promising for tumor therapy and regenerative medicine, leveraging the biocompatibility of polysaccharides with the precision of magnetic responsiveness.

4.6 *Multi-stimuli Responsiveness*

Polysaccharide-based multi-stimuli responsive have emerged as advanced biomaterials capable of simultaneous or sequential responses to multiple environmental stimuli, significantly enhancing their therapeutic precision. Through innovative copolymerization strategies, alginate backbones grafted with thermosensitive poly(N-isopropylacrylamide-co-N,N-diethylacrylamide) side chains and photothermal semicoke particles form ionic-crosslinked hydrogel beads that dynamically modulate water absorption and glyphosate release in response to pH, temperature, and light. Similarly, HA hydrogels engineered with aldehyde groups and disulfide-containing crosslinkers (3,3′-dithiobis(propionic hydrazide)) exhibit redox-, enzyme-, and UV-responsive behaviors, enabling stimulus-triggered drug release [76]. Cellulose derivatives demonstrate particular versatility through the strategic incorporation of enamine and disulfide bonds, achieving: (1) reversible sol–gel transitions under pH/redox changes, (2) exceptional stability in physiological conditions, and (3) controlled release of payloads with tunable kinetics. These multi-stimuli systems leverage the inherent biocompatibility of polysaccharides while introducing programmable responsiveness through covalent modifications, offering unprecedented spatiotemporal control for applications ranging from precision agriculture to targeted cancer therapy.

Wang et al. [77] developed smart hydrogels with dual pH- and thermoresponsive properties, along with UV-shielding capabilities, for diabetic wound healing applications. The researchers first oxidized pullulan to generate aldehyde-functionalized pullulan, which was then crosslinked via dynamic Schiff base bonds with Pluronic F127-modified polyethylenimine. This formulation exhibited a temperature-dependent sol–gel transition, with gelation occurring above 25 °C. The hydrogel's pH-sensitive behavior was leveraged for controlled drug delivery, as the acidic wound environment (pH ~ 5.5) triggered the breakdown of imine bonds, enabling sustained release of encapsulated adipose-derived mesenchymal stem cell exosomes. This multi-stimuli-responsive system not only facilitated on-demand exosome delivery but also demonstrated potential for adaptive wound dressings that respond dynamically to physiological changes.

The development of multi-stimuli-responsive wound dressings that dynamically adapt to infected wound environments. By chemically modifying sodium alginate with 3-aminophenyl boronic acid, researchers created a hydrogel crosslinked via pH- and ROS-sensitive boronic ester bonds, which remain stable under alkaline conditions (pH 8–9) but degrade in acidic, high-ROS wound microenvironments. To enhance therapeutic effects, hyaluronic acid was functionalized with cholesterol to form self-assembling micelles loaded with anti-inflammatory naproxen, while the antibiotic amikacin was embedded directly within the hydrogel network. Upon exposure to infected wound conditions (low pH, high ROS), the hydrogel undergoes controlled disintegration, triggering the simultaneous release of both drugs for combinatorial therapy [78].

5 Applications of 4D-Printed Polysaccharides

5.1 Biomedical Applications

4D printing of stimuli-responsive hydrogels has emerged as a promising strategy for creating dynamic structures in biomedicine, leveraging their ability to undergo programmed shape transformations. However, achieving high-precision 4D-printed hydrogels remains challenging due to limitations in printability, structural fidelity,

Table 2 4D printed stimuli-responsive polysaccharides with their reactive groups, active stimuli, and applications

Polysaccharide	Reactive Group	Active Stimuli	Applications	References
Chitosan	Thiol	pH	Wound healing	[79]
Chitosan	MnO_2	H_2O_2 (redox)/light	Wound healing	[80]
Chitosan	Polyaniline	Electric/pH responsive	Smart drug release	[81]
Chitosan	Gold nanoparticles	Near-infrared	Cancer treatment	[82]
Dextran	Boronic ester	Reactive oxygen species (ROS)	Drug delivery	[83]
Cellulose	Disulfide linkage	GSH (redox)/ temperature	Drug delivery	[84]
Glycol chitosan	Polyaniline	Near-infrared	Photothermal therapy	[85]
Carboxymethyl cellulose	Dithioerythritol	Photo- and redox-responsive	Shape memory hydrogel	[86]

and surface smoothness, particularly with micro-extrusion-based techniques. A 4D-printable polysaccharide hydrogel composed of alginate and methylcellulose (MC), combines superior rheological properties, extrudability, and shape retention. The optimized Alg/MC hydrogel enables the fabrication of intricate 2D patterns with anisotropic stiffness and controlled swelling behaviors, achieved by tuning network density gradients perpendicular to the printed strips. Upon immersion in a calcium chloride solution, these 2D structures undergo programmable morphing into complex 3D configurations, including tubes, helices, double helices, and floral architectures, driven by differential swelling across the hydrogel matrix. This polysaccharide-based system offers exceptional versatility for biomedical applications, where precise spatiotemporal control over material behavior is critical. Potential uses include tissue engineering scaffolds with dynamic topographies, adaptive biomedical devices, and soft robotic actuators that respond to physiological stimuli. By overcoming traditional printability challenges, our 4D-printed Alg/MC hydrogel unlocks new possibilities for designing high-resolution, stimuli-responsive structures in regenerative medicine and beyond [13]. A list of 4D printed stimuli-responsive polysaccharides with their reactive groups, active stimuli, and applications is represented in Table 2.

5.1.1 Dynamic Drug Delivery Systems

4D printing technology has revolutionized the field of drug delivery by enabling the fabrication of smart, stimuli-responsive systems that can change shape or properties over time. Alginate, a natural polysaccharide, has emerged as a promising material for 4D printed drug delivery systems due to its biocompatibility, biodegradability,

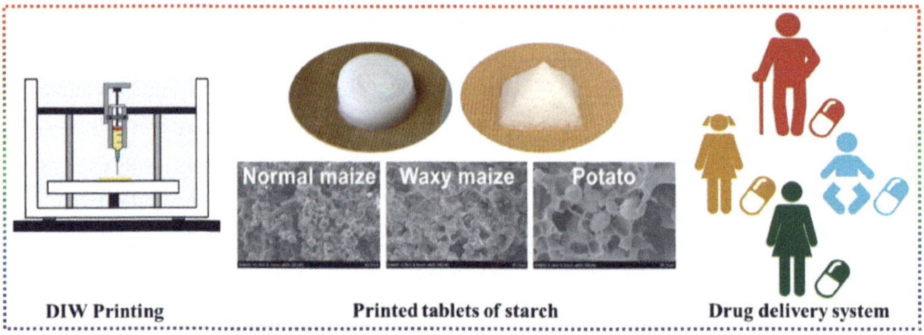

Fig. 3 3D printed starch (maize, waxy maize, and potato) tablets with combined release kinetics effect for drug delivery systems. Elsevier copyright 2022, reproduced from ref [87]

and responsiveness to various stimuli. Here, we summarize recent advancements and applications of 4D printed alginate in drug delivery.

4D-printed polysaccharide-based tablets were successfully fabricated using an efficient and straightforward 3D printing approach, enabling the production of customizable geometries loaded with a poorly water-soluble drug, as shown in Fig. 3. Initially, the gelatinization behavior of different polysaccharides was studied using DSC and rheological analysis, revealing that the gelatinization temperature varied depending on the polysaccharide's botanical source. Rheological characterization confirmed that all formulated inks exhibited optimal printability. Furthermore, the study demonstrated that the mechanical properties, swelling capacity, and drug release kinetics were influenced by factors such as microstructure, porosity, chemical composition (including amylose/amylopectin ratio and additional components), and the ability to form a stable polymeric network. By carefully selecting the polysaccharide type and tablet design, drug release profiles could be precisely controlled. These findings highlight the potential of 4D-printed polysaccharide matrices for personalized drug delivery applications, where dynamic structural changes in response to external stimuli could further enhance controlled release [87].

A study by Wang et al. demonstrated the use of a methotrexate-loaded pluronic diacrylate macromer and alginate-based hybrid hydrogel system. This system was fabricated using Direct Ink Writing (DIW) and UV-curing techniques. The hydrogels were designed to respond to aqueous calcium chloride (1% w/v), resulting in a folding mechanism that released approximately 80% of the drug over 12 h, with a fast release profile in the first 6 h [88]. Zhao et al. developed heparin-loaded gelatin methacryloyl hydrogels using DIW and UV-curing. These hydrogels folded into a tube-like structure when exposed to water, releasing about 70% of the drug over 30 h, with a fast release profile in the first 8 h. This system leverages the pH-responsive nature of alginate to achieve controlled drug release [89].

5.1.2 Tissue Engineering Scaffolds and Smart Implants

4D-printed polysaccharide-based hydrogels represent a transformative advancement in tissue engineering and smart implants, offering dynamic, stimuli-responsive scaffolds that closely mimic the native extracellular matrix. Unlike static 3D structures, these programmable materials adapt their shape, mechanical properties, or bioactivity in response to environmental cues (e.g., pH, temperature, or enzymes), enabling precise control over cell behavior and tissue regeneration [90]. This section will report current advancements in 4D-printed polysaccharide hydrogels, their design strategy, fabrication methods, and uses.

Polysaccharide-based vascularized bone scaffolds using 4D printing have facilitated new strategies of vascularized bone regeneration. While early techniques involved post-printing cell seeding, current strategies emphasize bioprinting cells directly within the scaffold using bioactive hydrogels. Polysaccharide-based bioinks, such as alginate, agarose, and methacrylated hyaluronic acid, are widely employed due to their biocompatibility, tunable mechanical properties, and support for cell viability. To enhance structural integrity and osteoconductivity, these hydrogels are often reinforced with nano-hydroxyapatite, mimicking the mineral phase of natural bone. Their stimuli-responsive behaviors, such as shape-morphing upon hydration or mechanical stiffening under physiological conditions, improve integration with host tissue. For instance, alginate-GelMA hybrid hydrogels can dynamically adapt to defect sites, while human umbilical vein endothelial cells-laden methacrylated hyaluronic acid constructs promote vascular network formation in situ. This cell-instructive, 4D-printed approach represents a paradigm shift in bone tissue engineering, enabling precise spatial control over vascularization and mineralization for enhanced regeneration.

4D-printed cellulose-based implants offer a promising approach for localized drug delivery in breast cancer treatment following breast-conserving surgery (BCS). Despite BCS being the standard for early-stage breast cancer, challenges such as tumor recurrence, tissue loss, and the side effects of adjuvant therapies (e.g., radiotherapy or systemic chemotherapy) highlight the need for personalized solutions. To address this, we developed a stimuli-responsive implant composed of carboxymethyl cellulose sodium salt (CMC) and cellulose nanocrystals (CNC), loaded with the chemotherapeutic agent doxorubicin (DOX). Rheological analysis ensured optimal printability, while the implant's 4D shape-morphing capability, triggered by swelling, allowed for a customized fit within the resection cavity, enhancing both functionality and aesthetic outcomes. Swelling studies demonstrated how formulation and printing parameters could be tailored to program the implant's dynamic shape transformation. In vitro drug release and cytotoxicity assays against MDA-MB-231 cells revealed sustained anticancer efficacy, reducing cell viability by 58% after 72 h, even when tested four weeks post-fabrication. These findings suggest that 4D-printed cellulose implants could serve a dual purpose: filling surgical voids to restore breast contour while delivering localized chemotherapy to minimize recurrence risks. This innovation underscores the potential of smart polysaccharide-based systems in improving

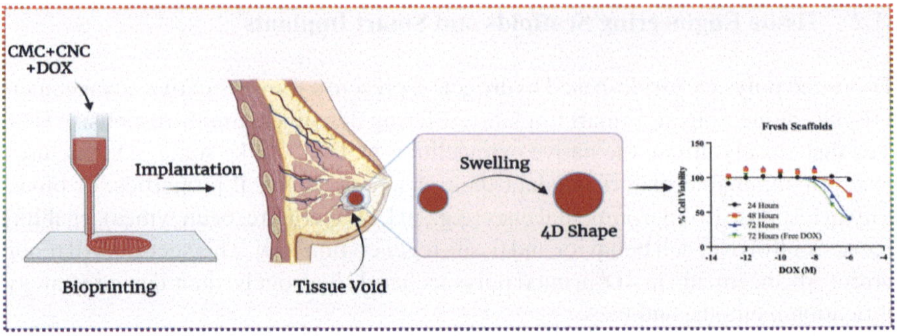

Fig. 4 CMC, CNC, and DOX-based 4D printed multipurpose smart implants for breast cancer management. Adapted from ref [91]

post-surgical cancer care through patient-specific design and controlled drug release, as shown in Fig. 4 [91].

Chitosan-based bioinks have been developed for 3D and 4D printing applications. These bioinks can be tailored to promote cell proliferation and function, making them suitable for creating advanced tissue engineering constructs. Chitosan has been combined with other materials like hydroxyapatite (HA) to create composite scaffolds with enhanced mechanical properties and biocompatibility. Li et al. demonstrated that chitosan-HA scaffolds significantly increased cell viability compared to pure chitosan scaffolds [92]. Gradient scaffolds made from chitosan and other biopolymers have been developed to mimic the natural gradient of tissues in the human body. These scaffolds can promote better tissue integration and regeneration [93]. For cartilage repair, 4D-printed hydrogels mimic the mechanical resilience of native cartilage, with shear-thinning or strain-stiffening behaviors that adapt to joint forces while supporting chondrocyte growth and ECM deposition [94]. By integrating stimuli-responsive properties with precise 3D architectures, these advanced scaffolds bridge the gap between static implants and living tissues. Polysaccharide-based scaffolds can be engineered to change shape in response to specific stimuli, such as temperature, pH, etc. [95]. This property is particularly useful for minimally invasive surgery, where the scaffold can be implanted in a compact form and then expand to fit the defect site.

5.2 *Soft Robotics and Actuators*

5.2.1 Design of Flexible, Responsive Actuation Systems

4D-printed polysaccharide-based hydrogels with programmable mechanical properties and dynamic responsiveness are increasingly sought after for applications

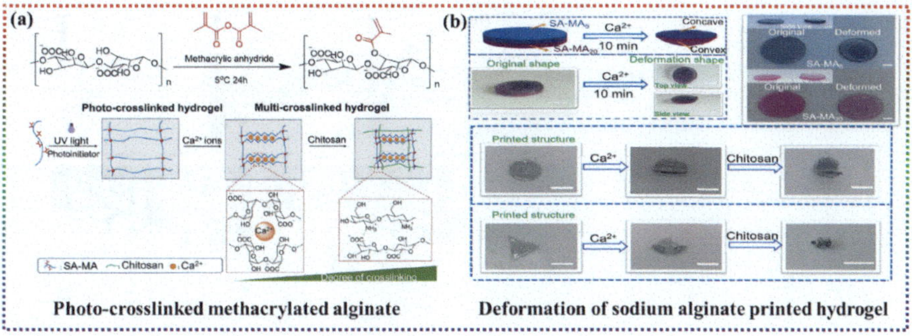

Fig. 5 **a** Synthetic scheme and distortion mechanism of photocrosslinked (SA-MA) hydrogel. **b** Schematic and graphic deformation of the SA-MA bilayer film, deformation of SA-MA5, SA-MA20, Step-wise SA-MA5 3D-printed hydrogels subjected to Ca^{2+} solution. ACS copyright 2021, reproduced from ref [96]

e.g., soft robotics, biosensors, and smart drug delivery. Methacrylated alginate (SA-MA) hydrogel was developed for 4D printing, and the influence of methacrylation degree on printability, crosslinking density, and multi-stimuli responsiveness (e.g., Ca^{2+}/chitosan crosslinking) was thoroughly investigated, as shown in Fig. 5a. The 4D-printed SA-MA hydrogel exhibited sequential volume contraction and mechanical strengthening when alternately exposed to Ca^{2+} and chitosan solution, driven by progressive crosslinking. By engineering a bilayer hydrogel with varying crosslinking densities, anisotropic shape-morphing was achieved upon Ca^{2+} immersion, as shown in Fig. 5b. Additionally, gradient crosslinking enabled complex, multi-stage deformations through controlled ionic stimulation. This approach offers a versatile platform for fabricating precisely defined, stimuli-responsive structures, advancing the design of adaptive actuation systems for biomedical and soft robotic applications [96].

Polysaccharide-based materials can be engineered to change shape in response to specific stimuli. This property is particularly useful for minimally invasive surgery, where the robot can be implanted in a compact form and then expand to fit the defect site. The application of cellulose-based hydrogel exhibits shape-memory features, allowing to undergo significant deformations and retain a temporary shape [97, 98].

A machine learning (ML)-guided 4D printing approach for fabricating cellulose-based soft pneumatic actuators (SPAs) with programmable deformation. A finite element model (FEM) was first developed to simulate the nonlinear hyperelastic behavior of the 4D-printed cellulose material under actuation, generating over a thousand training datasets. These data were used to train an ML model that predicts the optimal geometry and printing parameters required to achieve target bending motions in soft robotic applications. The trained ML model demonstrated high accuracy in forecasting both FEM and experimental results, validating its potential for designing adaptive, 4D-printed cellulose actuators, as shown in Fig. 6a. By integrating computational modeling and data-driven optimization, this work provides a framework for solving complex, nonlinear 4D printing challenges in soft robotics. The method

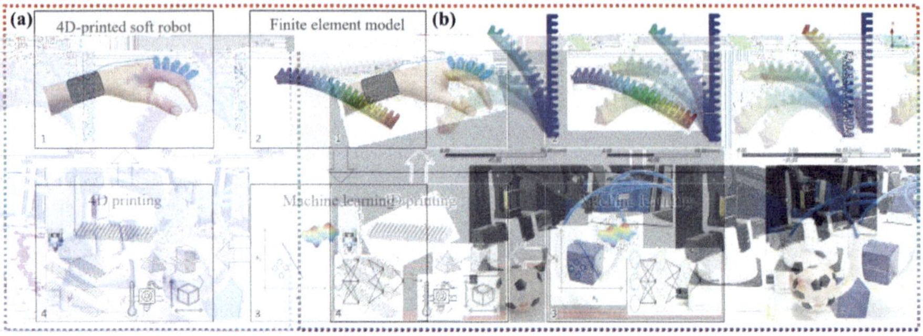

Fig. 6 **a** Schematic diagram of 4D printing soft robots guided by machine learning and finite element models for classification of different models. **b** FEM output compares 4D-printed SPAs bending, bending angles. 4D-printed SPAs-based experimental tests to grasp objects on the ML classification model representing various bending and geometry. Elsevier copyright 2021, reproduced from ref [99]

enables precise control over cellulose-based actuator deformation, paving the way for customizable, sustainable, and responsive soft robotic systems, as shown in Fig. 6b. This approach highlights the potential of 4D-printed polysaccharides in creating dynamically reconfigurable structures for flexible and eco-friendly robotic designs [99].

5.3 *Environmental Applications*

5.3.1 Smart Filtration Systems

Engineered living materials combine synthetic biology and smart material design to produce dynamic systems capable of sensing and responding to environmental cues. 4D-printed alginate hydrogels embedded with genetically modified cyanobacteria to develop stimuli-responsive biocomposites for on-demand bioremediation. Unlike conventional 3D-printed structures, these materials exhibit shape-morphing and self-adaptive properties upon chemical stimulation, enhancing their functionality in environmental applications. As a proof of concept, *Synechococcus elongatus* PCC 7942 was engineered with a synthetic riboswitch to control the expression of a fluorescent reporter, enabling real-time monitoring of cellular activity within the alginate matrix. Laccase-producing strain was incorporated into the 4D-printed hydrogel, demonstrating efficient decolorization of indigo carmine, a common textile pollutant. The alginate-based framework not only supports cyanobacterial viability but also responds dynamically to external stimuli, improving pollutant degradation efficiency. To address biocontainment concerns, an inducible cell death mechanism was integrated, ensuring controlled biodegradation of the material after use (Fig. 7). By combining 4D-printable alginate hydrogels with programmable cyanobacteria, this

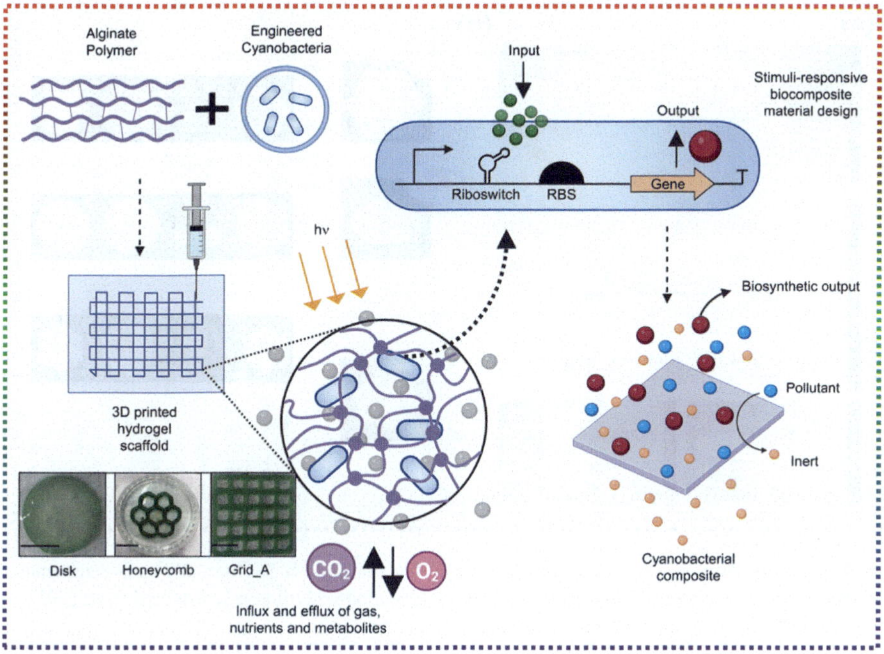

Fig. 7 Strains of *S. elongatus* have been genetically engineered embedded within a matrix made from the natural polymer alginate. This polymer scaffold supports cell viability and expand in aqueous environments. It facilitates nutrients and gases movement throughout the matrix and offers an optimal microenvironment for the growth of the embedded cyanobacteria and the activation of gene expression in response to external signals. Adapted from [100]

work establishes a scalable platform for next-generation bioremediation, offering precise spatial control, environmental adaptability, and on-demand functionality [100].

Diatoms, a diverse group of unicellular algae, are widely recognized as effective bioindicators for monitoring water quality in rivers and lakes. However, conventional diatom-based assessment methods are labor-intensive and rely on specialized equipment, as shown in Fig. 8a. 4D-printed cellulose-based hydrogels embedded with diatoms as innovative and low-cost biosensors for water quality evaluation. The hydrogel composites are fabricated using an extrusion-based 3D/4D printing technique, incorporating diatoms, cellulose nanocrystals (CNC), and alginate into a functional bio-ink, as shown in Fig. 8b. Rheological and mechanical analyses are conducted to optimize the bio-ink formulation, while cell culture experiments assess the viability and responsiveness of the encapsulated diatoms to water contaminants. The results demonstrate that 4D-printed diatom-cellulose hydrogels can serve as dynamic, self-sensing living materials, offering a scalable and efficient solution for real-time water quality monitoring [101].

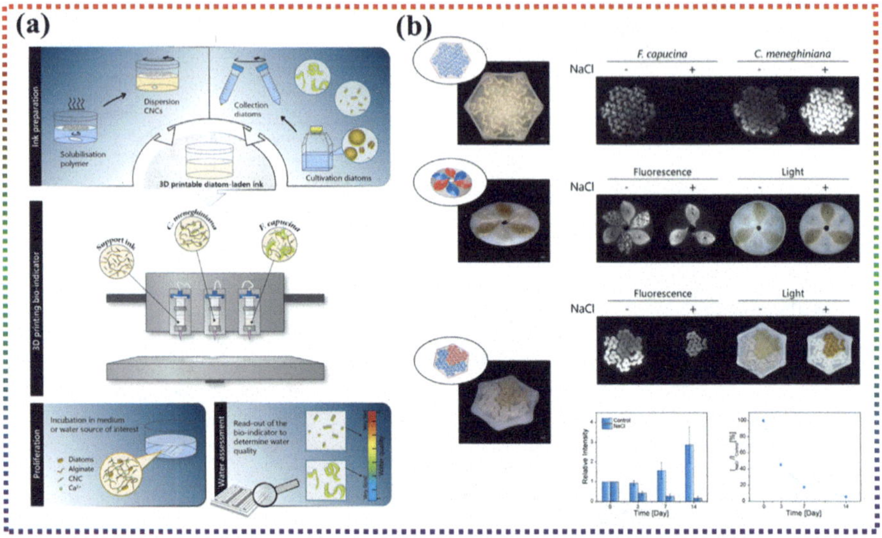

Fig. 8 **a** A production system for 3D-printing living diatoms. **b** Biologically active indicators produced via multi-material bioprinting of hydrogels loaded with diatoms. Adapted from [101]

5.4 Consumer Products

5.4.1 Adaptive Textiles and Wearable Devices

4D-printed polysaccharide hydrogels are revolutionizing regenerative medicine by providing dynamic, biomimetic scaffolds that actively promote tissue repair. For skin regeneration, these smart hydrogels create a moist, bioactive environment that not only accelerates wound closure but also responds to infection by releasing antimicrobial agents on-demand, enhancing healing while preventing complications [102]. In stem cell-based therapies, polysaccharide scaffolds (e.g., alginate or chitosan) are engineered to encapsulate and controllably release stem cells or growth factors in response to physiological triggers (e.g., enzymes or pH), optimizing tissue regeneration [103].

4D printed polysaccharide-based materials for wearable devices, showcasing their mechanical, thermal, self-healing, and shape memory capabilities. The thermoreversible network is produced using Diels–Alder reversible covalent bond, incorporating TPU and PCL, along with CNC for crosslinking, as shown in Fig. 9a. The addition of modified CNC significantly improves the mechanical features, thermal stability, and self-healing capabilities of the materials. Molecular dynamics simulations further analyze the mechanisms behind these performance enhancements. Notably, this mechanically tough and 4D-printed self-healing composite enables the formation of durable conductive devices and biomimetic skin applications, featuring

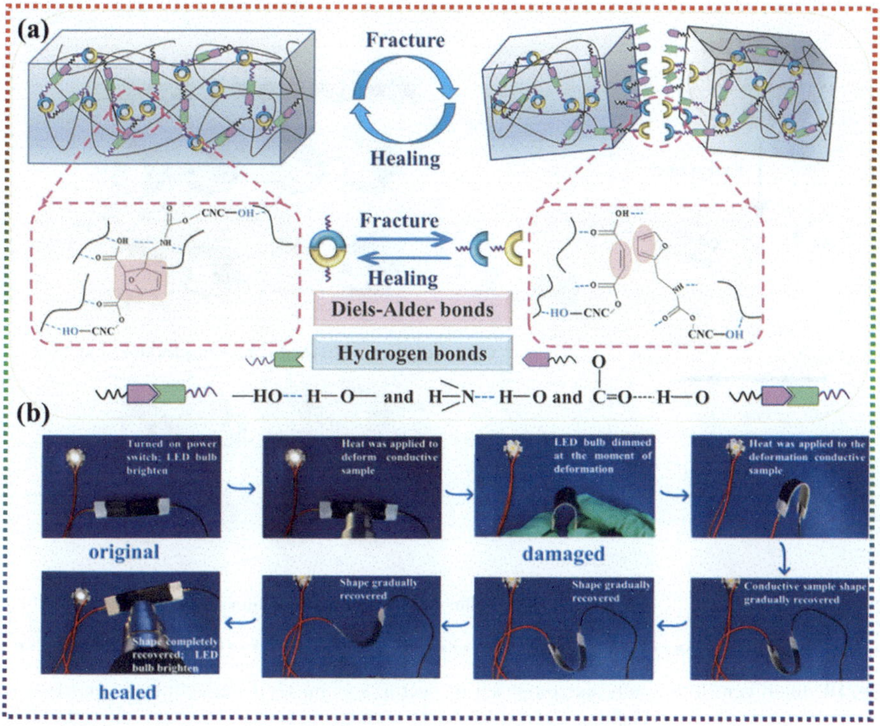

Fig. 9 **a** Design of self-healable, thermoreversible polymer composite featuring a dynamic cross-linking network. **b** Self-healing behavior of conductive devices assisted by shape memory properties. Elsevier copyright 2022, reproduced from ref [104]

higher mechanical ability, outstanding electrical healing, and remarkable strain-induced fluorescence, as shown in Fig. 9b. This innovation holds promise for next-generation flexible electronics, encryption devices, and electronic skins, expanding the range of applications for 4D-printable polysaccharide materials [104].

5.4.2 Customized Packaging Solutions

Indicators function as tools rely vital information to consumers about food spoilage, microbial contamination, and other important details. They enable the detection of specific substances, the degree of reactions among different substances, and provide information through noticeable color changes [105]. These indicators often trigger swift visual changes, like shifts in color or the movement of color boundaries. They play a key role in conveying information about a food product's microbial activity, quality, and other attributes.

To tackle consumer-level food waste and plastic pollution, an intelligent film was developed using sodium alginate, pectin, cellulose nanocrystals, and anthocyanins

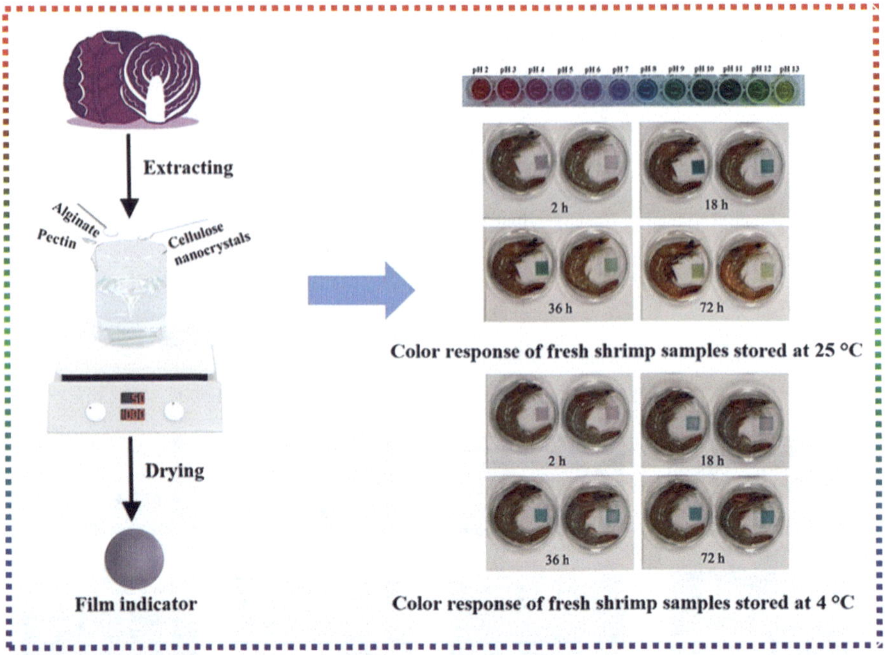

Fig. 10 Intelligent film synthesis composed of pectin, sodium alginate, cellulose nanocrystals, and anthocyanins in food packing as food freshness monitoring sensor. Elsevier copyright 2023, reproduced from ref [106]

from red cabbage. Cross-linking and adding CNCs enhanced their mechanical properties and thermal stability. The best-performing SA/PC/CNCs/RCA/CL films showed pH-dependent color changes and higher water resistance. Tested as colorimetric freshness indicators for shrimp, these films changed color from lilac to dark green to greenish-yellow after 72 h at 25 °C, but much slower at 4 °C, as shown in Fig. 10. This shows their potential to reduce food waste by signaling spoilage through pH changes in real-time [106].

5.5 Programmable, Adaptive 4D Printing with Polysaccharides

The history of additive manufacturing has acquired a transformational stage due to the introduction of 4D printing where a fourth dimension (time) confers a domain of moving, self-developing forms. Unlike 3D printing, 4D printing produces objects that can change their shape, property or functionality in a pre-programmed way depending on environmental cues. In this prospective domain, polysaccharides, including chitosan, alginate, cellulose, and starch, have also become especially attractive bio-based materials. This is on a very strong basis of their natural biocompatibility, biodegradability, and natural responsiveness to aqueous condition. However, what is really new in 4D printing with polysaccharides is not only their response to stimuli, but the complex control of dynamic behaviour with time, multi-stimuli responsiveness, and how these approaches can be combined with machine learning to achieve truly intelligent and programmable systems.

5.5.1 Intrinsic Stimuli-Responsiveness and Shape-Memory Effects

The core of 4D-printed polysaccharide systems is the stimuli-responsive property, which is mainly caused by the hydrophilic polymer chains. When these materials are exposed to certain triggers (in most cases, water or humidity) they swell or shrink. It is the volumetric change that is the key to transformation. This is done by printing design paths and material composition. Designers encode internal stresses by means of placing polysaccharide-based hydrogels of varying cross-linking density, polymer concentration, or fibre orientation near each other [107]. On hydration, areas of greater capacity to swell than those of smaller capacity swell further and cause the overall structure to bend, twist, or fold as dictated. It is the direct form of the shape-memory effect in which the preliminary shape printed in 3 dimensions is the transient shape, and the hydrated, equilibrated shape is the permanent shape, which was encoded in the printing and cross-linking process [108].

5.5.2 The Novelty of Multi-stimuli Responsiveness

Although multi-stimuli responsive polysaccharide systems are not yet the edge of research, water remains a major catalyst. Different stimuli directed to the response of a single structure at different levels, thus having advanced temporal control.

- **pH-Responsible Systems**: Chitosan is a cationic polysaccharide that swells in acidic conditions (because its amine functional groups protonate and repel one another) and de-swells in neutral or basic conditions. The opposite behaviour is seen with Alginate, which is an anionic polymer. Printing multi-material structures of chitosan and alginate allows a single object to be programmed such that one of the components activates in an acidic buffer and the other in a basic buffer to produce complex and sequential deformations [109].
- **Ion-Responsive Actuation**: The actuation of alginate with divalent ions, like Ca^{2+} is famous due to its quick gelation. Post-printing of structural locking or secondary actuation is possible by the use of this principle. The structure initially unfolded in pure water and then, when a certain concentration of a particular ion is added, the structure is subjected to a second-stage deformation or stiffens to the ultimate functional structure [22].

- **Enzyme-Specific and Biochemical Triggers**: It is the eventual aim of biomedical applications to be specific. It is possible to design polysaccharides as substrates of particular enzymes (e.g., cellulose with cellulase, chitosan with lysozyme). This will only enable degradation or transformation in the presence of a unique biological signal, and this will enable the targeted drug delivery systems releasing their payload on encountering a specific enzyme at a disease site [110].

5.5.3 The Role of Machine Learning

The dependency between the printing conditions (e.g., print speed, nozzle pressure, infill pattern), mixture of materials (e.g., polymer molecular weight, cross-linker density), and the ultimate dynamic behaviour is very complicated and non-linear. The manual methods of trial and error cannot be applied to formulate complex multi-stage deformations. Machine learning is then a life changer in this area.

- **Forward Prediction and Design Optimization**: ML models, specifically neural networks, trained to accept a data set of printing parameters and their corresponding shape changes. After training, the model predicts the ultimate morphed form of a fresh design without the need to physically print it. This radically shortens the design time of complicated structures, including those that resemble the botanical motions of the pine cone with its hygroscopic aperture and closure [111].
- **Inverse Design**: It is the strongest application of ML, an inverse model instead of forecasting the result of a design, accepting a favored final form or transformation route as input and returns the best printing material composition and printing parameters necessary to obtain it. As an illustration, a researcher might want to define a desired taper of a targeted gripper shape that a flat printed sheet should take when immersed and the ML algorithm would create the blueprint of the anisotropic print path to produce the desired grip [112].
- **Closed-Loop Control and Real-Time Adaptation**: ML used to correct errors in the printing process (e.g., cameras, sensors) in-situ, which cause the 4D transformation to derail. Moreover, the transformation process itself, learning and optimizing the application protocol of stimulus analyzed by ML to produce a more efficient or reliable shape change [113].

5.5.4 Emphasizing Temporal Transformation and Programmable Behaviors

The combination of these factors: multi-material polysaccharide printing and ML-based design brings the idea of shape-changing to the level of temporally programmable matter.

- **Sequential Self-Assembly**: A flat printed structure, which initially folds into a tube due to humidity, and as ions diffuse, begins to build helical ridges on its surface, and thus, after a time delay, changes its fluid dynamic characteristics [114].

- **Biomimetic Logic**: A drug delivery capsule made of chitosan and alginate is closed in the acidic environment of the stomach, opens through the neutral pH environment of the small intestine to release its therapeutic moiety, and is broken down only under the influence of the gastric enzymes found in the environment.
- **Adaptive Soft Robotics**: It is a soft gripper that is a polysaccharide-based, but the internal fluidic channels are optimized through machine-learning approaches. The gripper not only closed around a humidity trigger but also regulates its grip strength and conformation according to the evidence of integrated sensors, which is achieved by the programmed anisotropic swelling of its material matrix [115].

6 Challenges in 4D Printing of Polysaccharides

4D printing of polysaccharides presents several challenges that need to be addressed for successful implementation in various applications including:

6.1 Material Limitations and Stability

4D printing of polysaccharides presents several material limitations and stability concerns that must be addressed. These materials often exhibit lower mechanical strength compared to synthetic polymers, limiting their application in structural contexts [116]. Additionally, polysaccharides typically have low thermal stability, making them prone to degradation during printing or when exposed to elevated temperatures. Their hygroscopic nature leads to moisture absorption, resulting in dimensional changes that can impact accuracy and performance. Chemical stability is another concern, as some polysaccharides are susceptible to microbial degradation or chemical exposure, particularly in biomedical applications [41]. Achieving reliable shape memory effects is challenging, as variability in cross-linking and environmental conditions can affect their performance. Furthermore, the printability of polysaccharides can be hindered by their viscosity, necessitating optimization for consistent results [117]. Incorporating additional functionalities may compromise their inherent properties, while aging processes can affect their long-term mechanical and physical characteristics. Addressing these limitations is crucial for the successful integration of 4D printed polysaccharides into practical applications such as biomedical devices and smart materials.

6.2 Precision in Stimuli-Induced Transformations

Accuracy of stimuli-induced transformations is involved in the successful use of 4D printed materials, especially in a dynamic environment. However, such transformations, which react to external stimuli (temperature, humidity, light, or mechanical forces), should be very precise to provide the required functionality. The effectiveness of these transformations can be greatly affected by variability in the material properties i.e. sensitivity and responsiveness of the polymers being used [118]. Indicatively, uniform shape memory effects cannot be obtained without a controlled cross-linking density and thermomechanical characteristics of the materials. Also, the transformations may be affected by external factors such as environmental conditions that bring about instability. In order to increase precision, it is fundamental to refine the design of the materials and the print procedure, so that the stimuli-responsive properties can be clear and reproducible [20]. These transformations can be further enhanced by the advances in material engineering, such as the incorporation of nanomaterials and smart additives, which can be used to design more advanced applications, such as responsive actuators, adaptive structures, biomimetic systems. It is important that these transformations be accurate to ensure that the full potential of 4D printed materials will be achieved in the real world.

6.3 Scalability and Industrial Applications

Scalability is an important factor in the development of the 4D printing technologies, especially in the context of its implementation on the industrial level. With the increased need of innovative and adaptive materials the efficiency and uniform production of materials of large scale becomes important. The recent issues are the optimization of the printing process to preserve quality and accelerate the flow of production and the creation of cost-effective strategies of mass customization. They have the potential to be of great benefit in industry, e.g. in creating responsive components to the environment, improving some functions in aerospace, cars, and medicine. As an example, it is possible to enhance the durability and stability of elements with self-healing materials, and elements that change shape can respond to different working conditions and would result in more effective design. Nevertheless, to realize the scalability, one has to deal with technical challenges, including reproducibility of stimuli-induced changes and incorporation of various materials. Also, regulatory factors have to be fulfilled, particularly in such areas as healthcare, where safety and efficacy are of primary importance. These obstacles will be overcome and make 4D printing in industry widely adopted, allowing new opportunities of innovative products and solutions.

6.4 Regulatory Concerns for Biomedical Uses

Biomedical applications of 4D printed materials have regulatory issues that are paramount to maintaining safety, effectiveness, and adherence to the set standards. The materials should be evaluated thoroughly to address the needs of such regulatory organizations as FDA or EMA. Other critical issues are those of biocompatibility whereby materials incorporated in any medical equipment or implantation should not tend to generate negative reactions with biological tissues. Its manufacturing processes should prove to be consistent and reliable so that its products obtain safety requirements with each batch of products. This involves proper validation of the process of 4D printing, which can be replicated and with a good result. Long-term stability of these materials is also of interest especially in the way they behave under physiological conditions in the long run. Since 4D printed materials can alter their shape or functionality depending on stimuli, their predictability and behavior in applications are also questioned. The regulatory bodies need detailed information on these points in order to evaluate the possible dangers of their usage. Post-market surveillance is obligatory to watch the performance and safety of these materials when they are in operation, as to prevent any unanticipated problems to be addressed in a timely manner. These regulation issues are critical to the successful adoption of 4D printed materials to the biomedical subsector which will eventually lead to new solutions to the better care of patients.

7 Future Perspectives in 4D Printing with Polysaccharides

Material science and technology have facilitated the future of 4D printing with polysaccharides because of the promising future of this printing process. With the ongoing studies on the peculiarities of polysaccharides, it is probable that some trends and developments will influence the domain. Among the perspectives, one of them is the augmentation of the material formulations to enhance mechanical strength, thermal stability, and stimulus responsiveness. Advances in composite materials, such as incorporation of nanoparticles and bioactive compounds, have the potential to give them multifunctional functionality, and polysaccharides may be exploited in a broader application spectrum of biomedical devices, flexible electronics and smart textiles. The other major trend is that more elaborate printing methods have been developed where more accuracy and the control of the printing process is possible. This may include adopting state-of-the-art additive manufacturing techniques, allowing complex geometries and dynamic capabilities that react in an adaptive manner to the changes in the environment. The convergence of 4D printing, artificial intelligence and machine learning can enable design and optimization of materials and processes. The technologies may assist in predicting the behavior of structures made of polysaccharide in different conditions, which will result in more efficient and customized solutions. The future of 4D printing with polysaccharides will also be under sustainability. Polysaccharides will become more appealing to creating sustainable products as demand toward the use of environmentally-friendly materials grows. Further investigation in the area of 4D printing of polysaccharides has the potential to transform diverse industries and open the door to new applications that would increase functionality, flexibility, and sustainability of materials design.

8 Conclusion

This gives a detailed description of the new uses and novel methods of fabrication of 4D-printed polysaccharide materials. This technology combines time as the fourth dimension, enables the creation of adaptable, transformable, or modifiable structures in response to environmental stimuli, and this is very advantageous to the conventional 3D printing. summarizes the distinctive features of several polysaccharides, including alginate, cellulose, chitin, chitosan, gellan gum, hyaluronic acid, konjac glucomannan, pullulan and starch and their possible uses in 4D printing. These materials have been chemically, physically, and biologically engineered to be more printable, mechanically strong, and stimuli responsive to allow their use in a wide spectrum of applications, drug delivery, tissue engineering, soft robotics and environmental remediation. The functionality of polysaccharides in 4D printing depends on stimuli-responsive behavior, such as thermal, pH-sensitive, moisture-driven, light-induced and magnetic-sensitive behavior. Advanced fabrication techniques, such as extrusion-based printing and post-printing functionalization, enable the formation of complex structures with precise control over their shape and properties. Despite the numerous advantages, several challenges are addressed for the widespread adoption of 4D-printed polysaccharides. These include material limitations, precision in stimuli-induced transformations, scalability, and regulatory concerns for biomedical applications. Addressing these challenges successful integration of 4D-printed polysaccharides into practical applications. The future of 4D-printed polysaccharides, depends on material formulations, printing techniques, integration with artificial intelligence, and a focus on sustainability are expected to shape the future of biomedical field.

References

1. Kuang X, Roach DJ, Wu J, Hamel CM, Ding Z, Wang T, et al. Advances in 4D printing: materials and applications. Adv Func Mater. 2019;29(2):1805290.
2. Shokrani H, Shokrani A, Seidi F, Mashayekhi M, Kar S, Nedeljkovic D, et al. Polysaccharide-based biomaterials in a journey from 3D to 4D printing. Bioeng Transl Med. 2023;8(4): e10503.
3. Momeni F, M.Mehdi Hassani.N S, Liu X, Ni J. A review of 4D printing. Mater Design. 2017;122:42–79.

4. Lee KY, Mooney DJ. Alginate: properties and biomedical applications. Prog Polym Sci. 2012;37(1):106–26.
5. Klemm D, Heublein B, Fink H-P, Bohn A. Cellulose: fascinating biopolymer and sustainable raw material. Angew Chem Int Ed. 2005;44(22):3358–93.
6. Jayakumar R, Prabaharan M, Sudheesh Kumar PT, Nair SV, Tamura H. Biomaterials based on chitin and chitosan in wound dressing applications. Biotechnol Adv. 2011;29(3):322–37.
7. Anitha A, Sowmya S, Kumar PTS, Deepthi S, Chennazhi KP, Ehrlich H, et al. Chitin and chitosan in selected biomedical applications. Prog Polym Sci. 2014;39(9):1644–67.
8. Fakhri V, Janmaleki Dehchani A, Davoudi SA, Tavakoli Dare M, Jafari A, Nemati Mahand S, et al. Advancing biomedical frontiers with functionalized soybean oil: insights into tissue engineering and drug delivery. J Polym Environ. 2024;32(11):5516–43.
9. Liu S, Yu J-M, Gan Y-C, Qiu X-Z, Gao Z-C, Wang H, et al. Biomimetic natural biomaterials for tissue engineering and regenerative medicine: new biosynthesis methods, recent advances, and emerging applications. Mil Med Res. 2023;10(1):16.
10. Zhang Y, Dong L, Liu L, Wu Z, Pan D, Liu L. Recent advances of stimuli-responsive polysaccharide hydrogels in delivery systems: a review. J Agric Food Chem. 2022;70(21):6300–16.
11. Jin M, Shi J, Zhu W, Yao H, Wang DA. Polysaccharide-based biomaterials in tissue engineering: a review. Tissue Eng Part B Rev. 2021;27(6):604–26.
12. Zamboulis A, Michailidou G, Koumentakou I, Bikiaris DN. Polysaccharide 3D printing for drug delivery applications. Pharmaceutics. 2022;14(1):145.
13. Lai J, Ye X, Liu J, Wang C, Li J, Wang X, et al. 4D printing of highly printable and shape morphing hydrogels composed of alginate and methylcellulose. Mater Des. 2021;205: 109699.
14. Quiroz-Castillo JM, Rodríguez-Félix DE, Grijalva-Monteverde H, del Castillo-Castro T, PlascenciaJatomea M, Rodríguez-Félix F, et al. Preparation of extruded polyethylene/chitosan blends compatibilized with polyethylene-graft-maleic anhydride. Carbohyd Polym. 2014;101:1094–100.
15. Pien N, Bray F, Gheysens T, Tytgat L, Rolando C, Mantovani D, et al. Proteomics as a tool to gain next level insights into photo-crosslinkable biopolymer modifications. Bioactive Mater. 2022;17:204–20.
16. Sadraei A, Naghib SM. 4D printing of physical stimuli-responsive hydrogels for localized drug delivery and tissue engineering. Polym Rev. 2025;65(1):104–68.
17. Chan G, Mooney DJ. New materials for tissue engineering: towards greater control over the biological response. Trends Biotechnol. 2008;26(7):382–92.
18. Yang J, Chen J, Pan D, Wan Y, Wang Z. PH-sensitive interpenetrating network hydrogels based on chitosan derivatives and alginate for oral drug delivery. Carbohyd Polym. 2013;92(1):719–25.
19. Jiang Z, Diggle B, Tan ML, Viktorova J, Bennett CW, Connal LA. Extrusion 3D printing of polymeric materials with advanced properties. Adv Sci. 2020;7(17):2001379.
20. Antezana PE, Municoy S, Ostapchuk G, Catalano PN, Hardy JG, Evelson PA, et al. 4D printing: the development of responsive materials using 3D-printing technology. Pharmaceutics. 2023;15(12):2743.
21. Banerjee R, Kumar KJ, Kennedy JF. Structure and drug delivery relationship of acidic polysaccharides: a review. Int J Biol Macromol. 2023;243: 125092.
22. Tordi P, Ridi F, Samorì P, Bonini M. Cation-alginate complexes and their hydrogels: a powerful toolkit for the development of next-generation sustainable functional materials. Adv Func Mater. 2025;35(9):2416390.
23. Hamza M, Kanwal Q, Ali Z, Zargar S, Alshammari A, Wani T, et al. Exploring the molecular biology of zinc-doped hydroxyapatite nanocomposites as fillers for dental materials: a self-defensive approach targeting bacterial DNA. Mater Res Express. 2024;11.
24. Sharma A, Tewari R, Rana SS, Soni R, Soni SK. Cellulases: classification, methods of determination and industrial applications. Appl Biochem Biotechnol. 2016;179(8):1346–80.
25. Chang C, Zhang L. Cellulose-based hydrogels: present status and application prospects. Carbohyd Polym. 2011;84(1):40–53.

26. Dutta SD, Patel DK, Lim K-T. Functional cellulose-based hydrogels as extracellular matrices for tissue engineering. J Biol Eng. 2019;13(1):55.
27. Hamza M, Tariq M, Sabahat S, Ahmad A, Muhammad N, Rahim A. 4—Biological methods for fabricating nanomaterial-based metal–organic frameworks. In: Ahmad A, Verpoort F, Ahmad I, Tabassum S, Rahim A, editors. Nanomaterial-Based Metal Organic Frameworks for Single Atom Catalysis: Elsevier; 2023. p. 75–106.
28. Cohen E. Chitin synthesis and inhibition: a revisit. Pest Manag Sci. 2001;57(10):946–50.
29. Francis Suh JK, Matthew HWT. Application of chitosan-based polysaccharide biomaterials in cartilage tissue engineering: a review. Biomaterials. 2000;21(24):2589–98.
30. Park S-B, You J-O, Park H-Y, Haam SJ, Kim W-S. A novel pH-sensitive membrane from chitosan—TEOS IPN; preparation and its drug permeation characteristics. Biomaterials. 2001;22(4):323–30.
31. LogithKumar R, KeshavNarayan A, Dhivya S, Chawla A, Saravanan S, Selvamurugan N. A review of chitosan and its derivatives in bone tissue engineering. Carbohyd Polym. 2016;151:172–88.
32. Lalebeigi F, Alimohamadi A, Afarin S, Aliabadi HAM, Mahdavi M, Farahbakhshpour F, et al. Recent advances on biomedical applications of gellan gum: a review. Carbohyd Polym. 2024;334: 122008.
33. Moon HJ, Ko DY, Park MH, Joo MK, Jeong B. Temperature-responsive compounds as in situ gelling biomedical materials. Chem Soc Rev. 2012;41(14):4860–83.
34. Kuperkar K, Atanase LI, Bahadur A, Crivei IC, Bahadur P. Degradable polymeric bio(nano)materials and their biomedical applications: a comprehensive overview and recent updates. Polymers. 2024;16(2):206.
35. Tiwari S, Bahadur P. Modified hyaluronic acid based materials for biomedical applications. Int J Biol Macromol. 2019;121:556–71.
36. Zhou H, Zhang Z, Mu Y, Yao H, Zhang Y, Wang D-A. Harnessing nanomedicine for cartilage repair: design considerations and recent advances in biomaterials. ACS Nano. 2024;18(16):10667–87.
37. Zhang C, Chen J-d, Yang F-Q. Konjac glucomannan, a promising polysaccharide for OCDDS. Carbohydrate Polym. 2014;104:175–81.
38. Ma S, Zhu P, Wang M. Effects of konjac glucomannan on pasting and rheological properties of corn starch. Food Hydrocolloids. 2019;89:234–40.
39. Ahmed S, Soundararajan A. Pullulan: processing, properties, and applications: CRC Press; 2020.
40. Iravani S, Varma RS. Plants and plant-based polymers as scaffolds for tissue engineering. Green Chem. 2019;21(18):4839–67.
41. Quintanilla de Stéfano JC, Abundis-Correa V, Herrera-Flores SD, Alvarez AJ. pH-sensitive starchbased hydrogels: synthesis and effect of molecular components on drug release Behavior. Polymers. 2020;12(9):1974.
42. Fazal T, Murtaza BN, Shah M, Iqbal S, Rehman M-u, Jaber F, et al. Recent developments in natural biopolymer based drug delivery systems. RSC Adv. 2023;13(33):23087–121.
43. Jamil N, Saad Ali HM, Yasir M, Hamza M, Sagheer M, Ahmed T, et al. Biosynthesized metallic and bimetallic nanoparticles as effective biocides for plant protection: plausible mechanisms and challenges. J Chem. 2024;2024(1):3328223.
44. Luo Y, Lin X, Chen B, Wei X. Cell-laden four-dimensional bioprinting using near-infrared-triggered shape-morphing alginate/polydopamine bioinks. Biofabrication. 2019;11(4): 045019.
45. Pahlevanzadeh F, Emadi R, Valiani A, Kharaziha M, Poursamar SA, Bakhsheshi-Rad HR, et al. Threedimensional printing constructs based on the chitosan for tissue regeneration: state of the art, developing directions and prospect trends. Materials. 2020;13(11):2663.
46. Hamza M, Ahmad A, Tariq M, Riaz S, Rahim A. Metal–organic frameworks (an overview). Nanomaterial-based metal organic frameworks for single atom catalysis. 2023:1–38.
47. Neufurth M, Wang X, Schröder HC, Feng Q, Diehl-Seifert B, Ziebart T, et al. Engineering a morphogenetically active hydrogel for bioprinting of bioartificial tissue derived from human osteoblastlike SaOS-2 cells. Biomaterials. 2014;35(31):8810–9.

48. Li X, Li M, Tang L, Shi D, Lam E, Bae J. 3D shape morphing of stimuli-responsive composite hydrogels. Mater Chem Frontiers. 2023;7(23):5989–6034.
49. Nikolova M, Chavali M. Recent advances in biomaterials for 3D scaffolds: a review. Bioact Mater. 4 (2019) 271–92.
50. Thambi T, Phan VH, Lee DS. Stimuli-sensitive injectable hydrogels based on polysaccharides and their biomedical applications. Macromol Rapid Commun. 2016;37(23):1881–96.
51. Gholamali I. Stimuli-responsive polysaccharide hydrogels for biomedical applications: a review. Regenerative Eng Transl Med. 2021;7(1):91–114.
52. Anirudhan TS, Shainy F, Jefin PT. Effect of dual stimuli responsive dextran/nanocellulose polyelectrolyte complexes for chemophotothermal synergistic cancer therapy. Int J Biol Macromolecules. 2019;135:776–89.
53. Hamza M, Kanwal Q, Hussain MI, Khan K, Asghar A, Liu Z, et al. Recent progress in 3D printed piezoelectric materials for biomedical applications. Mater Sci Eng R Rep. 2025;164: 100962.
54. Hamza M, Kanwal Q, Hussain MI, Peng W, Wang B, Liu Z, et al. 3D printing of high-performance Ag/BaTiO3 ceramic-polymer piezoelectric sensors for biomedical applications. Chem Eng J. 2025;523: 168848.
55. Tran TS, Balu R, Mettu S, Roy Choudhury N, Dutta NK. 4D printing of hydrogels: innovation in material design and emerging smart systems for drug delivery. Pharmaceuticals. 2022;15(10):1282.
56. Wang X, He W, Huang H, Han J, Wang R, Li H, et al. Recent advances in hydrogel technology in delivering mesenchymal stem cell for osteoarthritis therapy. Biomolecules. 2024;14(7).
57. Pu X, Wu Y, Liu J, Wu B. 3D bioprinting of microbial-based living materials for advanced energy and environmental applications. Chem Bio Eng. 2024;1(7):568–92.
58. Zheng Z, Bian S, Li Z, Zhang Z, Liu Y, Zhai X, et al. Catechol modified quaternized chitosan enhanced wet adhesive and antibacterial properties of injectable thermo-sensitive hydrogel for wound healing. Carbohyd Polym. 2020;249: 116826.
59. Kazemi MS, Mohammadi Z, Amini M, Yousefi M, Tarighi P, Eftekhari S, et al. Thiolated chitosanlauric acid as a new chitosan derivative: Synthesis, characterization and cytotoxicity. Int J Biol Macromol. 2019;136:823–30.
60. He XY, Sun A, Li T, Qian YJ, Qian H, Ling YF, et al. Mussel-inspired antimicrobial gelatin/ chitosan tissue adhesive rapidly activated in situ by H(2)O(2)/ascorbic acid for infected wound closure. Carbohydr Polym. 2020;247: 116692.
61. Sanandiya ND, Lee S, Rho S, Lee H, Kim IS, Hwang DS. Tunichrome-inspired pyrogallol functionalized chitosan for tissue adhesion and hemostasis. Carbohydr Polym. 2019;208:77–85.
62. Lih E, Lee JS, Park KM, Park KD. Rapidly curable chitosan-PEG hydrogels as tissue adhesives for hemostasis and wound healing. Acta Biomater. 2012;8(9):3261–9.
63. Chen T, Chen Y, Rehman HU, Chen Z, Yang Z, Wang M, et al. Ultratough, self-healing, and tissue adhesive hydrogel for wound dressing. ACS Appl Mater Interfaces. 2018;10(39):33523–31.
64. Zhang Z, Wang X, Wang Y, Hao J. Rapid-forming and self-healing agarose-based hydrogels for tissue adhesives and potential wound dressings. Biomacromol. 2018;19(3):980–8.
65. Liu C, Liu X, Liu C, Wang N, Chen H, Yao W, et al. A highly efficient, in situ wet-adhesive dextran derivative sponge for rapid hemostasis. Biomaterials. 2019;205:23–37.
66. Gomes TD, Caridade SG, Sousa MP, Azevedo S, Kandur MY, Ö ner ET, et al. Adhesive free-standing multilayer films containing sulfated levan for biomedical applications. Acta Biomater. 2018;69:183–95.
67. Zheng K, Gu Q, Zhou D, Zhou M, Zhang L. Recent progress in surgical adhesives for biomedical applications. Smart Mater Med. 2022;3:41–65.
68. Xu X, Bizmark N, Christie KSS, Datta SS, Ren ZJ, Priestley RD. Thermo responsive polymers for water treatment and collection. Macromolecules. 2022;55(6):1894–909.
69. Srivastava N, Choudhury AR. Stimuli-responsive polysaccharide-based smart hydrogels and their emerging applications. Ind Eng Chem Res. 2023;62(2):841–66.

70. Wang X, Li C, Wang Y, Chen H, Zhang X, Luo C, et al. Smart drug delivery systems for precise cancer therapy. Acta Pharm Sin B. 2022;12(11):4098–121.
71. Ding L, Jiang Y, Zhang J, Klok H-A, Zhong Z. PH-sensitive coiled-coil peptide-cross-linked hyaluronic acid nanogels: synthesis and targeted intracellular protein delivery to CD44 positive cancer cells. Biomacromol. 2018;19(2):555–62.
72. Fernando IPS, Lee W, Han EJ, Ahn G. Alginate-based nanomaterials: fabrication techniques, properties, and applications. Chem Eng J. 2020;391: 123823.
73. Champeau M, Heinze DA, Viana TN, de Souza ER, Chinellato AC, Titotto S. 4D printing of hydrogels: a review. Adv Func Mater. 2020;30(31):1910606.
74. Kanwal Q, Ahmed M, Hamza M, Ahmad M, Atiq ur R, Yousaf N, et al. Curcumin nanoparticles: physicochemical fabrication, characterization, antioxidant, enzyme inhibition, molecular docking and simulation studies. RSC Adv. 2023;13(32):22268–80.
75. Saleem H, Jamil N, Ahmad A, Ali H, Zafar I, Majeed M, et al. Evaluation of anti-diabetic potential of Syzygium cumini seed (SCS) powder infused tortilla as therapeutic diet to mitigate hyperglycemia. Int J Chem Biochem Sci. 2024;25.
76. Hamza M, Qadeer A, Alsaiari M, Alsayari S. Recent advances in enzyme immobilization in nanomaterials. Nanomaterial-Supported Enzymes. 2022;126:1–66.
77. Wang M, Wang C, Chen M, Xi Y, Cheng W, Mao C, et al. Efficient angiogenesis-based diabetic wound healing/skin reconstruction through bioactive antibacterial adhesive ultraviolet shielding nanodressing with exosome release. ACS Nano. 2019;13(9):10279–93.
78. Hu C, Zhang F, Long L, Kong Q, Luo R, Wang Y. Dual-responsive injectable hydrogels encapsulating drug-loaded micelles for on-demand antimicrobial activity and accelerated wound healing. J Control Release. 2020;324:204–17.
79. Esmaeilzadeh P, Köwitsch A, Liedmann A, Menzel M, Fuhrmann B, Schmidt G, et al. Stimuli responsive multilayers based on thiolated polysaccharides that affect fibroblast cell adhesion. ACS Appl Mater Interfaces. 2018;10(10):8507–18.
80. Wang S, Zheng H, Zhou L, Cheng F, Liu Z, Zhang H, et al. Injectable redox and light responsive MnO(2) hybrid hydrogel for simultaneous melanoma therapy and multidrug-resistant bacteria-infected wound healing. Biomaterials. 2020;260: 120314.
81. Qu J, Zhao X, Ma PX, Guo B. Injectable antibacterial conductive hydrogels with dual response to an electric field and pH for localized "smart" drug release. Acta Biomater. 2018;72:55–69.
82. Reddy BPradeep K, Mishra SK, Ravichandran G, Chauhan DS, Srivastava R, De A. Preclinical evaluation of multi stimuli responsive core-plasmonic nanoshell for photo-triggered tumor ablation: a disintegrable nanohybrid. Appl Mater Today. 2020;20:100684.
83. Lv W, Xu J, Wang X, Li X, Xu Q, Xin H. Bioengineered boronic ester modified dextran polymer nanoparticles as reactive oxygen species responsive nanocarrier for ischemic stroke treatment. ACS Nano. 2018;12(6):5417–26.
84. Pan Y, Liu J, Yang K, Cai P, Xiao H. Novel multi-responsive and sugarcane bagasse cellulose-based nanogels for controllable release of doxorubicin hydrochloride. Mater Sci Eng C Mater Biol Appl. 2021;118: 111357.
85. Yan L-X, Chen L-J, Zhao X, Yan X-P. PH switchable nanoplatform for in vivo persistent luminescence imaging and precise photothermal therapy of bacterial infection. Adv Func Mater. 2020;30(14):1909042.
86. Li Z, Davidson-Rozenfeld G, Vázquez-González M, Fadeev M, Zhang J, Tian H, et al. Multi-triggered supramolecular DNA/bipyridinium dithienylethene hydrogels driven by light, redox, and chemical stimuli for shape-memory and self-healing applications. J Am Chem Soc. 2018;140(50):17691–701.
87. González K, Larraza I, Berra G, Eceiza A, Gabilondo N. 3D printing of customized all-starch tablets with combined release kinetics. Int J Pharm. 2022;622: 121872.
88. Wang Y, Miao Y, Zhang J, Wu JP, Kirk TB, Xu J, et al. Three-dimensional printing of shape memory hydrogels with internal structure for drug delivery. Mater Sci Eng, C. 2018;84:44–51.
89. Zhao Y-D, Lai J-H, Wang M. 4D printing of self-folding hydrogel tubes for potential tissue engineering applications. Nano LIFE. 2021;11(04):2141001.

90. Amirthalingam S, Rajendran AK, Moon YG, Hwang NS. Stimuli-responsive dynamic hydrogels: design, properties and tissue engineering applications. Mater Horiz. 2023;10(9):3325–50.
91. Moroni S, Bingham R, Buckley N, Casettari L, Lamprou DA. 4D printed multipurpose smart implants for breast cancer management. Int J Pharm. 2023;642: 123154.
92. Li T-T, Zhang Y, Ren H-T, Peng H-K, Lou C-W, Lin J-H. Two-step strategy for constructing hierarchical pore structured chitosan–hydroxyapatite composite scaffolds for bone tissue engineering. Carbohyd Polym. 2021;260: 117765.
93. Shi D, Shen J, Zhang Z, Shi C, Chen M, Gu Y, et al. Preparation and properties of dopamine-modified alginate/chitosan–hydroxyapatite scaffolds with gradient structure for bone tissue engineering. J Biomed Mater Res, Part A. 2019;107(8):1615–27.
94. Calafel MI, Criado-Gonzalez M, Aguirresarobe R, Fernández M, Mijangos C. From rheological concepts to additive manufacturing assessment of hydrogel-based materials for advanced bioprinting applications. Mater Adv. 2025;6(14):4566–97.
95. Yang Q, Peng J, Xiao H, Xu X, Qian Z. Polysaccharide hydrogels: functionalization, construction and served as scaffold for tissue engineering. Carbohydr Polym. 2022;278: 118952.
96. Cao P, Tao L, Gong J, Wang T, Wang Q, Ju J, et al. 4D printing of a sodium alginate hydrogel with step-wise shape deformation based on variation of crosslinking density. ACS Appl Polym Mater. 2021;3(12):6167–75.
97. Khalid MY, Arif ZU, Al Rashid A, Bukhari SMZS, Hossain M, Koç M. Shape-memory and self-healing properties of sustainable cellulosic nanofibers-based hybrid materials for novel applications. Giant. 2024;19: 100299.
98. Hamza M, Kanwal Q, Raza A, Zehra K, Yasir M, Iqbal D, et al. RP-HPLC-UV method validation for Ribavirin used in topical applications. Int J Chem Biochem Sci. 2024;26.
99. Zolfagharian A, Durran L, Gharaie S, Rolfe B, Kaynak A, Bodaghi M. 4D printing soft robots guided by machine learning and finite element models. Sens Actuators A. 2021;328: 112774.
100. Datta D, Weiss EL, Wangpraseurt D, Hild E, Chen S, Golden JW, et al. Phenotypically complex living materials containing engineered cyanobacteria. Nat Commun. 2023;14(1):4742.
101. Boons R, Siqueira G, Grieder F, Kim S-J, Giovanoli D, Zimmermann T, et al. 3D bioprinting of diatom-laden living materials for water quality assessment. Small. 2023;19(50):2300771.
102. Olteanu G, Neacşu SM, Joiţa FA, Musuc AM, Lupu EC, Ioniţă-Mîndrican CB, et al. Advancements in regenerative hydrogels in skin wound treatment: a comprehensive review. Int J Mol Sci. 2024;25(7).
103. Lee S-Y, Ma J, Khoo TS, Abdullah N, Nik Md Noordin Kahar NNF, Abdul Hamid ZA, et al. Polysaccharide-based hydrogels for microencapsulation of stem cells in regenerative medicine. Frontiers Bioeng Biotechnol. 2021;9.
104. Bi H, Ye G, Sun H, Ren Z, Gu T, Xu M. Mechanically robust, shape memory, self-healing and 3D printable thermoreversible cross-linked polymer composites toward conductive and biomimetic skin devices applications. Addit Manuf. 2022;49: 102487.
105. Sun X, Wang Y, Lei Y. Fluorescence based explosive detection: from mechanisms to sensory materials. Chem Soc Rev. 2015;44(22):8019–61.
106. Lei Y, Yao Q, Jin Z, Wang Y-C. Intelligent films based on pectin, sodium alginate, cellulose nanocrystals, and anthocyanins for monitoring food freshness. Food Chem. 2023;404: 134528.
107. Li Z, Lin Z. Recent advances in polysaccharide-based hydrogels for synthesis and applications. Aggregate. 2021;2(2): e21.
108. Srivastava N, Choudhury AR. Stimuli-responsive polysaccharide-based smart hydrogels and their emerging applications. Ind Eng Chem Res. 2022;62(2):841–66.
109. Qi X, Wu L, Su T, Zhang J, Dong W. Polysaccharide-based cationic hydrogels for dye adsorption. Colloids Surf B. 2018;170:364–72.
110. Poshina DN, Raik SV, Poshin AN, Skorik YA. Accessibility of chitin and chitosan in enzymatic hydrolysis: a review. Polym Degrad Stab. 2018;156:269–78.
111. Su J-W, Li D, Xie Y, Zhou T, Gao W, Deng H, et al. A machine learning workflow for 4D printing: understand and predict morphing behaviors of printed active structures. Smart Mater Struct. 2020;30(1): 015028.

112. Lee J, Park D, Lee M, Lee H, Park K, Lee I, et al. Machine learning-based inverse design methods considering data characteristics and design space size in materials design and manufacturing: a review. Mater Horiz. 2023;10(12):5436–56.
113. Snow Z, Scime L, Ziabari A, Fisher B, Paquit V. Scalable in situ non-destructive evaluation of additively manufactured components using process monitoring, sensor fusion, and machine learning. Addit Manuf. 2023;78: 103817.
114. Rudolf S, Sauer HM, Weißenseel S, Zembron J, Tone R, Doersam E, et al. Printing and processing techniques. Organic and printed electronics. Jenny Stanford Publishing; 2024. p. 47–124.
115. Cheng X, Shen Z, Zhang Y. Bioinspired 3D flexible devices and functional systems. Natl Sci Rev. 2024;11(3):nwad314.
116. Benalaya I, Alves G, Lopes J, Silva LR. A review of natural polysaccharides: sources, characteristics, properties, food, and pharmaceutical applications. Int J Mol Sci. 2024;25(2):1322.
117. Barrulas RV, Corvo MC. Rheology in product development: an insight into 3D printing of hydrogels and aerogels. Gels. 2023;9(12):986.
118. Montero de Espinosa L, Meesorn W, Moatsou D, Weder C. Bioinspired polymer systems with stimuli-responsive mechanical properties. Chemical Reviews. 2017;117(20):12851–92.

The manufacturer's authorised representative in the EU is Springer Nature Customer Service Centre GmbH, Europaplatz 3, 69115 Heidelberg, Germany. If you have any concerns regarding our products, please contact ProductSafety@springernature.com

Printed and bound by CPI Group (UK) Ltd, Croydon, CR0 4YY
07/10/2026
02229349-0017